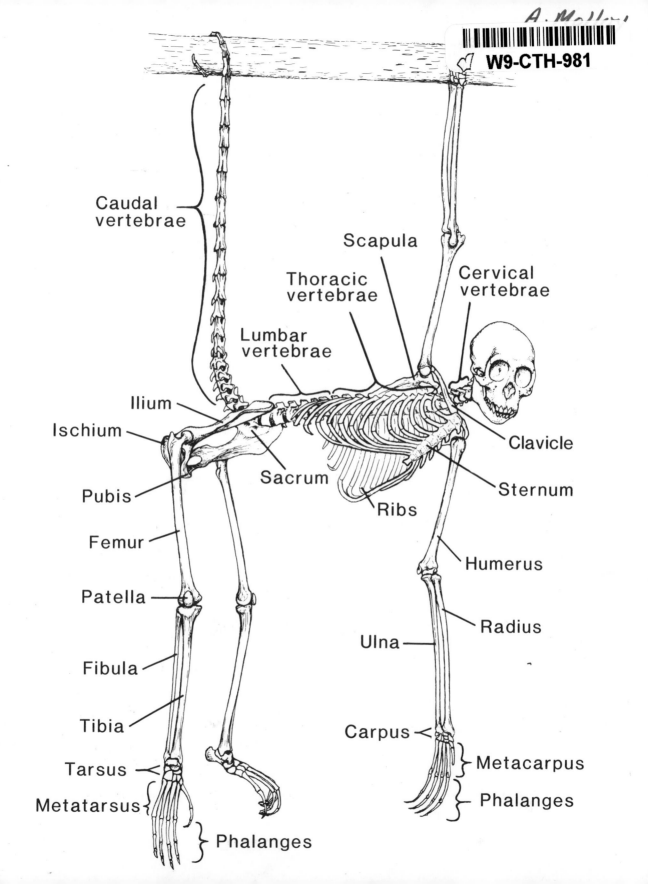

Caudal
vertebrae

Scapula

Thoracic
vertebrae

Cervical
vertebrae

Lumbar
vertebrae

Ilium

Clavicle

Ischium

Sternum

Pubis

Sacrum

Ribs

Femur

Humerus

Patella

Radius

Fibula

Ulna

Tibia

Carpus

Tarsus

Metacarpus

Metatarsus

Phalanges

Phalanges

Primate
Adaptation & Evolution

John G. Fleagle
State University of New York, Stony Brook

Academic Press, Inc.
HARCOURT BRACE JOVANOVICH, PUBLISHERS

San Diego   New York   Berkeley   Boston   London   Sydney   Tokyo   Toronto

Academic Press, Inc.
San Diego, California 92101

United Kingdom Edition published by
Academic Press Inc. (London) Ltd.
24–28 Oval Road, London NW1 7DX

Library of Congress Cataloging-in-Publication Data

Fleagle, John G.
  Primate adaptation and evolution.

  Includes index.
  1. Primates—Evolution. 2. Adaptation (Biology)
I. Title.
QL737.P9F57 1988   599′.0438   87-6508
ISBN 0-12-260340-0

PRINTED IN THE UNITED STATES OF AMERICA
88 89 90 91   9 8 7 6 5 4 3 2 1

CONTENTS

Tables

Illustrations

This book is an introduction to the biology of the mammalian order Primates. It is based on the contents of a course that has been offered to advanced undergraduate and beginning graduate students in anthropology and biology at the State University of New York at Stony Brook during the past ten years. It is designed for students with a general knowledge of basic biology and evolutionary theory who wish to examine the comparative anatomy, behavioral ecology, and paleontology of humans and their nearest relatives, a particularly well-studied and interesting group of animals. Anthropology textbooks beyond the freshman level have traditionally been devoted either to primate behavior and ecology or to primate and human evolution. This is unfortunate, since our understanding of the evolutionary history of primates hinges on our ability to interpret fossil bones and teeth from a comparison of these elements with those in the bodies of extant primates.

In this book, the major groups of living and extinct primates are presented as a series of adaptive radiations. For each radiation I examine those aspects of their biology that set them apart from other primates and those features they share with other members of the order. The book is divided into three sections. The first three chapters—references or primers on evolutionary biology, primate anatomy, and behavioral ecology—are designed to introduce the basic concepts and terminology used in later chapters. Chapters 4 through 7 cover the anatomy, ecology, and systematics of the major groups of extant primates—prosimians, New World monkeys, Old World monkeys, and hominoids. Each group is discussed genus by genus, with particular emphasis on diagnostic skeletal features and characteristic dietary and locomotor adaptations. Within each chapter are tables providing the species-level taxonomy of each group as well as common names, body weights, and limb proportions for each species. Each chapter includes more general discussions of the adaptive radiation of the group being considered as well as discussions of current issues concerning evolutionary relationships among the taxa. These chapters contain comparative anatomical drawings designed to illustrate the diagnostic features of each taxonomic group as well as summary charts reviewing the adaptive radiations. In addition, most genera are pictured in a series of drawings of animals in their natural environments which illustrate not only external appearance but also aspects of the typical habitat, diet, and locomotor and postural habits.

In Chapter 8, "Primate Adaptations," I examine common adaptive patterns in morphology and behavior that can be traced throughout the order Primates. This review provides a summary of adaptive themes from earlier chapters as well as a basis for interpretation of the adaptations of fossil taxa in later chapters.

In the remainder of the book, Chapters 9 through 16, we are concerned with the primate fossil record. Chapter 9, an introduction to paleontology, reviews the major differences between our knowledge of fossil primates and our understanding of living species. Chapters 10 through 15 are analyses of the fossil records of particular radiations of primates, beginning with the plesiadapiforms of the Paleocene and continuing through the evolution of hominids in the Pliocene and Pleistocene. As in the earlier chapters on living species, each radiation is considered in terms of its distinctive morphological characteristics and its adaptive diversity. Tables provide more detailed, species-level systematics, with estimated body weights based on regressions of dental dimensions to give the reader a comparative scale for visualizing the extinct primates. As in the chapters of living primates, there are discussions of the adaptive diversity of various extinct radiations as well as sections outlining current issues and unresolved problems on the evolutionary relationships of each group. In the final chapter, I survey 65 million years of primate evolution for evidence of general patterns in adaptive diversity and evolutionary mechanisms.

Although the book is designed as a single treatment of living and fossil primates, the arrangement is suitable for use in a less comprehensive course in either primate ecology or primate evolution. In addition, it should provide an introduction to primatology for biologists of all sorts.

This book has been many years in the making, and I have relied heavily on the good will and expertise of many colleagues and friends. The students of primate evolution at Stony Brook prompted me to write down my notes and provided me with numerous comments on early drafts of most chapters, as did students at the University of California, Berkeley, where I had the pleasure of teaching in 1986. Much of the material in these chapters is the result of interaction with my longtime friends and colleagues Russell Mittermeier, David Chivers, Elwyn Simons, Ken Rose, Phil Gingerich, Tom Bown, and especially Richard Kay. For the past thirteen years I have had the opportunity to work in the Department of Anatomical Sciences at Stony Brook with Gabor Inke, Jack Stern, Norman Creel, William Jungers, Randall Susman, David Krause, Sue Larson, Russell Mittermeier, Fred Grine, Lawrence Martin, and for all too brief a time Alfred Rosenberger and James Wells—a group of the most outstanding (and outrageous) primatologists ever assembled in one university. We owe a special debt to our chairman, Maynard Dewey, whose support has made this a truly enjoyable and productive place to work.

Much of the delight in putting together this book has come from the opportunity to work with several outstanding artists. Stephen Nash, Hugh Nachamie, Luci Betti, and Leslie Jungers all contributed greatly to the illustrations in this book, Jeff Meldrum helped with several maps and charts, and J. Muennig provided photographs. Many people and institutions generously provided copies of illustrations for use in the book, including Tom Bown, Eric Delson, Richard Kay, Gerald Eck, Ken Rose, Elwyn Simons, Leonard Ginsburg, Phil Gingerich, Brian Shea, W. von Koenigswald, Ronald Wolff, Kathy Schick, Jeanne Sept, Gunter Bräuer, Fred Grine, Vince Sarich, David Pilbeam, Tim White, B. Holly Smith, Russell Mittermeier, Meave Leakey, Russell Ciochon, Peter Andrews, Lawrence Martin, Wolfgang Maier, Mary Maas, Pan Yuerong, the Institute of Human Origins, and the National Museums of Kenya. Stephanie Rippel, Nancy Thompson Handler, and Mary Maas were invaluable in helping me go through many drafts of the text, sort out the bibliographies, and complete the index.

Many people have provided helpful comments on one or more chapters over the past four years, including Russell Ciochon, Susan Larson, David Krause, Ken Rose, Elizabeth Watts, Charles Janson, Frances White, Steve Redhead, Roderick Moore, F. Clark Howell, Nancy Handler, Todd Olson, Marc Godinot, Chris Beard, Tom Naimen, Tim Cole, Elizabeth Dumont, Greg Buckley, Liza Shapiro, Fred Grine, Mary Maas, and Randall Susman. William Jungers and Suzanne Strait provided both comments on the manuscript and invaluable assistance in compiling the tables of body weights. At Academic Press, Kerry Pinchbeck and John Thomas helped turn a ragged manuscript into a book. I am grateful for all their time, patience, and good humor. Finally, I owe extra thanks to Richard Kay, Lawrence Martin, and especially Patricia Wright, who contributed far more assistance with this book than one should normally expect from any colleague. I thank all of these people for their help and encouragement.

Adaptation, Evolution, and Systematics

ORDER PRIMATES

The subject of this book is the order Primates, the mammalian order that includes not only us humans but also a wide array of lemurs, lorises, galagos, tarsiers, monkeys, and apes. It also includes many extinct animals that are known to us only through fossilized remains and lack familiar names. Primates come in a variety of sizes and shapes, and this variety is matched by the diversity of behaviors primates have evolved to survive in equally various environments. This diversity in structure and behavior—and its evolution—is the theme of this book. Before considering this diversity, we review a few principles of evolutionary biology and discuss the mechanisms through which this array of creatures has come about. We also provide a brief review of biological classification and methods of reconstructing phylogeny.

Adaptation

Adaptation is a concept central to our understanding of evolution, but the term has proved very difficult to define in a simple phrase. One of the most succinct definitions has been offered by Vermeij (1978, p. 3): "An *adaptation* is a characteristic that allows an organism to live and reproduce in an environment where it probably could not otherwise exist." In the following chapters, we examine extant (living) and extinct (fossil) primates as a series of **adaptive radiations**— groups of closely related organisms that have evolved morphological and behavioral features enabling them to exploit different ecological niches. Adaptive radiations are central to our understanding of evolutionary processes. The adaptive radiation of finches on the Galapagos Islands of Ecuador played an important role in guiding Darwin's views on the origin of species.

"Adaptation" also refers to the process whereby organisms obtain their adaptive characteristics. The primary mechanism of adaptation is **natural selection**—the differential survival and reproductive success of individuals with different heritable characteristics. As Darwin argued, and subsequent generations of scientists have corroborated, natural selection ensures that any heritable features, either anatomical or behavioral, that increase the fitness of an individual relative to other individuals will be passed on to succeeding generations. In considering the evolution of behavioral traits in the following chapters, it is important to remember that natural selection acts primarily through differential reproductive success of individuals within a population. Through this differential reproductive success of different genotypes, the genetic composition of a population can change from generation to generation.

Evolution

Evolution is modification by descent, or genetic change in a population through time. Although biologists consider most evolution to be the result of natural selection, there are other, "non-Darwinian," mechanisms that can and do lead to genetic change within a population. **Genetic drift** is change in the genetic composition of a population from generation to generation due to chance sampling events independent of selection. **Founder effect** is a more extreme change in the genetic makeup of a population that occurs when a new population is established by only a few individuals. This new population may have a very different genetic composition than that found in the larger ancestral population. Thus the chance characteristics of a founder population can have dramatic effects on the subsequent evolution and evolutionary diversity of a group of organisms.

Evolutionary change within a population can ultimately lead to **speciation**—the appearance of a new species. Although biologists agree that the origin of new species is the result of evolution, there is considerable debate concerning the rate at which evolutionary change leading to formation of new species takes place and the actual mechanisms of species formation. According to the **phyletic gradualism** model, most evolutionary change takes place gradually. In contrast, the **punctuated equilibrium** model theorizes that populations are normally genetically stable, and that evolutionary change takes place primarily in abrupt genetic shifts too rapid to be preserved in the fossil record. We return to evaluate these theories in later chapters.

The origin of one species from another can take place in two ways. The change of a single species into another daughter species is called **anagenesis**; division of one species (or population) into two or more daughter species is called **cladogenesis**. Cladogenesis obviously has a more important function in the development of adaptive radiations. A reconstruction of the branching events in the evolution of a group of animals is called a **cladogram**.

Phylogeny

Because this book deals with the adaptive radiations of primates, we are interested in reconstructing the evolutionary branching sequence, or **phylogeny**, of various primate groups to see how they came to be the way they are. Although some of us can trace our own genealogies (or those of our pets) through several generations, tracing the genealogical relationships among all primates is a much more daunting undertaking. The evolutionary radiation of primates has taken place over geological time and has involved millions of generations, probably thousands of species, and billions of individuals. Moreover, the records available for reconstructing primate phylogeny are meager, consisting of individuals of about two hundred living species and occasional bony remains of several hundred extinct species drawn from various parts of the world at various times during the past 65 million years.

Morphology

The methods we use to reconstruct phylogeny are primarily based on identifying groups of related species through morphological similarities. Most biologists agree that organisms should be grouped together on the basis of **shared specializations** (or shared-derived features) that distinguish them from their ancestors. For example, body hair is a specialization that unites humans, apes, monkeys, and cats as mammals and distinguishes them from other types of vertebrates, whereas the common possession of a tail by many monkeys, lizards, and crocodiles is an ancestral feature that is of no particular value in assessing the evolutionary relationships among these organisms, since their common ancestor had a tail. On the other hand, the absence of a tail in apes and humans represents a derived specialization that sets them apart (Fig. 1.1). The common possession of a group of specializations by a cluster of species or genera is interpreted as indicating that this cluster shares a unique heritage relative to other related species.

Unfortunately, not all derived similarities among organisms are indicative of a unique heritage. Animals have frequently evolved morphological similarities independently—this is known as **parallel evolution**. In addition to apes and humans, for example, a few monkey species and a few prosimians have also lost their tail. The biologist's task in reconstructing phylogeny is to distinguish those specializations that are the result of a unique heritage from those that are the result of parallel evolution. In some cases, as in apes and tailless monkeys, the similarities may be only superficial; the underlying bone and muscle structure may be quite different, indicating different evolutionary histories for this feature. In other cases, as in amino

FIGURE 1.1

Shared specializations and ancestral features.

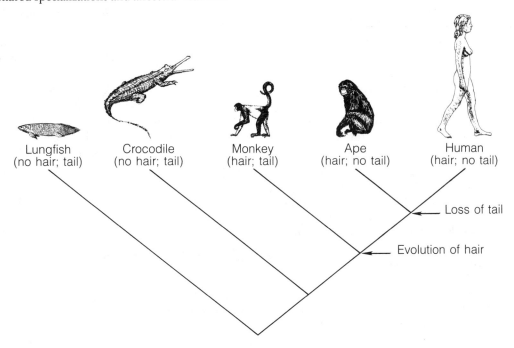

Lungfish
(no hair; tail)

Crocodile
(no hair; tail)

Monkey
(hair; tail)

Ape
(hair; no tail)

Human
(hair; no tail)

Loss of tail

Evolution of hair

acid sequences, this type of distinction cannot be made. Parallel evolution has been
very common in the evolution of mammals.
As a result, analyses of different morphological features often yield different evolutionary relationships among a group of animals.
In the following chapters we discuss many
aspects of primate phylogeny that cannot yet
be resolved because one set of data suggests
one phylogeny and other analyses suggest
another.

Biomolecular Phylogeny

Studies of primate genealogy have traditionally been based on gross morphology—tooth
and skull shape, patterns of blood supply,
and other anatomical features. These anatomical features still play a major role in our
understanding of the evolutionary relationships of living and fossil primates, but, for
investigating phyletic relationships among
living primates, studies in comparative immunology, molecular sequencing, and DNA
hybridization have become important tools.
Some biomolecular comparisons seem to
randomly sample a part of an organism's
genetic material—or, in the case of DNA
hybridization studies, all of it. Because they
presume that parallel evolution is less likely
to result in molecular similarities than in
similarities of gross morphology, many biologists consider the phyletic relationships indicated by such studies to more accurately
reflect genealogical relationships.

Numerous biomolecular studies of the
relationships among apes and humans have
yielded relatively consistent results (see
Chapter 7). For many other primate radiations (New World monkeys, prosimians),
however, the relationships indicated by
biomolecular studies are somewhat vague,
either because there are data from too few
species or because the resolution of the
methods is insufficient to distinguish the
branching sequence. In some cases bio-

molecular studies seem to have resolved
phyletic issues for which more traditional
morphological studies gave ambivalent results. In other cases they have supported
morphological studies, and in still others
there are dramatic contradictions between
the results of molecular studies and those
from more traditional methods (see, e.g.,
Lewin, 1987). In the following chapters, we
discuss the results of biomolecular studies of
phylogeny in conjunction with those based
on other data.

At present, biomolecular studies are limited to living species and a few recent fossils
(including woolly mammoths and some of
the recently extinct lemurs from Madagascar). Techniques have not yet been developed to allow analysis of fossils more than a
few thousand years old. However, some
biologists argue that immunological or DNA
hybridization comparisons reflect evolutionary changes that have taken place at a
relatively constant rate, and therefore that
these data can be used as "molecular clocks"
for determining the timing of evolutionary
divergences. The concept of the molecular clock, most avidly championed by Alan
Wilson and Vincent Sarich of the University
of California on the basis of their immunological comparisons, has had a stormy history. Current debate centers on whether the
rate, or rates, of molecular evolution are
strictly linear, and on what divergence dates
should be used to calibrate the clock. Disagreements over the timing of branching
events in primate evolution between scientists using molecular studies and those investigating the fossil record are numerous (see,
e.g., Houde, 1987). Both the theoretical
nature of the debates between molecular
biologists and paleontologists and the detailed differences in reconstruction of individual events in primate evolution are
beyond the scope of this book, but it is
noteworthy that the differences between the
results of these two approaches are narrow-

ing considerably. For many groups of primates, interpretations based on the two types of data are remarkably concordant (see, e.g., Gingerich, 1984; Andrews, 1986).

Taxonomy and Systematics

Taxonomy is a means of ordering our knowledge of biological diversity through a series of commonly accepted names for organisms. If scientists wish to communicate about animals and plants and to discuss their similarities and differences, they need a standard system of names both for individual types of organisms and for related groups of organisms. For example, the tufted capuchin monkey of South America, known to many people as the organ-grinder monkey, goes by over a dozen different names among the different tribal and ethnic groups of Surinam alone. To scientists around the world, however, this species is known by a single name, *Cebus apella*. The practice of assigning every biological species, living or fossil, a unique name composed of two Latin words was initiated by Carolus Linnaeus, a Swedish scientist of the eighteenth century whose system of biological nomenclature is universally followed today. Under the Linnean system, *Cebus* is the name of a **genus** (pl. *genera*), or group of animals, in this case all kinds of capuchin monkeys. (The name of a genus is always capitalized.) The word *apella*, the species name, refers to a particular type of capuchin monkey, the tufted capuchin monkey. (A species name always begins with a lower-case letter.) Each genus name must be unique, but species names need be unique only within a particular genus so that the combination of genus and species names is unique and refers to only one kind of organism. (The name is always written in italics—or underlined.) Somewhere in a museum there is a preserved skeleton (or skull, or skin) that has been designated as the **type specimen**

TABLE 1.1
A Classification of the Tufted Capuchin Monkey

Kingdom	Animal
Phylum	Chordata
Class	Mammalia (mammals)
Order	Primates (primates)
Suborder	Anthropoidea (higher primates)
Infraorder	Platyrrhini (New World monkeys)
Superfamily	Ceboidea (New World monkeys)
Family	Cebidae (capuchins, squirrel monkeys, and marmosets)*
Subfamily	Cebinae (capuchins and squirrel monkeys)*
Genus	*Cebus* (capuchins)
Species	*Cebus apella* (tufted capuchin monkey)

*Indicates only one of several common classifications (see Chapter 5).

for this species. The type specimen provides an objective reference for this species so that any scientist who thinks he or she may have discovered a different kind of monkey can examine the individual on which *Cebus apella* is based.

The Linnean system contains a hierarchy of levels for grouping organisms into larger and larger units (Table 1.1). Within the genus *Cebus*, for example, there are several species: *Cebus apella*, the tufted capuchin; *Cebus albifrons*, the white-fronted capuchin; *Cebus capucinus*, the caped capuchin; and others. Genera are grouped into **families**, families into **orders**, orders into **classes**, and classes into **phyla**. For particular lineages, these basic levels are often further subdivided or clustered into **suborders, infraorders, superfamilies, subfamilies, tribes, subgenera**, or **subspecies**. For convenience, names at different levels of the hierarchy are given distinctive endings. Family names usually end in *-dae*, superfamily names in *-oidea*, and subfamily names in *-inae*.

In the science of classifying organisms, **systematics**, we attempt to apply the tidy Linnean system to the untidy, unlabeled world of animals. Figure 1.2, the classifica-

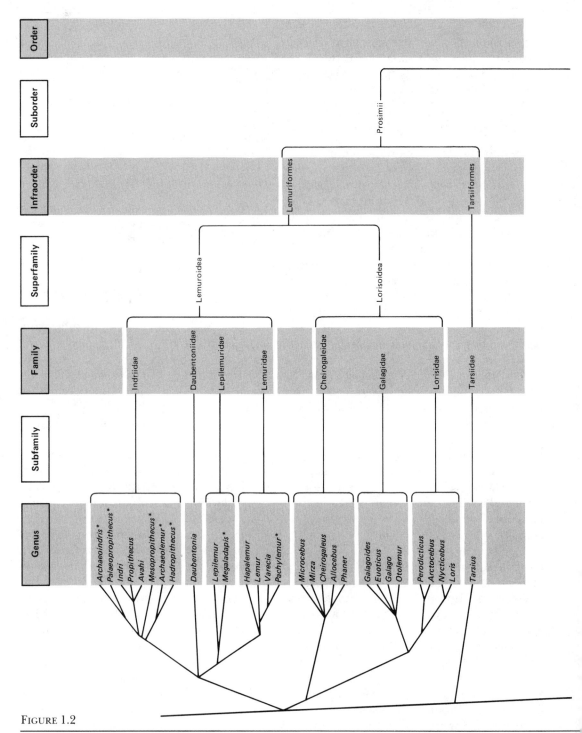

FIGURE 1.2

A classification of extant and subfossil (= recently extinct) (*) primate genera.

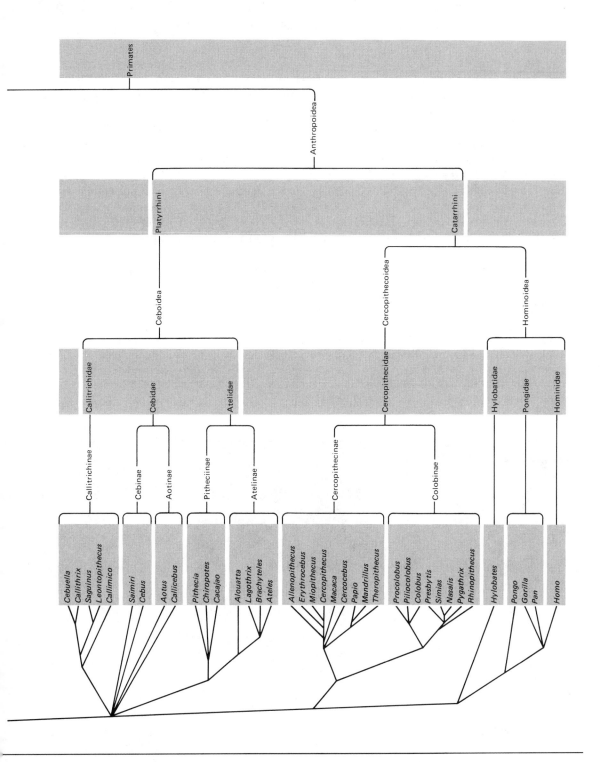

tion used in this book, is the result of one such attempt. But although biologists agree to use the Linnean framework for naming organisms, they frequently disagree about the proper classification of particular creatures. They may disagree as to whether each of the gibbon types on different islands in Southeast Asia is a distinct species or only a subspecies of a single species. Some authorities may feel that gibbons and great apes should be placed in a single family, others that they should be placed in separate families. Once they have learned the Linnean hierarchy, many students are understand-

ably frustrated and annoyed to find that textbooks often do not agree on the classification of different species. There are, however, usually good reasons for the disagreements about primate classification, as we see in the following chapters.

One reason for disagreement in primate classification is that the rules for distinguishing a genus, a family, or a superfamily are somewhat arbitrary. Scientists usually set their own standards. The only generally accepted rules are for species. A biological species is usually taken to be a group of organisms capable of interbreeding among

FIGURE 1.3

A strictly phyletic classification recognizes that humans, chimpanzees, and gorillas are more closely related to each other than any of them are to orangutans; the latter are thus grouped separately as the only pongids. A more traditional classification recognizes adaptive differences; in

this case, chimpanzees and gorillas are classified with orangutans (pongids), and humans are grouped separately (hominids) because of the great degree of adaptation that distinguishes humans from even their closest primate relatives.

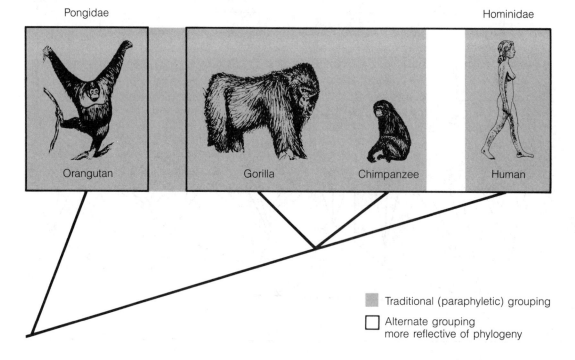

Pongidae   Hominidae

Orangutan   Gorilla   Chimpanzee   Human

▦ Traditional (paraphyletic) grouping

☐ Alternate grouping
more reflective of phylogeny

themselves but unable to interbreed with any other species without significant loss of fertility. This definition is, of course, impossible to apply to extinct primates, and it is often difficult to apply to living populations. A more practical approach to the identification of species is to examine the **metric variability**—a statistical measure of variations in the details of size, weight, or body dimension—among the individuals in question. Living species of mammals are remarkably consistent in their metric variability (Gingerich and Schoeninger, 1979), and we can use this standard to identify species in the fossil record. The limits for genera and families are, however, much more arbitrary.

It is generally agreed that classification should reflect phylogeny, and that taxonomic groups such as families, superfamilies, and suborders should be **monophyletic** groups; that is, that they should have a single common ancestor that gave rise to all members of the group. Many also feel that taxonomic groups should be **holophyletic** groups as well—they should contain all the descendents of their common ancestor, not just some of them. But it is often not practical or possible to achieve this unambiguously, and classifications are often compromises compatible with several possible phylogenies. In addition, many biologists feel that classification should reflect not only phylogeny but also major adaptive differences, even among closely related species. For example, most biologists now agree that humans are much more closely related to chimpanzees and gorillas than to orangutans. Thus a true phyletic taxonomy would group humans with the African apes in a single family and the orangutan in a separate family. In spite of this, most still place humans in a separate family, the Hominidae, and all living great apes in a common family, the Pongidae, because humans have departed further from

the common ancestor of humans and great apes than have chimpanzees and gorillas. The family Pongidae is called a **paraphyletic grouping** because some of its members (chimpanzees and gorillas) are more closely related to a species (humans) placed in another family than they are to other members (orangutans) of their own family (Fig. 1.3). The taxonomy used in this book (Fig. 1.2) contains several such departures from a strictly phyletic classification. In all cases, the evolutionary relationships are discussed in the text.

BIBLIOGRAPHY

Andrews, P. (1986). Fossil evidence on human origins and dispersal. *Cold Spring Harbor Symp. Quant. Biol.*, 1986.

———. (1987). Aspects of hominoid phylogeny. In *Molecules and Morphology in Evolution: Conflict or Compromise*, ed. C. Patterson, pp. 21–53. Cambridge: Cambridge University Press.

Ayala, F.J., and Valentine, J.W. (1979). *Evolving: The Theory and Processes of Organic Evolution*. Menlo Park, Ca.: Benjamin-Cummings.

Bendall, D.S., ed. (1983). *Evolution from Molecules to Men*. Cambridge: Cambridge University Press.

Cracraft, J., and Eldredge, N., eds. (1979). *Phylogenetic Analysis and Paleontology*. New York: Columbia University Press.

Darwin, C. (1859). *On the Origin of Species by Means of Natural Selection, or the Preservation of Favoured Races in the Struggle for Life*. London: John Murray.

Dawkins, R. (1976). *The Selfish Gene*. New York and Oxford: Oxford University Press.

———. (1987). *The Blind Watchmaker*. New York: W.W. Norton.

Dodson, E.O., and Dodson, P. (1976). *Evolution: Process and Product*, 2d ed. New York: Van Nostrand.

Eldredge, N., and Cracraft, J. (1980). *Phylogenetic Patterns and the Evolutionary Process: Method and Theory in Comparative Biology*. New York: Columbia University Press.

Eldredge, N., and Stanley, S.M., eds. (1984). *Living Fossils*. New York: Springer Verlag.

Futuyma, D.J. (1979). *Evolutionary Biology*. Sunderland, Mass.: Sinauer.

Gingerich, P.D. (1984). Primate evolution: Evidence from the fossil record, comparative morphology, and molecular biology. *Yrbk. Phys. Anthropol.* **27**:57–72.

Gingerich, P.D., and Schoeninger, M.J. (1979). Patterns of tooth size variability in the dentition of primates. *Am. J. Phys. Anthropol.* **51**:457–566.

Goodman, M., Tashian, R.E., and Tashian, J.H., eds. (1976). *Molecular Anthropology: Genes and Proteins in the Evolutionary Ascent of the Primates*. New York: Plenum Press.

Houde, P. (1987). Histological evidence for the systematic position of *Hesperornis* (Odontornithes: hesperornithiformes). *Auk* **104**:125–129.

Laporte, L.F., ed. (1978). *Readings from Scientific American: Evolution and the Fossil Record*. San Francisco: W.H. Freeman.

Lewin, R. (1987). Research news: My close cousin, the chimpanzee. *Science* **238**:273–275.

Lowenstein, J.M. (1981). Immunological reactions from fossil material. *Phil. Trans. Royal Soc. London*, B292, pp. 143–149.

———. (1985). Radioimmunoassay of extinct and extant species. In *Hominid Evolution: Past, Present, and Future*, ed. P.V. Tobias, pp. 401–410. New York: Alan R. Liss.

Milkman, R., ed. (1982). *Perspectives on Evolution*. Sunderland, Mass.: Sinauer.

Patterson, C. (1978). *Evolution*. London: British Museum (Natural History).

Pilbeam, D. (1984). The descent of hominoids and hominids. *Sci. Am.*, March 1984:84–96.

Simpson, G.G. (1953). *The Major Features of Evolution*. New York: Columbia University Press.

———. (1961). *Principles of Animal Taxonomy*. New York: Columbia University Press.

Smith, J. Maynard, ed. (1982). *Evolution Now: A Century after Darwin*. San Francisco: W.H. Freeman.

Vermeij, G.J. (1978). *Biogeography and Adaptation: Patterns of Marine Life*. Cambridge, Mass.: Harvard University Press.

Williams, G.C. (1966). *Adaptation and Natural Selection: A Critique of Some Current Evolutionary Thought*. Princeton, N.J.: Princeton University Press.

The Primate Body

PRIMATE ANATOMY

Fossil and living primates are an extraordinarily diverse array of species. Some are among the most generalized and primitive of all mammals; others show morphological and behavioral specializations unmatched in any other mammalian order. This diversity in structure, behavior, and ecology is our topic of study in this book. The purpose of this chapter is to establish an anatomical frame of reference—a survey of features common to all (or almost all) primates. This chapter, then, provides pictures and descriptions of primate anatomy and preliminary indications of those anatomical features that have undergone the greatest changes in primate evolution.

Compared to most other mammals, we primates have retained relatively primitive bodies. Some of us are specialized in that we have lost our tails, and many have a relatively large braincase. But no primates have departed so dramatically from the common mammalian body plan as bats, whose hands have become wings; as horses, whose fingers and toes have reduced to a single digit; or as baleen whales, who have lost their hindlimbs altogether, adapted their tails into flippers, and replaced their teeth with great hairlike combs. The anatomical features that distinguish the bones and teeth of primates from those of many other mammals are the result of subtle changes in the shape and proportion of homologous elements rather than major rearrangements, losses, or additions of body parts. We generally find the same bones and teeth in all species of primates, with only minor differences reflecting different diets or locomotor habits. The fact that humans are constructed of the same bony elements as other primates (and generally other mammals) is a major piece of evidence demonstrating our evolutionary origin.

Size

Size is a basic aspect of an organism's anatomy and plays a major role in its ecological adaptations. It is a feature that can be readily compared, both among living species and between living and fossil primates. Adult living primates range in size from mouse lemurs and pygmy marmosets, which weigh less than 100 g, to male gorillas, which reach weights of over 200 kg (Fig. 2.1). The fossil record provides evidence of a few extinct primates from the beginning of the age of mammals that were much smaller (probably as small as 20 g) and at least one, *Gigantopithecus blacki* from the Pleistocene of

FIGURE 2.1

A mouse lemur (*Microcebus*) and a gorilla (*Gorilla*), the smallest and largest living primates.

China (see Chapter 13), that was much larger (probably over 300 kg). In their range of body sizes, primates are one of the more diverse orders of living mammals. As a group, however, primates are rather medium-size mammals (Fig. 2.2)—larger than most insectivores and rodents and smaller than most ungulates, elephants, and marine mammals.

Cranial Anatomy

The anatomy of the head, or cranial region, plays a particularly important role in studies of primate adaptation and evolution. Many

of the anatomical features that have traditionally been used to delineate the systematic relationships among primates are cranial features, and most of our knowledge of fossil primates is based on this region.

Bones of the Skull

The adult primate skull (Fig. 2.3) consists of many different bones that together form a hollow, bony shell that houses the brain and special sense organs and also provides a base for the teeth and chewing muscles. Only the lower jaw, the mandible, and the three bones of the middle ear are separate, movable elements; the others are fused into a single

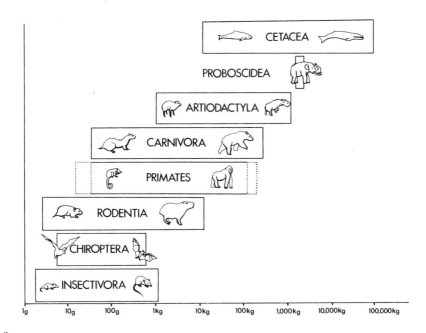

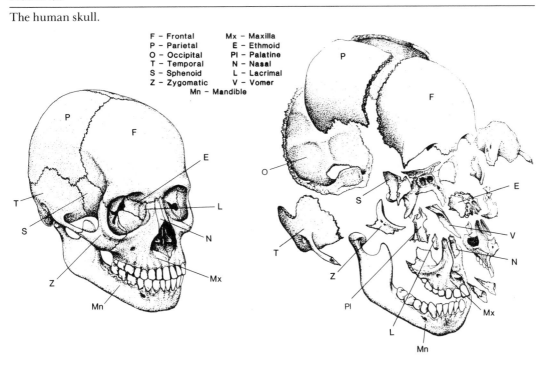

unit, the **cranium**. This unit can be roughly divided into two regions: a more posterior braincase, or neurocranium, and a more anterior facial region, or splanchnocranium.

The braincase serves as a protective bony case for the brain, a housing for the auditory region, and an area of muscle attachment for the larger chewing muscles and the muscles that move the head on the neck. Three paired flat bones—the **frontal, parietal,** and **temporal** bones—make up the top and sides of the braincase. (The temporal bone is a relatively complicated bone with several distinct parts.) The posterior and inferior surfaces of the braincase are formed by a single bone, the **occipital**, which also has a number of distinct parts. A complex, butterfly-shaped bone, the **sphenoid**, forms the anterior surface of the braincase and joins it with the facial region.

The facial region is formed by the **maxillary** and **premaxillary bones**, which contain the upper teeth; the **zygomatic bone**, which forms the lateral wall of the orbit, or eye

FIGURE 2.4

Skulls of a capuchin monkey (*Cebus*) and a lemur (*Lemur*), showing how differences in the size and shape of individual bones contribute to overall differences in skull form.

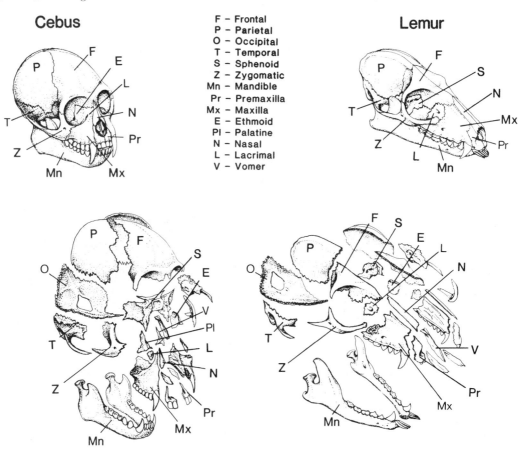

F – Frontal
P – Parietal
O – Occipital
T – Temporal
S – Sphenoid
Z – Zygomatic
Mn – Mandible
Pr – Premaxilla
Mx – Maxilla
E – Ethmoid
Pl – Palatine
N – Nasal
L – Lacrimal
V – Vomer

socket; the **nasal bones**, which form the bridge of the nose; and numerous small bones that make up the orbit and the internal nasal region. The lower jaw, or **mandible**, contains the lower teeth. In many mammals, and in most prosimian primates, the two halves of the mandible are loosely connected anteriorly in such a way that they can move somewhat independently of one another. This joint is called the **mandibular symphysis**. In higher primates, including humans, the two sides of the lower jaw are fused to form a single bony unit.

Although all primate skulls are made up of these same components, they can have very different appearances depending on the relative size and shape of individual bones (Fig. 2.4). The skull functions as a base and structural framework for the first part of the digestive system and as a housing for the brain and special sense organs of sight, smell, and hearing. Much of the diversity in primate skull shape reflects the need for this single bony structure to serve numerous, often conflicting functions. For example, although the size of the orbits is most directly related to the size of the eyeball and to whether a species is active during the day or night, it influences the shape and position of the nasal cavity and the direction of chewing forces in the face.

Teeth and Chewing

Many parts of the head and face are important in the acquisition and initial preparation of food. The lips, cheeks, teeth, mandible, tongue, hyoid bone (a small bone suspended in the throat beneath the mandible), and muscles of the throat all participate in this complex activity, and many of these same parts also play a role in communication and sound production. The two parts of the skull that can be linked most clearly to dietary habits are the teeth and the chewing muscles that move the lower jaw.

Teeth, more than any other single part of the body, provide the basic information underlying much of our understanding of primate evolution. Because of their extreme hardness and compact shape, teeth are the most commonly preserved identifiable remains of most fossil mammals. But teeth are more than just plentiful; they are also very complex organs that provide considerable information about both the phyletic relationships and the dietary habits of their owners. Because of the importance of teeth in evolutionary studies, there is an extensive but fairly simple terminology for dental anatomy.

All primates have teeth in both the upper jaw (maxilla) and the lower jaw (mandible), and, like most features of the primate skeleton, primate teeth are **bilaterally symmetrical**—the teeth on one side are mirror images of those on the other. Each primate jaw normally contains four types of teeth (Fig. 2.5). These are, from front to back, **incisors, canines, premolars**, and **molars**. The number of teeth a particular species possesses is usually expressed in a dental formula. The human dental formula is $\frac{2.1.2.3}{2.1.2.3}$, indicating that we normally have two incisors, one canine, two premolars, and three molars in each side of both the upper and the lower jaw for a total of thirty-two adult teeth. In most primate species, formulas for the upper and lower dentition are the same. In addition to adult (or permanent) teeth, primates have an earlier set of teeth, the **milk** (or **deciduous**) **dentition**, which precedes the adult incisors, canines, and premolars and occupies the same positions in the jaws. The human milk dentition, for example, contains two deciduous incisors, one deciduous canine, and two deciduous premolars (often called "milk molars") in each quadrant for a total of twenty deciduous teeth.

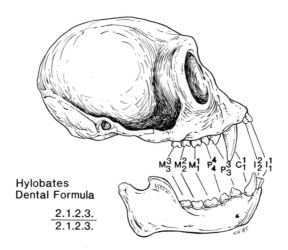

Hylobates
Dental Formula

$$\frac{2.1.2.3.}{2.1.2.3.}$$

FIGURE 2.5

The dentition of a siamang (*Hylobates*), showing two incisors (I), one canine (C), two premolars (P), and three molars (M) in each dental quadrant for a dental formula of $\frac{2.1.2.3.}{2.1.2.3.}$.

In discussions of primate dentition, individual teeth are usually denoted by a single-letter abbreviation with subscripts and superscripts (Fig. 2.5). For example, P^1 is the first upper premolar, M_2 is the second lower molar, and dP^3 is the third upper deciduous premolar. In addition to this shorthand for describing the position of teeth in a jaw, there is a widely accepted terminology for describing the shape and features of individual teeth, especially molars. The front of a tooth, in the direction of the central incisor, is the **mesial** end, and opposite is the **distal** end. The cheek side or outside of a tooth is the **buccal**, or **labial** side, and the inside or tongue side is the **lingual** side. The length of a tooth is commonly measured from the mesial end to the distal end, and the breadth from the buccal side to the lingual side.

Primate molars, and often premolars as well, have a series of cusps or bumps of enamel connected by crests or ridges. Because homologous cusps and crests can be identified in a wide range of species, they have been given names for reference (Fig.

2.6). The current terminology is based on an early idea that mammal molars evolved from a series of triangles that point toward the tongue in upper molars and away from it in lower molars.

The three main cusps of an upper molar are the **paracone**, the **metacone**, and the **protocone**. The triangle formed by these cusps is called the **trigon**. Many primates have evolved a fourth cusp distal to the protocone, called the **hypocone**. Small cusps adjacent and lingual to these major cusps are called **conules** (the paraconule and the metaconule). Accessory folds of enamel on the buccal surface of the tooth are called **styles**, and an enamel belt around the tooth is referred to as a **cingulum**. Shallow areas between crests are called **basins**.

The basic structure of a lower molar in a generalized mammal is another triangle, this one pointing toward the cheek side. The cusps have the same names as those of the upper molars, but with the suffix *-id* added (for example, protoconid, paraconid, and metaconid). This basic triangle in the front

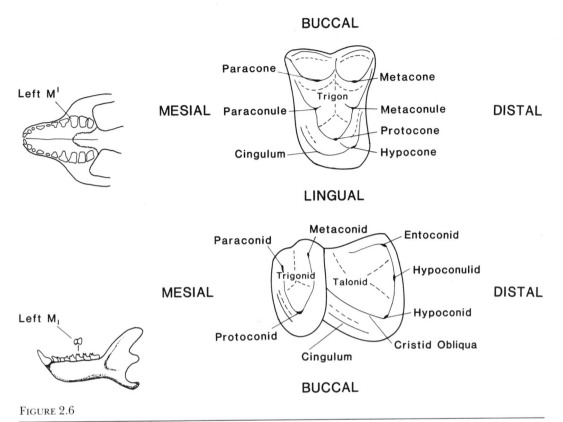

FIGURE 2.6

Major parts of the upper and lower teeth of a primitive primate.

of a lower molar is called the **trigonid**. In primates and all but the most primitive mammals there is an additional area added to the distal end of this primitive trigonid. This extra part, the **talonid**, is formed by two or three additional cusps: the hypoconid on the buccal side, the entoconid on the lingual side, and, in many species, a small, distalmost cusp between these two, the hypoconulid.

Primate dentitions are involved in two different aspects of feeding. The anterior part of the dentition, the incisors and often the canines (together with the lips and often the hands), is primarily concerned with ingestion—the transfer of food from the outside world into the oral cavity in manageable pieces that can then be further prepared by the cheek teeth (the molars and premolars).

The molars and premolars of primates break down food mechanically in three ways: (a) by puncture-crushing or piercing the food with sharp cusps, (b) by shearing the food into small pieces, that is, by trapping particles between the blades of enamel that are formed by the crests that link cusps, and (c) by crushing or grinding food in mortar-and-pestle fashion between rounded cusps and flat basins. Different types of food require different types of dental preparation before swallowing, and it is possible to relate

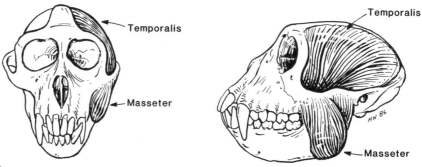

FIGURE 2.7

Anterior and lateral views of a primate skull showing the major chewing muscles.

the various characteristics of both the anterior teeth (for obtaining and ingesting objects) and the cheek teeth (for puncturing, shearing, or grinding) to diets with different consistencies (as we discuss in Chapter 8).

The movement of the lower jaw relative to the skull in both ingestion and chewing (mastication) is brought about by four chewing muscles that originate on the skull and insert on different parts of the lower jaw (Fig. 2.7). The largest is the **temporalis**,

FIGURE 2.8

Muscles of facial expression in a macaque (*Macaca*), a young orangutan (*Pongo*), and a human (*Homo sapiens*). Note the increasing differentiation of individual muscles, which enables finer control of expressions.

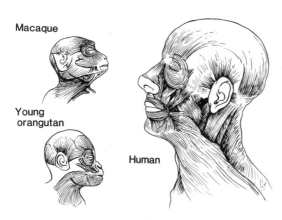

which has a fan-shaped origin on the side of the skull and inserts onto the coronoid process of the mandible. The second large muscle is the **masseter**, which originates from the zygomatic arch and inserts on the lateral surface of the mandible. Both of these muscles close the jaw when they contract. There are two smaller muscles on the inside of the jaw: the **medial** and **lateral pterygoids**. Much of the bony development of the primate skull seems to be related to the size and shape of these muscles and to the magnitude and direction of the forces generated in the skull during chewing. These muscular differences have in turn evolved to meet the different chewing requirements associated with dietary differences.

Muscles of Facial Expression

One additional aspect of cranial anatomy in primates that deserves special consideration is facial musculature (Fig. 2.8). Among primates, and especially in humans, the muscles of facial expression are more highly developed and differentiated into separate units than among any other groups of mammals (Hüber, 1931). It is these muscles that make possible the range of visual expressions that characterizes the behavior of primates.

The Brain and Senses

The structural shape of the skull—the development of bony buttresses and crests as well as the relative positioning of the face and the neurocranium—seems to be greatly influenced by the size and functional requirements of the masticatory system. However, the relative sizes of many parts of the skull, such as the neurocranium and the orbits, as well as the size and position of various openings in the skull, seem more directly related to the skull's role in housing the brain and the sense organs responsible for smell, vision, and hearing.

The Brain

The brain is the largest organ in the head, and its relative size is an important determinant of skull shape among primates. Relative to body weight, primates have the largest brains of any terrestrial mammals; only marine mammals are comparably brainy. There are, however, differences in relative brain size among primates. Lemurs, lorises, and *Tarsius* all have relatively smaller brains than do monkeys and apes, and human brains are relatively enormous. Still, the brain is a complex organ with many parts, and although some parts of primate brains are relatively large by mammalian standards, others are relatively small. In gross morphology, a primate brain can be divided into three parts (Fig. 2.9)—the brainstem, the cerebellum, and the cerebrum. Each part has very different functions and each, in turn, is made up of many different functionally distinct sections.

The **brainstem** forms the lower surface

FIGURE 2.9

Brains of a lemur (*Lemur*), a tarsier (*Tarsius*), a chimpanzee (*Pan*), and a human (*Homo sapiens*), showing differences in relative size of the parts of the brain. Note especially the differences in size of the olfactory bulb and size and development of convolutions on the cerebral hemispheres.

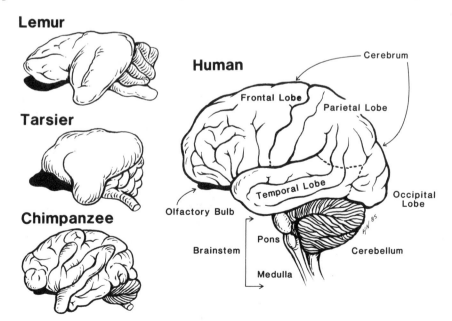

and base of the brain. It is an enlarged and modified continuation of the upper part of the spinal cord and is the part of the primate brain that differs least from that found among other mammals and lower vertebrates. The brainstem is concerned with basic physiological functions such as reflexes, control of heartbeat and respiration, and temperature regulation, as well as the integration of sensory input before it is relayed to "higher centers" in the cerebrum. All of the cranial nerves, which are responsible for innervation of such things as the organs of sight, smell, and hearing and the orbital muscles, arise from the brainstem. Very little of the primate brainstem is visible in either a lateral or superior view; it is covered by two areas that have become so large and specialized that they are recognized as separate parts—the cerebellum and the cerebral hemispheres.

The **cerebellum**, which lies between the brainstem and the posterior part of the cerebrum, is a developmental outgrowth of the caudal region of the brainstem. It is primarily concerned with control of voluntary movement and with motor coordination. Among primates there are few major differences in the relative size of the cerebellum, suggesting that this region has remained fairly conservative during primate evolution.

The paired **cerebral hemispheres** are the part of the brain that has undergone the greatest change during primate evolution. It is in this part that we find the greatest differences between primates and other mammals and the greatest differences among living primates (Fig. 2.9). Gigantic cerebral hemispheres are one of the hallmarks of human evolution. Anatomically, this part of the brain is divided into lobes named for the bones immediately overlying them—frontal, parietal, temporal, and occipital. In most primates the surface of the cerebral hemi-

spheres is covered with convolutions made up of characteristic folds, or **gyri**, which are separated by grooves, or **sulci**. The development of these convolutions is most apparent in larger species and reflects the fact that the most functionally significant part of the cerebrum, the gray matter, lies at the surface. The convolutions or foldings of the brain surface provide a greater increase in the surface area of the cerebral hemispheres with respect to brain or body volume than would be provided by a smooth spherical surface.

Overall, the cerebral hemispheres are involved with recognition of sensations, with voluntary movements, and with mental functions such as memory, thought, and interpretation. Different regions of the cerebrum (i.e., specific gyri) can be related to different functions. In many cases particular areas of the cerebral cortex can be related to different functions (Fig. 2.10). The central sulcus, for example, separates an anterior area related to voluntary movement from a more posterior area concerned with sensation. Within each of these areas, it is possible to identify more specific regions concerned with voluntary movement or sensory control of particular parts of the body. In addition, there are other parts of the cerebral hemispheres, called **association areas**, which are related to the integration of input from several different senses (such as hearing and vision) and to specific tasks such as speech. Two particularly well-developed association areas in the human brain are those related to speech, Broca's area in the frontal lobe and Wernike's area in the parietal lobe.

Although the brain is a soft structure, primate brains often leave their mark on the bony morphology of the skull. Size (in particular, volume) is an obvious feature of a primate brain that can be determined from a skull. Furthermore, in many species, sulci and gyri also leave impressions on the inter-

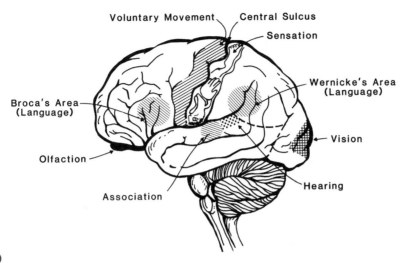

Voluntary Movement   Central Sulcus
                              Sensation

Wernicke's Area
(Language)

Broca's Area
(Language)

Vision

Olfaction

Hearing

Association

FIGURE 2.10

Important functional areas of the human brain.

nal surface of the cranium. Such impressions on fossil skulls can provide information about the development of different functional regions on the cerebral hemispheres of extinct primates.

All of the nerves that take signals to and from the brain enter and leave the cranial cavity through various holes, called **foramina**, in the skull bones. The largest of these holes is the **foramen magnum**, through which the spinal cord passes. The many smaller foramina vary considerably in size and position among living primates, and in a few cases it seems possible to correlate the size of a foramen carrying a specific nerve to the development of a particular function or anatomical region.

Cranial Blood Supply

Foramina also serve as passages for the arteries that supply blood to the brain and other cranial structures and for the veins that drain those same structures. The pathway of the blood supply to the brain shows a number of distinctly different patterns among living primates (Fig. 2.11). Although we know little about the functional significance of these differences, they have proved useful in sorting the phyletic relationships among many living and fossil primate species. The major blood supply to the head in primates comes from two branches of the common carotid artery at the base of the neck. The **external carotid** is primarily responsible for supplying structures in the neck and face, while the **internal carotid** (along with the smaller vertebral arteries) supplies the brain. The internal carotid artery enters the cranial cavity as two distinct arteries, a **stapedial artery** passing through the stapes bone and a **promontory artery** that generally lies medial to the stapedial artery and crosses the **promontorium**, a raised surface in the middle ear, to enter the cranial cavity further anteriorly. In most lemurs, for example, the stapedial is the larger artery; in tarsiers, New World monkeys, Old World monkeys, apes, and humans, the promontory provides most of the

Generalized Diagram of Cranial Blood Supply

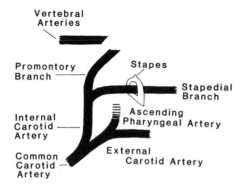

Lemur

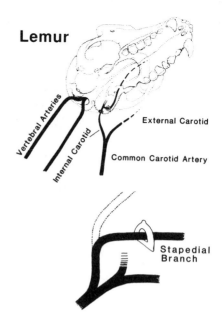

Slow loris

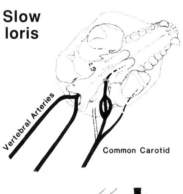
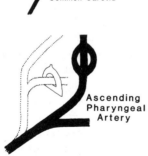

Macaque

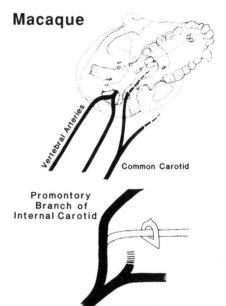

FIGURE 2.11

Cranial blood supply in several types of living primates. In all living primates, the vertebral arteries supply blood to the brain; however, species differ considerably in the relative contribution of the stapedial and promontory branches of the internal carotid artery and of the ascending pharyngeal branch of the external carotid artery. In a lemur (*Lemur*), the stapedial branch provides the major arterial supply to the brain; in a slow loris (*Nycticebus*), the intracranial blood supply comes from a large ascending pharyngeal artery; in a macaque (*Macaca*) and all higher primates, the promontory branch of the internal carotid provides the major arterial blood supply.

blood supply to the brain. In lorises, galagos, and cheirogaleids, a branch of the external carotid artery, the **ascending pharyngeal**, provides the major blood supply to the brain (Fig. 2.11).

Olfaction

In many mammals, smell is the dominant sensory mode. It provides much of the information on which animals rely to find their way around, locate their food, locate potential predators, communicate with their kin and neighbors, and determine the sexual status of potential mates. Among more diurnal (active during the day) higher primates, smell seems to be less important for some of these functions than other senses, such as vision. But even for these species this most basic of senses has not been abandoned. It still plays an important, but relatively poorly understood, role in reproduction, communication, and food evaluation in most primate species.

The sensation of smell is carried by the **olfactory nerves**, which lie under the large frontal lobes of most primates and end in paired swellings, the **olfactory bulbs** (Fig. 2.9). The bulbs receive their input from the special sensory membranes lining the scroll-like **turbinates** of the internal nasal cavity. The development of the nasal part of the olfactory system and its position with respect to the orbits shows two distinctly different arrangements among primates (Fig. 2.12). In lemurs and lorises, as well as in most other mammals, the nerves responsible for olfaction pass from the brain into the internal nasal cavity between the orbits. Within the nasal cavity, large numbers of turbinates are attached to several different bones, including several derived from the ethmoid bone that lies in a special cul-de-sac, the **sphenoid recess**. In tarsiers, monkeys, apes, and humans, the structure of this region is greatly simplified. The olfactory nerves pass over the interorbital septum, rather than between the orbits, and the sphenoid recess is missing, as are the posteriormost two turbinates. In apes and humans this region is even further reduced.

Although primate noses and the tissue-lined passages that make up their internal structure are primarily associated with olfaction, they also play important roles in respiration and temperature regulation by warming and humidifying the air that passes over them.

In addition to their sense of smell, lemurs, lorises, tarsiers, and many New World monkeys (but not Old World monkeys, apes, or humans) have an additional sense that

FIGURE 2.12

Structure of the interior nasal region of a lemur (*Lemur*), a tarsier (*Tarsius*), and a squirrel monkey (*Saimiri*). Note the reduction in number and relative size of the turbinates in *Tarsius* and *Saimiri*. M, maxilloturbinate; N, nasoturbinate; E, ethmoturbinates (numbered).

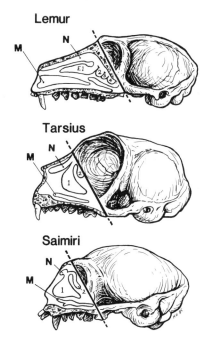

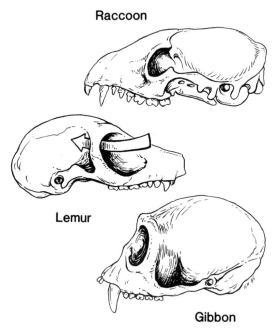

Raccoon

Lemur

Gibbon

FIGURE 2.13

The bony structure of the orbit in a raccoon, a lemur, and a gibbon. In the raccoon skull, the orbit is open laterally. In the lemur, the eye is surrounded by a bony ring which is open posteriorly. In the gibbon, the posterior opening of the orbit is closed off so that the eye is surrounded by a bony cup.

seems to be particularly important in sexual communication. The **vomeronasal organ** (or Jacobson's organ) is a chemical-sensing organ that lies in the anterior part of the roof of the mouth in many mammals. It is stimulated by substances found in the urine of female primates and permits other individuals to determine chemically the reproductive status of a female.

Vision

Primates rely extensively on vision to understand the world around them. Nevertheless, there are considerable differences among primate species in many aspects of their visual systems, both in the bony structure of the orbit and in the soft anatomy of the eye and the parts of the brain related to sight. Primate eyes vary strikingly in relative size. Nocturnal (active during the night) species have relatively larger eyes and bony orbits than do diurnal species.

In addition to their size, the bony orbits of primate skulls show important differences in construction (Fig. 2.13). In most mammals, and among the primitive, primatelike plesiadapiforms, each eye lies nestled in a pocket of tough but flexible connective tissue on the side of the skull, medial to the zygomatic bone. The lateral side of the orbit is formed by a fibrous ligament rather than by bone. In all living primates, however, the zygomatic bone and the frontal bones join to form a lateral strut, or **postorbital bar**, so that the eye is surrounded by a complete bony ring. In higher primates, and to a lesser extent in *Tarsius*, the orbit is further walled off behind by a bony partition, the **postorbital plate**; thus the eyeball lies within a bony cup. This condition is described as **postorbital closure**. The functional significance of the postorbital bar and postorbital plate is regularly debated, with no clear solution. Each is probably related to the stresses imposed on the orbital region during chewing. In addition to these major differences in the mechanical structure of primate orbits, there is considerable variation among primate species in the arrangement of the mosaic of small bones forming the medial wall of the orbit.

The overall structure of most primate eyeballs is similar; the main differences lie in the structure of the retina, the filmlike sheet of light-sensitive cells that lines the back of the eye. Two types of cells make up the retina in most primates: rods, which are very sensitive to light but do not distinguish color, and cones, which are sensitive to color. In many nocturnal primates, the retina is composed totally of rods. Furthermore, in lemurs and lorises, as well as in many other primates, we find an additional feature characteristic of many nocturnal mammals: the retina contains an extra layer that reflects light. This layer, the **tapetum lucidum**, seems to reduce visual acuity but enhances an animal's ability to see at night by "recycling" all incoming light. In tarsiers, monkeys, apes, and humans (all of which lack a tapetum), we find a different modification of the retina—a specialized area of the retina, called the **fovea**, in which the light-sensitive cells are packed extremely closely together, allowing very good visual acuity.

Hearing

Hearing plays an important role in many aspects of primate life. Many species, especially those active at night, use hearing to locate insect prey, and most use their ears to listen for approaching predators and to receive the vocal signals emitted by their family and neighbors. Although we know much about the anatomy of the auditory system, the physiological significance of many anatomical differences among primate ear regions is poorly understood.

Anatomically, the primate ear can be divided into three parts—the outer ear, the middle ear, and the inner ear (Fig. 2.14). The outer ear is composed of the **external ear**, or **pinna**, and a tube leading from that structure to the eardrum, or **tympanic membrane**. Primate pinnae are extremely variable in size, shape, and mobility. In many nocturnal primates that rely extensively on hearing to locate prey, the outer ear is often a large, membranous structure that can be moved in many directions by a distinct set of muscles. In other species it is smaller, often only slightly movable (as in humans), and may even be totally hidden under fur. The outer ear collects sounds, localizes them with respect to direction, and funnels them into the auditory canal, where they set the tympanic membrane in motion.

The eardrum, a sheet of connective tissue spread over a bony ring formed by the tympanic bone, forms the boundary between the outer ear and the middle ear and

FIGURE 2.14

A primate ear, showing the three major parts and the individual elements in each part.

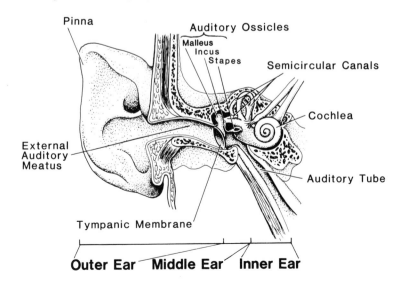

changes the moving air that makes up sound into mechanical movements that are passed along the three ossicles of the middle ear (the malleus, incus, and stapes). The last of these, the stapes, transfers this motion to the fluid-filled inner ear.

The inner ear contains three functionally different parts within the petrous portion of the temporal bone. The part concerned with hearing is the **cochlea**, a coiled, snail-shaped bony tube. Within the fluid-filled cochlea is a pressure-sensitive organ that registers the

FIGURE 2.15

The structure of the tympanic bone surrounding the eardrum and its position in relation to the bones surrounding the middle ear cavity vary considerably among living primate species (inferior view above, cross-sectional view of the middle ear below). In a lemur (*Lemur*), the tympanic bone is ring-shaped and is suspended within the bony bullar cavity. In lorises, the tympanic bone lies at the edge of the middle ear cavity and is connected to the wall of the bulla; note, also, that the bulla cavity is divided. In tarsiers (*Tarsius*), the tympanic bone is elongated to form a bony tube at the lateral edge of the bullar cavity. In New World monkeys, the tympanic bone is a ringlike structure fused against the lateral wall. In catarrhines, represented by an Old World monkey and a human, the tympanic bone is extended to form a bony tube.

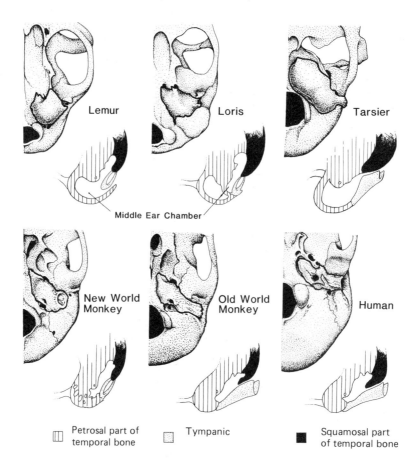

Lemur

Loris

Tarsier

Middle Ear Chamber

New World Monkey

Old World Monkey

Human

▥ Petrosal part of temporal bone　　▦ Tympanic　　■ Squamosal part of temporal bone

movement of the fluid and sends impulses to the brain through the acoustic nerve. The other two parts of the inner ear, three semicircular canals and two other fluid-filled chambers (the utricle and saccule), are responsible for sensing movement and for orientation with respect to gravity.

Apart from differences in relative sensitivity to particular frequencies, all primate ears seem to function in much the same way. There are, however, considerable architectural differences in the way the bony housing of the ear is constructed (Fig. 2.15). In all living primates, the inferior surface of the middle ear is covered by a thin sheet of bone, called the **auditory bulla**, derived from the petrous part of the temporal bone. In some primates, this bulla is inflated or balloonlike and is often divided into many compartments; in others, it is flatter. The physiological significance of the different types of ear architecture is poorly understood. The inflated auditory bullae of many small nocturnal primates seem to increase perception of low-frequency sounds and may be associated with nocturnal predation of flying insects.

The spatial relationship between the tympanic ring and the auditory bulla differs considerably among major groups of living primates (Fig. 2.15). In lemurs the ring lies within the cavity formed by the bulla, in lorises the ring is attached to the inside wall of the bulla, and in New World monkeys it is attached to the outside wall of the bulla. In tarsiers and catarrhines the ring is also attached to the wall of the bulla, but it extends laterally to form a bony tube, the **external auditory meatus**.

The Trunk and Limbs

Whereas the skull is primarily concerned with sensing the environment, with communication, and with the ingestion and prepa-

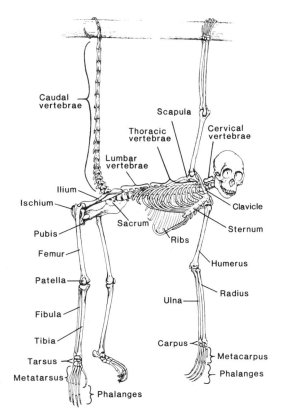

FIGURE 2.16

The skeleton of a spider monkey (*Ateles*). This species is unusual in having a very small thumb and a prehensile tail.

ration of food, the part of the skeleton behind the skull, the **postcranial skeleton**, as it is often called, serves quite different functions. Obviously it provides support and protection for the organs of the trunk, but its primary functions and those that seem to best account for the major differences in skeletal shape are related to locomotion. In this capacity, the postcranial skeleton provides both a structural support and a series of attachments and levers to aid in movement. The primate postcranial skeleton (Figs. 2.16, 2.17) is relatively generalized by

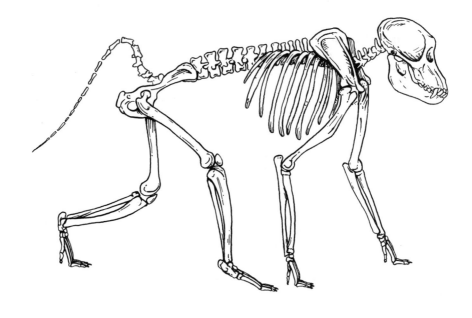

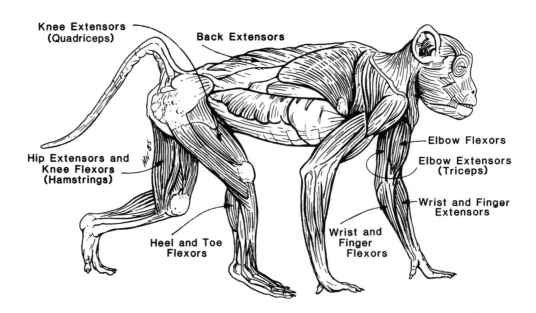

FIGURE 2.17

The skeleton of a baboon (*Papio*) and the superficial limb musculature of the same species showing the major muscle groups responsible for locomotion.

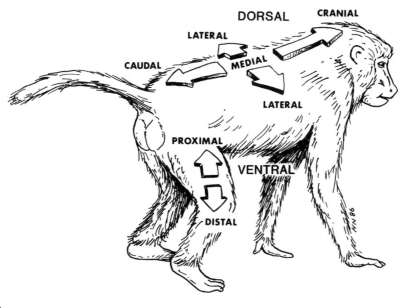

FIGURE 2.18

Terminology for anatomical orientation.

mammalian standards. Primates have retained many bones from their early mammalian ancestors that other mammals have lost. For example, most primates have a primitive limb structure with one bone in the upper (or proximal) part of each limb (the humerus or femur), a pair of bones in the lower (distal) part (the radius and ulna or tibia and fibula), and five digits on the hands and feet. Primate skeletons can be divided into three parts: the **axial skeleton** (the backbone and ribs), the **forelimbs**, and the **hindlimbs**. To facilitate descriptions of anatomical features, we use a standard terminology for directions with respect to an animal's body (Fig. 2.18).

Axial Skeleton

The backbone, which is made up of individual bones called **vertebrae**, is divided into four regions. The **cervical** or neck region

contains seven vertebrae in almost all mammals. The first two vertebrae, the atlas and the axis, are specialized in shape and serve as a support and pivot for the skull. The second region of the backbone is the **thorax**. Most of the rotational movements of the trunk involve movements between thoracic vertebrae. Primates have between nine and thirteen thoracic vertebrae, each of which is attached to a rib. The ribs are connected anteriorly with the sternum to enclose the thoracic cage, within which lie the heart and lungs. The outside of the thorax is covered by the muscles of the upper limbs. The thoracic vertebrae are followed by the **lumbar** vertebrae. There are no ribs attached to the lumbar vertebrae, but there are very large transverse processes for the attachment of the large back muscles that extend the back. Most of the flexion and extension of the back takes place in the lumbar region. The next lower region of the backbone is the

sacrum, a single bone composed of three to five fused vertebrae. The **pelvis**, or hipbone, is attached to the sacrum on its two sides, and the tail joins it distally. The last region of the spine, the **caudal region**, or tail, varies from a few tiny bones fused together (the **coccyx**, in humans) to a long, grasping organ of as many as thirty bones in some species (Fig. 2.16).

Upper Limb

The primate upper limb, or forelimb, is divided into four regions, most of which contain several bones. The most proximal part, nearest the trunk, is the **shoulder girdle**, which is composed of two bones—the **clavicle** anteriorly and the **scapula** posteriorly. All primates have a clavicle, in contrast to many other mammals—particularly, fast terrestrial runners such as dogs, cats, horses, and antelopes, which have lost this bone. The clavicle is one of the primitive skeletal characteristics of primates. This small S-shaped bone, attached to the sternum anteriorly and the scapula posteriorly, provides the only bony connection between the upper limb and the trunk.

The flat, triangular scapula is attached to the thoracic wall only by several broad muscles. It articulates with the single bone of the upper arm, the **humerus**, by a very mobile ball-and-socket joint. Most of the large propulsive muscles of the upper limb originate on the chest wall or the scapula and insert on the humerus. The muscles responsible for flexing and extending the elbow originate on the humerus (or just above, on the scapula) and insert on the forearm bones.

There are two forearm bones that articulate with the humerus—the **radius**, on the lateral or thumb side, and the **ulna**, on the medial side. The elbow joint is a complex region that involves the articulation of three bones. The articulation between the ulna and the humerus is a hinge joint that functions as a simple lever. The radius forms a more complex joint; this rodlike bone not only flexes and extends but also rotates about the end of the humerus. There are two articulations between the radius and the ulna, one at the elbow and one at the wrist.

FIGURE 2.19

The bony skeleton of a baboon hand (dorsal view).

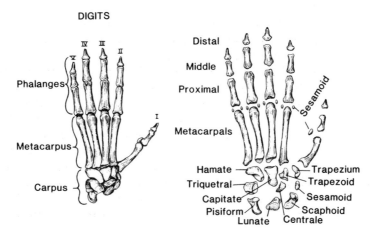

Because of its rotational movement, the radius can roll over the ulna. The movement of the radius and ulna is called **pronation** when the hand faces down and **supination** when the hand faces up. The muscles responsible for movements at the wrist and for flexion and extension of the fingers originate on the distal end of the humerus and on the two forearm bones. Distally, the radius and the ulna articulate with the bones of the wrist. The radius forms the larger joint between the forearm and the wrist, and in some primates (lorises, humans, and apes) the ulna does not even contact the wrist bones.

Primate hands (Figs. 2.19, 2.20) are divided into three regions—the **carpus**, or wrist, the **metacarpus**, and the **phalanges**. The wrist is a complicated region consisting of eight or nine separate bones aligned in two rows. The proximal row articulates with the radius, and the distal row articulates with the metacarpals of the hand. Between the two rows of bones is another composite joint, the midcarpal joint, which has considerable mobility in flexion, extension, and rotation.

The five rodlike metacarpals form the skeleton of the palm and articulate distally with the phalanges, or finger bones, of each digit. The joints at the base of most of the metacarpals are formed by two flat surfaces, offering little mobility, but the joint at the base of the first digit, the **pollex**, or thumb, is more elaborate and allows the more complex movements associated with grasping. The joints between the metacarpal and the proximal phalanx of each finger allow mainly flexion and extension and a small amount of side-to-side movement (abduction and adduction) for spreading the fingers apart. There are three phalanges (proximal, middle, and distal) for each finger except the thumb, which has only two (proximal and distal). The interphalangeal joints are purely flexion and extension joints.

As noted above, the muscles mainly responsible for flexing and extending the fingers and thumb lie within the forearm and send long tendons into the hand which insert on the middle and distal phalanges. The only muscles that lie completely within the hand are those forming the ball of the thumb, which are responsible for fine move-

FIGURE 2.20

Dorsal views of the left hand skeleton and palmar views of the right hand of six primate species.

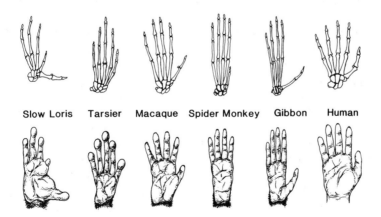

Slow Loris   Tarsier   Macaque   Spider Monkey   Gibbon   Human

ments of that digit, a smaller group forming the other side of the palm, and a series of small muscles within the palm which aid in complex movements of the digits. The palmar (relating to the palm) surfaces of primate hands and feet are covered with friction pads, a special type of skin covered with dermatoglyphics (fingerprints), and sweat glands. In most living primates, the tips of the distal phalanges have flattened nails, in contrast with the claws on the digits of most primitive mammals or the hooves of ungulates. A few primates have specialized claws on some of their digits.

Although primate hands usually have approximately the same numbers of bones, the relative sizes of the hand elements can vary greatly in conjunction with particular needs for locomotion or manipulation (Fig. 2.20). The slow-climbing loris, for example, has a robust thumb and long lateral digits for grasping branches; the more suspensory, hanging primates such as the gibbon or spider monkey have very long, slender fingers. Primates that use their hands for manipulating food, such as the macaque, or tools, such as humans, have well-developed thumbs that can be opposed to the fingers.

Lower Limb

The primate lower limb, or hindlimb, can be divided into four major regions: pelvic girdle, thigh, leg, and foot. These regions are comparable to the shoulder girdle, arm, forearm, and hand of the forelimb.

The primate **pelvic girdle** is composed of three separate bones on each side (the **ilium, ischium**, and **pubis**) which fuse to form a single rigid structure, the bony pelvis. In contrast with the pectoral girdle, which is quite mobile and loosely connected to the trunk, the pelvic girdle is firmly attached to the backbone through a nearly immobile joint between the sacrum and the paired ilia.

The primate pelvis, like that of all mammals, serves many roles. Forming the bottom of the abdomino-pelvic cavity, the internal part supports and protects the pelvic viscera, including the female reproductive organs, the bladder, and the lower part of the digestive tract. The bony pelvis also forms the birth canal through which the newborn must pass. In conjunction with this requirement, most female primates (including women) have a bony pelvis that is relatively wider than that of males. Finally, the pelvis plays a major role in locomotion; it is the bony link between the trunk and the hindlimb bones, and it is the origin for many large hindlimb muscles that move the lower limb.

The ilium is the largest of the three bones forming the bony pelvis. A long, relatively flat bone in most primates, it lies along the vertebral column and is completely covered with large hip muscles, primarily those responsible for flexing, abducting, and rotating the hip joint. The rodlike ischium lies posterior to the ilium; most of the muscles responsible for extending the hip joint and flexing the knee (hamstrings) arise from its most posterior surface, the **ischial tuberosity**. This tuberosity also forms the primate sitting bone. The pubis lies anterior to the other two bones and gives rise to many of the muscles that adduct the hip joint. The ischium and pubis join inferiorly and surround a large opening, the **obturator foramen**. The relative sizes and shapes of the ilium, ischium, and pubis vary considerably among different primate species in conjunction with different locomotor habits.

The part of the bony pelvis that articulates with the head of the femur, the **acetabulum**, lies at the junction of the three bones. The hip joint is a ball-and-socket joint that allows mobility in many directions.

The single bone of the thigh is the **femur**. The prominent features of this long bone

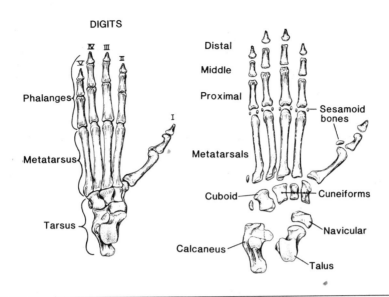

FIGURE 2.21

Dorsal views of the skeleton of a left baboon foot.

are a round head that articulates with the pelvis, the greater trochanter where many hip extensors and abductors insert, the shaft, and the distal condyles, which articulate with the tibia to form the knee joint. Most of the surface of the femur is covered by the quadriceps muscles, which are responsible for extension of the knee. Attached to the tendon of this set of muscles is the third bone of the knee, the small **patella**.

Two bones make up the lower leg, the **tibia** medially and the **fibula** laterally. The tibia is larger and participates in the knee joint; distally, it forms the main articulation with the ankle. The fibula is a slender, splintlike bone that articulates with the tibia both above and below and also forms the lateral side of the ankle joint. Arising from the surfaces of the tibia and fibula (and also from the distalmost part of the femur) are the large muscles responsible for movements at the ankle and those that flex and extend the toes during grasping.

Like the hand, the primate foot (Figs. 2.21, 2.22) is made up of three parts: tarsus,

metatarsus, and phalanges. The most proximal two tarsal bones are part of the ankle—the **talus** above and the **calcaneus** below. The head of the talus articulates with the **navicular bone**. This boat-shaped bone articulates with three small **cuneiform bones**, which in turn articulate with the first three metatarsals. The body of the talus sits roughly on the center of the calcaneus, the largest of the tarsal bones. The tuberosity of the calcaneus extends well posterior of the rest of the ankle and forms the heel process, to which the Achilles tendon from the calf muscles attaches. This process acts as a lever for the entire foot. Anteriorly, the calcaneus articulates with the **cuboid**, which in turn articulates with the metatarsals of the fourth and fifth digits.

In nonhuman primates, the digits of the foot resemble those of the hand (Figs. 2.19, 2.21). Each of the four lateral digits has a long metatarsal with a flat base and a rounded head, followed by three phalanges. The shorter first digit, the **hallux**, is opposable, like the thumb, and has a mobile joint

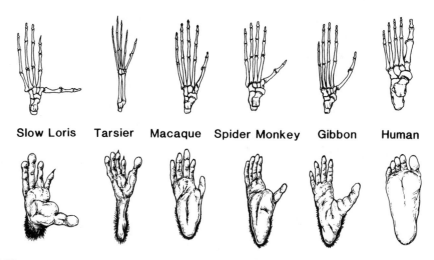

Slow Loris   Tarsier   Macaque   Spider Monkey   Gibbon   Human

FIGURE 2.22

Dorsal views of the left foot skeleton and plantar views of the right foot of six primate species.

at its base for grasping. Primate feet, like primate hands, show considerable differences from species to species in the relative proportions of different elements in association with different locomotor abilities. The climbing loris has a grasping foot, the tarsier has a very long ankle region for rapid

leaping, and the suspensory gibbon and spider monkey have long, slender digits for hanging. With their short phalanges and lack of opposable hallux, human feet are stiff, propulsive levers most suitable for walking on flat surfaces.

Limb Proportions

Primates vary dramatically in their overall body proportions. Some species have forelimbs longer than hindlimbs; others have hindlimbs longer than forelimbs. Some have limbs relatively long for the length of their trunk; others have relatively short limbs. These proportional differences are often described by a limb index, a ratio of the length of one part to the length of another part of the same animal. Table 2.1 gives the formula for some of the most commonly used indices. Of these, the **intermembral index**, a ratio of forelimb length to hindlimb length, is especially useful for describing the body proportions of a species and also seems to be correlated with locomotor differences in many primates (see Chapter 8). In gen-

TABLE 2.1
Skeletal Proportions

Intermembral index

$$\frac{\text{Humerus length} + \text{radius length}}{\text{Femur length} + \text{tibia length}} \times 100$$

Humerofemoral index

$$\frac{\text{Humerus length}}{\text{Femur length}} \times 100$$

Brachial index

$$\frac{\text{Radius length}}{\text{Humerus length}} \times 100$$

Crural index

$$\frac{\text{Tibia length}}{\text{Femur length}} \times 100$$

eral, leapers have a low intermembral index (longer hindlimbs), suspensory species have a high intermembral index (longer forelimbs), and quadrupedal species have intermediate indices (forelimbs and hindlimbs similar in size).

Soft Tissues

Primates are composed of more than just bones and teeth, but these are the parts usually preserved in the fossil record and in most museum collections of extant primate species. For extinct species, our knowledge of other aspects of anatomy must be based on inferences derived from our knowledge of the relationships between bony anatomy and the softer structures associated with that bony anatomy. For example, we can often reconstruct details of muscular attachments in extinct species from scars on bones. However, for understanding the adaptations and phylogenetic relationships of living primates, details of "soft" anatomy are often very important. Several organic systems have been well studied and provide insight into the evolution and adaptations of living primates.

Digestive System

In a previous section of this chapter we discussed the first part of the digestive system, the dentition and structures of the oral cavity; these cranial parts are involved in ingestion and the initial mechanical and chemical preparation of food items. The remainder of the digestive system (Fig. 2.23) lies primarily in the abdominal cavity and is concerned with further chemical preparation of food, absorption of nutrients, and excretion of wastes. The primate digestive system, like that of all vertebrates, is basically a long tube with some enlarged areas

(stomach and large intestine), some coiled loops (small intestine), one cul-de-sac (caecum), and two developmental outgrowths of the digestive tract, the liver and pancreas, which produce various digestive enzymes. Although there is considerable variation among primate species in the relative size and shape of individual organs in this system, largely associated with their different diets (see Chapter 8), the organs themselves and their functions are relatively similar throughout the order.

After food is prepared in the oral cavity, it is passed through the **esophagus**, a narrow muscular tube that traverses the thoracic cavity, into the abdominal cavity where it empties into the **stomach**. Here the food undergoes chemical preparation by digestive juices. The most specialized primate stomachs are those of the colobine monkeys; in

FIGURE 2.23

Diagram of the orangutan digestive system.

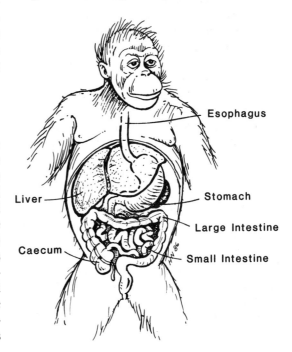

these primates, this organ is divided into several sections that function as fermenting chambers in which bacterial colonies break down cellulose.

From the stomach, the food passes to the **small intestine**, where further chemical preparation takes place. Here digestive juices from the liver and pancreas are mixed with the food, and much of the nutrient absorption takes place in this part of the gut. The small intestine is normally the longest part of the digestive tract. In most primates, it is several times as long as the animal's body and is usually folded into a series of loops within the abdominal cavity. At the end of the small intestine, the unabsorbed food and wastes are passed to the large intestine.

The **large intestine** is larger in diameter than the small intestine but usually shorter in length. It is involved primarily with further absorption of nutrients and water and, in its final parts, with excretion of solid wastes. At the beginning of the large intestine is the **caecum**, a cul-de-sac that varies considerably in size among different primate species and serves several special digestive functions. Like the colobine stomach, this out-of-the-way segment of the digestive tract is an ideal place for harboring the bacteria used to break down food items that primates can not normally digest, such as leaves or gums. The remainder of the large intestine, the **colon**, is usually divided into several parts on the basis of position within the abdominal cavity. From this last part of the large intestine, solid wastes leave the body through the rectum and anus.

Many of the adaptive differences in the digestive system of living primates are discussed further in Chapter 8. It is worth noting here, however, that different groups of primates have frequently evolved quite different visceral adaptations for similar digestive functions. Leaf-eating colobines,

for example, have evolved an enlarged stomach for digesting leaves, whereas leaf-eating primates of Madagascar have evolved an enlarged colon. Like all parts of primate anatomy, this system often reflects the interaction of evolutionary history and adaptation.

Reproductive System

All primates have a characteristically mammalian reproductive system in which the egg is fertilized internally and the embryo develops within the female's uterus for many months before it is born. This basic mammalian pattern of extensive investment by the mother during development and of infant nourishment for months or years after birth has important implications for the evolution of primate social behavior (discussed in Chapter 3). In this chapter we briefly review the anatomy underlying primate reproduction.

The anatomical structures associated with primate reproduction are similar to those found among other mammals (Fig. 2.24). The male reproductive system shows less variability among different species than does that of females. Like other mammals, male primates have paired **testicles** that normally lie suspended in a pouch, the **scrotum**, at the lower end of the anterior abdominal wall. Male primates differ from species to species in the position of the scrotum, which is usually behind the penis but may be in front, and in the timing of the descent of the testes from their fetal position within the abdomen to the reproductive position in the scrotum. There are also considerable differences in the relative size of the testes, which seem related to breeding systems (see Chapter 8), and in the size and external appearance of the penis. In most nonhuman primates there is a bone in the penis, the **baculum**.

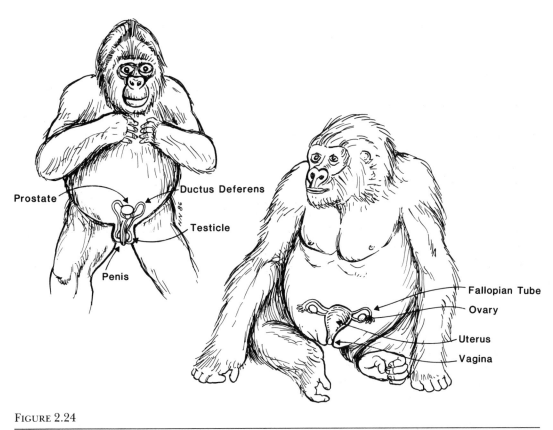

FIGURE 2.24

Diagram of the male and female reproductive organs in gorillas.

Like other mammals, female primates have paired **ovaries** and paired **fallopian** or **uterine tubes** extending laterally toward the ovaries from the midline uterus (Fig. 2.24). There is considerable variation among primate species in the relative size of the fallopian tubes and the body of the uterus. Among lemurs and lorises, the fallopian tubes are large relative to the body of the uterus, a condition normally found among mammals that have multiple births. Among tarsiers, monkeys, apes, and humans, the fallopian tubes are relatively slender and the body of the uterus is much larger, the condition normally found among mammals that give birth to single offspring.

The **vagina** lies below the uterus and opens onto the **perineum**, where the external genitalia are found. The external genitals of female primates generally consist of two sets of **labia** on either side of the vaginal opening and the **clitoris** anterior to the vagina. The clitoris of female primates varies in size and shape: in some species it is small and hidden beneath a hood; in others it is large and pendulous, in some cases larger than the male's penis. In addition, many female primates have areas of sexual skin surrounding the external genitalia which change color and size during the sexual cycle. In some species, such as baboons and chimpanzees, these sexual swellings are extremely large and provide a rather spectacular advertisement of an individual's reproductive condition.

Primates vary considerably in the periodicity of their reproductive physiology. At the extremes are Malagasy lemurs, in which reproductive activity in both males and females is limited to one day per year, and most higher primates, in which male sperm production seems to be relatively constant throughout the year and female ovulation occurs regularly at approximately monthly intervals. There are also numerous intermediate species in which both male and female reproductive activity (sperm production and ovulation) is limited to one or two seasons each year, often in response to environmental cues such as food availability or day length.

Compared to many mammals, primates have very small litters. Most species are characterized by single births. Only a few groups (some Malagasy prosimians and the New World marmosets and tamarins) regularly bear twins.

An aspect of primate reproduction that shows considerable differences among living primate species is the form of the placenta and other structures associated with the developing fetus within the mother's womb (Fig. 2.25). In most lemurs and lorises, the placental membranes are spread diffusely

FIGURE 2.25

Fetal membranes in three primates. In the lemur, several layers of tissue separate the uterus of the mother from the diffuse epitheliochorial membrane of the fetus. In the tarsier and the macaque, the developing embryo forms one or two placental disks that invade the lining of the uterus to become embedded in the uterine wall, providing a more intimate interchange between fetal and maternal circulation.

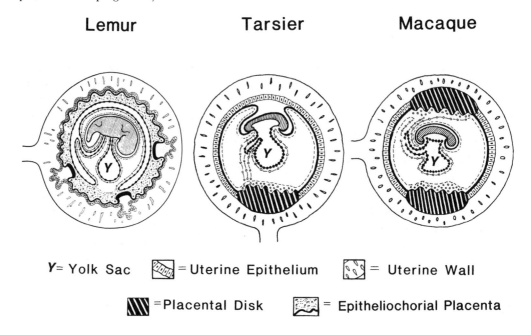

Lemur   **Tarsier**   **Macaque**

Y = Yolk Sac   = Uterine Epithelium   = Uterine Wall

= Placental Disk   = Epitheliochorial Placenta

throughout the uterine cavity, and fetal circulation is separated from maternal circulation by several tissue layers—a condition called **epitheliochorial placentation**. In *Tarsius* and in all higher primates, the placenta is localized into one or two discrete disks, and there is a much closer approximation between fetal and maternal blood supplies—a condition called **hemochorial placentation**. In great apes and humans, the intimacy of fetal and maternal circulation reaches its greatest degree and provides the most efficient transfer of nutrients to the fetus.

Growth, Development, and Aging

Compared to most other mammals of a similar size, primates are characterized by a long period of growth and development and concomitant large amounts of parental investment in immature offspring. An extreme in this regard are the great apes and humans, in which individuals become sexually mature only after ten or more years of growth. In most species, rapid growth during infancy is followed by a long childhood in which growth is relatively slow; just prior to sexual maturity, there is a phase of rapid growth, called the **adolescent growth spurt** (Fig. 2.26).

The slow growth and development of primates seems to be associated with a relatively long life span compared to other species of similar size. These primate characteristics of a long period of growth coupled with a relatively long life span seem causally associated with the primate emphasis on learning.

Systematically collected data on primate

FIGURE 2.26

A human growth curve (dashed line) and a generalized growth curve of a nonprimate mammal (solid line). In humans and other primates, there is a long period of slow childhood growth followed by the adolescent growth spurt. In contrast, most animals have a growth curve that decreases in rate from birth onward (modified from Watts, 1986a).

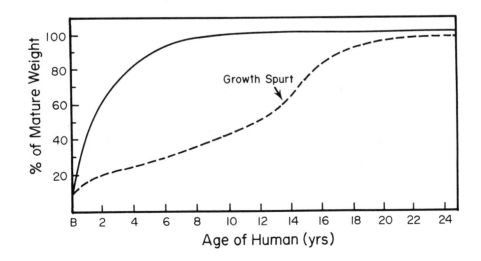

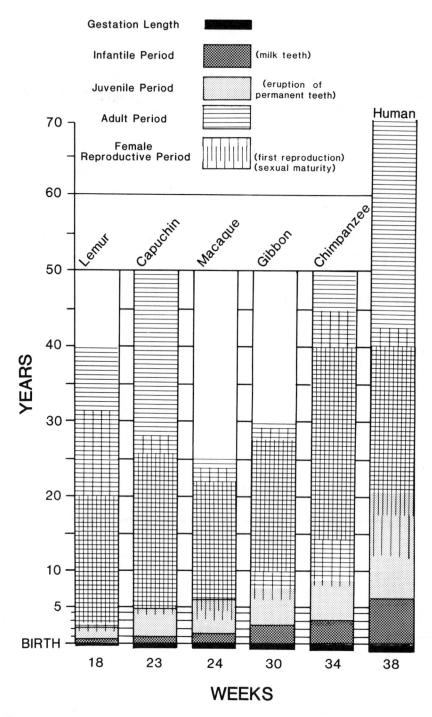

F<small>IGURE</small> 2.27

Life history parameters for several primates, showing lengths of different parts of the life cycle (modified from Schultz, 1960).

growth and development are available for only a few species (Fig. 2.27), and most of our knowledge of primate growth and development comes from isolated, anecdotal observations. Skeletal maturation has been studied in detail only in humans, macaques, chimpanzees, and capuchins. As a result, primate-wide comparisons on rates of skeletal maturation are almost impossible to make (Watts, 1986b).

One aspect of growth and development which has been studied in many primates and has played an important role in studies of primate and human evolution is the sequence and timing of the eruption of teeth (Fig. 2.28). There are a number of significant differences in the sequence in which the permanent teeth of primates appear in the jaw. In general, larger, longer-lived species are characterized by later eruption of the last two molar teeth, and humans are unusual in the early eruption of their canine teeth. These differences have been important in providing clues to the lifestyle of our early human ancestors (see Chapter 15 and, e.g., Mann, 1975; Smith, 1986).

Growth, development, and aging are often referred to collectively as **life history parameters**, and there is considerable interest in the ways these maturational and reproductive characteristics are related to ecological differences among species (see Harvey *et al.*, 1986). The most obvious correlations seem to be with size: larger species live longer and take longer to mature. There are also a few detailed differences among species in growth and development which can be related to particular ecological features.

FIGURE 2.28

Dental eruption sequences in a variety of primates. Arrows highlight eruption of the molar teeth to illustrate the sequence differences between genera (redrawn and modified from Schultz, 1960). C, canine; I, incisor; M, molar; P, premolar.

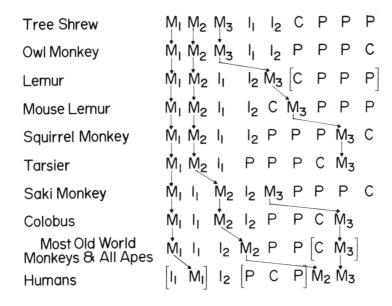

BIBLIOGRAPHY

GENERAL

Hartman, C.G., and Straus, W.L., Jr. (1933). *The Anatomy of the Rhesus Monkey*. Baltimore: Williams and Wilkins.

Hill, W.C.O. (1953–1970). *Primates*, vols. 1–8. Edinburgh: Edinburgh University Press.

———. (1972). *Evolutionary Biology of the Primates*. New York: Academic Press.

Hofer, H., Schultz, A.H., and Starck, D. (1956–1973). *Primatologia*, vols. 1–4. Karger: Basel.

Le Gros Clark, W.E. (1959). *The Antecedents of Man*. Edinburgh: Edinburgh University Press.

Schultz, A.H. (1969). *The Life of Primates*. New York: Universe Books.

Swindler, D.R., and Erwin, J., eds. (1986). *Comparative Primate Biology*, vol. 1: *Systematics, Evolution, and Anatomy*. New York: Alan R. Liss.

SIZE

Eisenberg, J. F. (1981). *The Mammalian Radiations*. Chicago: University of Chicago Press.

Jungers, W.L. (1985). *Size and Scaling in Primate Biology*. New York: Plenum Press.

CRANIAL ANATOMY

Gregory, W.K. (1922). *Origin and Evolution of the Human Dentition*. Baltimore: Williams and Wilkins.

Hiiemae, K.M., and Kay, R.F. (1973). Evolutionary trends in the dynamics of primate mastication. *Symp. Fourth Int. Cong. Primatol.* **3**:28–64.

Hüber, E. (1931). Evolution of facial musculature and cutaneous field of trigeminus, pt. II. *Q. Rev. Biol.* **5**:389–437.

Kay, R.F. (1975). The functional adaptations of primate molar teeth. *Am. J. Phys. Anthropol.* **43**:195–216.

———. (1984). On the use of anatomical features to infer foraging behavior in extinct primates. In *Adaptations for Foraging in Nonhuman Primates*, ed. P.S. Rodman and J.G.H. Cant, pp. 21–53. New York: Columbia University Press.

Moore, W.J. (1981). *The Mammalian Skull*. Cambridge: Cambridge University Press.

The Brain

Armstrong, E., and Falk, D., eds. (1982). *Primate Brain Evolution: Methods and Concepts*. New York: Plenum Press.

Falk, D. (1982). Primate neuroanatomy: An evolutionary perspective. In *A History of American Physical Anthropology*, ed. F. Spencer. New York: Academic Press.

Noback, C.R., and Montagna, W., eds. (1970). *The Primate Brain*. New York: Appleton-Century-Crofts.

Radinsky, L.B. (1975). Primate brain evolution. *Am. Sci.* **63**:656–663.

Cranial Blood Supply

Bugge, J. (1980). Comparative anatomical study of the carotid circulation in New and Old World primates: Implications for their evolutionary history. In *Evolutionary Biology of the New World Monkeys and Continental Drift*, ed. R. L. Ciochon and A.B. Chiarelli, pp. 293–316. New York: Plenum Press.

Conroy, G.C. (1982). A study of cerebral vasculature evolution in primates. In *Primate Brain Evolution*, ed. E. Armstrong and D. Falk, pp. 246–261. New York: Plenum Press.

MacPhee, R.D.E., and Cartmill, M. (1986). Basicranial structures and primate systematics. In *Comparative Primate Biology*, vol. 1: *Systematics, Evolution, and Anatomy*, ed. D.R. Swindler and J. Erwin, pp. 219–275. New York: Alan R. Liss.

Olfaction

Cave, A.J.E. (1973). The primate nasal fossa. *J. Linn. Soc.* **5**:377–387.

Fobes, J.L., and King, J.E. (1982). Auditory and chemoreceptive sensitivity in primates. In *Primate Behavior*, ed. J.L. Fobes and J.E. King, pp. 245–270. New York: Academic Press.

Vision

Cartmill, M. (1980). Morphology, function, and evolution of the anthropoid postorbital septum. In *Evolutionary Biology of the New World Monkeys and Continental Drift*, ed. R.L. Ciochon and A.B. Chiarelli, pp. 243–274. New York: Plenum Press.

Hearing

MacPhee, R.D.E., and Cartmill, M. (1986). Basicranial structures and primate systematics. In *Comparative Primate Biology*, vol. 1: *Systematics, Evolution, and Anatomy*, ed. D.R. Swindler and J. Erwin, pp. 219–275. New York: Alan R. Liss.

THE TRUNK AND LIMBS

Schultz, A.H. (1969). *The Life of Primates*. New York: Universe Books.

Schultz, M. (1986). The forelimb of the Colobinae. In *Comparative Primate Biology*, vol. 1: *Systematics, Evolution, and Anatomy*, ed. D.R. Swindler and J. Erwin, pp. 559–670. New York: Alan R. Liss.

Sigmon, B.A., and Farslow, D.L. (1986). The primate hindlimb. In *Comparative Primate Biology*, vol. 1: *Systematics, Evolution, and Anatomy*, ed. D.R. Swindler and J. Erwin, pp. 671–718. New York: Alan R. Liss.

Stern, J.T., Jr. (1971). Functional myology of the hip and thigh of cebid monkeys and its implications for the evolution of erect posture. *Bibl. Primatol.* **14**:1–318.

SOFT TISSUES

Digestive System

Chivers, D.J., and Hladik, C.M. (1980). Morphology of the gastrointestinal tract in primates: Comparisons with other mammals in relation to diet. *J. Morphol.* **166**:337–386.

Hill, W.C.O. (1958). Pharynx, oesophagus, stomach, small and large intestine: Form and position. *Primatologia* **3**:139–147.

———. (1972). *Evolutionary Biology of the Primates*. New York: Academic Press.

Hladik, C.M. (1967). Surface relative du tractus digestif de quelques primates, morphologie des villosities intestinales et correlations avec le regime alimentaire. *Mammalia* **31**:120–147.

Reproductive System

Dukelow, W.R., and Erwin, J., eds. (1986). *Comparative Primate Biology*, vol. 3: *Reproduction and Development*. New York: Alan R. Liss.

Hill, W.C.O. (1972). *Evolutionary Biology of the Primates*. New York: Academic Press.

Luckett, W.P. (1974). *Reproductive Biology of the Primates, Contributions to Primatology*, vol. 3. Basel: Karger.

———. (1975). Ontogeny of the fetal membranes and placenta. In *Phylogeny of the Primates*, ed. W.P. Luckett and F.S. Szalay, pp. 157–182. New York: Plenum Press.

GROWTH, DEVELOPMENT, AND AGING

Harvey, P.H., Martin, R.D., and Clutton-Brock, T.H. (1986). Life histories in comparative perspective. In *Primate Societies*, ed. B.B. Smuts, D.L. Cheney, R.M. Seyfarth, R.W. Wrangham, and T.T. Struhsaker, pp. 181–196. Chicago: University of Chicago Press.

Mann, A. (1975). *Paleodemographic Aspects of the South African Australopithecines*. Philadelphia: University of Pennsylvania Press.

Schultz, A.H. (1956). Postembryonic age changes. *Primatologia* **1**:887–964.

———. (1960). Age changes in primates and their modification in man. In *Human Growth*, ed. J.M. Tanner. *Symposium for the Study of Human Biology* **3**:1–20.

———. (1969). *The Life of Primates*. New York: Universe Books.

Smith, B.H. (1986). Dental development in *Australopithecus* and early *Homo*. *Nature* **323**:327–330.

Watts, E.S. (1985). *Nonhuman Primate Models for Human Growth and Development*. New York: Alan R. Liss.

———. (1986a). Evolution of the human growth curve. In *Human Growth*, ed. F. Falkner and J.M. Tanner, pp. 153–165. New York: Plenum Press.

———. (1986b). Skeletal development. In *Comparative Primate Biology*, vol. 3: *Reproduction and Development*, ed. W.R. Dukelow and J. Erwin, pp. 415–439. New York: Alan R. Liss.

THREE

Primate Life

In the previous chapter we discussed physical characteristics of primates. The purpose of this chapter is to enliven the primate body by introducing general aspects of primate behavior and ecology—where primates live, what they eat, how they move, and how they organize their social life. In later chapters we see how these parameters vary from species to species; in this one we introduce terminology and general principles.

Primate Habitats

Nonhuman primates today are found naturally on five of the seven continents (Fig. 3.1). There are no living primates other than humans on either Antarctica or Australia and no evidence that primates ever inhabited either continent before the relatively recent arrival of humans. Although primates occupy only marginal areas of Europe (Gibraltar) and North America (Central America and southern Mexico), they were formerly much more widespread on both continents. For the present, however, Africa, Asia, South America, and their nearby islands are the home of most living nonhuman primates.

A few hardy primate species live in temperate areas, where the winters are cold (South Africa, Nepal, and Japan), but these are exceptional. The vast majority of primate species and individuals are found in tropical climates, where daily fluctuations in temperature between day and night far exceed the average temperature changes from season to season. In these climates, seasonal changes in rainfall have a much greater effect on the vegetation and on the primates than do any seasonal temperature changes.

Forest Habitats

Within their geographic range, living primates are found in a variety of habitats ranging from deserts to tropical rain forests. Only a few hardy types such as chimpanzees, baboons, and Senegal bushbabies manage to successfully ply their primate trade year after year in the drier, more poorly

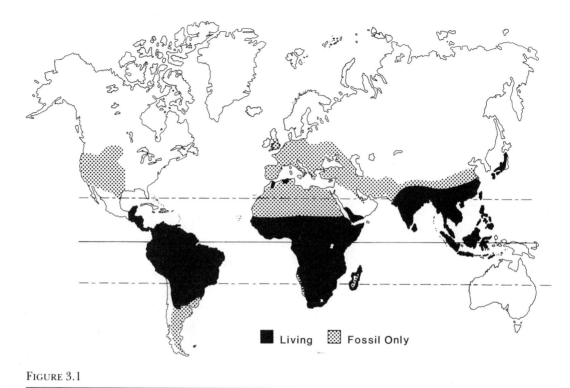

■ Living   ▦ Fossil Only

FIGURE 3.1

The geographic distribution of extant nonhuman primates and extinct primate species.

vegetated areas. The majority of primate species and individuals live in tropical forests of one sort or another. The forests come in many shapes, with variations in climate, altitude, topography, and soil type, as well as the characteristic flora of each particular continental area. A few of the more distinctive forest types are illustrated in Figure 3.2.

Primary rain forests are usually characterized by the height of the trees (up to 80 m) and the relatively continuous canopy that results from intense competition between many tree species for access to light. The dark **understories** of primary rain forests, which are usually quite open, are made up primarily of trunks and vines. The **canopies** of primary rain forests are punctuated by occasional **emergent trees**, which stand above the rest, and by gaps resulting from tree falls. It is through these gaps that light reaches the forest floor, enabling the forests to renew themselves.

Secondary forests, like the areas around tree falls, are characterized by denser, more continuous vegetation because of the availability of light. The canopy structure is less distinct and is often characterized by an abundance of vines and short trees. Because of the high levels of light, leaves and fruit can be very abundant in secondary forests.

African **woodlands** are made up of relatively shorter, deciduous trees. Between individual trees are continuous growths of grasses and low bushes. As the trees be-

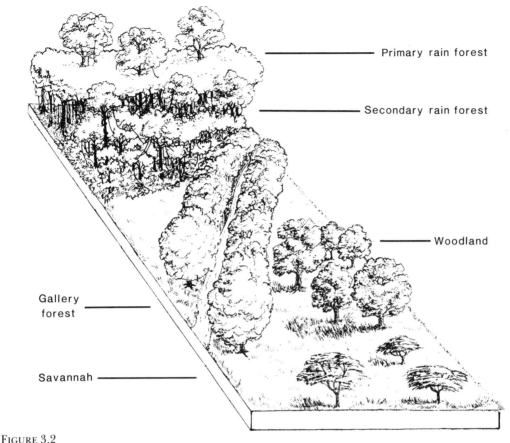

Primary rain forest

Secondary rain forest

Woodland

Gallery forest

Savannah

FIGURE 3.2

The diversity of habitats occupied by extant primates.

come more sparse, woodland gives way to **savannah**.

In many relatively dry tropical regions, forests are concentrated around rivers. These **gallery forests** can contrast strongly with surrounding areas in the types of animals they support.

There are other ways of categorizing forests. We find highland rain forests and lowland rain forests, as well as swamp forests, montane forests, and bamboo forests. Each of these environments presents a primate with a different array of trees on which

to move, different places to sleep, and different things to eat from season to season. The primates that inhabit these forests must meet these different demands. Many of the behavioral differences among living primate populations reflect adaptations to this diversity of habitats.

Habitats within the Forest

Equally diverse as the types of forests primates inhabit are the different niches primate species may occupy within a single

FIGURE 3.3

A rain forest scene from Surinam showing the different levels of a tropical rain forest, each with different types of substrates and each occupied by different primate species.

forest (Fig. 3.3). In a tropical forest, which reaches heights of 80 m, the temperature and humidity, the shapes of the branches, the kinds of plant foods, and the types of other animals a species encounters are usually quite different on the ground level, 20 m above the ground, or 40 m up in the canopy. Near the ground, there is little light, there are many vertical supports (such as small lianas and young trees), and there are terrestrial predators. Higher in the canopy, there are more horizontally continuous supports, which provide convenient highways for arboreal travel, and a greater abundance of leaves and fruits. Still higher, in the emergent layer, the canopy again becomes discontinuous, heat from the sun may be quite intense, and individuals are exposed to aerial predators. Primates, as well as many other arboreal animals, often move and feed in specific forest levels; they have adapted to these different demands and opportunities.

Primates also specialize on different types of trees within a forest—trees which may have distinctive structures or produce foods with unique characteristics. Some species rely on bamboo patches within the forest, others on palms, and still others on vines. Primates often seem to specialize on trees with characteristic sizes and productivity; some species seem to feed primarily on small trees that produce small quantities of fruit, while others concentrate on the forest giants that produce huge bonanzas of fruit. In sum, there are many different niches within any single forest habitat, each of which offers a slightly different way of making a living for a primate.

Primates in Tropical Ecosystems

The tropical forests inhabited by most primates are the most complex ecosystems on earth, often containing thousands of plant species, hundreds of vertebrates, and innumerable insects and other invertebrate species. It is important to remember that the evolution of living primates has taken place in conjunction with the evolution of other members of these complex environments. Plants, for example, are not the passive structural elements of the forest they might appear to be. Natural selection in plants has led to the evolution of elaborate and complex mechanisms for obtaining needed resources of light and nutrients, including defending leaves from herbivores, attracting pollinators, and ensuring that seeds are adequately dispersed and prepared for germination. The brightly colored, juicy fruits that form the diet of many primates have probably evolved those attributes for the purpose of attracting primates: once the fruits have been eaten, the seeds they con-

tain will be scattered about the forest. At the same time that many plants have evolved ways of enticing animals to help them disperse their seeds, they have also evolved mechanisms to protect their leaves and immature fruits from predators; they may cover them with spines, for example, or fill them with undigestible materials or toxic substances. Thus the dietary and foraging adaptations of living primates have evolved hand in hand with features in the tropical plants that affect their dietary choices.

In their roles as competitors, predators, and prey, the other animals of the forest have also had an important influence on the evolutionary history of primates. In Manu National Park, a pristine rain forest environment in the upper Amazon basin of Peru, frugivorous (fruit-eating) monkeys account for only about one-third of the biomass of frugivorous vertebrates (Terborgh, 1983). Birds, bats, various carnivores, and numerous rodents eat many of the same fruits as the primates and are often found in the same trees at the same time. There has certainly been competition among these different animals for access to the various food items in the forest. Many of the animals that inhabit the same forests as primates are not interested in the monkey's food, but in the monkeys as food. Large felids (including lions, tigers, leopards, jaguars, and pumas) prey on primates, as do many large birds and snakes. The presence of predators has probably had an important influence on the evolution of many aspects of primate ecology and behavior, including activity patterns, social organization, choice of sleeping sites, vocalizations, and coloration patterns (but see Cheney and Wrangham, 1986).

Unlike the large carnivores, few primates rely exclusively on other animals for their food, but many, especially the smaller species, do include various invertebrates and

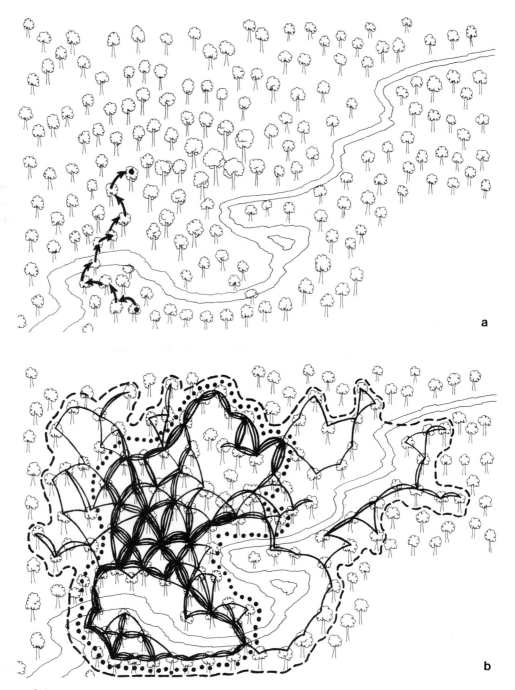

FIGURE 3.4

Primate land use: (a) The path an individual or group travels in a day is called a day range; (b) If all day ranges are combined, the total area utilized by the group is its home range (dashed line). The part of the home range that is most heavily used is called the core area (dotted line).

small vertebrates such as lizards as a regular part of their diet. As we shall see, primate species have evolved a number of unique predation strategies to exploit different types of prey in distinct parts of the forest structure. Capturing flying insects requires keen eyes and quick hands, and locating cryptic insects that live beneath the bark of trees or in leaf litter requires a keen sense of smell or hearing. Often such prey can only be reached by gnawing through the bark with specialized teeth, by ripping it open with strong hands, or by probing in crevices with slender fingers. Again, the evolution of primate adaptations reflects an interaction with the evolutionary history of other organisms in the forest.

Land Use

Primates live in a complex environment with many constantly changing variables. One way groups of primates deal with this complexity is to restrict their activities to a limited area of forest that they know well. Thus we find that primates are very conscious of real estate. In contrast with many birds or other mammals that have seasonal migrations, most primates spend their days, years, and often their entire lives in a single, relatively small patch of forest. To exploit this patch effectively, they must know many things about it—the different food trees and their seasonal cycles, the best pathways for moving, the best water sources, and the safest places for sleeping. Many researchers have suggested that it is this need for knowledge of their environment that is responsible for the evolution of primate mental abilities (MacKinnon, 1974; Milton, 1981).

There is a standard terminology used to describe the normal patterns of land use by primates and other animals (Fig. 3.4). The distance an individual or group moves in a single day (or night) is called a **day range** (Fig. 3.4a, arrows). If we map all the day ranges for a primate group, we can see the total area of land used over a longer period of time, for example, a year. This area of land (or forest) is called the **home range** (Fig. 3.4b, dashed line). Often a group uses one part of its home range intensively with only occasional, usually seasonal, forays into other parts. This heavily used area is called the **core area** (Fig. 3.4b, dotted line). Frequently the home ranges of neighboring groups of the same species overlap. In other instances there is almost no overlap and adjacent groups actively defend the boundaries of their home ranges with actual fighting or vocal battles. Such defended areas are called **territories**.

Activity Patterns

Most primates restrict their environment by limiting their activities to one particular segment of each twenty-four-hour day. Most mammals are **nocturnal**; they are active primarily at night and sleep during the day. By contrast, most birds are **diurnal**; they are active during the hours of light and sleep when it is dark. Some mammals are **crepuscular**; they are most active in the hours around dawn and dusk, when the light is at low levels. Nocturnality seems to be the primitive condition for primates; nearly three-quarters of the more primitive prosimians (lemurs, lorises, and tarsiers) are nocturnal, but there is only one nocturnal monkey. The majority of living primates are, however, diurnal. Many primate species show peaks of activity at dawn and dusk and have a rest period at either midday or midnight, but few, if any, are only crepuscular. There are also primates with quite variable activity

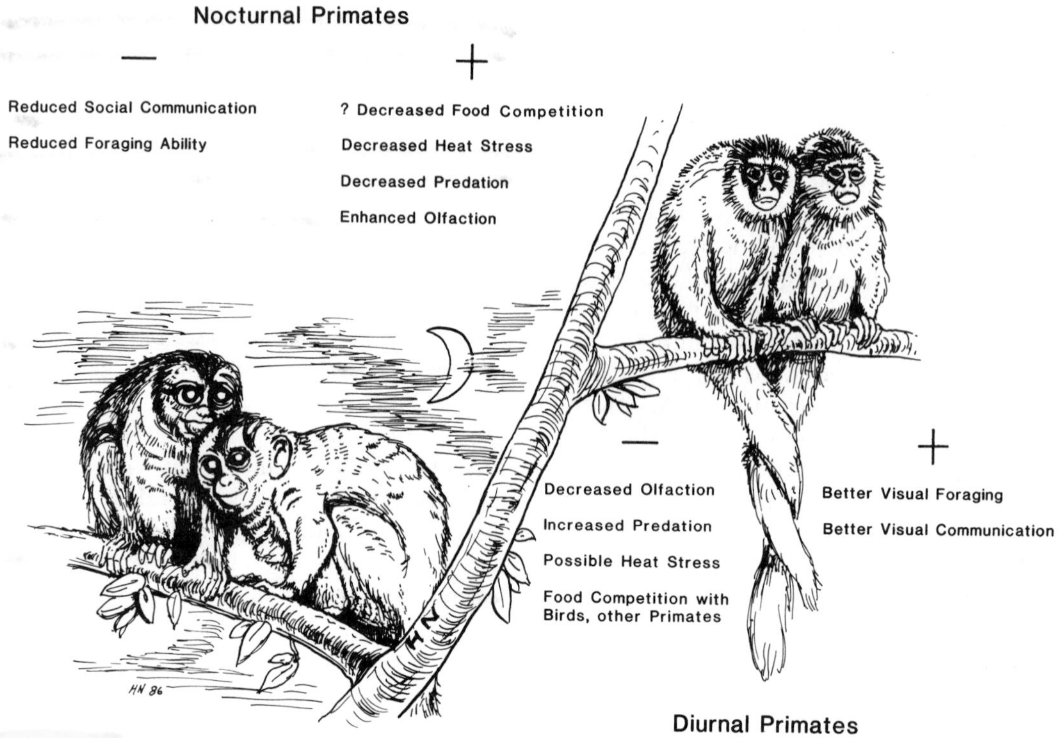

FIGURE 3.5

Potential benefits ($+$) and costs ($-$) of diurnality and nocturnality for two New World monkeys— the dusky titi monkey (*Callicebus*) and the owl monkey (*Aotus*).

patterns. Rather than being strictly diurnal or nocturnal, they seem to be sporadically active throughout a twenty-four-hour day, an activity pattern that we can call **cathemeral** (I. Tattersall, personal communication). Several lemur species show this cathemeral activity pattern.

Each of these ways of life has its advantages and disadvantages (Fig. 3.5). Diurnal species presumably have a better view of where they are going, of available food, and of potential mates, friends, competitors, and predators. At the same time, they have a greater risk of being seen by predators. Nocturnal species are better concealed from many predators, and they have fewer direct primate or avian competitors. They avoid heat stress due to sunlight, and they may even avoid diurnal parasites. They have the difficulties in feeding and social communication associated with restricted visual abilities, but their vocal communication may be better during hours of darkness, and olfactory communication seems to be enhanced by the humid night air (Wright, 1985). It is thus not surprising that nocturnal primates tend to live in small groups or alone and to communicate primarily through smells and sounds. A cathemeral activity pattern enables a species to exploit the advantages of both diurnality and nocturnality in conjunction with changes in temperature or food availability. The mongoose lemur, for example, is most active during daylight hours for

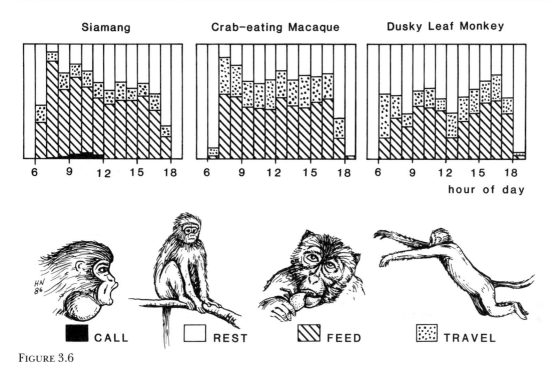

FIGURE 3.6

Primate activity histograms showing the portion of each hour of the day spent calling, resting, feeding, and traveling by three Asian primates.

the part of the year in which it feeds of fruits and new leaves; in the dry season, however, when these food items are scarce, it becomes more active at night and feeds on nectar (see Chapter 4).

A Primate Day

In addition to such drastic differences as diurnality and nocturnality, primates also show differences in the way they spend each day or night. For most primates the day is generally divided among three main activities: feeding, moving, and resting. Such activities as sex, grooming, and territorial displays usually occupy a relatively small part of each day. There are exceptions, of course. During their short breeding season, males of some lemur species may spend half

of their waking hours engaged in fighting with other males for the opportunity of mating with females during their brief period of sexual activity. For most primates, however, these activities are just occasional punctuations of long sequences of resting, feeding, and travel.

The distribution of activities throughout the day is usually not random (Fig. 3.6). Many primates generally travel early and late in the day and rest in the middle of the day. Most also begin and end each day with a long feeding period. Gibbons, and perhaps many other species as well, show a temporal pattern in food preference; they eat fruit in the morning and leaves in the evening. Their preference for fruit early in the day reflects a need for the quick energy available from fruits because of the high sugar and low fiber composition. Their choice of leaves in the evening perhaps reflects an attempt to maximize available digestion time (overnight) and to obtain the highest potential sugar content in leaves. Because plants cannot photosynthesize in the dark, leaves have higher sugar levels late in the day than they have early in the morning.

Primate Diets

Variation in the choice of foods on a daily, seasonal, and yearly basis is one of the greatest differences among living primates and one that has far-reaching effects on virtually all aspects of their life and morphology. Primate diets have generally been divided into three main food categories—fruit, leaves, and fauna (including both vertebrates and invertebrates, but usually insects and arachnids). Species that specialize on one of these dietary types are sometimes referred to as **frugivores, folivores**, and **insectivores** (or **faunivores**), respectively. These dietary categories accord well with the structural and nutritional characteristics of primate foods, and thus frugivores, folivores, and faunivores have characteristic features of teeth and guts (see Chapter 8). These gross dietary categories are also correlated with aspects of primate daily activity patterns such as home range size, day range size, and group size.

Like any categorization, however, this one glosses over many subtle differences in the types of foods primates eat and the different problems they must overcome to obtain a balanced diet from day to day. For example, new leaves and mature leaves often have very different chemical, textural, and nutritional compositions and may be available during different seasons of the year (Ganzhorn, 1988). Some fruits appear in large clumps; others are more evenly scattered in small numbers over a large area. As noted above, flying

insects must be hunted differently than burrowing insects. In addition, foods such as gums, seeds, and nectar, which are important in the diets of many primates and often require unique adaptations, do not easily fit into these three categories.

The many intricate ways primates obtain their food are usually referred to as **foraging strategies**. They are called "strategies" because many factors are involved, and the behavior of any species is probably the result of compromises and decisions among an array of potential behaviors, each with unique costs and benefits. Thus, within any one dietary category, such as frugivory, different species may have quite different foraging strategies. One species specializes on fruits that are regularly available in small amounts throughout the forest, while another species may specialize on fruits that are found in more irregularly spaced, but larger, clumps. We would expect two such species to be similar in their dentition and digestive system but to have very different ranging patterns. Many of the descriptions of individual primate species in later parts of this book emphasize the subtle differences in foraging strategies that have been found among primate species within the same general habitat. These subtle differences in feeding habits demonstrate the richness of primate adaptations that have evolved over the past 60 million years.

Locomotion

A major aspect of the foraging strategy of any species, and an aspect of behavior that shows considerable variation among primates, is **locomotion**, the way animals move. No other order of mammals displays the diversity of locomotor habits seen among primates. Like diet, primate locomotor habits can be crudely divided into several major

Arboreal quadrupedalism

Terrestrial quadrupedalism

Knuckle-walking quadrupedalism

Leaping

Suspensory climbing

Bipedalism

FIGURE 3.7

Examples of primate locomotor behavior.

categories (Fig. 3.7), each characterized by different patterns of limb use: leaping, arboreal and terrestrial quadrupedalism, suspensory behavior, and bipedalism. Each of these ways of moving may provide a primate with better access to a particular type of forest structure or may be more efficient for traveling on a particular type of substrate.

Leaping (**saltation**) allows arboreal species to move between discontinuous supports, for example, between separate trees or between tree trunks in the understory. **Arboreal quadrupedalism** is more suitable for movement on a continuous network of branches and is probably less hazardous than leaping, especially for larger species.

Terrestrial quadrupedalism enables a primate to move rapidly on the ground. **Suspensory behavior** allows larger species to spread their weight among small supports and also to avoid the problem of balancing their body above a support. Finally, **bipedalism** allows a species to progress on a continuous, level substrate while freeing the hands for other tasks.

As with dietary categories, the locomotor categories do considerable injustice to the actual diversity of primate movements. Some species leap from a vertical clinging position, others from more horizontal supports. Quadrupedal walking and running may involve different gaits in the trees or on the

FIGURE 3.8

A variety of primate feeding postures.

ground. Suspensory behavior includes many different activities, including brachiation (swinging by two arms), climbing, and bridging. As with dietary groups, this lumping of behaviors is mainly for the purpose of examining general patterns in either morphology or ecology.

In addition to locomotion, primatologists also pay careful attention to differences in primate postures—the way primates sit, hang, cling, or stand while they obtain their food, rest, or sleep (Fig. 3.8). In many instances, feeding postures may be as important in the evolution of the species as locomotion. Primates that feed on gums or other tree exudates, for example, often must cling to the side of a large trunk. This clinging ability may be more important than the method by which the tree is reached. Likewise, the suspensory locomotion of many

NOYAU   MONOGAMY   POLYANDRY

MULTIMALE GROUP   ONE-MALE GROUP

FISSION-FUSION SOCIETY   HAMADRYAS BABOONS

FIGURE 3.9

Common types of primate social group.

primates may be just a by-product of their need to hang below supports to feed on food sources at the end of small branches.

Social Life

The size and composition of the groups in which primates carry on their daily activities and methods they use to explore the area of land they inhabit are the most extensively studied aspects of primate behavior and ecology. All primates are social animals; they interact regularly with other members of their species (Charles-Dominique, 1971). Most diurnal species and some nocturnal ones are also **gregarious**—they feed, travel, and sleep in groups. The composition of these social groups differs considerably, however, from species to species.

Several distinct types of groups are particularly common among living primates (Fig. 3.9). The simplest, and certainly most primitive, social group is the **noyau**, which seems to characterize most primitive, nocturnal mammals (Charles-Dominique, 1983). The basic unit of this arrangement is the individual female and her offspring. In the noyau, adult males and females do not form permanent mixed-sex groups; rather, individual males have ranges that overlap several different female ranges. Thus, even though the two sexes do not travel together regularly, they interact often enough for males to monitor the reproductive status of the females and for females to have a choice of potential mates.

The next simplest grouping, at least in terms of numbers, is the **monogamous family** consisting of one adult female, one male, and their offspring. Nonhuman primates that live in these families appear to mate for life, and there is usually intense territorial competition between adjacent groups. Most of this competition appears to be intersexual—males compete to exclude other males and females compete to exclude other females.

One of the most interesting revelations in the study of primate societies in recent years has been the discovery that many marmosets and tamarins (small New World monkeys) live in **polyandrous** groups consisting of a single reproducing female and several sexually active males (Terborgh and Goldizen, 1985). In these groups, several of the males, as well as many other group members, participate in the care of offspring, and some authors have suggested that these groups are better viewed as a "communal breeding system" (Sussman and Garber, 1987).

Many primate species live in groups consisting of a single adult male along with several females and their offspring. Adult males not living with females band together to form separate all-male groups in some species, and in others they live alone. These **one-male groups** are almost invariably characterized by repeated instances of takeovers in which outside males oust the resident male, kill dependent infants, and mate with the females.

In contrast to these one-male groups, many species live in large bisexual groups that include several adult males, numerous females, and offspring, all of which forage as a group. Such groups are characterized by complex intratroop politics and competition. The distinction between one-male groups and such **multi-male** groups is very difficult to make for many species of living primates. As the young males mature, many one-male groups seem to become multi-male groups. Also, the composition of primate groups within a species may depend on factors such as troop size or population density. Because of this blurry distinction, Eisenberg and his colleagues (1972) have introduced an intermediate type of social group, the **age-graded group**.

There are many primate species for which social organization cannot be characterized so easily in terms of the numbers of males and females because the groupings change for different activities, or perhaps in different seasons. Hamadryas baboons, for example, forage all day in small one-male harem groups consisting of one male, one or a few females, and their offspring. Each evening, however, dozens of these small groups congregate on a single sleeping cliff, and sometimes the troop travels as a unit from one area to another. Among common chimpanzees, the social units are even more fluid—in what is called a **fission-fusion society**. Adult females usually forage alone, or with their offspring, whereas adult males are more frequently found in groups. Still, these subgroups of a single community frequently join to feed at a particularly rich food source, and individuals may associate temporarily for various reasons.

These categories of social organization, like all such classifications, mainly provide us with a convenient framework for comparing different species. The ultimate goal of such classification is to facilitate investigation of the factors that have given rise to this diversity in social organization. Primate social groupings are the result of many selective factors, each of which influences in a different way the size, composition, and dynamics of the social group. It is the dynamics of interindividual interaction, rather than just the numbers of males and females, that provides the real clues to understanding primate social systems (see, e.g., Rubenstein and Wrangham, 1986; Strum, 1987). It is important to note, too, that these selective factors affect individual group members in somewhat different ways. Factors that are of critical importance for one individual may be less significant for another of a different gender, age, or kinship. Most of our early knowledge of primate social behavior came from studies that lasted only one or two years. There are, however, increasing numbers of long-term studies indicating that individuals regularly move from troop to troop much more commonly than earlier workers suspected. Primate groups should not be considered stable, permanent units; rather, many are dynamic associations that constantly change as individuals are born, mature, emigrate, immigrate, mate, reproduce, and die. Many studies demonstrate that the social structure within a single species frequently changes with differences in resource availability or even chance demographic fluctuations. This dynamic nature of primate groups reenforces the notion that primate social groups are the result of many selective factors acting on each individual.

Why Primates Live in Groups

Compared with most other types of mammals, primates seem to be extremely social animals. This behavior is evident, not only in the diverse types of social groups described above, but also in the elaborate systems of scents, postures, facial expressions, and vocalizations that primates have evolved for communicating with their conspecifics. Primate social behavior has evolved through natural selection. Like all other primate adaptations, social behavior can be viewed as the result of a complex and often dynamic balance of selective advantages and disadvantages. From an evolutionary perspective, the fitness of an individual animal, or its evolutionary success, is equivalent to its reproductive success—the number of reproductively successful offspring it contributes to the next generation. Thus all aspects of a primate's life—feeding behavior, locomotion, defense against predators, as well as social behavior—have evolved to enable survival and successful reproduction. Only recently have biologists begun to explore social behavior from this evolutionary perspective, but the results are impressive in their utility for explaining details of behavior.

From the point of view of the individuals that make up primate groups, there are four potential advantages to group living: improved access to food, greater protection from predators, better access to mates, and assistance in caring for offspring. Each of these potential advantages is likely to have greater selective value for some individuals than others, depending on the individual's gender and age, the reproductive physiology of the species, and the ecological environment. Each potential advantage also must be balanced against the likely disadvantages of group living: increased competition with other individuals for these same resources of food, mates, and assistance in rearing offspring. The behavioral and physiological adaptations individual primates have evolved for maximizing their survival and that of their offspring in this maze of advantages and disadvantages are referred to as **reproductive strategies**.

Improved Access to Food

The reproductive strategies of all individuals depend ultimately on their ability to obtain enough food for themselves and their offspring. As noted above, the ways animals obtain and select their diets from the array of potential food sources are often referred to as foraging strategies. Foraging strategies are, in a larger sense, just one part of reproductive strategies for group-living animals—like most primates, which often breed throughout the year and have slowly maturing offspring that are dependent for many years after their birth.

The most important factor in determining the size of groups in which primates live seems to be the distribution of food resources in time and space. Primate species relying on foods that are found in small, evenly scattered patches, such as gums or many small forest fruits, usually live in small groups. Those that specialize on foods such as figs, which are usually found in gigantic but erratically spaced patches, tend to live in large groups. It is easy to see how the distribution of food resources can limit the size of groups that are able to feed on any single resource patch.

There are also several ways group feeding seems to provide individuals with better access to food than they might be able to obtain by foraging alone. Many primate species actively defend food sources—in some cases individual food trees, in others the troop's entire range. In general, disputes over food resources are often resolved by group size; larger groups can displace smaller groups in preferred food trees or in preferred areas. By joining a group, individuals gain access to its resources. There is obviously a fine balance between a group size that is small enough to subsist on a particular resource or set of resources and one that is still large enough to defend those resources from other groups. It should not be surprising that many primate groups that defend their food resources are composed of closely related individuals, usually females and their offspring.

Living in groups may also help primates locate food. Individuals may benefit in several ways from communal knowledge about the location of food sources, either through the memory of other individuals or through food calls given by other group members who are foraging semi-independently. There are also suggestions that primates feeding on insects may benefit from the disturbance caused by other troop members, who inadvertently flush out insects as they move.

Protection from Predators

Individuals living in groups seem to gain increased protection from predators; each individual benefits from the eyes, ears, and warning calls of every other individual. Furthermore, groups can gang up on an attacking predator, whereas an individual can only run away. Although comparative data are admittedly difficult to obtain, these potential advantages seem to outweigh the group's obvious disadvantage of being more visible and noisy (see Terborgh, 1983).

Access to Mates

Sexual reproduction requires that each reproductively successful male and female find a mate of the opposite sex. The reproductive strategies of males and females are, however, quite different for virtually all sexually reproducing animals. A critical aspect of primate reproduction that influences individual reproductive strategies is the marked asymmetry in the roles played by males and females during the early development of offspring. Female primates, like all female mammals, nourish and carry developing young for many months before birth and also provide milk for the newborn infant for months or years after birth. In contrast, the investment by a male primate to its offspring during this part of development is much less—and theoretically could be as little as a single sperm cell.

There are several consequences of this dramatic difference in the time and energy required of male and female primates. First, because of the time required by gestation, the maximum number of potential offspring a female primate can have in a lifetime is far less than the number that can be sired by a male, and the female's offspring must necessarily be more evenly spaced in time. With unlimited food resources, a single female with a twenty-year period of reproductive fertility, a litter size of one, and a six-month gestation period can theoretically (but not actually) produce forty offspring in her lifetime. To achieve this reproductive success, she will (again theoretically) need to associate with a male for mating purposes only briefly every six months. A male of many primate species can theoretically father the same number of offspring in a week (or even a day) if that number of receptive, fertile females are available. Thus, in the number of offspring they can physiologically produce, females are limited primarily by time, whereas males are limited by their access to females.

There are other consequences of this asymmetry in required investment. One is that females are always sure that the offspring they bear are their own. An individual male, on the other hand, can never be sure that he is the father of a newborn. Only by limiting the access of his mates to other males can he increase the likelihood that the offspring they produce are his own.

From these physiological differences in the relative minimal investment required to produce offspring and the relative certainty of parentage, we can predict that the theoretically optimal strategies of males and females for maximizing their reproductive success will be very different. The most successful male is one that mates with the greatest number of females and excludes other males from mating with these same females in order to ensure that all offspring are his own progeny. Females, on the other hand, have fewer obvious strategies for producing greater numbers of offspring. Female reproductive strategies seem to emphasize the quality rather than the quantity of offspring. Because every offspring involves such a large investment in time and energy, female strategies are concerned with ensuring that the male that sires the offspring is likely to engender healthy, strong progeny through paternal investment in such forms as protection from predators and access to food resources.

From these considerations we would expect male reproductive behavior to involve more intensive competition with other males for access to reproductively active females. The relatively greater intensity of male–male competition for access to mates over that expected in females is generally regarded as a major cause of sexual dimorphism in overall size and in the size of canine teeth, which are important in fighting and in dominance displays (see Chapter 8). Thus it seems that access to mates plays an important role in the reproductive strategies of males and a major selective factor in males joining groups.

Access to many potential mates seems to be a less important factor favoring group living for females of many primate species. Indeed, adult females outnumber males in most primate groups. However, among species that live in groups with numerous males and females, there is certainly competition among females for access to mates, and females frequently mate with numerous males. For females, a major important factor is the contribution that one or more of these males can make toward the survival of the offspring.

Assistance in Rearing Offspring

Mating is only the first step in successful reproduction. An individual's reproductive success is determined by the number of offspring that live to reproduce themselves, not by the number of conceptions. Offspring that do not survive to themselves reproduce are, from an evolutionary perspective, a wasted effort. For primates that give birth to relatively helpless young that require a relatively long time to reach adulthood, parental investment in the growing offspring is a particularly important aspect of reproductive behavior.

Because of their greater initial investment in offspring and the certainty of maternity,

females always make a substantial contribution toward the upbringing of infants in a primate group. Milk is expensive to produce and females may eat twice as much food when they are lactating. Thus it is not surprising that female primates receive help in raising offspring from other troop members. There is, however, considerable variability among primates in the contributions of males, females, and other, less closely related troop members in the care and rearing of immature animals. Investment in infants and dependent young seems to be correlated with the degree to which individuals are likely to be related to the offspring. In monogamous species, in which the male is virtually assured of paternity, males often contribute as much or more to the care of infants as do females. In larger, more complex social groups, adults of both sexes often assist the mother in caring for infants. In many primate societies, the adult females in the group are probably related, so infants are the "nieces" and "nephews" of other troop members. In addition, female primates have evolved many behavioral strategies to ensure assistance in rearing infants. By mating with several males, for example, females in multi-male troops can confuse the issue of paternity and perhaps elicit some investment from all of the males, since none can exclude the possibility that an infant is his offspring. Likewise, a male's willingness or ability to care for offspring may be a prerequisite for future matings. One worker has suggested that baboon females are more likely to mate with males who have helped care for her offspring in the previous year (Smuts, 1985). This access to help in rearing offspring from other individuals of all genders and ages is probably a more important factor favoring group living by females than is access to mates.

Primate Communities

The ability of primate species to specialize on different canopy levels or different types of food within a single forest habitat—and to do so during different parts of each day—often permits several species to thrive in the same habitat. Two species that are found in the same area are said to be **sympatric**; species whose ranges do not overlap are **allopatric**. Studies of groups of sympatric species are particularly important for our understanding of primate adaptations because they allow direct comparison of ecological variables (locomotion, diet, social organization) within one environment. Such studies of primate communities are essentially natural experiments in which the climate, the forest, and the competing species (such as predators, parasites, or other arboreal mammals or birds) are held constant. Thus the observer can see how changes in one ecological variable are correlated with changes in other variables.

In the following chapters we generally discuss primate species individually, but we often also compare related species that are sympatric. In this way we not only highlight the diversity of related animals but also see how differences in one parameter, such as diet, change in relation to other parameters, such as locomotion or social organization. Many of this book's illustrations show the ecological features of several sympatric species, enabling comparisons both within selected communities and across various communities (see, e.g., Madagascar, Fig. 4.8; Africa, Figs. 4.20, 6.12; South America, Fig. 5.7; Southeast Asia, Fig. 6.6). In each taxonomic chapter we discuss the diversity of adaptations within a particular evolutionary radiation, and in Chapter 8 we look more broadly across these radiations to examine general correlations among anatomical and ecological features that hold true for all species, regardless of their evolutionary history. Only by looking at primate communities from several perspectives can we appreciate how adaptation and evolution have produced the diversity of species we see today.

BIBLIOGRAPHY

PRIMATE HABITATS

Cheney, D.L., and Wrangham, R.W. (1986). Predation. In *Primate Societies*, ed. B.B. Smuts, D.L. Cheney, R.M. Seyfarth, R.W. Wrangham, and T.T. Struhsaker, pp. 227–239. Chicago: University of Chicago Press.

Napier, J.R. (1966). Stratification and primate ecology. *J. Anim. Ecol.* **35**:411–412.

Napier, J.R., and Napier, P.H. (1967). *A Handbook of Living Primates*. New York: Academic Press.

Terborgh, J. (1983). *Five New World Primates*. Princeton, N.J.: Princeton University Press.

Wolfheim, J.H. (1983). *Primates of the World: Distribution, Abundance, and Conservation*. Seattle: University of Washington Press.

LAND USE

Chevalier-Skolnikoff, S., Galdikas, B.M.F., and Skolnikoff, A.Z. (1982). The adaptive significance of higher intelligence in wild orang-utans: A preliminary report. *J. Hum. Evol.* **11**:639–652.

MacKinnon, J. (1974). The behaviour and ecology of wild orangutans (*Pongo pygmaeus*). *Anim. Behav.* **22**:3–74.

Milton, K. (1981). Distribution patterns of tropical plant foods as an evolutionary stimulus to primate mental development. *Am. Anthropol.* **83**:534–548.

Mitani, J.C., and Rodman, P.S. (1979). Territoriality: The relation of ranging patterns and home range size to defendability, with an analysis of territoriality among primate species. *Behav. Ecol. Sociobiol.* **5**:241–251.

Oates, J.F. (1986). Food distribution and foraging behavior. In *Primate Societies*, ed. B.B. Smuts, D.L. Cheney, R.M. Seyfarth, R.W. Wrangham, and T.T. Struhsaker, pp. 197–209. Chicago: University of Chicago Press.

Pollock, J.I. (1974). Spatial distribution and ranging behavior in lemurs. In *The Study of Prosimian Behavior*, ed. G.A. Doyle and R.D. Martin, pp. 359–409. New York: Academic Press.

ACTIVITY PATTERNS

Charles-Dominique, P. (1975). Nocturnality and diurnality: An ecological interpretation of these two modes of life by an analysis of the higher vertebrate fauna in tropical forest ecosystems. In *Phylogeny of the Primates: A Multidisciplinary Approach*, ed. W.P. Luckett and F.S. Szalay, pp. 69–88. New York: Plenum Press.

Raemaekers, J. (1978). Changes through the day in the food choice of wild gibbons. *Folia Primatol.* **30**:194–205.

Sussman, R.W., and Tattersall, I. (1976). Cycles of activity, group composition, and diet of *Lemur mongoz mongoz* Linnaeus 1766 in Madagascar. *Folia Primatol.* **26**:270–283.

Wright, P.C. (1982). Adaptive advantages of nocturnality in *Aotus*. *Am. J. Phys. Anthropol.* **57**(2):242.

———. (1985). *The Costs and Benefits of Nocturnality for Aotus trivirgatus (the Night Monkey)*. Ph.D. Dissertation, City University of New York.

PRIMATE DIETS

Chivers, D.J., and Hladik, C.M. (1980). Morphology of the gastrointestinal in primates: Comparisons with other mammals in relation to diet. *J. Morphol.* **166**:337–386.

Chivers, D.J., Wood, B.A., and Bilsborough, A. (1984). *Food Acquisition and Processing in Primates*. New York: Plenum Press.

Ganzhorn, J.U. (1988). Food partitioning among Malagasy primates. *Oecologia* **75**:436–450.

Kay, R.F. (1984). On the use of anatomical features to infer foraging behavior in extinct primates. In *Adaptations for Foraging in Nonhuman Primates: Contributions to an Organismal Biology of Prosimians, Monkeys and Apes*, ed. P.S. Rodman and J.G.H. Cant, pp. 21–53. New York: Columbia University Press.

Milton, K. (1978). The quality of diet as a possible limiting factor on the Barro Colorado Island howler monkey population. In *Recent Advances in Primatology*, vol. 1: *Behaviour*, ed. D.J. Chivers and K.A. Joysey, pp. 387–389. London: Academic Press.

Oates, J.F. (1986). Food distribution and foraging behavior. In *Primate Societies*, ed. B.B. Smuts, D.L. Cheney, R.M. Seyfarth, R.W. Wrangham, and T.T. Struhsaker, pp. 197–209. Chicago: University of Chicago Press.

Rodman, P.S., and Cant, J.G.H., eds. (1984). *Adaptations for Foraging in Nonhuman Primates: Contributions to an Organismal Biology of Prosimians, Monkeys and Apes*. New York: Columbia University Press.

LOCOMOTION

Fleagle, J.G. (1979). Primate positional behavior and anatomy: Naturalistic and experimental approaches. In *Environment, Behavior and Morphology: Dynamic Interactions in Primates*, ed. M.E. Morbeck, H. Preuschoff, and N. Gomberg, pp. 313–325. New York: Gustav Fischer.

Fleagle, J.G., and Mittermeier, R.A. (1980). Locomotor behavior, body size and comparative ecology of seven Surinam monkeys. *Am. J. Phys. Anthropol.* **52**:301–322.

Mittermeier, R.A., and Fleagle, J.G. (1976). The locomotor and postural repertoires of *Ateles geoffroyi* and *Colobus guereza*, and a reevaluation of the locomotor category semibrachiation. *Am. J. Phys. Anthropol.* **45**(2):235–251.

Prost, J. (1965). A definitional system for the classification of primate locomotion. *Am. Anthropol.* **67**:1198–1214.

Ripley, S. (1967). The leaping of langurs: A problem in the study of locomotor adaptation. *Am. J. Phys. Anthropol.* **26**:149–170.

Rose, M.D. (1974). Postural adaptations in New and Old World monkeys. In *Primate Locomotion*, ed. F.A. Jenkins, pp. 201–222. New York: Academic Press.

SOCIAL LIFE

Altmann, S.A., and Altmann, J. (1979). Demographic constraints on behavior and social organization. In *Primate Ecology and Human Origins*, ed. I.S. Bernstein and E.O. Smith, pp. 47–63. New York: Gartland STPM Press.

Charles-Dominique, P. (1971). Sociologie chez les lemuriens. *La Recherche* **15**:780–781.

———. (1983). Ecology and social adaptations in didelphid marsupials: Comparison with eutharians of similar ecology. In *Advances in the Study of Mammalian Behavior*, ed. J.F. Eisenberg and D. Kleiman, pp. 395–422. Special Publication no. 7, American Society of Mammalogists.

Eisenberg, J.F., Muckenhirn, N.A., and Rudran, R. (1972). The relationship between ecology and social structure in primates. *Science* **176**:863–874.

Garber, P.A. (1984). A preliminary study of the moustached tamarin monkey (*Saguinus mystax*) in northeastern Peru: Questions concerned with the evolution of a communal breeding system. *Folia Primatol.* **42**:17–32.

Jay, P.C. (1968). *Primates: Studies in Adaptation and Variability*. New York: Holt, Rinehart and Winston.

Leighton, D.R. (1986). Gibbons: Territoriality and monogamy. In *Primate Societies*, ed. B.B. Smuts, D.L. Cheney, R.M. Seyfarth, R.W. Wrangham, and T.T. Struhsaker, pp. 135–145. Chicago: University of Chicago Press.

McFarland, M.J. (1986). Ecological determinants of fission-fusion sociality in *Ateles* and *Pan*. In *Primate Ecology and Conservation*, ed. J.G. Else and P.C. Lee, pp. 181–190. Cambridge: Cambridge University Press.

Moore, J. (1984). Female transfer in primates. *Int. J. Primatol.* **5**(6):537–589.

Rubenstein, D.I., and Wrangham, R.W. (1986). *Ecological Aspects of Social Evolution*. Princeton, N.J.: Princeton University Press.

Smuts, B.B., Cheney, D.L., Seyfarth, R.M., Wrangham, R.W., and Struhsaker, T.T., eds. (1986). *Primate Societies*. Chicago: University of Chicago Press.

Strum, S.C. (1987). *Almost Human*. New York: Random House.

Sussman, R.W., and Garber, P.A. (1987). A new interpretation of the social organization and mating system of the *Callitrichidae*. *Int. J. Primatol.* **8**:73–92.

Terborgh, J., and Goldizen, A. Wilson. (1985). On the mating system of the cooperatively breeding saddle-backed tamarin (*Saguinus fuscicollis*). *Behav. Ecol. Sociobiol.* **16**:293–299.

WHY PRIMATES LIVE IN GROUPS

Alexander, R.K. (1974). The evolution of social behavior. *Ann. Rev. Ecol. Syst.* **5**:325–383.

Janson, C.H. (1986). Capuchin counterpoint. *Nat. Hist.* **95**(2):45–53.

Schaik, C.P. Van. (1983). Why are diurnal primates living in groups? *Behaviour* **87**:120–144.

Schaik, C.P. Van, and Van Hoof, J.A.R.A.M. (1983). On the ultimate causes of primate social systems. *Behaviour* **85**:91–117.

Terborgh, J. (1983). *Five New World Primates*. Princeton, N.J.: Princeton University Press.

Wilson, E.O. (1975). *Sociobiology, the New Synthesis*. Cambridge, Mass.: Belknap Press.

Wrangham, R.W. (1979). On the evolution of ape social systems. *Social Science Information* **18**:335–368.

———. (1980). An ecological model of female-bonded primate groups. *Behaviour* **75**:262–300.

———. (1986). Evolution of social structure. In *Primate Societies*, ed. B.B. Smuts, D.L. Cheney, R.M. Seyfarth, R.W. Wrangham, and T.T. Struhsaker, pp. 282–296. Chicago: University of Chicago Press.

Improved Access to Food

Wrangham, R.W. (1980). An ecological model of female-bonded primate groups. *Behaviour* **75**:262–300.

Protection from Predators

Cheney, D.L., and Wrangham, R.W. (1986). Predation. In *Primate Societies*, ed. B.B. Smuts, D.L. Cheney, R.M. Seyfarth, R.W. Wrangham, and T.T. Struhsaker, pp. 227–239. Chicago: University of Chicago Press.

Schaik, C.P. Van, Van Noordwijk, M.A., Warsono, B., and Sutriono, E. (1983). Party size and early detection of predators in Sumatran forest primates. *Primates* **24**(2):211–221.

Terborgh, J. (1983). *Five New World Primates*. Princeton, N.J.: Princeton University Press.

———. (1986). The social systems of New World primates: An adaptationist view. In *Primate Ecology and Conservation*, ed. J.G. Else and P.C. Lee, pp. 199–212. Cambridge: Cambridge University Press.

Access to Mates

Milton, K. (1985). Mating patterns of woolly spider monkeys, *Brachyteles arachnoides*: Implications for female choice. *Behav. Ecol. Sociobiol.* **17**:53–59.

Wrangham, R.W. (1979). On the evolution of ape social systems. *Social Science Information* **18**:335–368.

Assistance in Rearing Offspring

Janson, C.H. (1984). Female choice and mating system of the brown capuchin monkey *Cebus apella* (Primates: Cebidae). *Z. Tierpsychol.* **65**:177–200.

Kleiman, D. (1977). Monogamy in mammals. *Q. Rev. Biol.* **52**:39–69.

Leutenegger, W. (1980). Monogamy in callitrichids: A consequence of phyletic dwarfism? *Int. J. Primatol.* **1**:163–176.

Smuts, B.B. (1985). *Sex and Friendship in Baboons.* Hawthorne, N.Y.: Aldine.

Whitten, P.L. (1986). Infants and adult males. In *Primate Societies*, ed. B.B. Smuts, D.L. Cheney, R.M. Seyfarth, R.W. Wrangham, and T.T. Struhsaker, pp. 343–357. Chicago: University of Chicago Press.

FOUR

Prosimians

SUBORDER PROSIMII

In many respects, the living members of the primate order seem to form a natural ladder from primitive to more advanced, or specialized, types. This remarkable array provides us with living species that preserve some of the conditions through which early primates must have evolved. These living links can give us some idea of the pathways of primate evolution. Unfortunately, this series of linking forms is not easily classified.

The order Primates is commonly divided into two major groups or suborders, Prosimii (lemurs, lorises, and tarsiers) and Anthro-poidea (monkeys, apes, and humans). The prosimians are the more primitive of the two suborders and in many respects preserve a morphology similar to that found in primates of the Eocene epoch, 50 to 40 million years ago. Their English name, *pro-simian* ("before apes"), suggests this primitive nature, and the German *halbaffen* ("half-ape") is even more evocative.

There are eight families of living prosimians, all from the Old World (Fig. 4.1). Five of these are from the island of Madagascar: the cheirogaleids, lemurids, indriids, and

FIGURE 4.1

Geographic distribution of extant prosimians.

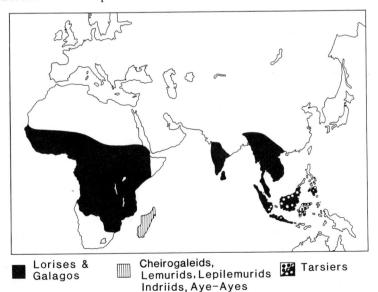

Lorises & Galagos

Cheirogaleids, Lemurids, Lepilemurids Indriids, Aye–Ayes

Tarsiers

two families that contain only a single living genus each, the lepilemurids and the daubentoniids. Closely related to the Malagasy families are the lorises of Africa and Asia and the galagos of Africa. Finally, there are the tarsiers of Southeast Asia. The diverse groups contained in the suborder Prosimii are united by their retention of primitive primate features and their lack of the features characteristic of the other suborder, the Anthropoidea, or higher primates.

This division of living primates into two suborders (Fig. 4.2) is a **gradistic** or **horizontal classification**; that is, the two groups, prosimians and anthropoids, are grades or, in a rough sense, stages of evolution. Such a classification provides no indication of which group of living prosimians may be closer to the origin of anthropoids, nor does it emphasize the derived characteristics, if any, that may be used to group prosimians. It simply expresses the fact that prosimians are primitive primates that lack anthropoid features.

An alternate grouping of living primates is by presumed lines of descent into a **phyletic** or **vertical classification** (Fig. 4.2). In this approach, the first seven groups of prosimians mentioned above (cheirogaleids, lemurids, lepilemurids, indriids, daubentoniids, galagids, and lorisids) are grouped together in one suborder, Strepsirhini. Tarsiers are then grouped with anthropoids as

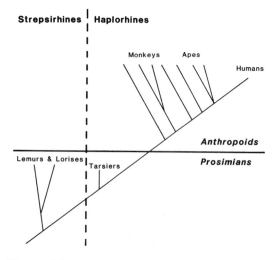

FIGURE 4.2

The gradistic division of primates into prosimians and anthropoids contrasted with the phyletic division into strepsirhines and haplorhines.

the Haplorhini, since they share a number of derived anatomical features with anthropoids which suggests that anthropoids are derived from a tarsier-like prosimian. In this scheme, strepsirhines and haplorhines are both monophyletic groups. Neither grouping scheme is totally satisfactory. Because of the difficulties of fitting many fossil primates into a strepsirhine–haplorhine dichotomy, we follow the more traditional classification while still discussing the phyletic grouping of the prosimian families.

Malagasy Strepsirhines

Living strepsirhines are united by two specialized features of "hard anatomy" that could be identified in fossils—their unusual dental tooth comb (and associated small upper incisors) and the grooming claw on the second digit of their feet (Figs. 4.3, 4.4). Their skull (Fig. 4.5) is characterized by the retention of primitive primate features such as a postorbital bar (without postorbital closure), a relatively small braincase, and a primitive mammalian nasal region with an ethmoid recess. Many of the distinctive soft structures of the strepsirhine cranial region, such as the well-developed nasal rhinarium

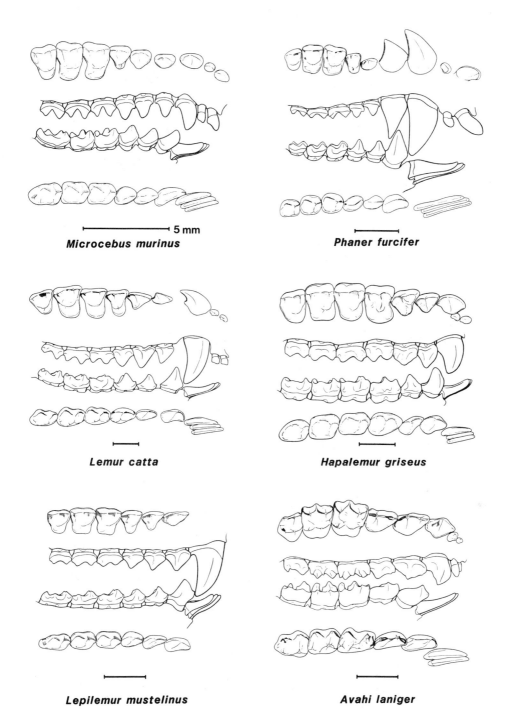

5 mm

Microcebus murinus

Phaner furcifer

Lemur catta

Hapalemur griseus

Lepilemur mustelinus

Avahi laniger

FIGURE 4.3

Dentition of representative Malagasy strepsirhines. For each species, occlusal view of upper
dentition (above; lateral view of upper and lower
dentition (center); and occlusal view of lower
dentition (below) (from Maier, 1980).

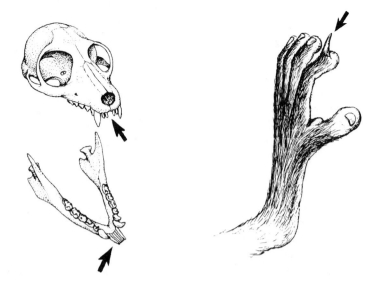

FIGURE 4.4

Distinctive skeletal features of strepsirhine primates: small upper incisors, separated by a large cleft, dental tooth comb composed of lower incisors and canines, and grooming claw on the second digit of the foot.

FIGURE 4.5

Skulls of a variety of extant strepsirhines.

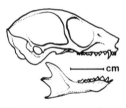

Microcebus murinus

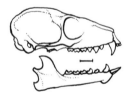

Varecia variegata

Lepilemur mustelinus

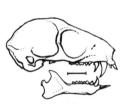

Phaner furcifer

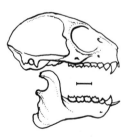

Hapalemur griseus

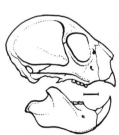

Daubentonia madagascariensis

and the reflecting tapetum lucidum in the eye, are also primitive mammalian features.

The reproductive system of all strepsirhines is characterized by at least two pairs of nipples, a bicornuate uterus, and an epitheliochorial type of placentation. Although these features distinguish strepsirhines from other primates, they are probably primitive primate characteristics, for they are found in many other groups of mammals.

Cheirogaleids

The greatest abundance and diversity of extant strepsirhines occurs on the island of Madagascar, off the eastern coast of Africa (Fig. 4.6). Five families live on this large and ecologically diverse island. The smallest strepsirhines and perhaps the most primitive of the Malagasy families are the cheirogaleids (Fig. 4.7, Table 4.1). They are all nocturnal, nest-building animals weighing less than 1 kg. They share with lemurids the primitive strepsirhine dental formula of $\frac{3.1.3.3}{3.1.3.3}$ and, in all but one genus, a primitive ear structure with the tympanic ring lying free within the bulla (see Fig. 2.15). The arrangement of the cranial blood supply in cheirogaleids shows the same unique pattern as that of the lorises and galagos. In both groups, the ascending pharangeal artery enters the skull near the center of the cranial base to form the internal carotid artery supplying the brain (see Fig. 2.11). The reproductive system is unusual among primates in that females have three pairs of nipples and normally give birth to twins. There are five extant genera.

The **mouse lemurs** (*Microcebus murinus* and *Microcebus rufus*) are among the smallest of all living primate species. They have a fairly short, pointed snout and large membranous ears. Their limbs are short relative to the length of their trunk, and their forelimbs are slightly shorter than their

Savanna and steppes
Dense Rainforest
Savoka
Mountain Forest
Dry Deciduous Forest
Spiny Desert

FIGURE 4.6

Madagascar and the distribution of different forest types.

FIGURE 4.7

Five genera of cheirogaleids: upper left, two mouse lemurs (*Microcebus murinus*); to their right, Coquerel's dwarf lemur (*Mirza coquereli*); just below, a fat-tailed dwarf lemur (*Cheirogaleus me-dius*); below it, a greater dwarf lemur (*Cheirogaleus major*); on the right, a fork-marked lemur (*Phaner furcifer*) clings to a tree and licks exudates.

Table 4.1
Infraorder Lemuriformes
Family CHEIROGALEIDAE

Common Name	Species	Intermembral Index	Body Weight (g)
Gray mouse lemur	*Microcebus murinus*	72	70
Brown mouse lemur	*M. rufus*	71	60
Coquerel's dwarf lemur	*Mirza coquereli*	70	330
Greater dwarf lemur	*Cheirogaleus major*	72	450
Fat-tailed dwarf lemur	*C. medius*	68	140–300
Fork-marked lemur	*Phaner furcifer*	68	440
Hairy-eared dwarf lemur	*Allocebus trichotis*	—	?100

hindlimbs. Their hands are very humanlike in proportion. Their tail is approximately the same length as their body.

The slightly larger, gray species (*M. murinus*) is found throughout the drier forests of the western, northern, and southern coastal areas, while the brown species (*M. rufus*) lives in the more humid forests of the east coast as well as in patches of humid forest in the north and on the central plateau. Mouse lemurs are particularly abundant in secondary forests and in the undergrowth and lower levels of virtually all forest types, including cultivated areas. They are arboreal quadrupeds that move primarily by walking and running along very small branches and leaping between terminal twigs.

Mouse lemurs are the most faunivorous of the cheirogaleids (Fig. 4.8). They eat invertebrates as well as small vertebrates (tree

Figure 4.8

Diet and forest height preference for five sympatric prosimians in the dry forest of western Madagascar (data from Hladik *et al.*, 1980).

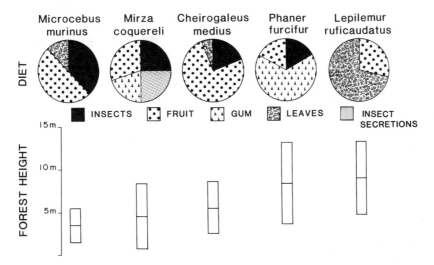

frogs, chameleons), which they catch by quick hand grasps. Although they are mainly arboreal, they frequently prey on terrestrial insects by leaping on them from low perches. The greater portion of their diet, however, consists of fruits, flowers (nectar), buds, and leaves. In feeding, they use a wide range of postures, including hindlimb suspension.

Mouse lemurs are nocturnal and seem to be most active just after nightfall and before sunrise. During the day they sleep in nests, which they make among small branches or in hollow trees. On the west coast, their activity shows considerable seasonal variation. In the relatively lush, wet season they increase their weight from 50 to 80 g, largely by storing fat in their tail, which increases fourfold in volume. During the long dry season their activity decreases, and they may spend several days without feeding, using the stored fat for sustenance. In captivity, mouse lemurs show a similar seasonal change in dietary preference, with a relatively greater protein intake during the wet season.

Mouse lemurs are basically solitary, with individuals foraging in separate stable ranges. Adult females occupy small, distinct ranges, but neighboring individuals frequently nest together in common sleeping beds or "dormitories." The home ranges of male mouse lemurs are much larger and usually overlap with several (about four) female ranges in a noyau type of social structure. Young, probably nonreproductive males appear to occupy adjacent, less optimal areas that do not overlap with female ranges.

Like all Malagasy species, mouse lemurs are seasonal breeders. Individual females are receptive for only one day at the end of the dry season (September–October), and their birth season coincides with the wet season (November–February). Female mouse lemurs usually have litters of two or three infants, which remain in the nest while the mother forages. The infants do not cling to the mother's fur. To move them, she carries them in her mouth.

Often placed in the same genus with the mouse lemurs, **Coquerel's dwarf lemur** (*Mirza coquereli*) is a substantially larger species. It shares a number of dental features with the dwarf lemurs (*Cheirogaleus*) but has a very long tail and limb proportions similar to those of the mouse lemurs. Like mouse lemurs, *Mirza* has a pointed snout and large, membranous ears. *Mirza* is found only on the western and northwestern coast of Madagascar. It is sympatric with *Microcebus murinus* but seems to prefer thicker and taller forests near rivers or ponds and is found in slightly higher parts of the canopy (Fig. 4.8). Like *M. murinus*, Coquerel's dwarf lemur moves mainly by quadrupedal running, with some leaping.

Microcebus mirza feeds on insects and vertebrate prey, fruits, nectar, and some gums but seems to specialize on secretions from colonial insects. During the dry season, these insect secretions account for up to 60 percent of all feeding time. Like mouse lemurs, *Mirza* uses a variety of feeding postures, including clinging postures on the trunks of trees.

Coquerel's dwarf lemurs construct large, elaborate circular nests of leaves for their daytime resting. After leaving their nests at nightfall, these primates seem to devote the first half of the night to feeding and the second half to social interaction with conspecifics. They live in a noyau social system, but the overlapping female and male ranges appear to be similar in size. In the dry season, male–female interactions (including contact calls, grooming, play, and chasing) are very common and take place in the central areas of overlapping ranges. These lemurs show no indication of reduced activity during the dry season. Like mouse lemurs, they give birth to twins or triplets, which stay in the nests during the first three weeks.

There are two allopatric species of **dwarf lemurs**, which differ mainly in size: *Cheirogaleus medius*, the fat-tailed dwarf lemur, and *Cheirogaleus major*, the greater dwarf lemur. They have pointed snouts and moderate-size ears that are often partly hidden by their fur. In both species the tail is slightly shorter than the long trunk and the arms are much shorter than the legs.

The smaller species (*C. medius*) is abundant in the dry forest of the west and south, while the larger (*C. major*) is found in the more humid forests of the east and on the plateau. Both are arboreal quadrupeds that move more slowly than *Microcebus* or *Mirza* and are much less agile leapers.

Cheirogaleus medius (and presumably *C. major*) are predominantly frugivorous but opportunistically eat small amounts of insects, small vertebrates, gums, and nectar. Dwarf lemurs are less versatile than mouse lemurs in their feeding postures, and they usually maintain a quadrupedal posture for their body as they move in and out of hollow logs in search of prey. During the night they intersperse periods of activity with periods of rest.

Dwarf lemurs adapt to the dry winter of the west and south of Madagascar by hibernating for six to eight months of each year. During this time, they metabolize the enormous fat reserves stored in their tails during the wet season. In this period, the mean adult body weight of *C. medius* drops from 217 g in March to 142 g in November (Hladik et al., 1980). Dwarf lemurs generally nest in hollow trees, during both normal daytime sleeping and hibernation. All that is known about the social grouping of dwarf lemurs is that several individuals are frequently found hibernating in the same nest. Presumably

their social system resembles that of *Microcebus*. Mating is from September to October and litters of two or three are born in December or January.

The **fork-marked lemur** (*Phaner furcifer*) is one of the largest and ecologically most specialized of the cheirogaleids (Fig. 4.7). Its facial features are characterized by large membranous ears and dark rings around the eyes which join on the top of the skull to form a stripe down the back. It has relatively long hindlimbs and a very long bushy tail.

Fork-marked lemurs are widely distributed in Madagascar but are most common in the west. These lemurs forage in all levels of the forest and specialize on gum. They have a number of distinctive anatomical adaptations commonly found among primates with this unusual diet (Martin, 1972b; Charles-Dominique, 1977). They have very large hands and feet with expanded digital pads, and their fingernails are keeled like claws for clinging to the trunks of trees. For obtaining gums, they have very procumbent incisors, both above and below, long canines and anterior upper premolars (see Fig. 4.3), and a long and narrow tongue. Their gut is characterized by a large caecum in which the gums are chemically broken down. Their locomotion is rapid quadrupedalism interspersed with leaps from branch to branch.

In contrast with other cheirogaleids, male and female fork-marked lemurs seem to live in more or less permanent groups, many of which contain one male and one female. The male follows between 1 and 10 m behind the female during the night, and the two stay in constant vocal contact. They forage one at a time at the gum sites, with the females appearing to have first choice. During the day, a pair of fork-marked lemurs normally sleeps in a tree hole or in a nest built by *Mirza*.

The **hairy-eared dwarf lemur** (*Allocebus trichotis*) is known only from a handful of

FIGURE 4.9

Three lemurid species from different parts of Madagascar: above, a pair of brown lemurs (*Lemur fulvus*); center, a pair of ruffed lemurs (*Varecia variegata*); below, three ring-tailed lemurs (*Lemur catta*), on the ground.

osteological specimens and may well be extinct. It is similar in size to *Microcebus murinus*, but in its dentition it is more like *Phaner* (Schwartz and Tattersall, 1985). The most distinctive feature of the genus, which separates it from other cheirogaleids, is the construction of its auditory region. Rather than having a free tympanic ring within the auditory bulla, *Allocebus* resembles lorises and galagos in having a tympanic ring fused to the wall of the bulla (Cartmill, 1975, 1982). Nothing is known of its natural behavior, but the dental similarities to *Phaner* and the presence of keeled nails on its digits suggest a diet of gums.

Lemurids

The lemurids (Fig. 4.9, Table 4.2) are the better-known, typical, Malagasy lemurs. They share the same dental formula with the cheirogaleids, $\frac{3.1.3.3}{3.1.3.3}$, and the tympanic ring lies free within the auditory bulla as in most cheirogaleids and indriids (Fig. 2.15). Their cranial blood supply, however, is largely through the stapedial artery rather than through the ascending pharangeal artery as in cheirogaleids. They are medium-size (1–4 kg), generally diurnal, group-living prosimians and do not build nests. The genus *Lemur* contains at least six species and displays a wide range of behavioral and ecological characteristics.

Lemur catta, the **ring-tailed lemur** (Fig. 4.9), is one of the larger species of the genus. It is a gray animal with a long striped tail; sexes look alike.

Ring-tailed lemurs are diurnal, live in the dry south of Madagascar, and feed both on the ground and in the trees. They are the most terrestrial of living strepsirhines, spending 30 percent of each day and 65 percent of their traveling time on the ground. They are primarily quadrupedal walkers and runners. Their diet varies from region to region, depending on both habitat

TABLE 4.2
Infraorder Lemuriformes
Family LEMURIDAE

Common Name	Species	Intermembral Index	Body Weight (g)
Ring-tailed lemur	*Lemur catta*	70	2,670
Brown lemur	*L. fulvus*[a]	72	2,500
Mongoose lemur	*L. mongoz*	72	2,025
Black lemur	*L. macaco*	71	2,401
Red-bellied lemur	*L. rubriventer*	68	2,350
Crowned lemur	*L. coronatus*	69	?2,000
Ruffed lemur	*Varecia variegata*	72	3,800
	(!)*Pachylemur insignis*	98	—
	(!)*P. jullyi*	94	—
Gentle bamboo lemur	*Hapalemur griseus*	64	880
Greater bamboo lemur	*H. simus*	65	2,500
Golden bamboo lemur	*H. aureus*	—	1,200

[a]simplified systematics.
(!)extinct.

and competition from other lemur species, and contains large amounts of both
fruit and leaves.

Ring-tailed lemurs live in large social
groups of about twenty individuals which
contain approximately equal numbers of
males and females. The groups travel almost
a kilometer a day and occupy a home range
of between 5 and 10 ha (hectares). *Lemur
catta* societies, like those of most strepsirhines, seem to center around a group of
females who are dominant over the males in
the troop. There is very little competition
among the males except during and just
preceding the annual breeding season—in
contrast with the situation seen in most
higher primates.

The **brown lemur** (*Lemur fulvus*) is similar
in size to *L. catta* (Fig. 4.9) but quite different
in many aspects of its ecology. Many subspecies of brown lemurs are found in forests
throughout Madagascar. Despite a considerable variability in chromosome number, all
seem to be capable of interbreeding. Many
of the subspecies of *L. fulvus* are sexually
dichromatic—that is, males and females
have different pelage patterns. Brown lemurs are primarily diurnal, but there are
some indications of cathemeral activity (Conley, 1975). In contrast to ring-tailed lemurs,
brown lemurs are totally arboreal. They
move primarily by quadrupedal walking and
running and by leaping. Their diet is largely
of leaves.

Compared with *L. catta*, brown lemurs live
in somewhat smaller groups, averaging
about a dozen individuals, with equal numbers of males and females. In the southwest
of Madagascar they travel less than 50 m
each day within their tiny (less than 1 ha)
home ranges, but subspecies in the eastern
rain forest have much larger day ranges and
home ranges of up to 80 ha.

The **mongoose lemur** (*Lemur mongoz*) is a
most unusual primate in many aspects of
its behavior and ecology. These small lemurs
(2 kg) live in forested areas both in the north
of Madagascar and on the nearby Comoro
Islands of Anjouan and Moheli. They are
exclusively arboreal. Mongoose lemurs show
extreme variability in their activity pattern,
both between different populations and, in
at least one area, from season to season.
Tattersall (1976) found them active at night
on the island of Moheli and in the lowlands
of Anjouan. Yet in the cold, wet highlands of
Anjouan they are active in the daytime. On
Madagascar, Tattersall and Sussman (1975)
found mongoose lemurs active only in the
night in July and August (the dry season),
whereas Harrington (1978) saw the same
population active during the day in the rainy
months of February through July. Similar
cathemeral activity cycles have been reported
for wild populations of *Lemur rubriventer*
(Overdorff, 1987) and also in captive populations of both *L. catta* and *L. fulvus*. The
field observations suggest that mongoose
lemurs, at least, tend to be nocturnal in
dry conditions and diurnal in cold, wet conditions.

Oddly enough, for *L. mongoz* only the
nocturnal diet and ranging patterns are well
known. During the dry season in Madagascar, according to one study, when they are
nocturnal, they fed 80 percent of the time on
a single species and from a total of only five
species during the entire observation period
(Tattersall and Sussman, 1975). The major
component of their diet was flowers and
nectar (and some fruit), for which their main
competitors were bats (Sussman, 1978).
They ate virtually no leaves.

The social organization of this dichromatic species seems to be as flexible as its
activity period, but the two are not clearly
correlated (Tattersall, 1978). Most populations live in monogamous family groups

composed of an adult male, an adult female, and their offspring. Other groups seem to have more adults (possibly older offspring?).

Very little is known about the ecology and behavior of other species of the genus *Lemur* (*L. macaco*, *L. coronatus*, and *L. rubriventer*).

The beautiful **ruffed lemur** (*Varecia variegata*), from the east coast forests, is the largest lemurid and seems to be the most primitive in many aspects of its behavior (Fig. 4.9). Ruffed lemurs have a long, doglike snout, thick fur, and a long tail. Compared with other lemurs, they have relatively short limbs. Ruffed lemurs are totally arboreal and are reported to be almost exclusively quadrupedal. In captivity they frequently adopt hindlimb suspensory postures. They are reported to be frugivorous.

Nothing is known of the social organization of ruffed lemurs, but most sitings are of small groups (2–5 individuals). In some features of their reproductive behavior, ruffed lemurs resemble cheirogaleids more than other lemurids. As in cheirogaleids, female *Varecia* have three pairs of nipples, regularly give birth to twins, and are reported to build nests. Like other Malagasy species, they are seasonal breeders, and females are fertile on only one day per year.

Genus *Hapalemur* (Fig. 4.10) is a periph-

FIGURE 4.10

A family of gentle bamboo lemurs (*Hapalemur griseus*) in a typical bamboo habitat.

eral member of the lemurids which is sometimes placed in a separate family. There are
three species: a small one, the **gentle lemur**
(*H. griseus*), with a fairly broad distribution;
and two larger, rarer species—the **greater** or
broad-nosed bamboo lemur (*H. simus*) and
the **golden bamboo lemur** (*H. aureus*), both
from restricted areas on the southeastern
rain forest. All *Hapalemur* species have relatively short faces and small, hairy ears. They
have relatively short arms and long legs
compared with other lemurids (Jungers,
1979). All show a preference for bamboo
forests and live almost entirely on bamboo
shoots and leaves, but they seem to eat
different parts of the plant. Their locomo-

tor and postural behavior in this vertical habitat is mainly clinging and leaping, but they
also move quadrupedally along bamboo
branches when feeding. They are diurnal
and possibly crepuscular in their activity. All
Hapalemur species are regularly found in
small (presumably family) groups and have
single births.

Lepilemurids

Lepilemur, the **sportive** or **weasel lemur** (Fig.
4.11, Table 4.3), is often grouped with the
lemurids or indriids, but it is distinctive
enough to deserve its own family (Petter *et
al.*, 1977). This small, drab-colored lemur is

FIGURE 4.11

Several sportive lemurs (*Lepilemur mustelinus*) in a dry forest of Didiereaceae bush.

TABLE 4.3
Infraorder Lemuriformes
Family LEPILEMURIDAE

Common Name	Species	Intermembral Index	Body Weight (g)
Subfamily Lepilemurinae			
Sportive lemur, Weasel lemur	*Lepilemur mustelinus[a]*	59	500–1,000
(!)Subfamily Megaladapinae			
	(!)*Megaladapis edwardsi*	120	140,000
	(!)*M. grandidieri*	115	65,000
	(!)*M. madagascariensis*	114	60,000

[a]simplified systematics.
(!)extinct, weight estimated from dental dimensions.

characterized by a lack of permanent upper incisors (Fig. 4.3), an unusual articulation between the mandible and the skull (Tattersall and Schwartz, 1974), large digital pads on its hands and feet, and a large caecum. There is considerable variability in the chromosomes among the widespread populations of this genus, and some authors (Petter *et al.*, 1977) recognize up to seven distinct species.

Sportive lemurs are found in all types of forests throughout Madagascar and are often locally abundant. They prefer vertical postures and travel mainly by leaping.

These lemurs are predominantly nocturnal and folivorous (Fig. 4.8). Like other folivorous primates, they are unable to digest cellulose, the main structural component of leaves, so they rely on the bacteria they maintain in their digestive tract for this task. In sportive lemurs, as in horses and rabbits, the bacteria live in the caecum, at the base of the large intestine; therefore the cellulose is broken down very near the end of the digestive tract. *Lepilemur* (like rabbits) overcomes this difficulty by reingesting its feces (containing the broken-down cellulose) during the day.

Like many of the cheirogaleids, sportive lemurs have been reported to live as solitary individuals with overlapping home ranges in a noyau arrangement or in small groups (especially in the mating season). Because of the low nutritional content of leaves and their small size, these lemurs are on a particularly tight energy budget, and their nightly activity pattern reflects this condition. Individuals have very small home ranges (0.2–0.5 ha) and they are intensely territorial, as their name suggests. Males give loud crowlike calls throughout the night and have been reported to engage in fisticuffs in defense of their tiny leafy domains. Most of their evening, however, is spent in resting and guarding their territory, presumably letting their leaves digest. Sportive lemurs have single births.

Indriids

The living indriids (Fig. 4.12, Table 4.4) are a very uniform family, consisting of three similar genera that differ most obviously in size and activity pattern. They have a reduced dental formula, with two premolars, and four rather than six teeth in their tooth

FIGURE 4.12

Three indriids from different parts of Madagascar: left, the sifaka (*Propithecus verreauxi*); right, a pair of indris (*Indri indri*); below, a family of the nocturnal woolly lemurs (*Avahi laniger*) in a typical sleeping posture.

TABLE 4.4
Infraorder Lemuriformes
Family INDRIIDAE

Common Name	Species	Intermembral Index	Body Weight (g)
Subfamily Indriinae			
Woolly lemur	*Avahi laniger*	58	920
White sifaka	*Propithecus verreauxi*	59	3,780
Diademed sifaka	*P. diadema*	64	6,500
Indris	*Indri indri*	64	?10,000
	(!)*Mesopropithecus pithecoides*	—	11,000
	(!)*M. globiceps*	?100	—
(!)Subfamily Archaeolemurinae			
	(!)*Archaeolemur edwardsi*	92	29,000
	(!)*A. majori*	92	17,000
	(!)*Hadropithecus stenognathus*	100	47,000
(!)Subfamily Palaeopropithecinae			
	(!)*Palaeopropithecus ingens*	138	—
	(!)*P. maximus*	144	97,000
	(!)*Archaeoindris fontoynonti*	—	200,000

(!)extinct, weight estimated from dental dimensions.

comb. Indriids are remarkably similar in their cranial morphology (Fig. 4.13) and, like lemurids, have a tympanic ring that lies free in the bulla and a large stapedial artery. All extant indriids are specialized leapers with long limbs; their hindlimbs are especially long (Jungers, 1979). They tend to have single births and most are monogamous, although some populations are also found in larger groups.

The smallest of the indriids is the **woolly lemur** (*Avahi laniger*), a mottled brown, white, and beige, thickly furred animal with a long tail. It is the only nocturnal indriid and is common in the wet forests of eastern Madagascar and the moister areas of the northwest. Woolly lemurs move mainly by leaping. Their diet consists mostly of leaves, which they procure by bringing the branches to their mouth rather than by climbing out on small supports (Albignac and Dorst, 1981). They live in monogamous family groups and issue long, high-pitched whistles. They have single births. Woolly lemurs do not build nests; they huddle together during the day among tangles of vines and leaves (Fig. 4.12).

The **sifaka** (*Propithecus*) is a medium-size indriid. There are two species: the smaller *P. verreauxi*, from the dry forest on the west coast, and the larger, lesser known diademed sifaka, *P. diadema*, from the rain forests in the east. Like *Avahi*, both species of *Propithecus* have very long limbs, especially long legs, and a long tail. Both species are diurnal. They are found in a variety of forest types, and both species are primarily vertical clingers and leapers and travel by leaping between vertical supports. When they come to the ground, they progress with bipedal

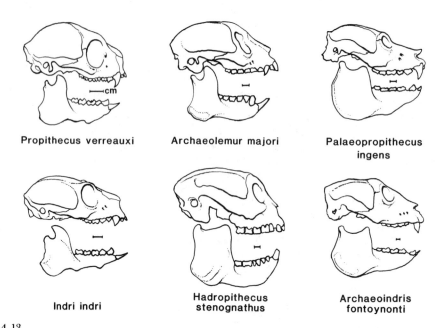

Propithecus verreauxi   Archaeolemur majori   Palaeopropithecus
                                                          ingens

Indri indri   Hadropithecus   Archaeoindris
                         stenognathus   fontoynonti

FIGURE 4.13

Skulls of a variety of extant and recently extinct indriids.

hops. During feeding they also use a variety of suspensory postures, often hanging upside down among small branches.

The diet of *P. verreauxi* varies from locality to locality but always includes large amounts of fruit (particularly in the wet season) and leaves (especially in the dry season). The two species have similar day ranges, but *P. diadema* has a home range fifteen times the size of that used by the dry forest species. Neither species of *Propithecus* is monogamous, and both live in moderate-size groups (3–9 individuals) with more than one breeding female. Females give birth to one infant every two years. Males regularly change groups, and there is intense male–male competition in the breeding season.

The largest of the indriids are the **indris** (*Indri indri*), a diurnal, tailless species from the hilly rain forests of the east coast. Like *Propithecus*, *Indri* has a short face and thick fur. These largest indriids have extremely long hands, long feet, and slender arms as well as very long legs (Fig. 4.14). They move primarily by leaping in lower levels of the forest but also hang below more horizontal supports, especially during feeding. Like sifakas, indris eat both fruit and leaves, with the proportions varying seasonally. They seem to avoid mature leaves, however, and specialize on young leaves and shoots. The small family groups, averaging about four individuals, defend relatively large territories from other groups. They are extremely vocal and give long, haunting morning calls to advertize their presence to neighboring groups, which answer sequentially.

Daubentoniids

The **aye-aye** (*Daubentonia madagascariensis*) is about as improbable a primate as one could

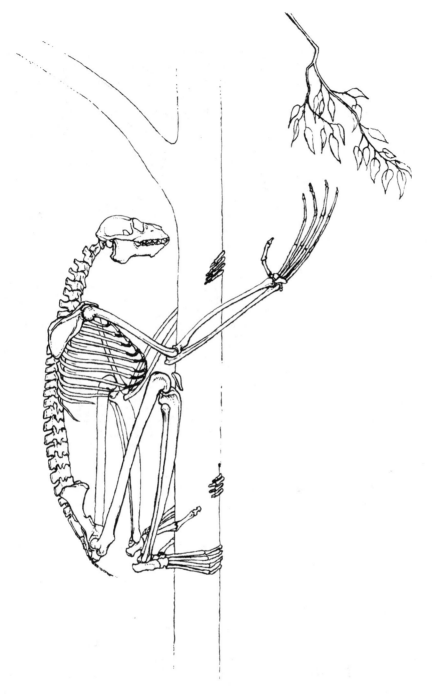

FIGURE 4.14

The skeleton of an indris (*Indri indri*). Note the long trunk, the extremely long hindlimbs, and the slender forelimbs.

FIGURE 4.15

The solitary and nocturnal aye-aye (*Daubentonia madagascariensis*) probing the bark of a tree with its slender third digit. In the background is an aye-aye nest made of leaves.

imagine (Fig. 4.15, Table 4.5). This moderate-size (3 kg), black animal with coarse, shaggy fur, enormous ears, and a large bushy tail has more extreme morphological specializations than any other living primate (Oxnard, 1981) but retains many features that clearly link it with lemurs (Tattersall, 1982b). The aye-aye has a greatly reduced dental formula of $\frac{1.0.1.3}{1.0.0.3}$. The most distinctive feature of its dentition, indeed of the entire skull (Fig. 4.5), is the pair of large, ever-growing, rodentlike incisors. The skull has a relatively large, globular braincase compared with other lemurs, but the auditory bulla and the cranial arteries are like those found in lemurids and indriids. Aye-ayes have relatively large, clawed digits on both hands and feet (except for the hallux), and the third digit of each hand is extremely long and slender.

Aye-ayes are nocturnal, and most of their odd anatomical features are adaptations for their dietary specialization of grubs and larvae. They locate their prey within a log by sound and then obtain the food by gnawing away the bark with their large incisors and by using the slender finger as a probe. There are no woodpeckers on Madagascar, but the

TABLE 4.5
Infraorder Lemuriformes
Family DAUBENTONIIDAE

Common Name	Species	Intermembral Index	Body Weight (g)
Aye-aye	*Daubentonia madagascariensis*	71	2,800
	(!)*D. robusta*	85	?4,000

(!)extinct.

aye-aye has been described as a woodpecker avatar (Cartmill, 1974).

The aye-aye's locomotion is mainly quadrupedal, and it is reported to be a solitary forager. Aye-ayes build large round nests of sticks and leaves.

Subfossil Malagasy Prosimians

Despite their diversity, the living strepsirhines of Madagascar are a small part of the primate fauna that inhabited the island in the very recent past. An extensive fauna of larger genera and species is known from fossil deposits about one thousand years old (Fig. 4.16), and there are reports (but not sightings) of large lemurs as recently as three hundred years ago (Flacourt, 1661, in Tattersall, 1982). The extinction of many species seems to coincide with the first appearance of humans on Madagascar, and bones of the extinct species are often found in conjunction with human artifacts or in sites indicating human activity. Most of the extinct species were large, diurnal, and probably terrestrial. In discussing the adaptive radiation of the Malagasy strepsirhines, these subfossil taxa are most appropriately considered with the living species, since their demise is a very recent and in some ways an unnatural tragedy. Perhaps the most striking evidence that the large extinct Malagasy forms are an integral part of the present radiation is that all of the extinct species are related to the living families and are often found in association with fossil remains of living species and genera. With the possible exception of *Allocebus*, there are no known extinct cheirogaleids, and the fossil aye-aye species was larger (approximately 30 percent in linear dimensions) but not known to be otherwise different from the living one. The extinct members of other families greatly extend our knowledge of the adaptive radiation of those groups. This is most dramatic in the case of the indriids, of which five extinct genera are known.

Subfossil Indriids

The genus **Mesopropithecus**, with three species, is the fossil indriid most similar to the living genera (Table 4.4). Dentally and cranially *Mesopropithecus* is similar to *Propithecus* but is larger, more robust, and has larger upper incisors. Like the living indriids, it seems to have been largely folivorous. There are only a few skeletal remains of this genus, but the similar-size humerus and femur suggest that *Mesopropithecus* was probably an arboreal quadruped (Jungers, 1980).

The genus **Palaeopropithecus** is one of the largest of the indriids, with an estimated weight of nearly 100 kg for the larger northern population and somewhat less for the southern population. *Palaeopropithecus* is dentally similar to *Propithecus*, with long

FIGURE 4.16

Artistic reconstruction of the fossil site of Ampa-zambazimba, Madagascar (ca. 8,000–1,000 B.P.), showing a variety of subfossil prosimians from that locality. At the upper left a *Megaladapis* feeds on leaves while clinging to a trunk. Below are two individuals of *Pachylemur insignis* and to the right a family of slothlike *Palaeopropithecus*. On the ground, an *Archaeoindris* feeds in the background while another individual of *Megaladapis* ambles along to another tree. In the foreground a group of *Archaeolemur* feed on tamarind pods while a group of *Hadropithecus* wander in from the right.

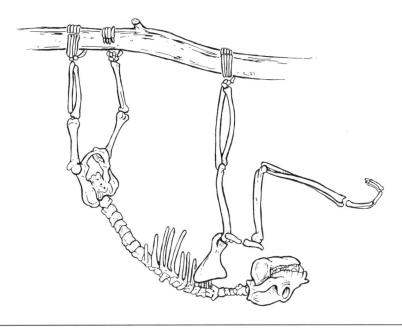

FIGURE 4.17

A skeleton of *Palaeopropithecus ingens* restored in a suspensory feeding position.

narrow molars and well-developed shearing crests, but it has small vertical lower incisors with no tooth comb. It was almost certainly folivorous. The skull is similar to that of living indriids (Fig. 4.13) but more robustly built, with a longer snout and a heavily buttressed nasal region suggesting more prehensile lips. The auditory region is superficially quite different from that of living indriids in that it has a tubular meatus extending laterally from the tympanic ring, apparently related to the extreme development of the mastoid region (Saban, 1975; Szalay and Delson, 1979; Tattersall, 1982).

Unlike the living indriids, which have relatively long legs and extraordinary leaping abilities, *Palaeopropithecus* has considerably longer forelimbs than hindlimbs (Fig. 4.17). It has a short thumb, long curved phalanges, and very mobile joints. It was the most suspensory of all known strepsirhines, with locomotor abilities that have generally been compared with that of either sloths or orangutans. Although the proportions and skeletal adaptations of this large extinct indriid were extraordinary, suspensory feeding postures are not unusual in the larger living indriids and provide evidence for a behavioral continuity between the living leapers and this extinct suspensory giant.

The genus *Archaeoindris* is closely related to *Palaeopropithecus* but was substantially larger, with an estimated weight (based on dental dimensions) of nearly 200 kg, the size of a male gorilla. Dentally and cranially (Fig. 4.12) *Archaeoindris* is similar to *Palaeopropithecus* and was probably folivorous. The few limb bones attributed to this giant indriid, together with its great size, suggest that it was probably terrestrial (Fig. 4.16). Its limb bones suggest that the closest locomotor analogues for this giant prosimian are the extinct ground sloths of North and South America.

Whereas *Palaeopropithecus* and *Archaeoindris* seem to have been the Malagasy equivalents of orangutans or sloths, **Archaeolemur** and its relative *Hadropithecus* evolved remarkable morphological similarities to cercopithecoid monkeys such as macaques or baboons. Although small compared with other extinct indriids (Table 4.4), they were larger than animals of any living genera. One species of *Archaeolemur* (*A. edwardsi*) has an estimated weight of between 25 and 30 kg, and a second species (*A. majori*) was slightly smaller.

Both *Archaeolemur* and *Hadropithecus* have one more premolar than other indriids, and they have large upper central incisors, expanded but slightly procumbent lower incisors and canines, and a fused mandibular symphysis. In *Archaeolemur*, the anterior premolar is caniniform and the entire premolar row forms a long cutting edge. The broad molars have low, rounded cusps arranged in a bilophodont pattern similar to that characterizing Old World monkeys. The dental similarities between the extinct indriid and living Old World monkeys such as macaques suggest that *Archaeolemur* was probably frugivorous. Cranially, it is similar to living indriids (Fig. 4.13), with no striking development of monkeylike features (Tattersall, 1973).

In skeletal anatomy *Archaeolemur* has further similarities to Old World monkeys in its limb proportions and in the configuration of individual limb elements and joint surfaces (Walker, 1974; Jungers, 1980). While many details of its limbs suggest a terrestrial habitat, the relatively short limbs compared with trunk length are characteristic of arboreal quadrupeds. It was a quadrupedal indriid that probably exploited both arboreal and terrestrial supports.

The single species of **Hadropithecus** was the largest (nearly 50 kg) and the most specialized of the monkeylike indriids. It has smaller incisors, reduced anterior premolars, and expanded, molarized posterior premolars compared with *Archaeolemur*. The molars have additional foldings of enamel that form a complex array of dentine and enamel crests and develop extremely flat wear. Because of similarities to the graminivorous (grass-feeding) gelada baboons, several authors have suggested that *Hadropithecus* fed on small grass seeds (Fig. 4.16). The reduced anterior dentition has been suggested as evidence that in *Hadropithecus*, as in gelada baboons, the hands played a major role in feeding (Jolly, 1970). *Hadropithecus* has a relatively short face and a robust skull with well-developed sagittal and nuchal crests (Fig. 4.13). The few available limb bones of *Hadropithecus* suggest that it was a terrestrial quadruped.

Subfossil Lemurids

Pachylemur insignis (Fig. 4.16, Table 4.2), a subfossil lemurid, is dentally similar to the living ruffed lemur, *Varecia variegata*, and probably had a frugivorous diet (Seligsohn and Szalay, 1974). This fossil genus is more robustly built in its limb skeleton and has forelimbs and hindlimbs that are similar in length (Jouffroy and Lessertisseur, 1979). It was a slow, arboreal quadruped with even less leaping ability than the living *Varecia*.

Subfossil Lepilemurids

One of the largest and most unusual of the extinct lemurs was **Megaladapis**, known from three species (Table 4.3), the largest of which had an estimated body size of 150 kg. This extinct giant shares several dental and cranial features with the living *Lepilemur*, and the two are usually placed in the same family. Like *Lepilemur*, *Megaladapis* has long narrow molars with well-developed crests, indicating a folivorous diet, and no upper incisors. The lack of incisors suggests to some authors that *Megaladapis* may have had a pad covering the premaxillary region (as in living artiodactyls) which occluded with the lower incisors for cropping herbivorous foods. In addition, the pronounced nasal region suggests the possibility of large prehensile lips, perhaps for cropping tough, thorny vegetation. The mandibular condyle on *Megaladapis* is like that of *Lepilemur* and distinct from that of other strepsirhines. Like many of the larger fossil lemurs, *Megaladapis* has a fused mandibular symphysis. It has a long, flat cranium, a long snout, and a small braincase.

Megaladapis has long forelimbs relative to the size of its hindlimbs, a very long trunk, and long curved phalanges. In contrast to the slender limbs of *Palaeopropithecus*, those of *Megaladapis* are extremely robust. The locomotor behavior of this genus seems to have been that of a vertical clinger and climber similar to the living koala of Australia. *Megaladapis* probably clung to vertical trunks while cropping vegetation with its snout (Fig. 4.16). On the ground it moved quadrupedally between trees.

ADAPTIVE RADIATION OF MALAGASY PRIMATES

Like the Galapagos finches, the Malagasy strepsirhines were a natural experiment in evolution. Isolated from repeated faunal invasions and from ecological competition with other primates (until humans arrived) and with many other groups of mammals, this lineage evolved an extremely diverse array of species with dietary and locomotor adaptations for exploiting a wide range of ecological conditions (Fig. 4.18). In size,

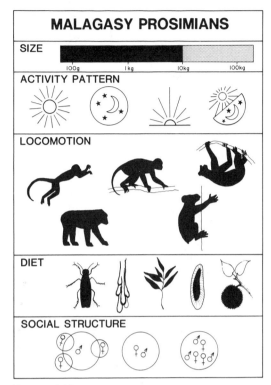

FIGURE 4.18

The adaptive diversity of Malagasy prosimians, including both extant and subfossil species.

these animals range from the smallest living primate, the mouse lemur (60 g), to *Megaladapis* and *Archaeoindris*, which must have weighed as much as living gorillas. In their gross dietary preferences there are species specializing on insects, on gums, on fruits, on nectar, on leaves, and probably on seeds, and they show dental adaptations in accordance with these dietary differences. In locomotor abilities they include prodigious leapers, arboreal and probably terrestrial quadrupeds, long-armed suspensory species, and some, such as the koala-like *Megaladapis* and the ground sloth-like *Archaeoindris*, which have no analogue among other primates. Likewise in their limb proportions and many other aspects of the musculature and skeletal anatomy they show considerable anatomical diversity that is functionally related to the behavioral differences.

The Malagasy radiation includes numerous diurnal, nocturnal, and cathemeral species. The remarkably flexible activity period of the mongoose lemur is apparently not uncommon among lemurs, but it is not known for other primates. Generalizations about the social organization of the Malagasy primates are more difficult, both because many aspects are poorly understood and because we have no information about the behavior of the larger (diurnal and in some cases probably terrestrial) subfossil species. Many species (notably the nocturnal ones) have very primitive, relatively simple noyau social structures, while the species living in larger groups with numerous males and females have a social organization that is clearly quite different from that characterizing most living anthropoids (see, e.g., Jolly, 1984; Richard, 1986). Troops seem to be dominated by females, with very little male–male competition except during breeding seasons. Accordingly, and in contrast to most higher primates and many other groups of mammals, the Malagasy primates show very little sexual dimorphism. It is unclear at present how the unique features in the social organization of the diurnal Malagasy strepsirhines are related to the extreme seasonality of breeding, reduced predation pressure, or some unique features in their evolutionary past.

Galagos and Lorises

In addition to the diverse radiation of strepsirhines on Madagascar, there is a smaller, mainland radiation represented by the galagos in Africa and the lorises in Africa and Asia (Figs. 4.19, 4.20). These families share with the Malagasy families the strepsirhine characteristics of a dental tooth comb and a grooming claw on the second digit (Fig. 4.4), but they also have cranial features that separate them from most of the prosimians of Madagascar. The main blood supply to the brain comes through the ascending pharyngeal artery (Fig. 2.11) rather than through the stapedial. In the ear region, the tympanic ring is fused to the lateral wall rather than being suspended within the bulla. In both of these features, the lorises and galagos show similarities with the cheirogaleids. The overall cranial morphology of lorises and galagos is very similar and distinct from that seen in indriids and most lemurids. Galagos and lorises are nocturnal and arboreal, but the two families are extremely different in their postcranial morphology and locomotor behavior. The former are primarily leapers; the latter are slow climbers.

Galagids

The galagos (Table 4.6) are a far more diverse group than many early systematists realized, and the exact arrangement of species and genera is a subject of some disagreement. Most current authorities recognize at least ten species and three or four genera, and many galago species have been the subject of ecological field studies.

Otolemur crassicaudatus, the **thick-tailed bushbaby**, is the largest of the galagos. It has a wide distribution throughout eastern and southern Africa. Like all galagos, thick-tailed bushbabies have large ears, a long tail, relatively long lower limbs, and an elongated calcaneus and navicular. These proportions of both the limbs and the ankle are less extreme in this species than in other galagos (Berge and Jouffroy, 1984).

Perodicticus potto

Galagoides demidovii

Euoticus elegantulus

Galagoides alleni

Arctocebus calabarensis

FIGURE 4.19

Five sympatric lorises and galagos from Gabon.

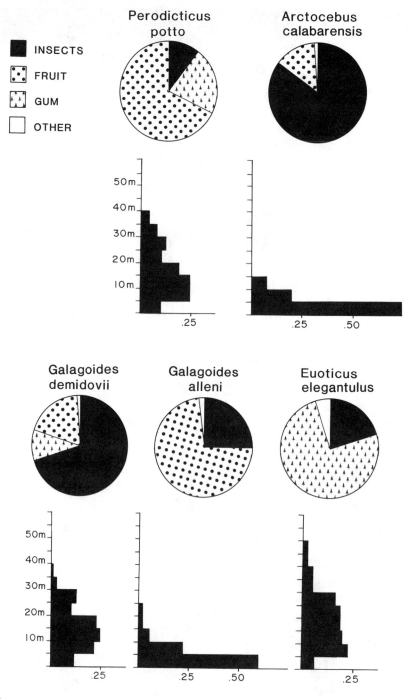

FIGURE 4.20

Diet and forest height preference for five sympatric lorises and galagos from Gabon (data from Charles-Dominique, 1977).

TABLE 4.6
Infraorder Lemuriformes
Family GALAGIDAE

Common Name	Species	Intermembral Index	Body Weight (g)
Large-eared greater bushbaby	*Otolemur crassicaudatus*	70	1,151
Small-eared greater bushbaby	*O. garnettii*	69	760
Senegal bushbaby	*Galago senegalensis*	52	215
Somali bushbaby	*G. gallarum*	—	—
South African lesser bushbaby	*G. moholi*	54	161
Needle-clawed bushbaby	*Euoticus elegantulus*	64	274
Matschie's needle-clawed bushbaby	*E. matschiei*	60	210
Demidoff's bushbaby	*Galagoides demidovii*	68	70
Thomas' bushbaby	*G. thomasi*	67	110
Zanzibar bushbaby	*G. zanzabaricus*	60	150
Allen's bushbaby	*G. alleni*	64	295

Like all galagos, thick-tailed bushbabies are nocturnal. They are found in relatively low forests between 6 and 12 m high and move mainly by quadrupedal walking and running, less often by leaping. Their diet consists primarily of fruits and gums and varies considerably from season to season.

Like cheirogaleids, thick-tailed bushbabies are solitary foragers that live in noyau social organization. Females build leaf nests for their twin or even triplet offspring and carry their infants in their mouth if they must move them.

Galago senegalensis, the **Senegal bushbaby**, is the most widespread species of galago, extending from Senegal in the west through eastern and southern Africa. It inhabits a wide range of forests, savannahs, open woodlands, and isolated thickets. It is smaller (215 g) than *Otolemur crassicaudatus*, has relatively long legs and ankle bones, and occupies the opposite extreme of the galago locomotor spectrum (McArdle, 1981). This species prefers the lower forest levels and smaller supports and travels almost exclu-

sively by leaping. Its diet is mainly insects, but gum is a major component in the dry season. During gum feeding, these small galagos cling to the rough bark of acacia trees by grasping.

Senegal bushbabies forage separately in large individual home ranges, but they often group together in daytime sleeping nests. Among the adult males there appears to be a social dominance hierarchy related to age and weight. Because of the harsh, unpredictable nature of their environment and high mortality rates, these primates seem to have a "boom or bust" reproductive strategy, with females capable of having up to two litters of twins per year. They are not particular about their choice of nest sites, frequently choosing tangles as well as holes in trees.

Allen's bushbaby (*Galagoides alleni*) is one of three west African galagos whose behavior has been studied by Pierre Charles-Dominique in Gabon (Fig. 4.19). It is similar in size to *G. senegalensis* and is often placed in the same genus with that species, but it may be more closely related to another group of bushbabies. *Galagoides alleni* moves by leaping between small vertical supports in the understory and between these small trees and the ground. Its diet has been reported to be quite different in different localities, and possibly during different seasons. In the primary forests of Gabon, Allen's bushbabies eat 25 percent animal matter, 75 percent fruit, much of which is from the ground, and some gums (Fig. 4.20), but in a secondary forest locality they eat a much higher number of insects. Individuals forage alone and males have large home ranges overlapping the ranges of several females. Allen's bushbabies are much less prolific than Senegal bushbabies. In their relatively stable rain forest habitat, females give birth to only a single infant per year.

The smallest galago is *Galagoides demidovii*, the **dwarf galago** (Fig. 4.19). Dwarf galagos rival mouse lemurs for the title of smallest living prosimian; they also resemble *Microcebus* in many behavioral features (Charles-Dominique and Martin, 1970). Their range extends in a band across central Africa from west to east. Throughout this region, they are very common in dense vegetation of either the canopy of primary forests or the understory of secondary forests. They are less specialized leapers than either *Galago senegalensis* or *Galagoides alleni* but move mainly by quadrupedal walking and running with short leaps between branches. Their diet (Fig. 4.20) in western Africa is predominantly insects (70 percent), with lesser amounts of fruit (19 percent) and gums (10 percent). Their social structure is a noyau with overlapping male and female ranges and common day sleeping nests for groups of females and occasional visiting males. They seem to have single births once a year in some parts of their range, but they frequently have twins in other areas.

The most specialized of the west African

galagos is *Euoticus elegantulus*, the **needle-clawed galago** (Fig. 4.19). This medium-size species resembles the cheirogaleid *Phaner* in having numerous morphological specializations such as procumbent upper incisors, caniniform upper anterior premolars, and laterally compressed, clawlike nails related to its gum-eating habits. These galagos use all levels of the canopy and move both quadrupedally and by leaping. They are particularly adept at clinging to large trunks and branches, which are the source of their main food—gums (Fig. 4.20). Most of their foraging is solitary, and their social behavior is largely unknown.

Lorisids

In contrast with the rapid running and leaping of galagos, lorises (Table 4.7) are best known for their slow, stealthy habits. They have smaller ears than galagos, limbs that are more similar in length, and no long tails. Despite their slowness, they have the broadest geographic distribution (between endpoints) of any prosimian family, with two genera in Africa and two in Asia.

The **potto** (*Perodicticus potto*) is the larger of the two African lorises and the more widespread (Fig. 4.19). Its range extends from Liberia in the west to Kenya in the east. Pottos prefer the main continuous canopy of both primary and secondary forests and move on relatively large supports. Their diet in western Africa has been reported to be largely frugivorous and to include much smaller amounts of animal matter and gums (Fig. 4.20). In eastern Africa, however, they reportedly eat much more gum (60 percent) and less fruit (10 percent) and animal material. The potto has been described as an animal that specializes in olfactory foraging, and its most characteristic activity is slow climbing along large supports with its nose to the branch.

TABLE 4.7
Infraorder Lemuriformes
Family LORISIDAE

Common Name	Species	Intermembral Index	Body Weight (g)
Potto	*Perodicticus potto*	88	1,150
Angwantibo, Golden potto	*Arctocebus calabarensis*	89	265
Slender loris	*Loris tardigradus*	90	275
Common slow loris	*Nycticebus coucang*[a]	88	920
Pygmy slow loris	*N. pygmaeus*	91	300

[a]simplified systematics.

Like galagos, pottos seem to be solitary foragers and to have overlapping male and female ranges. The females have single births. Pottos do not build nests but rather sleep curled up on branches. During the night, females do not carry their infant but "park" it for the evening and return for it later. The infants are white at birth and turn a drab brown within a few months.

Arctocebus calabarensis, the **golden potto or angwantibo,** is a smaller, more slender African lorisid with a restricted distribution in west central Africa. Angwantibos prefer the understory of primary and secondary forests, where they are usually found less than 5 m above the ground. They move by slow quadrupedal climbing on very small branches and lianas. They are predominantly insectivores and eat relatively little fruit. Socially, they appear very similar to pottos.

Asian lorises (Fig. 4.21), like their African counterparts, come in two shapes—thin and plump—but they are not sympatric. The thin species is the **slender loris** (*Loris tardigradus*), from Sri Lanka and southern India. This species has been most accurately described as a banana on stilts. It is found in the understory of dry forests and in the canopy of wetter forests, where it moves about among the fine twigs. It is mainly insectivorous. Virtually nothing is known of its social behavior.

The **slow loris** (*Nycticebus*), from Southeast Asia, is the stockier Asian genus and contains two allopatric species: *N. coucang* and *N. pygmaeus*. Slow lorises are found primarily in the main canopy, but any preference for primary, secondary, or deciduous forests is unclear (Barrett, 1981). Like all lorises, they move mainly by slow quadrupedal climbing and prefer larger supports, 3–6 cm in diameter. All reports indicate that they are frugivorous. There are no detailed studies of their social organization.

ADAPTIVE RADIATION OF GALAGOS AND LORISES

Compared with that of their Malagasy relatives, the radiation of galagos and lorises is limited. They are all small and nocturnal, a likely correlate of their geographic overlap with the larger, diurnal catarrhine primates throughout their range. Within their nocturnal habitus, however, the galagos and lorises have evolved sufficient adaptive diversity to permit as many as five sympatric species. There are many different things to do at night in a tropical forest. Both families include some species that specialize on

FIGURE 4.21

The two Asian lorisids: left, a slow loris (*Nycticebus coucang*) from Southeast Asia; right, a slender loris (*Loris tardigradus*) from India and Sri Lanka.

fruits, gums, or insects, as well as many that show seasonal specializations but a more diverse diet over a full year. Even though the lorises seem to have a rather stereotyped, stealthy locomotor behavior, their different sizes permit sympatric species (in Africa) to use different types of supports in different parts of the canopy. The galagos show more diversity in locomotor abilities and range, from extremely saltatory species to largely quadrupedal species. In contrast with the Malagasy strepsirhines, the nocturnal lorises and galagos show little diversity in social organization. They all appear to be solitary foragers and vary only in their tendencies to sleep in groups.

PHYLETIC RELATIONSHIPS OF STREPSIRHINES

There is fairly general, though not universal,

agreement that the lemurids, indriids, lepilemurids, and the aye-aye are distinct groups of strepsirhines, and that the first three are more closely related to one another than to any other group of strepsirhines. The phyletic relationships of the aye-aye and cheirogaleids to the other families of large Malagasy strepsirhines are less clear (Figs. 4.22, 4.23). Dental and cranial studies place the aye-aye with indriids, but molecular studies suggest that this odd primate is more distantly related. The difficulty in reconstructing the phyletic relationships of strepsirhines results largely from the considerable amount of parallel evolution that occurred during the radiation of the group. Between 20 and 40 percent of all features used to distinguish the different taxonomic groups must have evolved independently in at least two radiations to account for their distribution within living taxa (Eaglen, 1983). Thus,

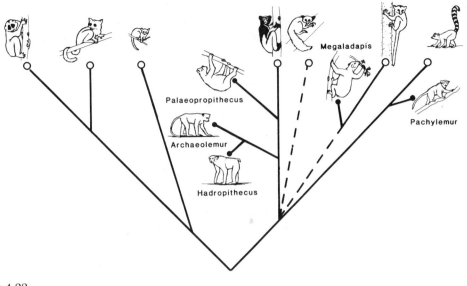

LORISIDS   GALAGIDS   CHEIROGALEIDS   INDRIIDS   AYE-AYE   LEPILEMURIDS   LEMURIDS

Megaladapis

Palaeopropithecus

Archaeolemur

Hadropithecus

Pachylemur

FIGURE 4.22

A phylogeny of strepsirhines, including representative subfossil genera.

FIGURE 4.23

Two biomolecular phylogenies of strepsirhines: left, Sarich and Cronin (1976); right, Dene *et al.* (1976).

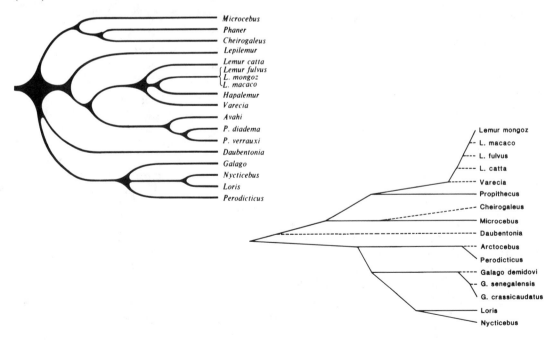

Microcebus
Phaner
Cheirogaleus
Lepilemur
Lemur catta
Lemur fulvus
L. mongoz
L. macaco
Hapalemur
Varecia
Avahi
P. diadema
P. verrauxi
Daubentonia
Galago
Nycticebus
Loris
Perodicticus

Lemur mongoz
L. macaco
L. fulvus
L. catta
Varecia
Propithecus
Cheirogaleus
Microcebus
Daubentonia
Arctocebus
Perodicticus
Galago demidovi
G. senegalensis
G. crassicaudatus
Loris
Nycticebus

depending on which anatomical characters are used, we can reconstruct a different phylogenetic branching sequence for many of the groups.

A particularly vexing issue in strepsirhine phylogeny concerns the relationship of the cheirogaleids to both the other Malagasy families and to the lorises and galagos. Historically, the cheirogaleids have been grouped with the other Malagasy families rather than with the galagos and lorises, a division that is supported by the structure of the ear region (Fig. 2.15) as well as by some molecular studies (Fig. 4.23). Cheirogaleids, galagos, and lorises are, however, unique among living primates in their cranial arterial system, and one genus of cheirogaleid (*Allocebus*) has a galago-like ear region, so many recent authorities have argued that cheirogaleids should be grouped with the galagos and lorises (Fig. 4.22). This grouping of one Malagasy family with the mainland strepsirhines raises a number of more interesting evolutionary and biogeographical questions. Are the mainland galagos and lorises descended from a cheirogaleid or are the cheirogaleids evolved from galagos that found their way to Madagascar? To what extent do small cheirogaleids and galagos represent primitive strepsirhines? There is evidence from both traditional and molecular studies to support each of these alternatives and insufficient data to reject any of them outright (Cartmill, 1975).

Tarsiers

The tarsiers (genus *Tarsius*) of Southeast Asia (Table 4.8) are among the smallest and most unusual of all living primates. In many respects they are anatomically intermediate between the strepsirhine prosimians and the anthropoid monkeys and apes. Their similarities to other prosimians are primitive features: an unfused mandibular symphysis, molar teeth with high cusps, grooming claws on their second (and third) toes, multiple nipples, and a bicornate uterus. Their similarities to higher primates seem to be derived specializations indicative of a phyletic relationship. In addition, tarsiers have many distinctive features all their own (Fig. 4.24).

A most striking feature of a tarsier is the size of its eyes, each of which is actually larger than the animal's brain, not to mention its stomach. Tarsiers differ from other nocturnal primates, and resemble all diurnal primates, in having a retinal fovea and lacking the reflective tapetum found in all lemurs and lorises as well as many other groups of mammals. Their large eyes are protected by a bony socket that is nearly complete posteriorly, similar to higher primates. The nose of tarsiers resembles that of higher primates as well, both externally in the lack of an attached upper lip with a median fold, and internally in the greatly reduced turbinates and in the absence of an ethmoid recess. In tarsiers, as in higher primates, the major blood supply to the brain comes through the promontory branch of the internal carotid (Fig. 2.11). The tympanic ring lies external to the auditory bulla and extends laterally to form a bony tube, the external auditory meatus (Fig. 2.15). The tarsier dental formula, $\frac{2.1.3.3}{1.1.3.3}$, is unique among primates, but tarsier teeth resemble those of anthropoids in overall proportions, with large upper central incisors, small lower incisors, and large canines. Their high-cusped, simple molar teeth with conules on the upper molars superficially look very primitive, but there are indications that they may have been modified from a previously more complex molar type in their ancestry.

The postcranial skeleton of tarsiers is striking in many of its proportions. The hands and feet of these tiny animals are

TABLE 4.8
Infraorder Tarsiiformes
Family TARSIIDAE

Common Name	Species	Intermembral Index	Body Weight (g)
Philippine tarsier	*Tarsius syrichta*	58	122
Bornean tarsier	*T. bancanus*	52	123
Spectral tarsier	*T. spectrum*	—	?140
Pygmy tarsier	*T. pumilus*	—	—

FIGURE 4.24

The skull, dentition, and skeleton of *Tarsius*, showing some of the distinctive features of the genus.

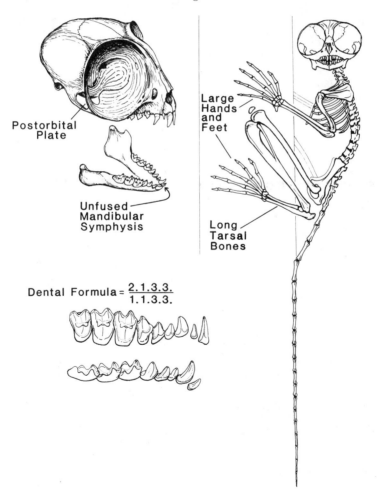

Postorbital Plate

Unfused Mandibular Symphysis

Large Hands and Feet

Long Tarsal Bones

Dental Formula = $\dfrac{2.1.3.3.}{1.1.3.3.}$

FIGURE 4.25

A pair of tarsiers (*Tarsius spectrum*) from the island of Sulawesi.

relatively enormous, reflecting both their clinging abilities and their predatory habits. They have extremely long legs and many more specific adaptations for leaping, including a fused tibia and fibula and the very long ankle region responsible for their name.

In their reproductive physiology, tarsiers show several similarities to higher primates. They have a hemochorial placenta like that of monkeys, apes, and humans rather than the epitheliochorial type found in lemurs (see Fig. 2.25), and they produce relatively large offspring. Female tarsiers undergo monthly sexual cycles with swellings reminiscent of some Old World monkeys.

There are four distinct species of *Tarsius*: *T. bancanus* from Borneo, Java, and Sumatra, *T. spectrum* from Sulawesi (Fig. 4.25), *T.*

syrichta from the Philippines, and the smaller *T. pumilus* from montane mossy forests in Sulawesi. Both *T. bancanus* and *T. spectrum* seem to be most common in the lower levels of all types of forests but are particularly abundant in secondary forests and scrub. They are totally nocturnal and spend their days sleeping in the grass or on vines. Both the Bornean and spectral tarsiers travel and feed very near the ground. They move mainly by rapid leaps of up to 3 m. In Sulawesi, other activities such as calling and resting take place higher in the canopy. *Tarsius pumilus* is unusual in having keeled nails for clinging to moss-covered trees. Tarsiers are totally faunivorous; they eat insects, arachnids, and small vertebrates such as snakes and lizards.

Tarsiers seem to live in either families

consisting of a mated pair and their off-spring or a noyau system with individual but overlapping ranges. There is also evidence from captive studies of considerable differences in social behavior among the species. Spectral tarsiers live in families, sleep together, give complex territorial duet calls early every evening, and then forage together all night. They actually chase other tarsiers out of their territories. In the Bornean tarsier (*T. bancanus*), on the other hand, males and females have separate, overlapping ranges and forage independently (Crompton and Andau, 1987). All tarsiers have a six-month gestation period and single births. Infants are 30 percent of the mother's weight at birth and suckle for two months. Females are largely responsible for care of the infant. In contrast to many monogamous species, there is no indication of male infant care in tarsiers.

PHYLETIC RELATIONSHIPS OF TARSIERS

The relationship of tarsiers to other primates, both living and fossil, has been a source of debate for many years and remains a topic of lively controversy and widely divergent views today. On the basis of their many primitive primate features as well as the morphology of their chromosomes, tarsiers are often grouped with the strepsirhines as prosimians. But many authorities feel that *Tarsius* and the higher primates share a large enough number of distinctive cranial, dental, and reproductive features to justify grouping them in a single lineage, the haplorhines.

Regardless of the classification one adopts, the evolutionary history of tarsiers is far from being resolved, and increasing evidence of morphological and behavioral differences among tarsier species further complicates the picture. In the past, most researchers have regarded tarsiers as a group of prosimians that gave rise to anthropoids. Still, many of their similarities to anthropoids, such as their lack of a tapetum, their partial postorbital closure, and their very large teeth for their body size, are equally compatible with a view of tarsiers as a group of anthropoids that have "reinvaded" the "prosimian adaptive zone" (Cartmill, 1980). Perhaps tarsiers are dwarfed anthropoids. The phyletic history of tarsiers and the adaptive story behind their similarities with anthropoids are important issues in primate evolution that remain unresolved at present. We discuss these topics again in later chapters when we examine fossil prosimians.

BIBLIOGRAPHY

GENERAL

Bourliere, F. (1974). How to remain a prosimian in a simian world. In *Prosimian Biology*, ed. R.D. Martin, G.A. Doyle, and A.C. Walker, pp. 17–22. London: Duckworth.

Cartmill, M. (1982b). Basic primatology and prosimian evolution. In *Fifty Years of Physical Anthropology in North America*, ed. F. Spencer, pp. 147–186. New York: Academic Press.

Cartmill, M., and Kay, R.F. (1978). Cranio-dental morphology, tarsier affinities and primate suborders. In *Recent Advances in Primatology*, vol. 3, ed. D.J. Chivers and K.A. Joysey, pp. 205–213. London: Academic Press.

Hladik, A. (1980). The dry forest of the west coast of Madagascar: Climate, phenology and food available for prosimians. In *Nocturnal Malagasy Primates*, ed. P. Charles-Dominique, H.M. Cooper, A. Hladik, C.M. Hladik, E. Pages, G.F. Pariente, A. Petter-Rousseaux, and A. Schilling, pp. 3–40. New York: Academic Press.

Jouffroy, F.K. (1962). La musculature des membres chez les lémuriens de Madagascar. Etude descriptive et comparative. *Mammalia* **26** (Supple. 2):1–326.

Luckett, W.P. (1975). Ontogeny of the fetal membranes and placenta: Their bearing on primate phylogeny.

In *Phylogeny of the Primates: A Multidisciplinary Approach*, ed. W.P. Luckett and F.S. Szalay, pp. 157–182. New York: Plenum Press.

Maier, Wolfgang (1980). Konstruktion morphologishe Untersuchungen am Gebis der rezenten. Prosimiae (Primates). *Abh. senckenb. naturforsch. Ges.* **538**:1–158.

Martin, R.D. (1979). Phylogenetic aspects of prosimian behavior. In *The Study of Prosimian Behavior*, ed. G.A. Doyle and R.D. Martin, pp. 45–78. New York: Academic Press.

Pariente, G. (1979). The role of vision in prosimian behavior. In *The Study of Prosimian Behavior*, ed. G.A. Doyle and R.D. Martin, pp. 411–460. New York: Academic Press.

Petter, J.J., and Petter-Rousseaux, A. (1979). Classification of the prosimians. In *The Study of Prosimian Behavior*, ed. G.A. Doyle and R. D. Martin, pp. 1–44. New York: Academic Press.

Petter, J.J., Albignac, R., and Rumpler, Y. (1977). *Faune de Madagascar*, vol. 44: *Mammiferes lémuriens (Primates, Prosimiens)*. Paris: Orstrom, CNRS.

Schwartz, J.H., and Tattersall, I. (1985). Evolutionary relationships of living lemurs and lorises (Mammalia, primates) and their potential affinities with European Eocene Adapidae. *Anthropol. Papers Am. Mus. Nat. Hist.* **60**(1):1–100.

Walker, A. (1979). Prosimian locomotor behavior. In *The Study of Prosimian Behavior*, ed. G.A. Doyle and R.D. Martin, pp. 543–566. New York: Academic Press.

STREPSIRHINES

Charles-Dominique, P., and Martin, R.D. (1970). Evolution of lorises and lemurs. *Nature* **227**:257–260.

Conroy, G.G., and Packer, D.J. (1981). The anatomy and phylogenetic significance of the carotid arteries and nerves in strepsirhine primates. *Folia Primatol.* **35**:237–247.

Martin, R.D. (1972b). Adaptive radiation and behavior of the Malagasy lemurs. *Philos. Trans. R. Soc. London B:* **264**:295–352.

Richard, A. (1986). Malagasy prosimians: Female dominance. In *Primate Societies*, ed. B.B. Smuts, D.L. Cheney, R.M. Seyfarth, R.W. Wrangham, and T.T. Struhsaker, pp. 25–33. Chicago: University of Chicago Press.

Tattersall, I. (1982). *The Primates of Madagascar*. New York: Columbia University Press.

———. (1986). Systematics of the Malagasy strepsirhine primates. In *Comparative Primate Biology*, vol. 1:
Systematics, Evolution, and Anatomy, ed. D.R. Swindler and J. Erwin, pp. 43–72. New York: Alan R. Liss.

Cheirogaleids

Cartmill, M. (1975). Strepsirhine basicranial structures and the affinities of the Cheirogaleidae. In *Phylogeny of the Primates: A Multidisciplinary Approach*, ed. W.P. Luckett and F.S. Szalay, pp. 313–356. New York: Plenum Press.

———. (1982). Assessing tarsier affinities: Is anatomical description phylogenetically neutral? In *Phylogenie et paleobiogeographie*, ed. E. Buffetaut, P. Janvier, J.C. Rage, and P. Tassy, pp. 279–287. *Geobios, Mem. Spec.* **6**.

Charles-Dominique, P. (1977). *Ecology and Behavior of Nocturnal Primates*. New York: Columbia University Press.

Charles-Dominique, P., and Petter, J.J. (1980). Ecology and social life of *Phaner furcifer*. In *Nocturnal Malagasy Primates*, ed. P. Charles-Dominique, H.M. Cooper, A. Hladik, C.M. Hladik, E. Pages, G.F. Pariente, A. Petter-Rousseaux, and A. Schilling, pp. 75–96. New York: Academic Press.

Hladik, C.M., Charles-Dominique, P., and Petter, J.J. (1980). Feeding strategies of five nocturnal prosimians in the dry forest of the west coast of Madagascar. In *Nocturnal Malagasy Primates*, ed. P. Charles-Dominique, H.M. Cooper, A. Hladik, C.M. Hladik, E. Pages, G.F. Pariente, A. Petter-Rousseaux, and A. Schilling, pp. 41–74. New York: Academic Press.

Martin, R.D. (1972a). A preliminary field study of the lesser mouse lemur (*Microcebus murinus* J.F. Miller, 1777). *Z. Comp. Ethol. Suppl.* **9**:43–89.

———. (1972b). Adaptive radiation and behavior of the Malagasy lemurs. *Philos. Trans. R. Soc. London B:* **264**:295–352.

———. (1973). A review of the behavior and ecology of the lesser mouse lemur (*Microcebus murinus* J.F. Miller, 1777). In *Comparative Ecology and Behavior of Primates*, ed. R.P. Michael and J.H. Crook, pp. 1–68. London: Academic Press.

Pages, E. (1980). Ethoecology of *Microcebus coquereli* during the dry season. In *Nocturnal Malagasy Primates*, ed. P. Charles-Dominique, H.M. Cooper, A. Hladik, C.M. Hladik, E. Pages, G.F. Pariente, A. Petter-Rousseaux, and A. Schilling, pp. 3–40. New York: Academic Press.

Petter, J.J., Albignac, R., and Rumpler, Y. (1977). *Faune de Madagascar*, vol. 44: *Mammiferes lémuriens (Primates, Prosimiens)*. Paris: Orstom, CNRS.

Petter, J.J., Schilling, A., and Pariente, G. (1971).

Observations eco-ethologiques sur deux lémuriens malgaches nocturnes: *Phaner furcifer* et *Microcebus coquereli*. *Terre Vie* **25**:287–327.

Schwartz, J.H., and Tattersall, I. (1985). Evolutionary relationships of living lemurs and lorises (Mammalia, Primates) and their potential affinities with European Eocene Adapidae. *Anthropol. Papers Am. Mus. Nat. Hist.* **60**(1):1–100.

Szalay, F.C., and Katz, C.C. (1973). Phylogeny of lemurs, galagos and lorises. *Folia Primatol.* **19**:88–103.

Tattersall, I. (1982). *The Primates of Madagascar*. New York: Columbia University Press.

Lemurids

Conley, J.M. (1975). Notes on the activity pattern of *Lemur fulvus*. *J. Mammal.* **56**:712–715.

Harrington, J. (1974). Olfactory communication in *Lemur fulvus*. In *Prosimian Biology*, ed. R.D. Martin, G.A. Doyle, and A.C. Walker, pp. 331–346. London: Duckworth.

———. (1975). Field observation of social behavior of *Lemur fulvus fulvus* E. Geoffroyi, 1812. In *Lemur Biology*, ed. I. Tattersall and R.W. Sussman, pp. 259–279. New York: Plenum Press.

———. (1978). Diurnal behavior of *Lemur mongoz* at Ampijoroa, Madagascar. *Folia Primatol.* **29**:291–302.

Jolly, A. (1966). *Lemur Behavior—A Madagascar Field Study*. Chicago: University of Chicago Press.

Jungers, W.L. (1979). Locomotion, limb proportions and skeletal allometry in lemurs and lorises. *Folia Primatol.* **32**: 8–28.

Meier, B., Albignac, R., Peyrierns, A., Rumpler, Y., and Wright, P. (1987). A new species of *Hapalemur* (Primates) from South East Madagascar. *Folia Primatol.* **48**:211–215.

Overdorff, D.J. (1987). Activity rhythms and ecology of *Lemur rubriventer* in Madagascar. *Am. J. Phys. Anthropol.* **72**:239.

Petter, J.J., Albignac, R., and Rumpler, Y. (1977). *Faune de Madagascar*, vol. 44: *Mammiferes lémuriens (Primates, Prosimiens)*. Paris: Orstom, CNRS.

Petter, J.J., and Peyrieras, A. (1975). Preliminary notes on the behavior and ecology of *Hapalemur griseus*. In *Lemur Biology*, ed. I. Tattersall and R.W. Sussman, pp. 281–286. New York: Plenum Press.

Pollack, J.I. (1979). Spatial distribution and ranging behavior in lemurs. In *The Study of Prosimian Behavior*, ed. G.A. Doyle and R.D. Martin, pp. 359–410. New York: Academic Press.

———. (1986). A note of the ecology and behavior of *Hapalemur griseus*. *Primate Conservation* **6**:97–101.

Sussman, R.W. (1974). Ecological distinctions in sympatric species of *Lemur*. In *Prosimian Biology*, ed. R.D. Martin, G.A. Doyle, and A.C. Walker, pp. 75–108. London: Duckworth.

———. (1975). A preliminary study of the behavior and ecology of *Lemur fulvus fulvus* Audebert, 1880. In *Lemur Biology*, ed. I. Tattersall and R.W. Sussman, pp. 237–258. New York: Plenum Press.

———. (1978). Nectar feeding by prosimians and its evolutionary implications. In *Recent Advances in Primatology*, vol. 3, ed. D.J. Chivers and K.A. Joysey, pp. 119–125. London: Academic Press.

Sussman, R.W., and Tattersall, I. (1976). Cycles of activity, group composition and diet of *Lemur mongoz mongoz* (Linnaeus, 1766) in Madagascar. *Folia Primatol.* **26**:270–283.

Tattersall, I. (1976). Group structure and activity rhythm in *Lemur mongoz* (Primates, Lemuriformes) on Anjouan and Moheli Islands, Comoro Archipelago. *Anthropol. Papers Am. Mus. Nat. Hist.* **53**:367–380.

———. (1978). Behavioral variation in *Lemur mongoz* (= *L. m. mongoz*). In *Recent Advances in Primatology*, vol. 3, ed. D.J. Chivers and K.A. Joysey, pp. 127–132. London: Academic Press.

———. (1982). *The Primates of Madagascar*. New York: Columbia University Press.

Tattersall, I., and Sussman, R.W. (1975). Observations on the ecology and behavior of the mongoose lemur, *Lemur mongoz mongoz* Linnaeus (Primates, Lemuriformes) at Ampyoroa, Madagascar. *Anthropol. Papers Am. Mus. Nat. Hist.* **52**:193–216.

Wright, P.C. (1987). The greater bamboo lemur in Madagascar. *From the Forest* **2**(1):1–4.

———. (1988). A lemur's last stand. *Animal Kingdom* **91**:12–25.

Lepilemurids

Charles-Dominique, P., and Hladik, C.M. (1971). Le *Lepilemur* du sud de Madagascar: Ecologie, alimentation et vie sociale. *Terre Vie* **25**:3–66.

Hladik, C.M. (1979). Diet and ecology of prosimians. In *The Study of Prosimian Behavior*, ed. G.A. Doyle and R.D. Martin, pp. 307–358. New York: Academic Press.

Hladik, C.M., and Charles-Dominique, P. (1974). The behavior and ecology of the sportive lemur (*Lepilemur mustelinus*) in relation to its dietary peculiarities. In *Prosimian Biology*, ed. R.D. Martin, G.A. Doyle, and A.C. Walker, pp. 24–38. London: Duckworth.

Hladik, C.M., Charles-Dominique, P., and Petter, J.J.

(1980). Feeding strategies of five nocturnal prosimians in the dry forest of the west coast of Madagascar. In *Nocturnal Malagasy Primates*, ed. P. Charles-Dominique, H.M. Cooper, A. Hladik, C.M. Hladik, E. Pages, G.F. Pariente, A. Petter-Rousseaux, and A. Schilling, pp. 41–74. New York: Academic Press.

Hladik, C.M., Charles-Dominique, P., Valdebouze, P., Delort-Lavale, J., and Flanzy, J. (1971). La caecotrophie chez un primate phyllophage du genre *Lepilemur* et les correlations avec les particularites de son appareil digestif. *C. R. Acad. Sci. (Paris)* **272**:3191–3194.

Petter, J.J., Albignac, R., and Rumpler, Y. (1977). *Faune de Madagascar*, vol. 44: *Mammiferes lémuriens (Primates, Prosimiens)*. Paris: Orstom, CNRS.

Tattersall, I., and Schwartz, J.H. (1974). Craniodental morphology and the systematics of the Malagasy lemurs (Primates, Prosimii). *Anthropol. Papers Am. Mus. Nat. Hist.* **52**(3):141–192.

Indriids

Albignac, R., and Dorst, J. (1981). Zoologies des vertebres variabilite dans l'organisation territoriale et l'ecologie de *Avahi laniger* (Lemurien nocturne de Madagascar). *C. R. Acad. Sci. (Paris)* **292** (ser. III):331–334.

Jolly, A. (1966). *Lemur Behavior—A Madagascar Field Study*. Chicago: University of Chicago Press.

Jungers, W.L. (1979). Locomotion, limb proportions and skeletal allometry in lemurs and lorises. *Folia Primatol.* **32**:8–28.

Petter, J.J., Albignac, R., and Rumpler, Y. (1977). *Faune de Madagascar*, vol. 44: *Mammiferes lémuriens (Primates, Prosimiens)*. Paris: Orstom, CNRS.

Pollock, J.I. (1975). Field observations on *Indri indri*: A preliminary report. In *Lemur Biology*, ed. I. Tattersall and R.W. Sussman, pp. 287–312. New York: Plenum Press.

———. (1979). Spatial distribution and ranging behavior in lemurs. In *The Study of Prosimian Behavior*, ed. G.A. Doyle and R.D. Martin, pp. 359–410. New York: Academic Press.

———. (1986). The song of the indris (*Indri indri*; Primates: Lemuroidea): Natural history, form, and function. *Int. J. Primatol.* **7**:225–264.

Richard, A.F. (1974). Patterns of mating in *Propithecus verreauxi*. In *Prosimian Biology*, ed. R.D. Martin, G.A. Doyle, and A.C. Walker, pp. 49–74. London: Duckworth.

———. (1978). *Behavioral Variation: Case Study of a Malagasy Lemur*. Lewisburg, Pa.: Bucknell University Press.

Richard, A.F., and Heimbuch, R. (1975). An analysis of the social behavior of three groups of *Propithecus verreauxi*. In *Lemur Biology*, ed. I. Tattersall and R.W. Sussman, pp. 313–334. New York: Plenum Press.

Wright, P.C. (1987). Diet and ranging patterns of *Propithecus diadema edwardsi* in Madagascar. *Am. J. Phys. Anthropol.* **72**:271.

Daubentoniids

Cartmill, M. (1974). *Daubentonia*, *Dactylopsila*, woodpeckers and kinorhynchy. In *Prosimian Biology*, ed. R.D. Martin, G.A. Doyle, and A.C. Walker, pp. 655–672. London: Duckworth.

Constable, I.D., Mittermeier, R., Pollock, J.I., Ratsirarson, J., and Simons, H. (1985). Sightings of aye-ayes and red ruffed lemurs on Nosy Mangabe and the Masoala Peninsula. *Primate Conservation* **5**:59–62.

Oxnard, E. (1981). The uniqueness of *Daubentonia*. *Am. J. Phys. Anthropol.* **54**:1–21.

Petter, J.J., and Peyrieras, A. (1970). Nouvelle contribution a l'etude d'un lémurien malgache, le aye-aye (*Daubentonia madagascarensis* E. Geoffroyi). *Mammalia* **34**:167–193.

Petter, J.J., Albignac, R., and Rumpler, Y. (1977). *Faune de Madagascar*, vol. 44: *Mammiferes Lémuriens (Primates, Prosimiens)*. Paris: Orstom, CNRS.

Tattersall, I. (1982a). *The Primates of Madagascar*. New York: Columbia University Press.

———. (1982b). Two misconceptions of phylogeny and classification. *Am. J. Phys. Anthropol.* **57**:13.

SUBFOSSIL MALAGASY PROSIMIANS

Dewar, R.E. (1984). Recent extinctions in Madagascar: The loss of the subfossil fauna. In *Quaternary Extinctions: A Prehistoric Revolution*, ed. P.S. Martin and R.G. Klein, pp. 574–593. Tucson: University of Arizona Press.

Flacourt, E. de. (1661). Histoire de la grande Isle Madagascar. Avec une relation de ce qui s'est passe des annees 1655, 1656 et 1657, non encor veue par la premiere impression. pp. 1–202 (Histoire); 203–471 (Relation). Troyes, N. Oudot: Paris, Pierre L' Amy.

Jungers, W.L. (1980). Adaptive diversity in subfossil Malagasy prosimians. *Z. Morphol. Anthropol.* **71**(2):177–186.

MacPhee, R.D.E. (1986). Environment, extinction, and Holocene vertebrate localities in southern Madagascar. *National Geographic Research* **2**:441–455.

MacPhee, R.D.E., Burney, D.A., and Wells, N.A. (1985).

Early Holocene chronology and environment of Ampazambazimba, a Malagasy subfossil lemur site. *Int. J. Primatol.* **6**:463–489.

Szalay, F.S., and Delson, E. (1979). *Evolutionary History of the Primates.* New York: Academic Press.

Tattersall, I. (1982). *The Primates of Madagascar.* New York: Columbia University Press.

Walker, A. (1967). Patterns of extinction among the subfossil Madagascan lemuroids. In *Pleistocene Extinctions: The Search for a Cause*, ed. P.S. Martin and H.E. Wright, Jr., pp. 425–532. New Haven, Conn.: Yale University Press.

———. (1974). Locomotor adaptations in past and present prosimians. In *Primate Locomotion*, ed. F.A. Jenkins, pp. 349–381. New York: Academic Press.

Mesopropithecus

Vuillaume-Randriamanantena, M. (1982). Contribution a l'etude des os longs des lémuriens subfossiles malagaches: Leurs particularites au niveau des proportions. Dissertation de 3ed cycle, Universite de Madagascar, Antananarivo.

Palaeopropithecus

Saban, R. (1975). Structure of the ear region in living and subfossil lemurs. In *Lemur Biology*, ed. I. Tattersall and R.W. Sussman, pp. 83–109. New York: Plenum Press.

Tattersall, I. (1975). Notes on the cranial anatomy of the subfossil Malagasy lemurs. In *Lemur Biology*, ed. I. Tattersall and R.W. Sussman, pp. 111–124. New York: Plenum Press.

Archaeoindris

Carlton, A. (1936). The limb bones and vertebrae of the extinct lemurs of Madagascar. *Proc. Zool. Soc. London* **110**:281–307.

Lamberton, C. (1936). Fouilles faites en 1936. *Bull. Acad. Malgache, n.s.* **19**:1–19.

Archaeolemur

Godfrey, L.R. (1977). Structure and function in *Archaeolemur* and *Hadropithecus* (Subfossil Malagasy lemurs): The postcranial evidence. Ph.D. Dissertation. Harvard University.

Jolly, C.J. (1970). The seed-eaters: A new model of hominid differentiation based on baboon analogy. *Man* **5**:5–26.

Tattersall, I. (1973). Cranial anatomy of the *Archaeolemurinae* (Lemuroidea, Primates). *Anthropol.*

Papers Am. Mus. Nat. Hist. **52**:1–110.

———. (1974). Facial structure and mandibular mechanics in *Archaeolemur*. In *Prosimian Biology*, ed. R.D. Martin, G.A. Doyle, and A.C. Walker, pp. 563–578. London: Duckworth.

Hadropithecus

Jolly, C. (1970). *Hadropithecus*, a lemuroid small-object feeder. *Man n.s.* **5**:525–529.

Jungers, W.L. (1980). Adaptive diversity in subfossil Malagasy prosimians. *Z. Morphol. Anthropol.* **71**(2):177–186.

Tattersall, I. (1973). Cranial anatomy of the *Archaeolemurines* (Lemuroidea, Primates). *Anthropol. Papers Am. Mus. Nat. Hist.* **52**:1–110.

Pachylemur

Jouffroy, F.K., and Lessertisseur, J. (1979). Relationships between limb morphology and locomotor adaptations among prosimians: An osteometric study. In *Environment, Behavior and Morphology: Dynamic Interactions in Primates*, ed. M.E. Morbeck, H. Preuschoft, and N. Gomberg, pp. 143–181. New York: Gustav Fischer.

Seligsohn, D., and Szalay, F.S. (1974). Dental occlusion and the masticatory apparatus in *Lemur* and *Varecia*: Their bearing on the systematics of living and fossil primates. In *Prosimian Biology*, ed. R.D. Martin, G.A. Doyle, and A.C. Walker, pp. 543–562. London: Duckworth.

Megaladapis

Jungers, W.L. (1977). Hindlimbs and pelvic adaptations to vertical climbing and clinging in *Megaladapis*, a giant subfossil prosimian from Madagascar. *Yrbk. Phys. Anthropol.* **20**:508–524.

Major, C.J.F. (1893). On *Megaladapis madagascariensis*, an extinct gigantic lemuroid from Madagascar. *Philos. Trans. R. Soc. London* **185**:15–38.

Tattersall, I. (1972). The functional significance of airorhynchy in *Megaladapis*. *Folia Primatol.* **18**:20–26.

Zapfe, H. (1963). Lebensbild von *Megaladapis edwardsi* (Grandidier): Ein Rekonstruktionversuch. *Folia Primatol.* **1**:178–187.

ADAPTIVE RADIATION OF MALAGASY PRIMATES

Jolly, A. (1984). The puzzle of female feeding priority. In *Female Primates: Studies by Women Primatologists*, ed.

M.F. Small, pp. 197–215. New York: Alan R. Liss.

Jouffroy, F.K. (1963). Contribution a la connaissance du genre *Archaeolemur*. *Ann. Paleontol.* **49**:3–29.

———. (1975). Osteology and myology of the lemuriform postcranial skeleton. In *Lemur Biology*, ed. I. Tattersall and R.W. Sussman, pp. 149–192. New York: Plenum Press.

Jouffroy, F.K., and Lessertisseur, J. (1979). Relationships between limb morphology and locomotor adaptations among prosimians: An osteometric study. In *Environment, Behavior and Morphology: Dynamic Interactions in Primates*, ed. M.E. Morbeck, H. Preuschoft, and N. Gomberg, pp. 143–181. New York: Gustav Fischer.

Jungers, W.L. (1979). Locomotion, limb proportions and skeletal allometry in lemurs and lorises. *Folia Primatol.* **32**:8–28.

———. (1980). Adaptive diversity in subfossil Malagasy prosimians. *Z. Morphol. Anthropol.* **71**(2):177–186.

Kay, R.F. (1975). The functional adaptations of primate molar teeth. *Am. J. Phys. Anthropol.* **43**:195–215.

Martin, R.D. (1972). Adaptive radiation and behavior of the Malagasy lemurs. *Philos. Trans. R. Soc. London B:* **264**:295–352.

Richard, A. (1986). Malagasy prosimians: Female dominance. In *Primate Societies*, ed. B.B. Smuts, D.L. Cheney, R.M. Seyfarth, R.W. Wrangham, and T.T. Struhsaker, pp. 25–33. Chicago: University of Chicago Press.

Seligsohn, D. (1977). Analysis of species-specific molar adaptations in strepsirhine primates. *Contributions to Primatology*, vol. 11, ed. F.S. Szalay. Basel: S. Karger

GALAGIDS

Bearder, S.K. (1986). Lorises, bushbabies, and tarsiers: Diverse societies in solitary foragers. In *Primate Societies*, ed. B.B. Smuts, D.L. Cheney, R.M. Seyfarth, R.W. Wrangham, and T.T. Struhsaker, pp. 11–24. Chicago: University of Chicago Press.

Bearder, S.K., and Doyle, G.A. (1974). Ecology of bushbabies, *Galago senegalensis* and *G. crassicaudatus*, with some notes on their behavior in the field. In *Prosimian Biology*, ed. R.D. Martin, G.A. Doyle, and A.C. Walker, pp. 109–130. London: Duckworth.

Berge, C.H., and Jouffroy, F.K. (1984). Morpho-functional study of *Tarsius'* foot as compared to the galagines': What does an "elongate calcaneus" mean? *Ninth Congress, International Primatological Society*.

Charles-Dominique, P. (1971). Eco-ethologie des prosimiens du Gabon. *Biol. Gabonica* **7**:121–228.

———. (1972). Ecologie et vie sociale de *Galago demidovii* (Fischer 1808, Prosimii). *Z. Tierpsychol. Beih.* **9**:7–41.

———. (1977a). *Ecology and Behavior of Nocturnal Primates*. New York: Columbia University Press.

———. (1977b). Urine marking and territoriality in *Galago alleni* (Waterhouse 1837–Lorisoidea, Primates), a field study by radio telemetry. *Z. Tierpsychol.* **43**:113–138.

Charles-Dominique, P., and Martin, R.D. (1970). Evolution of lorises and lemurs. *Nature (London)* **227**: 257–260.

Crompton, R.H. (1983). Age differences in locomotion of two subtropical Galaginae. *Primates* **24**(2):241–259.

———. (1984). Foraging, habitat structure and locomotion in two species of *Galago*. In *Adaptations for Foraging in Non-human Primates*, ed. P.S. Rodman and J.G.H. Cant, pp. 73–111. New York: Columbia University Press.

Harcourt, C.S. (1983). Bright eyes at night—bushbabies in Diani Forest. *Swara* 6:26–27.

Harcourt, C.S., and Nash, L.T. (1986). Social organization of galagos in Kenyan coastal forests: I, *Galago zanzibaricus. Am. J. Primatol.* **10**:339–356.

Jungers, W.L., and Olson, T.R. (1984). Relative brain size in galagoes and lorises. *Fortschr. Zool.* **30**:6–9.

Martin, R.D., and Bearder, S.K. (1979). Radio Bush-baby. *Nat. Hist.* **88**(8):77–81.

McArdle, J.E. (1981). Functional morphology of the hip and thigh of the lorisiformes. *Contributions to Primatology*, vol. 7, ed. F.S. Szalay. Basel: S. Karger.

Molez, N. (1976). Adaptation alimentaire du Galago D' Allen aux milieux forestiers secondaires. *Terre Vie* **30**:210–228.

Nash, L.T., and Harcourt, C.S. (1986). Social organization of galagos in Kenyan coastal forests: II, *Galago garnettii. Am. J. Primatol.* **10**:357–370.

Olson, T.R. (1981). Abstract: Systematics and zoogeography of the greater galago. *Am. J. Phys. Anthropol.* **54**(2):259.

LORISIDS

Amerasinghe, F.B., Van Cuylenberg, B.W.B., and Hladik, C.M. (1971). Comparative histology of the alimentary tract of Ceylon primates in correlation with diet. *Ceylon J. Sci., Biol. Sci.* **9**:75–87.

Barrett, E. (1981). The present distribution and status of the slow loris in peninsular Malaysia. *Malaysian Applied Biol.* **10**(2):205–211.

Bearder, S.K. (1986). Lorises, bushbabies, and tarsiers:

Diverse societies in solitary foragers. In *Primate Societies*, ed. B.B. Smuts, D.L. Cheney, R.M. Seyfarth, R.W. Wrangham, and T.T. Struhsaker, pp. 11–24. Chicago: University of Chicago Press.

Charles-Dominique, P. (1971). Eco-ethologie des prosimiens du Gabon. *Biol. Gabonica* **7**:121–228.

———. (1977). *Ecology and Behavior of Nocturnal Primates*. New York: Columbia University Press.

Charles-Dominique, P., and Bearder, S.K. (1979). Field studies of lorisid behavior: Methodological aspects. In *The Study of Prosimian Behavior*, ed. G.A. Doyle and R.D. Martin, pp. 567–630. New York: Academic Press.

Eliot, O., and Eliot, M. (1967). Field notes on the slow loris in Malaysia. *J. Mammal.* **48**:497–498.

Hladik, C.M. (1975). Ecology, diet and social patterning in Old and New World primates. In *Socioecology and Psychology of Primates*, ed. R.H. Tuttle, pp. 3–35. The Hague: Mouton.

———. (1979). Diet and ecology of prosimians. In *The Study of Prosimian Behavior*, ed. G.A. Doyle and R.D. Martin, pp. 307–358. New York: Academic Press.

Jolly, C.J., and Gorton, C. (1974). Proportions of the extrinsic foot muscles in some lorisid prosimians. In *Prosimian Biology*, ed. R.D. Martin, G.A. Doyle, and A.C. Walker, pp. 801–816. London: Duckworth.

Jungers, W.L., and Olson, T.R. (1984). Relative brain size in galagos and lorises. *Fortschr. Zool.* **30**:6–9.

Oates, J.F. (1984). The niche of the potto, *Perodicticus potto. Int. J. Primatol.* **5**(1):51–61.

Walker, A. (1969). The locomotion of the lorises, with special reference to the potto. *E. Afr. Wildlife J.* **8**:1–5.

PHYLETIC RELATIONSHIPS OF STREPSIRHINES

Cartmill, M. (1975). Strepsirhine basicranial structures and the affinities of the Cheirogaleidae. In *Phylogeny of the Primates: A Multidisciplinary Approach*, ed. W.P. Luckett and F.S. Szalay, pp. 313–356. New York: Plenum Press.

Dene, H., Goodman, M., Prychodko, W., and Moore, G.W. (1976). Immunodiffusion systematics of the primates. *Folia Primatol.* **25**:35–61.

Eaglen, R.H. (1983). Parallelism, parsimony and the phylogeny of the Lemuridae. *Int. J. Primatol.* **4**(3):249–273.

Sarich, V.M., and Cronin, J.E. (1976). Molecular systematics of the primates. In *Molecular Anthropology*, ed. M. Goodman and R. Tashkan, pp. 141–170. New York: Plenum Press.

Tattersall, I., and Schwartz, J.H. (1985). Evolutionary relationships of living lemurs and lorises (Mammalia, Primates) and their potential affinities with European Eocene Adapidae. *Anthropol. Papers Am. Mus. Nat. Hist.* **60**:1–110.

TARSIERS

Aiello, L.C. (1986). The relationships of the tarsiformes: A review of the case for the Haplorhini. In *Major Topics in Primate and Human Evolution*, ed. B. Wood, L. Martin, and P. Andrews, pp. 47–65. Cambridge: Cambridge University Press.

Cartmill, M. (1980). Morphology, function and evolution of the anthropoid post orbital septum. In *Evolutionary Biology of the New World Monkeys and Continental Drift*, ed. R.L. Ciochon and A.B. Chiarelli, pp. 243–274. New York: Plenum Press.

———. (1982). Assessing tarsier affinities—is anatomical description phylogenetically neutral? In *Phylogenie et paleobiogeographie*, ed. E. Buffetaut, P. Janvier, J.C. Rage, and P. Tassy, pp. 279–287. *Geobios, Mem. Spec.* **6**.

Crompton, R.H., and Andau, P.M. (1986). Locomotion and habitat utilization in free-ranging *Tarsius bancanus*: A preliminary report. *Primates* **27**:337–355.

———. (1987). Ranging, activity rhythms, and sociality in free-ranging *Tarsius bancanus*: A preliminary report *Int. J. Primatol.* **8**:43–72.

Fogden, M. (1974). A preliminary field study of the western tarsier, *Tarsius bancanus* Horsefield. In *Prosimian Biology*, ed. R.D. Martin, G.A. Doyle, and A.C. Walker, pp. 151–166. London: Duckworth.

Gebo, D.L. (1987). Functional anatomy of the tarsier foot. *Am. J. Phys. Anthropol.* **73**:9–31.

Gingerich, P.D. (1978). Phylogeny reconstruction and the phylogenetic position of *Tarsius*. In *Recent Advances in Primatology*, vol. 3, ed. D.J. Chivers and K.A. Joysey, pp. 249–255. London: Academic Press.

Hubrecht, A.A.W. (1908). Early ontogenetic phenomena in mammals and their bearing on our interpretation of the phylogeny of the vertebrates. *Quan. J. Microsc. Sci.* **53**:1–181.

Luckett, W.P., and Maier, W. (1982). Development of deciduous and permanent dentition in *Tarsius* and its phylogenetic significance. *Folia Primatol.* **37**:1–36.

Mackinnon, J., and Mackinnon, K. (1980). The behavior of wild spectral tarsiers. *Int. J. Primatol.* **1**:361–379.

Martin, R.D. (1975). Ascent of the primates. *Nat. Hist.* **74**(3):52–61.

Musser, G.G., and Dagosto, M. (1987). The identity of

Tarsius pumilus, a pygmy species endemic to the montane mossy forests of central Sulawesi. *Am. Mus. Nov.*, no. 2867, pp. 1–53.

Niemitz, C. (1977). Zur functionellen Anatomie der Papillarleisten und ihrer Muster bei *Tarsius bancanus borneanus* Horsefield, 1821. *Z. Saugetierk.* **42**:321–346.

———. (1979). Outline of the behavior of *Tarsius bancanus*. In *The Study of Prosimian Behavior*, ed. G.A. Doyle and R.D. Martin, pp. 631–660. New York: Academic Press.

———. (1984). *Biology of Tarsiers*. Stuttgart and New York: G. Fischer Verlag.

Pollock, J.I., and Mullin, R.J. (1987). Vitamin C biosynthesis in prosimians: Evidence for the anthropoid affinity of *Tarsius*. *Am. J. Phys. Anthropol.* **73**:65–70.

Schwartz, J. (1984). What is a tarsier? In *Living Fossils*, ed. N. Eldredge and S.M. Stanley, pp. 38–49. New York: Springer Verlag.

Wright, P.C., Izard, M.K., and Simons, E.L. (1986). Reproductive cycles in *Tarsius bancanus*. *Am. J. Primatol.* **11**:207–215.

New World Anthropoids

PRIMATE GRADES AND CLADES

Extant primates have traditionally been divided into two suborders, Prosimii and Anthropoidea. Prosimians are a gradistic unit; that is, they are characterized by their retention of primitive anatomical features. Extant anthropoids are also a grade—characterized by numerous derived anatomical features with respect to the more primitive conditions found among prosimians.

But, in contrast to prosimians, they are also a **clade**, or natural phyletic unit—because all anthropoids have a single anthropoid ancestor. As we discussed in the last chapter, among extant prosimians *Tarsius* shares a number of derived anatomical features with anthropoids, including partial postorbital closure, absence of a stapedial artery, loss of the ethmoid recess in the nasal region, loss of the nasal rhinarium, and loss of the tapetum and development of a retinal fovea. As a result, in a phyletic classification tarsiers and anthropoids are normally grouped in a single clade, the Haplorhini.

The features that *Tarsius* and anthropoids share are, however, considerably fewer and more anatomically restricted than those distinguishing anthropoids from all other, more primitive primates, including tarsiers. This relative magnitude of differences, together with the difficulty of fitting many fossil prosimians into a strepsirhine–haplorhine division, has led many primatologists to retain the more gradistic classification of

anthropoids and prosimians while still recognizing that a strepsirhine–haplorhine dichotomy probably more accurately reflects the phylogenetic relationships of living primates.

There are three major radiations of higher primates: the platyrrhines, or New World monkeys, from South and Central America, and two groups of catarrhines, from Africa, Europe, and Asia—the cercopithecoids, or Old World monkeys, and the hominoids, or apes and humans. The origins and evolutionary divergence of these groups, a topic of current debate, is discussed in later chapters. In this chapter and in Chapters 6 and 7 we consider the evolutionary diversity of the living higher primates. After a general characterization of the anthropoid suborder, we begin with the platyrrhines of the New World.

Anatomy of Higher Primates

Several anatomical features distinguish anthropoids as a group from prosimians (Fig. 5.1). Although some of these features are found in various other mammals (and quite a few in tarsiers), the suite as a whole is diagnostic of higher primates and presumably reflects a unique ancestry for the various groups of primates belonging to that suborder.

The dentition of anthropoids is more conservative than that of extant prosimians in both dental formula and shape of the teeth. All higher primates have two incisors in each quadrant. Although these may be somewhat procumbent, as in pithecines, they are never markedly so, as in strepsirhines. The anthropoid canine is always larger in caliber and usually taller in height than the incisors. There is considerable diversity in the size and shape of anthropoid canines, both among different species and between sexes within many species, but the canine is never absent, drastically reduced, or markedly procumbent, as in strepsirhines, and it usually retains a large root even when the crown is reduced, as in *Homo sapiens*.

Old World higher primates have two premolars, and New World monkeys have three. The morphology of higher primate premolars varies considerably, but the anteriormost premolar is often a simple tooth with a broad buccal surface that sharpens the upper canine, and the posteriormost premolar is always a semimolariform tooth with a differentiated trigonid and talonid.

All living higher primates have three molars, except for marmosets and tamarins, which have two. Higher primate upper molars have a moderate-size hypocone, making a relatively square tooth. The lower molars of higher primates are relatively broad with reduced trigonids (usually no paraconid) and an expanded talonid basin.

In most prosimians and in placental mammals in general the two halves of the mandible are joined at the front by a mobile symphysis that permits some degree of independent movement by the two halves of the jaw during chewing. In higher primates the two halves of the mandible are fused together. This condition seems to be functionally associated with the development of vertical incisors in anthropoids.

In conjunction with their diurnal habits, higher primates as a group have decreased their reliance on smell and have emphasized their reliance on sight. This change is reflected in several diagnostic characteristics of skull morphology (Fig. 5.1). In prosimians the frontal bone joins with the zygomatic bone to form a postorbital bar and thus to complete a bony ring around the orbit. In higher primates the frontal, zygomatic, and sphenoid bones combine to form a bony cup that surrounds the eye and separates it from the temporal fossa behind, a condition known as postorbital closure. In anthropoids the paired frontal bones usually fuse into a single unit early in ontogeny.

The orbits of anthropoids face forward in conjunction with their well-developed stereoscopic vision. Like tarsiers, all higher primates lack a reflecting tapetum and have a retinal fovea on the posterior surface of their eyeball; all higher primates, including the nocturnal owl monkey (*Aotus*), have color vision.

Most higher primates have a relatively short snout, and the lacrimal bone lies within the orbit rather than external to it on the snout. Internally, the nasal region of anthropoids lacks an olfactory recess and has a reduced number of turbinates, as in tarsiers.

In higher primates, as in tarsiers, the

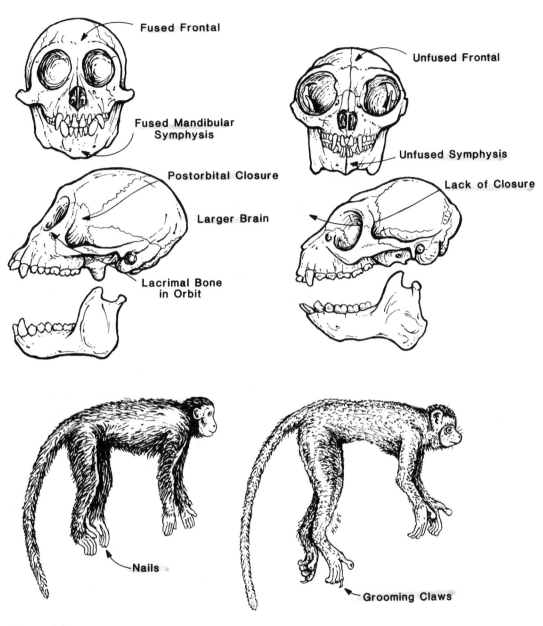

Anthropoids

Fused Frontal

Fused Mandibular
Symphysis

Postorbital Closure

Larger Brain

Lacrimal Bone
in Orbit

Nails

Prosimians

Unfused Frontal

Unfused Symphysis

Lack of Closure

Grooming Claws

FIGURE 5.1

Characteristic anatomical features of anthropoids and prosimians.

blood supply to the brain is primarily through the promontory branch of the internal carotid artery; the stapedial artery is usually absent in adult anthropoids (Fig. 2.11). The tympanic ring is fused to the lateral wall of the auditory bulla in all anthropoids, but the shape of the tympanic bone differs among major living groups (Fig. 2.15).

In skeletal anatomy, anthropoids share few diagnostic features that clearly characterize them as a group. In general, most anthropoids are larger than most extant prosimians and have relatively shorter trunks. Anthropoid forelimbs and hindlimbs are more similar in length, or the forelimbs are longer. With our relatively long legs,

we humans have very unusual higher primate proportions. Anthropoids do not have grooming claws.

Platyrrhines

The living anthropoids of the tropical areas of Central and South America (Fig. 5.2)—the platyrrhines, or ceboids—have an evolutionary history extending back over 30 million years and are a much more diverse group than their common name, New World monkeys, suggests. They are "monkeys" only in that they are not apes (tailless, close relatives of humans). Evolving in South America in the absence of other primates

FIGURE 5.2

Geographic distribution of extant and extinct platyrrhines.

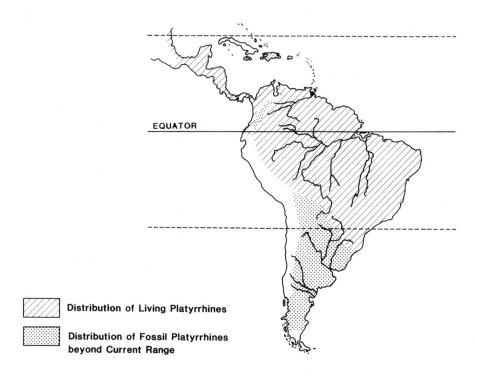

EQUATOR

▨ Distribution of Living Platyrrhines

▦ Distribution of Fossil Platyrrhines beyond Current Range

PLATYRRHINES   **CATARRHINES**

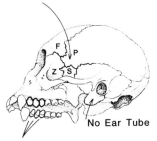
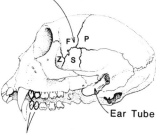

Zygomatic–Parietal Contact   Frontal–Sphenoid Contact

No Ear Tube   Ear Tube

Three Premolars   Two Premolars

FIGURE 5.3

Skulls of a platyrrhine and a catarrhine, showing some of the features distinguishing these two major groups of anthropoids.

(including prosimians), platyrrhines have evolved some species that are prosimian-like in habits, some that are apelike, and many that have no close analogy among other groups of living primates.

Several distinctive anatomical features define the platyrrhine group. They are all small- to medium-size primates ranging from about 100 g to just over 10 kg. Their name is derived from the rounded shape of their external nostrils, which often, but not always, distinguishes them from the Old World anthropoids, which often have narrow, or catarrhine, nostrils. In dental and cranial anatomy (Figs. 5.3, 5.4), platyrrhines have many primitive features that have been

FIGURE 5.4

Upper and lower dentitions of four platyrrhines, illustrating structural differences in the teeth associated with ingesting and processing different types of food.

Ateles   Alouatta   Chiropotes   Callimico

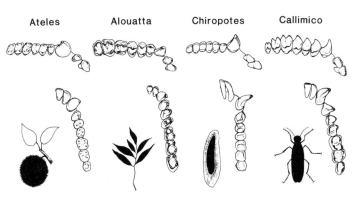

lost in the evolution of Old World catarrhines. For example, they have three premolars, and the tympanic ring is fused to the side of the auditory bulla but does not extend laterally as a bony tube. They also share some unique specializations of their own. The first two molar teeth of all platyrrhines lack hypoconulids, and on the lateral wall of the skull (the pterion region) the parietal and zygomatic bones join to separate the frontal bone above from the sphenoid below. In addition, the cranial sutures of platyrrhines fuse relatively late and many species have relatively long, narrow skulls (Fig. 5.5).

Platyrrhine limb proportions are relatively conservative, with intermembral indices ranging between 70 and 100. Platyrrhines lack the extremes found among either prosimians or catarrhines; most have a relatively short forearm, and most lack an opposable thumb. All have a tail of some sort, and in five genera the tail is a prehensile, fifth limb.

The sixteen genera of extant platyrrhines can be most conveniently grouped into five subfamilies: pitheciines, aotines, cebines, atelines, and callitrichines. Each of these probably represents a relatively old lineage, but—as among the Malagasy strepsirhines—platyrrhine subfamilies vary considerably in both size and cohesiveness. Some of the ecological diversity of platyrrhines is illustrated and characterized ecologically in Figures 5.6 and 5.7.

FIGURE 5.5

Skulls of several extant platyrrhines.

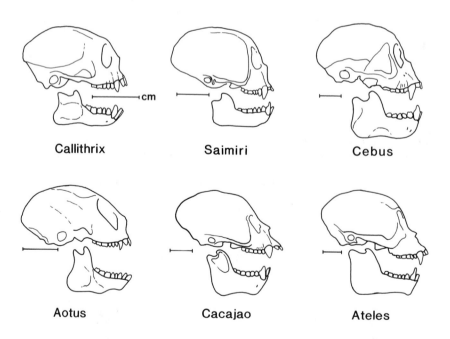

Callithrix   Saimiri   Cebus

Aotus   Cacajao   Ateles

FIGURE 5.6

Seven sympatric platyrrhine species in a Surinam rain forest, showing typical locomotor and postural behavior as well as use of different heights in the forest. At the highest levels are the bearded saki (*Chiropotes satanas*) and the black spider monkey (*Ateles paniscus*); below them are tufted capuchins (*Cebus apella*) on the left and red howling monkeys (*Alouatta seniculus*) on the right; in the lower levels are squirrel monkeys (*Saimiri sciureus*) on the left and golden-handed tamarins (*Saguinus midas*) and white-faced sakis (*Pithecia pithecia*) on the right.

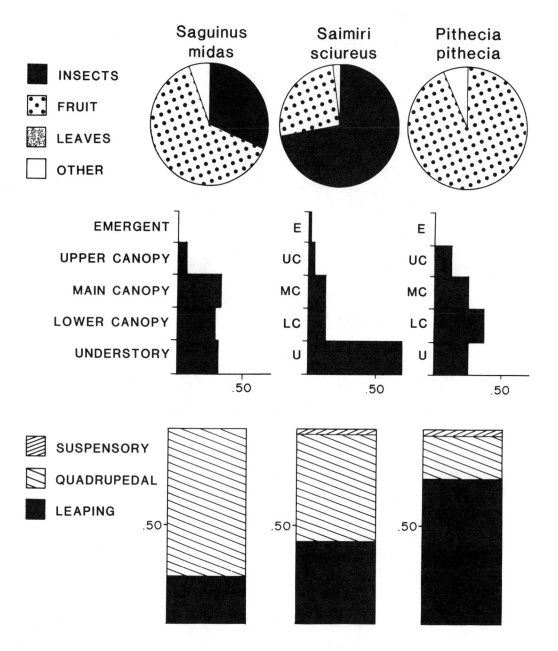

FIGURE 5.7

Diet, forest height preference, and locomotor behavior for seven sympatric platyrrhines from Surinam.

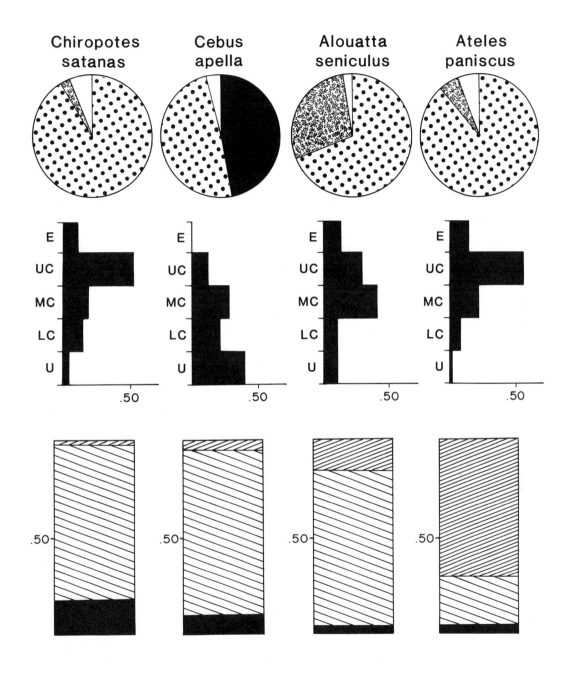

TABLE 5.1
Infraorder Platyrrhini
Subfamily PITHECIINAE

Common Name	Species	Intermembral Index	Body Weight (g)	
White-faced saki	*Pithecia pithecia*	75	MF	1,800
Bald-faced saki	*P. irrorator*	—	M	2,900
			F	2,100
Monk saki	*P. monachus*	77	M	2,800
			F	2,200
White saki	*P. albicans*	—		—
Equatorial saki	*P. aequatorialis*	—		—
Black-bearded saki	*Chiropotes satanas*	83	MF	2,980
White-nosed bearded saki	*C. albinasus*	—	M	3,125
			F	2,518
Bald uakari	*Cacajao calvus*	83	M	3,450
			F	2,875
Black-headed uakari	*C. melanocephalus*	—	F	2,800

Pitheciines

The pitheciines, a distinctive subfamily of platyrrhines, contain three genera of medium-size monkeys (Table 5.1). Pitheciines are characterized by unusual dental specializations including large procumbent incisors, robust canines, and relatively small, square premolar and molar teeth with low cusps (Fig. 5.4, *Chiropotes*). In association with their dental specializations, they have a slightly prognathic snout and enlarged nasal bones (Fig. 5.5, *Cacajao*). Their pelage varies considerably from species to species. New analyses suggest that many pitheciines have a fluid, fission-fusion social structure composed of multi-male, multi-female troops that often separate into smaller foraging parties and at other times forage as a troop, depending on available food resources.

With their broad noses, bushy fur, and long fluffy tails, **sakis** (*Pithecia*) are among the most distinctive of all New World pri-

mates (Fig. 5.8). They are the smallest pitheciine, averaging about 2 kg. They have a slightly more gracile skull and jaw than other pitheciines, a relatively longer trunk, and longer legs. There are four species of sakis. The best known is *Pithecia pithecia*, the white-faced saki, from the Guianas and northeastern Brazil. Males have a distinctive white face; females are totally gray. In addition, there are four Amazonian species, *P. albicans, P. irrorator, P. monachus,* and *P. aequatorialis* (Hershkovitz, 1987).

Species of *Pithecia* have been reported in a wide range of forest types, and the available data provide no clear evidence of habitat preference. White-faced sakis are most commonly seen in the understory and lower canopy levels, where they move primarily by leaping. They are among the most saltatory of all New World monkeys and frequently move by spectacular leaps. A dietary special-

FIGURE 5.8

Two pitheciine primates that live sympatrically in Surinam and other areas of northeastern South America: above, the bearded saki (*Chiropotes satanas*); below, the white-faced saki (*Pithecia pithecia*).

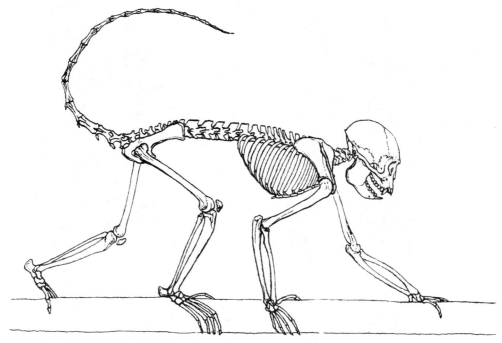

FIGURE 5.9

The skeleton of a bearded saki (*Chiropotes satanas*).

ization on fruit and especially seeds is probably responsible for their dental anatomy.
They use their robust incisors and canines to
open seed cases and their broad, flat molars
for crushing fruits and seeds. They rarely
eat insects.

Sakis have generally been reported to live
in monogamous family groups that travel
and sleep together but often separate while
feeding during the day. There are recent
reports, however, that in many species these
small groups often come together, suggesting that the smaller groups may not be as
distinct as previously thought. Saki social
organization may well be more a fission-
fusion structure than one of strict monogamy. Sakis have single births and the young
seem to be cared for primarily by the
females.

The **bearded saki** (*Chiropotes*) is larger
than *Pithecia* but lacks any obvious sexual
dichromatism. The skull and jaw of *Chiropotes* is more robust than that of *Pithecia* and
the limbs are more similar in length (Figs.
5.8, 5.9). There are two species: *Chiropotes
satanas*, a smaller (3 kg), chocolate brown
species with a black beard and bouffant
hairdo from the Guianas and northeastern
Brazil, and a larger species, *C. albinasus*, with
black fur and a white nose, from Brazil south
of the Amazon. *Chiropotes* prefers high rain
forests (also mountain savannah forests in
Surinam), where it is usually found in the
middle and upper levels of the main canopy
(Fig. 5.7). Bearded sakis are primarily arboreal quadrupeds that frequently use hindlimb suspension in feeding. They feed on
hard, often unripe fruits and on seeds with

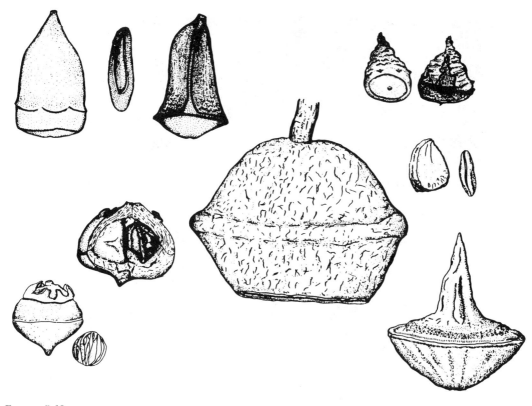

FIGURE 5.10

A few of the hard nuts included in the diet of *Chiropotes satanas* in Surinam.

hard shells (Fig. 5.10), which they open with their large canines. They occasionally eat insects.

Bearded sakis live in large groups with numerous males and females which split into smaller foraging units during the day. They seem to have extremely large home ranges, and groups may travel several kilometers in a day. They have single births and show no evidence of paternal care.

Uakaris (*Cacajao*) are the largest pithe- ciines and are easily distinguished from the other genera by their large size and short tail. There are three types: *C. melanocephalus*, the rare black uakari from the upper Ama- zon of Brazil, eastern Colombia, and south-

ern Venezuela, and two subspecies of *C. calvus*, the bald uakaris. Both subspecies of *C. calvus* have scarlet faces and bald heads. In one subspecies, the long shaggy fur is red; in the other, it is white.

The natural behavior of uakaris is largely unknown. They seem to prefer flooded forests and move primarily by quadrupedal walking and running, both in the trees and on the ground. Like *Chiropotes*, uakaris often feed by hindlimb suspension. They seem to be mainly frugivorous.

Like bearded sakis, uakaris live in large social groups, perhaps including as many as fifty animals. There are no data on their ranging behavior.

FIGURE 5.11

The night monkey, or owl monkey (*Aotus trivirgatus*), the only nocturnal higher primate.

Aotines

Aotus, the **owl monkey**, night monkey, or douracouli (Figs. 3.5, 5.11, Table 5.2), is the only nocturnal higher primate. There are many allopatric species. All are of medium size (about 1 kg) with no marked sexual dimorphism. They have relatively long legs and a long tail. They have large digital pads on the hands and feet, a slightly opposable thumb, and often a compressed clawlike grooming nail on the fourth digit of each foot. They have very large upper central incisors and small third molars. The most distinctive feature of their cranial anatomy is their large orbits, which are relatively larger than the orbits of any other anthropoid (see Fig. 5.5). Like other anthropoids, *Aotus* has

no tapetum lucidum but has a retina with rods and cones as well as a fovea. These are all unusual features in a nocturnal mammal, suggesting that *Aotus* evolved from diurnal ancestors.

Owl monkeys are found throughout South America from Panama to northern Argentina, but they are absent from the Guianas and southeastern Brazil. They live in a variety of forest habitats and there are no indications that they prefer any particular canopy level. They are predominantly quadrupedal but are adept leapers. Their diet is primarily frugivorous and is supplemented by both foliage and insects. They feed in small, evenly dispersed trees that produce a

TABLE 5.2
Infraorder Platyrrhini
Subfamily AOTINAE

Common Name	Species	Intermembral Index	Body Weight (g)	
Night monkey, Owl monkey	*Aotus trivirgatus*[a]	74	MF	1,220
Yellow-handed titi monkey	*Callicebus torquatus*	—	M	1,490
			F	1,265
Dusky titi monkey	*C. moloch*	74	MF	1,070
Masked titi monkey	*C. personatus*	73	MF	1,700

[a]simplified systematics.

small number of fruits on a regular basis. Unlike other small, monogamous platyrrhines, they also feed in larger trees—at night when more dominant species are asleep.

Owl monkeys live in monogamous families of two to four individuals which occupy home ranges of 6 to 10 ha. Night ranges are long in the wet season and very short (250 m) in the dry season, when available food is more clumped in its distribution. Their nightly activity consists mainly of feeding, with the remainder of the time divided between travel and resting. During the southern winters of Paraguay and northern Argentina, owl monkeys are active during both day and night. Each family has several daytime sleeping nests consisting of holes, vine tangles, or open branches. Solitary individuals give low, owl-like hoots, perhaps to attract mates.

In contrast with most monogamous monkeys, adult owl monkeys rarely groom one another. Nevertheless, pairs stay in close contact throughout the night and sleep huddled together. Females give birth to single offspring annually. During the first week of life the infant is increasingly entrusted to the male, who carries it through-

out much of the night and sleeps with it during the day. The infant returns to its mother only to nurse.

Titi monkeys (*Callicebus*) are most closely related to owl monkeys, but they also seem near the ancestry of pitheciines (Fig. 3.5). These small monkeys have short faces, fluffy bodies with long fluffy tails, and long legs. Compared with other platyrrhines, titi monkeys have very short canine teeth. There are three species (Table 5.2). Two of them, *C. torquatus*, the yellow-handed titi, and *C. moloch*, the dusky titi, are from Amazonian areas of Brazil, Venezuela, Colombia, and Peru. A third species, the beige, masked titi, *C. personatus*, is from southeastern Brazil. There is slight dental sexual dimorphism in *C. torquatus* but none reported for other species.

Although the ranges of the two Amazonian species overlap broadly, they seem to have different habitat preferences. *Callicebus torquatus* prefers the main canopy, high ground, and mature high forest, while *C. moloch* favors the understory and low levels in forest and bamboo thickets (Kinzey, 1981). They are both quadrupeds and leapers, but *C. moloch* leaps more frequently than does *C. torquatus*. Both occasionally use vertical

FIGURE 5.12

Two cebines that live sympatrically throughout much of South America: above, the tufted capuchin (*Cebus apella*); below, the squirrel monkey (*Saimiri sciureus*).

clinging positions when feeding, but *C. moloch* does so more frequently.

Titi monkeys are mainly frugivorous, but *C. moloch* eats large numbers of leaves (and bamboo shoots), whereas *C. torquatus* supplements its fruity diet with less foliage and more insects. Titi monkeys live in monogamous family groups that advertise their presence with elaborate dawn duets lasting five to fifteen minutes which are answered sequentially by neighboring pairs. In addition, they actively defend their relatively small territories. All three species have relatively small day ranges, similar to those of *Aotus* (600 m), and they seem to specialize on fruits that are found in small patches which they can harvest on a regular basis. All species are diurnal, with feeding peaks in the early morning and late afternoon and a long resting peak in late morning. They have single births and the young are carried by the male after their first week.

Cebines

The subfamily Cebinae contains two genera, *Saimiri* and *Cebus* (Fig. 5.12, Table 5.3),

which are united on the basis of their dental proportions, overall cranial morphology, and several skeletal features.

Squirrel monkeys (*Saimiri*) are less than a kilogram in average body mass. They are commonly used as laboratory animals and were once common as pets. Squirrel monkeys have a distinctive cranial morphology with a very long occipital region and a foramen magnum that lies well under the skull base (Fig. 5.5). The orbits are close together, so much so that the interorbital septum is perforated by a large opening. Squirrel monkeys are characterized by relatively broad quadrate upper molars with a large lingual cingulum and very small last molars. Their cheek teeth have very sharp cusps, which are associated with an insectivorous diet (Fig. 5.7). The canines are sexually dimorphic, with those of males larger than those of females.

The postcranial skeleton of squirrel monkeys (Fig. 5.13) is characterized by a relatively long trunk, long hindlimbs, and a long tail that is prehensile in infants but not in adults. Their hands have comparatively short fingers and a short but relatively unop-

TABLE 5.3
Infraorder Platyrrhini
Subfamily CEBINAE

Common Name	Species	Intermembral Index	Body Weight (g)	
Tufted capuchin	*Cebus apella*	82	M	3,300
			F	1,940
White-fronted capuchin	*C. albifrons*	82	M	3,260
			F	2,220
White-throated capuchin	*C. capucinus*	—	M	3,700
			F	2,500
Wedge-capped capuchin	*C. nigrivitattus* (= *olivaceus*)	—	M	3,500
Squirrel monkey	*Saimiri sciureus*[a]	80	M	960
			F	790

[a]simplified systematics.

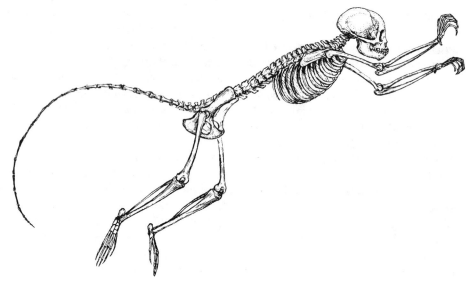

FIGURE 5.13

The skeleton of a squirrel monkey (*Saimiri sciureus*).

posable thumb. There is frequently a small amount of fusion between the tibia and fibula distally, making the ankle joint more restricted to simple hinge movements than in other platyrrhine species.

Traditionally, all squirrel monkeys have been placed in a single species, *Saimiri sciureus*. Hershkovitz (1984), however, recognizes numerous allopatric species from Panama and Costa Rica through Amazonian Brazil, Colombia, Ecuador, Peru, and Bolivia, the Guianas, and southern Venezuela. All are small, gray to yellow monkeys with short fur and a long, thin tail.

Squirrel monkeys occupy a variety of rain forest habitats but show a preference for riverine and secondary forests, where they are commonly found in the lower levels (Fig. 5.7). They are arboreal quadrupeds that frequently leap, especially when traveling in lower forest levels. During feeding, they are almost totally quadrupedal and occasionally come to the ground.

Squirrel monkeys are frugivores that seem to specialize on large fruit trees throughout the year. As they travel between fruit trees, they forage for insects (for almost half of each day) and frequently catch large arthropods on the wing. *Saimiri sciureus* often travels and forages in conjunction with *Cebus apella* (Fig. 5.12).

Squirrel monkeys live in large, continuously active groups that vary in size from about twelve to over one hundred individuals, with numerous adults of both sexes and offspring. They communicate frequently throughout the day with high-pitched whistles and chatter—a group is usually heard well before it comes into view. The estimated size of the overlapping home ranges varies dramatically from troop to troop. Squirrel monkeys use large day ranges (2.5–4 km), in keeping with the large group sizes and extensive insect foraging.

The intragroup social organization of *Saimiri* is unusual among anthropoids. Social interactions seem to center around a hierarchy of adult females, as in many prosimian groups. In the mating season, males put on extra fat and become more aggressive. Squirrel monkeys have restricted breeding seasons. Females give birth to relatively large single offspring at yearly intervals. In contrast with the males of most small New World monkeys, male squirrel monkeys do not play an important role in the care of infants; rather, infants are cared for by several females in addition to their mother.

Cebus, the **capuchin** or **organ-grinder monkeys**, are among the most well known of New World monkeys. These medium-size, sexually dimorphic primates all have large premolars and square molar teeth with thick enamel, which they use to open hard nuts. Their forelimbs and hindlimbs are more similar in size than those of many platyrrhines. Capuchins have a relatively short, fur-covered, prehensile tail. They have short fingers and an opposable thumb and are the most dexterous platyrrhines.

The genus is usually divided into four species (Table 5.3). *Cebus apella*, the black-capped or tufted capuchin (Fig. 5.12), is a relatively robust species found throughout most of the neotropics from northern South America to northern Argentina. Two allopatric species of untufted capuchins are sympatric with *C. apella* throughout much of South America. *Cebus nigrivittatus* (= *olivaceus*), the wedge-capped or brown capuchin, is found in Venezuela, the Guianas, and parts of Brazil, and *C. albifrons*, the white-fronted capuchin, is from the upper Amazon. *Cebus capucinus*, the white-capped capuchin, is the only species from Central America.

Capuchins are found in virtually all types of neotropical forest. They seem to prefer the main canopy levels (Fig. 5.7), but they frequently descend to the understory or to the ground during both travel and feeding. All four species are arboreal quadrupeds that use their prehensile tails mainly during feeding. The capuchin diet includes many types of fruit and animal matter (Fig. 5.7). Tufted capuchins are opportunistic monkeys that use their manipulative abilities and their strength to obtain foods that are unavailable to other species. They forage destructively for invertebrates in bark and leaf litter and are also able to break open hard palm nuts and the hard shells covering immature flowers. In contrast, the more gracile *C. albifrons* specializes on superabundant fruit sources in the same forests as *C. apella*.

All capuchins live in social groups of a dozen or so individuals with several adult males, several adult females, and offspring. There is greater sexual size dimorphism and a more obvious dominance hierarchy among males of *C. apella* than of *C. albifrons* (Janson, 1986a,b). Capuchins have large home ranges and relatively long day ranges. They have single births.

Atelines

The four ateline genera (Table 5.4) are the largest platyrrhines, and all have a long, prehensile tail covered on its ventral surface with friction ridges similar to the fingerprints found on our hands. In many aspects of their limb and trunk anatomy and in their use of suspensory behavior, they show similarities to the extant apes (Erikson, 1963). In dental and cranial anatomy, as well as diet and social structure, they are quite diverse.

The **howling monkeys** (*Alouatta*) have a broad distribution, ranging from southern Mexico to northern Argentina (Fig. 5.14).

TABLE 5.4
Infraorder Platyrrhini
Subfamily ATELINAE

Common Name	Species	Intermembral Index	Body Weight (g)	
Red howler	*Alouatta seniculus*	97	M	7,880
			F	5,500
Black-and-red howler	*A. belzebul*			
Brown howler	*A. fusca*			
Mantled howler	*A. palliata*	98	M	11,590
			F	6,290
Guatemalan howler	*A. villosa*	—		—
Black howler	*A. caraya*	97	M	8,280
			F	5,410
Common woolly monkey	*Lagothrix lagothricha*	98	M	8,770
			F	5,740
Yellow-tailed woolly monkey	*L. flavicauda*	—		—
Woolly spider monkey	*Brachyteles arachnoides*	104	MF	12,000–
				15,000
Black spider monkey	*Ateles paniscus*	105	MF	9,000
Black-handed spider monkey	*A. geoffroyi*	105	MF	7,730
Brown-headed spider monkey	*A. fusciceps*	103		—
Long-haired spider monkey	*A. belzebuth*	109		—

There are six allopatric species. All are large (6–10 kg) and sexually dimorphic by size and have prehensile tails, but individual species vary dramatically in color, ranging from red to brown, black, or blond. In some species both sexes are the same color; in others the sexes look strikingly different.

Howling monkeys have small incisors and large sexually dimorphic canines (Fig. 5.4). The lower molars have a narrow trigonid and a large talonid; the upper molars are quadrate, with well-developed shearing crests characteristic of folivorous primates.

The skull of *Alouatta* is distinguished by its relatively small cranial capacity and lack of cranial flexion. The mandible is quite large and deep, and the hyoid bone is expanded into a very large, hollow, resonating chamber.

Howling monkeys have forelimbs and hindlimbs that are similar in length and a long prehensile tail. Like many platyrrhines, howling monkeys have a poorly differentiated thumb and usually hold objects and branches between their second and third digits, a grasp known as schizodactyly ("between fingers").

Howling monkeys are found in a variety of habitats, including primary rain forests, secondary forests, dry deciduous forests, montane forests, and llanos habitats containing patches of relatively low trees in open savannah. Their distribution ranges from sea level to altitudes above 3,200 m. Within these diverse forest habitats, most species seem to prefer the main canopy and emergent levels (Fig. 5.7), but several species (especially *A. caraya*) that live in drier areas

FIGURE 5.14

A troop of red howling monkeys (*Alouatta seniculus*).

regularly come to the ground and cross open areas between patches of forest. Howlers are slow, quadrupedal monkeys that rarely leap (Fig. 5.7). During feeding, and less often during travel, they use suspensory locomotion, primarily climbing, in which all five limbs grasp supports opportunistically. During feeding they frequently use their tail in suspension.

Howling monkeys are one of the most folivorous of all New World monkeys (Fig. 5.7). There is considerable variation in their diet from month to month, but leaves, especially new leaves, constitute half or more of the yearly diet. Fruits and flowers are the next most common components.

Most howling monkeys live in groups containing a single adult male, several adult females, and their offspring, but the normal group composition seems to vary from one species to another. Groups of *A. palliata* may contain from twelve to thirty individuals, but in *A. seniculus* and *A. caraya* troops are usually smaller. Howler day ranges are very small, often less than 100 m, because of the howlers' ability to subsist on both a diversity of food items and on relatively common foods such as leaves. In keeping with their small home ranges, howling monkeys spend little time traveling each day and have long periods of resting and digesting. Home ranges generally vary from 4 to 20 ha, depending on group size—a small home range for New World monkeys of their biomass.

As their name indicates, howling monkey groups regularly advertise their presence with loud, lionlike roars given by both males and females. Vocal battles are often reinforced by actual physical combat between individuals, usually males, but there is disagreement regarding the extent to which howlers actively defend their territories or merely their daily positions. As with many

other animals that live in single-male groups, there is considerable competition between males for access to a troop and the females within it. Males taking over a troop have been reported to kill the dependent infants that were probably the offspring of the ousted male. Howlers have single births. Infants are frequently cared for by females other than their mother.

Little is known about the behavior and ecology of *Lagothrix*, the **woolly monkeys** (Fig. 5.15). There are two species: the more common *L. lagotricha* and the very rare yellow-tailed woolly monkey, *L. flavicaudata*, from Peru. Like howlers, woolly monkeys are sexually dimorphic in body size.

Woolly monkeys seem to be largely restricted to high rain forests. They are primarily arboreal quadrupeds that use their prehensile tail mainly during feeding. Their diet is mainly fruit. Their social structure and patterns of home range use are largely unknown.

The four allopatric species of **spider monkeys** (*Ateles*) range from the Yucatan peninsula of Mexico through Amazonia and are quite variable in details of coloration (Fig. 5.16). They are large, graceful, long-limbed monkeys with a long prehensile tail. Their color ranges from beige to black. Male and female spider monkeys are virtually identical in size and color in every species, and this similarity is strengthened by the long pendulous clitoris in the females, which is often mistaken for a penis.

The dentition of spider monkeys (Fig. 5.4) is characterized by relatively large, broad incisors and small molars with low rounded cusps. The skull has large orbits, a globular braincase, and a very gracile mandible (Fig. 5.5). Spider monkeys have relatively long slender limbs (see Fig. 2.16) that resemble those of gibbons and other suspensory species in many features. Their fingers and toes

FIGURE 5.15

A troop of woolly monkeys (*Lagothrix lagotricha*).

FIGURE 5.16

A group of black spider monkeys (*Ateles paniscus*).

are long and slender (see Figs. 2.20, 2.22), and most species lack an external thumb.

Spider monkeys are largely restricted to high primary rain forests, where they prefer the upper levels of the main canopy (Fig. 5.7). They have extremely diverse locomotor abilities. During travel, they use both arboreal quadrupedalism and suspensory behavior including brachiation and climbing. They move bipedally in the trees and occasionally leap. During feeding they are almost totally suspensory, and they use all five limbs to utmost advantage. Spider monkeys feed primarily on ripe fruit, but in some seasons they eat large amounts of new leaves (Fig. 5.7).

Spider monkey social organization is like the fission-fusion type found in chimpanzees. Groups are generally large, comprising a dozen or more individuals of both sexes and all ages. During the day, the large social group generally breaks down into smaller foraging units of two to five individuals which give loud, barking contact calls. These units are most frequently either adult females and their offspring or groups of young males. Spider monkeys have single births and the young are cared for by the mother.

Brachyteles arachnoides, the **woolly spider monkey** or **muriqui** (Fig. 5.17), is the largest nonhuman primate in the neotropics and one of the primate species closest to extinction. Its range is restricted to the few remaining patches of rain forest in southeastern Brazil. Although *Brachyteles* resembles the spider monkey in limb proportions and in the lack of a thumb, the dentition of the woolly spider monkey is more like that of *Alouatta*, with numerous shearing crests on the molar teeth. The canines of both sexes are small.

In their disappearing habitat, *Brachyteles*

are totally restricted to high forest areas, where they prefer canopy levels. Like *Ateles*, they are arboreal quadrupeds that rely extensively on suspensory behavior during travel and especially during feeding. Leaves make up the greater part of their diet, followed by fruit. Their social organization seems most comparable to that of spider monkeys, with large multi-male, multi-female groups that split into smaller foraging units during the day. Mating is extremely promiscuous, with females frequently copulating with several males successively. Males have extremely large testicles, suggesting that competition between males is largely by sperm competition rather than by interindividual aggression (Milton, 1985).

Callitrichines

The smallest and most distinctive of the New World anthropoids are the callitrichines (Table 5.5). There are three distinct groups among the callitrichines: Goeldi's monkey (*Callimico*), tamarins (*Saguinus* and *Leontopithecus*), and marmosets (*Cebuella* and *Callithrix*). They are all small (100–750 g), brightly colored monkeys with little if any sexual dimorphism in size or pelage coloration.

Marmosets and tamarins have a unique dentition (Fig. 5.18) with a dental formula of $\frac{2.1.3.2.}{2.1.3.2.}$. They have lost their third molars from the primitive platyrrhine condition. They have simple, tritubercular upper molars with no hypocone which differ from the tritubercular molars of most primitive mammals in also lacking conules. *Callimico* has a more primitive dentition, with three molars and a tiny hypocone. All callitrichines have very short snouts and long braincases (Fig. 5.5).

Callitrichines have skeletons with relatively long trunks, tails, and legs. All digits except

A troop of muriquis, or woolly spider monkeys (*Brachyteles arachnoides*), from southeastern Brazil, one of the most endangered living primates.

TABLE 5.5
Infraorder Platyrrhini
Subfamily CALLITRICHINAE

Common Name	Species	Intermembral Index	Body Weight (g)	
Goeldi's monkey	*Callimico goeldii*	69	MF	630
Black-and-red tamarin	*Saguinus nigricollis*	—	MF	465
Saddle-back tamarin	*S. fuscicollis*	79	MF	462
Moustached tamarin	*S. mystax*	76	MF	580
White-lipped tamarin	*S. labiatus*	76	MF	580
Emperor tamarin	*S. imperator*	75	MF	400
Red-handed tamarin	*S. midas*	74	MF	570
Inustus tamarin	*S. inustus*	—	MF	740
Barefaced tamarin	*S. bicolor*	—		—
Cottontop tamarin	*S. oedipus*	74	MF	490
White-footed tamarin	*S. leucopus*	74	MF	490
Lion tamarin	*Leontopithecus rosalia*	88	MF	500
Golden-headed lion tamarin	*L. chrysomelas*	—	MF	550
Golden-rumped lion tamarin	*L. chrysopygus*	—	MF	?550
Silvery marmoset	*Callithrix argentata*	76		—
Tassel-eared marmoset	*C. humeralifer*	—		—
Common marmoset	*C. jacchus*	75	MF	310
Buffy tufted-ear marmoset	*C. aurita*	—		—
Buffy-headed marmoset	*C. flaviceps*	—		—
White-faced marmoset	*C. geoffroyi*	—		—
Black tufted-ear marmoset	*C. penicillata*	75		—
Pygmy marmoset	*Cebuella pygmaea*	82	MF	135

the great toe end in tegulae, or claws, rather than the nails characteristic of other higher primates (Fig. 5.18), an adaptation that enables them to cling to the sides of large tree trunks to feed on gums, saps, and insects.

Despite having a simple unicornate uterus and a single pair of nipples, features usually found among mammals characterized by single births, marmosets and tamarins typically give birth to twins. The more primitive *Callimico* has single births. In all callitrichines, males play a major role in the care of infants and are primarily responsible for transporting them.

Marmoset and tamarin social groups are characterized by having a single breeding female in each group and by intensive care of infants by one or more adult males and often by other troop members as well. Although callitrichines regularly live in stable monogamous groups in captivity, in natural situations they are found in larger, more complicated groups that contain one breeding female but often several adult males that mate with the female. Marmoset and tamarin groups are subject to frequent immigrations and emigrations by members of both sexes. It appears that, whereas group size and home range often remain constant from year to year, the individuals constituting the groups usually change (Garber, 1984; Terborgh and Goldizen, 1985).

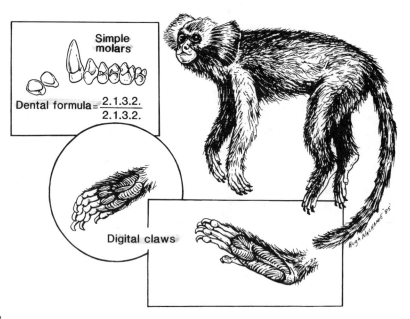

FIGURE 5.18

The unusual features of callitrichines.

The evolutionary history of the unique marmoset and tamarin (callitrichid) features is a topic of considerable debate. Hershkovitz (1977) has argued that the small size, claws, and simple molars of callitrichines are primitive features indicating an independent origin of platyrrhines from a very primitive (nonanthropoid) primate ancestor. Others (e.g., Ford, 1980; Leutenegger, 1980; Rosenberger, 1984; Sussman and Kinzey, 1984) have argued—more convincingly in my opinion—that the unique anatomical features of callitrichids are derived specializations related to their small size or unusual ecological adaptations for insectivory and exudate eating. For example, among platyrrhines, and primates in general, smaller species generally have larger infants (relative to the size of the mother). This results in considerable problems for the female in birthing and in early postnatal care, which callitrichids have overcome by giving birth to twins (rather than a single large infant) and by extensive caring for infants by many group members (Leutenegger, 1980). Likewise, their high-energy diet and small size permit a reduced dentition compared with that needed by larger, more folivorous or frugivorous species (see Chapter 8).

Goeldi's Monkey

Callimico goeldii (Fig. 5.19, Table 5.5) is a tufted, silky black monkey from the upper Amazon regions of Colombia, Ecuador, Peru, and Bolivia. It is intermediate in several anatomical features between other callitrichines and more "normal" platyrrhines. This species has a more primitive dentition than other callitrichines, with a full set of three molars and a small hypocone on the upper molars (Fig. 5.4). *Callimico* also has single births—the primitive condition

FIGURE 5.19

A group of Goeldi's monkeys (*Callimico goeldii*) in a typical bamboo habitat.

for all New World monkeys. In other respects, such as claws and limb proportions, *Callimico* resembles tamarins and marmosets. It is slightly smaller than the squirrel monkey. Males and females are virtually indistinguishable in size and coloration.

Goeldi's monkeys are animals of secondary forests and are most frequently found in low bushes and bamboo thickets of the understory. In this vertical milieu, they move mainly by leaping from trunk to trunk a few meters off the ground. The diet of *Callimico* includes large amounts of invertebrates and also fruits. There is no evidence that they ever feed on exudates.

In captivity, these monkeys live in family groups of one adult male and one adult female, but under natural conditions groups seem to contain more adults of both sexes. Their social organization has not been studied in detail. Compared with marmosets and tamarins, Goeldi's monkeys have relatively large home ranges and day ranges. They have single offspring twice a year. As in *Callicebus* and *Aotus*, males are largely responsible for carrying infants after birth.

Tamarins

The tamarins, *Saguinus* and *Leontopithecus*,

Saguinus bicolor   Saguinus imperator   Saguinus oedipus   Saguinus labiatus

FIGURE 5.20

The faces of four tamarins.

are the most diverse group of callitrichines, with ten to twelve distinct species displaying an extraordinary array of pelage color patterns and elaborations of facial hair (Fig. 5.20, Table 5.5). All are relatively small, with relatively long trunks, legs, and tail. Like most other anthropoid primates, tamarins have canines that are much larger than their incisors. Although many tamarins eat exudates, they seem to lack the marmoset's ability to gnaw holes in tree bark. Several tamarin species have been the subject of extensive field studies.

Saguinus oedipus, the **crested** or **cotton-top tamarin**, lives in low secondary forests of Panama and Colombia. These monkeys move primarily by quadrupedal walking and running on medium-size supports and less frequently by leaping between vertical trunks. Fruit (40 percent) and animal material (40 percent) seem to make up the bulk of their diet, and exudates are an important third component. In foraging for fruits, they range from the middle of the canopy to the ground. They forage for insects in the shrub layer and feed on exudates by clinging to the sides of relatively large trunks.

Cotton-top tamarin groups have a mean size of six individuals but vary considerably. The relationships among individuals in these groups are not known from present studies. Group composition changes frequently and there is considerable immigration and emigration between groups.

The **golden-handed tamarin** (*Saguinus midas*) has been briefly studied in both eastern Colombia and Surinam. In Surinam these tamarins are most common in primary forest, but they prefer the edge habitats between forest types (Fig. 5.7). They spend most of their time in the middle levels of the forest, where they move primarily by quadrupedal walking and running along medium-size supports and by leaping between the ends of branches. They seem to be largely frugivorous; insects and exudates are less important components of their diet.

Social groups of golden-handed tamarins average about six individuals, with a considerable range (2–12). There are no detailed studies of their composition or social dynamics. *Saguinus oedipus* and *S. midas* are the only tamarins living in their respective geographic areas.

Saguinus fuscicollis (Fig. 5.21), the **saddle-back tamarin** (named for the distinctive

<small>FIGURE 5.21</small>

Two sympatric tamarins from Bolivia: above, the white-lipped tamarin (*Saguinus labiatus*); below, the saddle-back tamarin (*Saguinus fuscicollis*).

patterns on its trunk), has over a dozen subspecies throughout Amazonian Colombia, Peru, Bolivia, and Brazil. It has been studied in several localities, often in conjunction with other tamarins. These small monkeys move and feed in the lower levels of the forest. Their locomotion involves frequent leaps between large vertical tree trunks and from branches to trunks. During the wet season they are primarily frugivores specializing on small, widely dispersed fruits. In the dry season, when fruits are rare, the herbivorous portion of their diet consists almost totally of nectar. Insect foraging accounts for nearly half of the daily feeding time in this species, which specializes on relatively large, cryptic insects that it locates by probing in the hollows, crevices, and bases of trees.

Saddle-back tamarins live in small groups of three to eight individuals and actively defend their territories against neighboring groups of the same species. The most common group structure is a polyandrous mating system with a single breeding female and two breeding males. Monogamous and polygynous groups are less common, and all monogamous groups studied failed to successfully raise their offspring without a second male caretaker. Although group territories remain stable from year to year, there is considerable turnover of individuals in a group as a result of predation and births as well as emigration and immigration. Groups often range more than a kilometer a day.

Throughout their distribution, saddle-back tamarins are normally found in association with one of the moustached tamarins. There are three members of this species-group whose behavior has been studied in recent years: the white-lipped tamarin, the emperor tamarin, and the moustached tamarin.

Saguinus labiatus, the **white-lipped tamarin** (Figs. 5.20, 5.21), has a relatively small distribution in the middle Amazonian region of western Brazil and eastern Bolivia, where it lives in sympatry with *S. fuscicollis*. Although similar in size to the saddle-back species, white-lipped tamarins differ in several aspects of behavior and ecology. They are found at higher levels in the forest, most commonly in the middle levels of the canopy, where they move by quadrupedal running and short leaps between branches. Like saddle-backs, they eat both fruits and insects, but their insect foraging is primarily for small insects, which they find among the leaves and terminal branches within the main canopy.

White-lipped tamarins and saddle-back tamarins have home ranges of similar size (30 ha) for their small groups (2–4 individuals). Most striking is the fact that groups of the two species overlap almost completely in their daily ranging behavior, not only when traveling and feeding but also during resting and sleeping. This seems to be primarily an adaptation for predator detection.

Saguinus imperator, the **emperor tamarin** (Fig. 5.20), was named for Emperor Franz Joseph of Austria because of its sweeping moustache. This medium-size tamarin from the upper Amazonian region of Peru and Bolivia is similar in many aspects of its ecology to the white-lipped tamarin. It relies on fruit from small trees throughout the year as its dietary staple, but it takes a substantial amount of nectar during the dry season, when fruits are less abundant. Like white-lipped tamarins, emperor tamarins forage for visible insects among the leaves and small branches of the forest canopy.

The small family groups of emperor tamarins share their moderate territories (30 ha) with a group of saddle-backed tamarins of a

A group of golden lion tamarins (*Leontopithecus rosalia*), one of the most beautiful and most endangered primates.

similar size, but they defend it against con-specifics. They often travel as much as a kilometer a day. Twin offspring are normally born at the beginning of the rainy season.

The **moustached tamarin** (*S. mystax*), from the middle Amazon region of northern Peru and western Brazil, has been the subject of several recent studies undertaken in conjunction with the trapping and transfer of wild populations onto a natural island in northern Peru. These studies provide some experimental evidence on habitat preferences and group formation. Among the available habitats, these tamarins seem to prefer the drier, upland forest and to avoid flooded forests. They travel and feed most commonly on thin, flexible supports.

The demography of this species has provided considerable insight into the unusual nature of tamarin social systems. Moustached tamarins live in groups of three to eight individuals which usually contain a single breeding female, up to three presumably reproductively active adult males, and several nonreproductive females and sub-adults. In the newly formed groups, there was frequent emigration and immigration of adults of both sexes. Usually more than one adult male participated in the care of infants and census data on numerous groups show a correlation between the number of male helpers and the number of surviving infants. The social organization of these callitrichines seems more like the communal breeding systems of birds such as woodpeckers (Stacy and Keonig, 1984) than the more stable, monogamous groups of other platyr-rhines, prosimians, or gibbons.

Lion tamarins (*Leontopithecus*) are the largest callitrichines, with an adult body weight of nearly 700 g (Fig. 5.22). There are three allopatric species, all from southeast Brazil: *L. rosalia*, the golden lion tamarin; *L. chrysomelas*, the golden-headed lion tamarin; and *L. chrysopygus*, the golden-rumped lion tamarin. All are on the verge of extinction in the wild because of extensive habitat destruction. Of the three, *L. chrysomelas* is the most distinctive in dental and cranial features, with particularly large anterior teeth. *Leontopithecus rosalia* has the most gracile anterior dentition (Rosenberger and Coimbra-Filho, 1984).

Lion tamarins are largely confined to the lowland primary rain forest of southeast Brazil and seem to fare poorly in secondary forests. All species are found primarily in the main canopy levels, where they move in a quadrupedal fashion. Lion tamarins feed on a wide range of invertebrates and small vertebrates as well as on fruit, but they have never been observed eating exudates. They use their long fingers to extract insects from holes and crevices and beneath tree bark. It has been suggested that the enlarged anterior teeth of *L. chrysomelas* are used to gnaw through bark to expose insects.

Lion tamarins live in moderate-size social groups and have twin births. At present there is no information available on either home range or day range. There are indications from both captive and field observations that lion tamarins use holes in trees for sleeping.

Marmosets

The marmosets (Table 5.5) are the smallest platyrrhines and also those with the most specialized dentition (Fig. 5.23). They are distinguished from tamarins and other platyrrhines by their uniquely enlarged incisors, which are similar in height to their canines. Furthermore, these large incisors have only a thin layer of enamel on the lingual surface which quickly wears away and causes the teeth to assume a chisel-like shape similar to the incisors of rodents. Marmosets use these chisel-like incisors for biting holes in trees to elicit the flow of gums, saps, and resins.

The larger marmosets, *Callithrix* (Fig. 5.24), are divided into three species groups: *C. humeralifer*, the **tassel-eared marmoset**, from Amazonia; *C. argentata*, the **silvery marmoset**, from Amazonia, western Brazil and eastern Bolivia; and *C. jacchus*, the **common marmoset**, a superspecies that includes five closely related allopatric species (or subspecies) from southeastern Brazil. *Callithrix* species seem to be most commonly found in dry rather than flooded forests, and especially in edge habitats and secondary forest environments. They forage for insects in the vine tangles of the understory and move by a combination of quadrupedal walking and running as well as by leaping. They use clinging postures on tree trunks when eating exudates. In contrast with the pygmy marmoset, however, *Callithrix* species are primarily frugivorous and insectivorous, relying on exudates primarily at the end of the wet season when fruits are scarce.

There have been no long-term studies of the social behavior of *Callithrix*. In captivity they live in strictly monogamous groups with a single adult male and a single adult female, each of which excludes other members of the same sex by fighting. They give birth to twins at six-month intervals, and the young are carried by the male from the first week of life. Wild groups of *Callithrix* generally consist of more individuals, suggesting that natural populations probably live in more complex social groups with several adult males and females. The number of reproducing individuals in these *Callithrix* groups is unknown. At night *Callithrix* often sleep in tree holes.

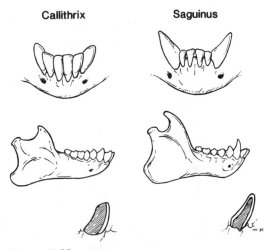

Callithrix   Saguinus

FIGURE 5.23

The lower jaw and teeth of a marmoset (left) and a tamarin (right), showing the differences in proportion of the canines and incisors and the thickness of the enamel on the lower incisors.

FIGURE 5.24

The faces of four marmosets, showing the diversity in facial ornamentation.

 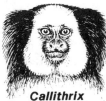 

Callithrix jacchus   *Callithrix argentata*   *Callithrix geoffroyi*   *Callithrix aurita*

FIGURE 5.25

A family of pygmy marmosets (*Cebuella pygmaea*).

Cebuella pygmaea, the **pygmy marmoset** (Fig. 5.25), has an adult body weight of approximately 100 g. It is the smallest marmoset, the smallest platyrrhine, and the smallest anthropoid. It is only slightly larger than *Microcebus murinus* and *Galago demidovii*, the smallest living primates. Because of its small size, Hershkovitz (1977) has argued that the pygmy marmoset is the most primitive platyrrhine. However, because of their unusual diet and the dental specializations they share with *Callithrix* (Fig. 5.23), most authorities feel that pygmy marmosets are extremely specialized platyrrhines. They are found in the Amazonian regions of Colombia, Peru, Ecuador, Brazil, and Bolivia, where their distribution and density seem to be linked to the abundance of special feeding trees.

Feeding on tree exudates occupies 67 percent of pygmy marmoset feeding time, part of which is devoted to actual feeding from a primary exudate tree and part to preparing new trees by gnawing holes in their bark. The remainder of the monkeys' feeding time consists of foraging for insects and occasional fruits. Because they are so dependent on tree exudates, which they obtain by gnawing holes in trunks and large branches, pygmy marmosets tend to be found in the lower levels of the forest. Their insect foraging takes place in vine tangles. During exudate eating, they frequently adopt clinging positions on the large trunks and move by leaping between vertical supports (Kinzey *et al.*, 1975).

Pygmy marmosets live primarily in small groups with a single adult male, a single adult female, and offspring of various ages. They have tiny home ranges centered around whatever the main food tree is at the time. Because these primary exudate trees change from year to year, so do the home ranges of pygmy marmoset groups. Pygmy marmosets give birth to dizygotic (fraternal) twins at approximately six-month intervals. The young are carried most frequently by the adult male. During the night pygmy marmosets sleep in vine tangles or in tree holes.

ADAPTIVE RADIATION OF PLATYRRHINES

Like the Malagasy prosimians, the platyrrhines of the neotropics arrived on an island continent tens of millions of years ago and have evolved into a diverse radiation with no competition from other groups of primates. The extent of their adaptive diversity (Fig. 5.26) is indicated by the presence of six or more sympatric species throughout most of South America and up to thirteen species at some Amazonian sites (see Figs. 5.6, 5.7).

In size, platyrrhines are small- to medium-size primates; they range from the pygmy marmoset at about 100 g to *Brachyteles*, which weighs over 10 kg. All of the genera but one are diurnal, but the single nocturnal genus (*Aotus*) is very widespread.

Although they lack the extremes in limb proportion or skeletal specialization seen in many other groups of primates, platyrrhines show a wide range of locomotor abilities. Some species are excellent leapers, many are arboreal quadrupeds, and the larger species frequently use suspensory postures. Platyrrhines are the only primates to have evolved a prehensile tail, an organ that adds considerably to the locomotor abilities of five genera. In addition to locomotor specializations, New World monkeys have evolved

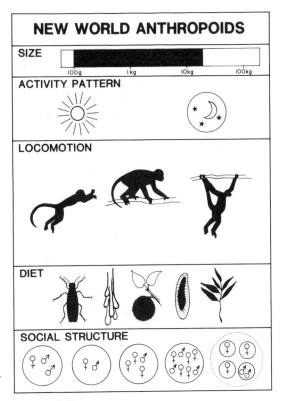

FIGURE 5.26

The adaptive diversity of the platyrrhines.

unique postural adaptations such as the abilities of the clawed callitrichines to cling to vertical trunks—an adaptation for feeding on exudates and cryptic prey on trunks and in tree holes.

A striking feature of the platyrrhine radiation is the absence of terrestrial species. A few species (*Alouatta caraya*, *Cebus apella*, and *Saimiri sciureus*) occasionally forage on the ground or travel for short distances between trees, but none spend a large portion of each day feeding on the ground.

The New World anthropoids include species that specialize on gums, on fruits, on leaves, and on seeds. Some of the smaller species rely heavily on nectar during the dry periods of the year. There are only two predominantly folivorous genera, *Alouatta* and *Brachyteles*. Only the woolly spider monkey seems to rely almost exclusively on leaves.

Platyrrhine social organization is much more diverse than that found among any other major radiation of primates and more complex than earlier studies anticipated. Many New World monkeys live in monogamous groups (*Aotus, Callicebus*, and perhaps *Pithecia pithecia*) that seem to be stable for several years. In several genera (*Ateles, Brachyteles*, and *Chiropotes*), the normal social structure is a large group of many adult males and females that fragments into smaller foraging units. One genus (*Alouatta*) lives in single-male groups in some environments and multi-male groups at different sites or when the population density changes. *Cebus* and *Saimiri* live in more complex groups of several adults of each sex. The social organization of *Saimiri* is reminiscent of that of Malagasy lemurs, with female dominance throughout the year and intense male–male competition for a few brief weeks in the breeding season. Social organization in the callitrichines is much more complex than the simple monogamy suggested by studies of captive monkeys. Social groups of many species in natural environments usually contain a single breeding adult female with several reproductively active males and often several nonbreeding adults as well as younger animals. These seem to be polyandrous mating systems.

PHYLETIC RELATIONSHIPS OF PLATYRRHINES

New World anthropoids have traditionally been divided into two families, the callitrichids (marmosets and tamarins) and the cebids (everything else). *Callimico* has always been a problem genus that does not fit cleanly into either group. Although the marmosets and tamarins are certainly the most distinctive group of platyrrhines, such a division does not resolve the relationships of the remaining genera, nor does it offer any insight into the problem of which other group of platyrrhines is most closely related to the callitrichids.

The morphological distinctiveness of the extant platyrrhine subfamilies suggests that each is the result of an early evolutionary diversification, an interpretation that accords well with biomolecular studies of their evolutionary relationships. The most recent attempts at reconstructing the phyletic relationships of the extant taxa using dental and skeletal anatomy are those of Rosenberger (1981, 1984) and of Ford (1986). The two phylogenies (Fig. 5.27) agree in their grouping of pitheciines with atelines and in the placement of *Callimico* with the marmosets. The main differences concern the relationships of cebines and aotines to one another and to either the pitheciine–ateline group or the callitrichines.

Rosenberger (1981) groups *Cebus* and *Saimiri* with the callitrichines, into the cebids, largely on the basis of their reduced third molars, "gracile masticatory apparatus," and genital similarities; he places the remaining genera in a second group, the atelids. He suggests that *Aotus* and *Callicebus* are most closely related to the pitheciines. Rosenberger further argues that this dichotomy reflects both a phyletic and an adaptive division of platyrrhines into those that are largely frugivore-insectivores (cebids) and those that are predominantly frugivore-folivores (atelids).

Ford's phylogeny does not link *Cebus, Saimiri, Aotus*, or *Callicebus* with either the pitheciines and atelines or the callitrichines;

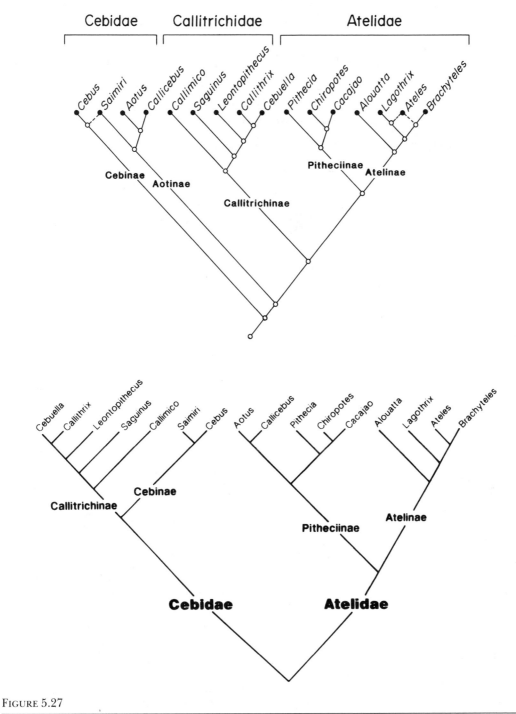

FIGURE 5.27

Two platyrrhine phylogenies based on teeth, skulls, and skeletons: top, Ford, 1986; bottom, Rosenberger, 1981.

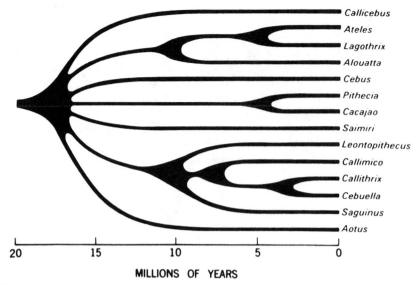

Callicebus
Ateles
Lagothrix
Alouatta
Cebus
Pithecia
Cacajao
Saimiri
Leontopithecus
Callimico
Callithrix
Cebuella
Saguinus
Aotus

20   15   10   5   0

MILLIONS OF YEARS

FIGURE 5.28

A biomolecular phylogeny of platyrrhines based on immunological comparisons (from Sarich and Cronin, 1980).

it sets them apart as a group of primitive genera. Her suggested taxonomy, based on an extensive quantitative analysis of dental, cranial, and skeletal anatomy, reflects this threefold division of platyrrhines.

Immunological studies (Fig. 5.28) provide another assessment of platyrrhine relationships. Atelines, pitheciines, and callitrichines cluster as distinct groups, but *Cebus, Saimiri, Callicebus,* and *Aotus*—the genera about which Rosenberger and Ford differ—do not seem to be particularly closely allied with any other taxa; rather, the immunological studies suggest a rapid diversification of many distinct lineages sometime between fifteen and twenty million years ago (Sarich and Cronin, 1980). More precise phylogenetic relationships among living platyrrhines have not been resolved and certainly deserve further study.

BIBLIOGRAPHY

GENERAL

Ford, S.M. (1986). Systematics of the New World monkeys. In *Comparative Primate Biology*, vol. 1: *Systematics, Evolution, and Anatomy,* ed. D.R. Swindler and J. Erwin, pp. 73–135. New York: Alan R. Liss.

Hershkovitz, P. (1977). *Living New World Monkeys (Platyrrhine), with an Introduction to the Primates*, vol. 1. Chicago: University of Chicago Press.

Mittermeier, R.A., and Coimbra-Filho, A.F., eds. (1981). *Ecology and Behavior of Neotropical Primates*. Rio de Janeiro: Academia Brasiliera de Ciencias.

PITHECIINES

Sakis

Buchannon, D.B., Mittermeier, R.A., and van Roosmalen, M.G.M. (1981). The saki monkeys, genus *Pithecia*. In *Ecology and Behavior of Neotropical Primates*, ed. A.F. Coimbra-Filho and R.A.

Mittermeier, pp. 391–417. Rio de Janeiro: Academia Brasiliera de Ciencias.

Happel, R.E. (1982). Ecology of *Pithecia hirsuta* in Peru. *J. Hum. Evol.* **11**:581–590.

Hershkovitz, P. (1979). The species of sakis, genus *Pithecia* (Cebidae, Primates) with notes on sexual dimorphism. *Folia Primatol.* **31**:1–22.

———. (1986). The piebald saki. *Field Museum Natural History Bulletin* **57** (2):24–25.

———. (1987). The taxonomy of South American sakis, genus *Pithecia* (Cebidae, Platyrrhini): A preliminary report and critical review with the description of a new species and a new subspecies. *Am. J. Primatol.* **12**:387–468.

Izawa, K. (1975). Foods and feeding behavior of monkeys in the Upper Amazon Basin. *Primates* **16**:295–316.

———. (1976). Group sizes and composition of monkeys in the Upper Amazon Basin. *Primates* **17**:367–399.

Johns, A. (1986). Notes on the ecological current status of the buffy saki, *Pithecia albicans. Primate Conservation* **7**:26–29.

Mittermeier, R.A., and van Roosmalen, M.G.M. (1981). Preliminary observations on habitat utilization and diet in eight Surinam monkeys. *Folia Primatol.* **36**:1–39.

Oliveira, J.M.S., Guerreiro de Lima, M., Bonvincino, C., Ayres, J.M., and Fleagle, J. (1985). Preliminary notes on ecology and behavior of the white-faced saki (*Pithecia pithecia*, Linnaeus, 1766; Cebidae, Primates). *Acta Amazonica* **15**:249–263.

Bearded Sakis

Ayres, J.M. (1981). *Observacoes sobre Ecologia e o Compartamento dos Cuxius (Chiropotes albinasus and Chiropotes satanus: Cebidae, Primates)*. Consuelho Nacional de Desenvolvimento Cientifico E Technologico Instituto Nacional de Pesquisas da Amazonia. Manaus: Fundacao Univ. do Amazonas.

Fleagle, J.G., and Mittermeier, R.A. (1980). Locomotor behavior, body size and comparative ecology of seven Surinam monkeys. *Am. J. Phys. Anthropol.* **52**:301–314.

Hershkovitz, P. (1985). A preliminary taxonomic review of the South American bearded saki monkeys genus *Chiropotes* (Cebidae, Platyrrhini) with the description of a new subspecies. *Fieldiana* **27**(NS):1–46.

Mittermeier, R.A., and van Roosmalen, M.G.M. (1981). Preliminary observations on habitat utilization and diet in eight Surinam monkeys. *Folia Primatol.* **36**:1–39.

van Roosmalen, M.G.M., Mittermeier, R.A., and Milton, K. (1981). The bearded sakis, genus *Chiropotes*. In *Ecology and Behavior of Neotropical Primates*, ed. A.F. Coimbra-Filho and R.A. Mittermeier, pp. 419–441. Rio de Janeiro: Academia Brasiliera de Ciencias.

van Roosmalen, M.G.M., Mittermeier, R.A., and Fleagle, J.G. (1988). Diet of the bearded saki (*Chiropotes satanas chiropotes*): A neotropical seed predator. *Am. J. Primatol.* **14**:11–35.

Uakaris

Ayres, J.M. (1986). *Uakaries and Amazonian Flooded Forest*. Ph.D. Dissertation. Cambridge University.

Fontaine, R. (1981). The uakaris, genus *Cacajao*. In *Ecology and Behavior of Neotropical Primates*, ed. A.F. Coimbra-Filho and R.A. Mittermeier, pp. 443–493. Rio de Janeiro: Academia Brasiliera de Ciencias.

Fontaine, R., and Dumond, F.V. (1977). The red ouakari in a seminatural environment: Potentials for propagation and study. In *Primate Conservation*, ed. HSH Prince Ranier III and G.H. Bourne, pp. 167–236. New York: Academic Press.

Hershkovitz, P. (1987). Uacaries, New World monkeys of the genus *Cacajao* (Cebidae, Platyrrhini): A preliminary taxonomic review with the description of a new subspecies. *Am. J. Primatol.* **12**:1–54.

Moynihan, M. (1976). *The New World Primates—Adaptive Radiation and the Evolution of Social Behavior, Language and Intelligence*. Princeton, N.J.: Princeton University Press.

Owl Monkeys

Hershkovitz, P. (1983). Two new species of night monkeys, genus *Aotus* (Cebidae, Primates): A preliminary report on *Aotus* taxonomy. *Am. J. Primatol.* **4**:209–243.

Jacobs, G.H. (1977). Visual sensitivity: Significant within-species variations in non-human primate. *Science* **197**:499–500.

Moynihan, M. (1976). *The New World Primates—Adaptive Radiation and the Evolution of Social Behavior, Language and Intelligence*. Princeton, N.J.: Princeton University Press.

Wright, P.C. (1978). Home range, activity pattern and agonistic encounters of a group of night monkeys (*Aotus trivirgatus*) in Peru. *Folia Primatol.* **29**:43–55.

———. (1981). The night monkeys, genus *Aotus*. In

Ecology and Behavior of Neotropical Primates, ed. A.F. Coimbra-Filho and R.A. Mittermeier, pp. 214–240. Rio de Janeiro: Academia Brasiliera de Ciencias.

———. (1982). Adaptive advantages of nocturnality in *Aotus. Am. J. Phys. Anthropol.* **57**(2):242.

———. (1983). Abstract: Day-active night monkeys (*Aotus trivirgatus*) in the Chaco of Paraguay. *Am. J. Phys. Anthropol.* **60**(2):272.

———. (1984). Biparental care in *Aotus trivirgatus* and *Callicebus moloch*. In *Female Primates: Studies by Women Primatologists*, ed. M.F. Small, pp. 59–75. New York: Alan R. Liss.

———. (1985). *The Costs and Benefits of Nocturnality for Aotus trivirgatus (the Night Monkey)*. Ph.D. Dissertation, City University of New York.

Titi Monkeys

Kinzey, W.G. (1981). The titi monkey, genus *Callicebus*. In *Ecology and Behavior of Neotropical Primates*, ed. A.F. Coimbra-Filho and R.A. Mittermeier, pp. 241–276. Rio de Janeiro: Academia Brasiliera de Ciencias.

Robinson, J.G., Wright, P.C., and Kinzey, W.G. (1986). Monogamous cebids and their relatives: Intergroup calls and spacing. In *Primate Societies*, ed. B.B. Smuts, D.L. Cheney, R.M. Seyfarth, R.W. Wrangham, and T.T. Struhsaker, pp. 44–53. Chicago: University of Chicago Press.

Wright, P.C. (1984). Biparental care in *Aotus trivirgatus* and *Callicebus moloch*. In *Female Primates: Studies by Women Primatologists*, ed. M.F. Small, pp. 59–75. New York: Alan R. Liss.

———. (1986). Ecological correlates of monogamy in *Aotus* and *Callicebus*. In *Primate Ecology and Conservation*, ed. J.G. Else and P.C. Lee, pp. 159–167. Cambridge: Cambridge University Press.

CEBINES

Squirrel Monkeys

Baldwin, J.D., and Baldwin, J.I. (1981). The squirrel monkey, genus *Saimiri*. In *Ecology and Behavior of Neotropical Primates*, ed. A.F. Coimbra-Filho and R.A. Mittermeier, pp. 277–330. Rio de Janeiro: Academia Brasiliera de Ciencias.

Fleagle, J.G., Mittermeier, R.A., and Skopec, A.L. (1981). Differential habitat use by *Cebus apella* and *Saimiri sciureus* in central Surinam. *Primates* **22**(3):361–367.

Hershkovitz, P. (1984). Taxonomy of squirrel monkeys, genus *Saimiri* (Cebidae, Platyrrhini): A preliminary report with description of a hitherto unnamed form. *Am. J. Primatol.* **7**:155–210.

Terborgh, J. (1983). *Five New World Primates*. Princeton, N.J.: Princeton University Press.

Thorington, R.W., Jr. (1967). Feeding and activity of *Cebus* and *Saimiri* in a Columbian forest. In *Progress in Primatology*, ed. D. Starck, R. Schneider, and H.J. Kuhn, pp. 180–184. Stuttgart: Gustav Fischer.

———. (1968). Observations of squirrel monkeys in a Columbian forest. In *The Squirrel Monkey*, ed. L.A. Rosenblum and R.W. Cooper, pp. 69–85. New York: Academic Press.

Capuchins

Freese, C., and Oppenheimer, J.R. (1981). The capuchin monkeys, genus *Cebus*. In *Ecology and Behavior of Neotropical Primates*, ed. A.F. Coimbra-Filho and R.A. Mittermeier, pp. 331–390. Rio de Janeiro: Academia Brasiliera de Ciencias.

Janson, C.H. (1984). Female choice and mating system of the brown capuchin monkey, *Cebus apella* (Primates: Cebidae). *Z. Tierpsychol.* **65**:177–200.

———. (1986a). Capuchin counterpoint. *Nat. Hist.* **95**(2):45–53.

———. (1986b). The mating system as a determinant of social evolution in capuchin monkeys (*Cebus*). In *Primate Ecology and Conservation*, ed. J.G. Else and P.C. Lee, pp. 169–179. Cambridge: Cambridge University Press.

Jungers, W.L., and Fleagle, J.G. (1981). Postnatal growth allometry of the extremities of *Cebus albifrons* and *Cebus apella:* A longitudinal and comparative study. *Am. J. Phys. Anthropol.* **53**:471–478.

Robinson, J.G. (1981). Spatial structure in foraging groups of wedge-capped capuchin monkeys, *Cebus nigrivittatus. Anim. Behav.* **29**:1036–1056.

Robinson, J.G., and Janson, C.H. (1986). Capuchins, squirrel monkeys, and atelines: Socioecological convergence with Old World monkeys. In *Primate Societies*, ed. B.B. Smuts, D.L. Cheney, R.M. Seyfarth, R.W. Wrangham, and T.T. Struhsaker, pp. 69–82. Chicago: University of Chicago Press.

Terborgh, J. (1983). *Five New World Primates*. Princeton, N.J.: Princeton University Press.

Thorington, R.W., Jr. (1967). Feeding and activity of *Cebus* and *Saimiri* in a Columbian forest. In *Progress in Primatology*, ed. D. Starck, R. Schneider, and H.J. Kuhn, pp. 180–184. Stuttgart: Gustav Fischer.

ATELINES

Howling Monkeys

Altmann, S.A. (1959). Field observations on howling monkey society. *J. Mammal.* **40**:317–330.

Carpenter, C.R. (1934). A field study of the behavior and social relations of howling monkeys. *Comp. Psychol. Monogr.* **10**:1–168.

Chivers, D.J. (1969). On the daily behavior and spacing of howling monkey groups. *Folia Primatol.* **10**:48–102.

Crockett, C.M., and Eisenberg, J.F. (1986). Howlers: Variations in group size and demography. In *Primate Societies*, ed. B.B. Smuts, D.L. Cheney, R.M. Seyfarth, R.W. Wrangham, and T.T. Struhsaker, pp. 54–68. Chicago: University of Chicago Press.

DaSilva, E.C., Jr. (1981). A preliminary survey of brown howler monkeys (*Alouatta fusca*) at the Cantareira Reserve (Sao Paulo, Brazil). *Rev. Brasil. Biol.* **41**(4):897–909.

Erikson, G.E. (1963). Brachiation in New World monkeys and in anthropoid apes. *Symp. Zool. Soc. London* **10**:135–164.

Fleagle, J.G., and Mittermeier, R.A. (1980). Locomotor behavior, body size and comparative ecology of seven Surinam monkeys. *Am. J. Phys. Anthropol.* **52**:301–314.

Gaulin, S.J.C., and Gaulin, C.K. (1982). Behavioral ecology of *Alouatta seniculus* in Andean Cloud Forest. *Int. J. Primatol.* **3**(1):1–52.

Glander, K.E. (1975). Habitat description and resource utilization: A preliminary report on mantled howling monkey ecology. In *Socioecology and Psychology of Primates*, ed. R.H. Tuttle, pp. 37–57. The Hague: Mouton.

———. (1978). Howling monkey feeding behavior and plant secondary compounds: A study of strategies. In *The Ecology of Arboreal Folivores*, ed. G.G. Montgomery, pp. 561–573. Washington, D.C.: Smithsonian Institution Press.

———. (1981). Feeding behavior in mantled howling monkeys. In *Foraging Behavior: Ecological, Ethological and Psychological Approaches*, ed. A.C. Kamil and T.D. Sargent, pp. 231–257. New York: Garland Press.

Leighton, M., and Leighton, D.R. (1982). The relationship of size of feeding aggregate to size of food patch: Howler monkeys (*Alouatta palliata*) feeding in *Trichelia cipo* fruit trees on Barro Colorado Island. *Biotropica* **14**(2):81–90.

Malinow, M.R., Pope, B., Depaoli, J.R., and Katz, S. (1968). Laboratory observations on living howlers.

In *Biology of the Howler Monkey*, ed. A. Caraya, pp. 224–230. Basel: S. Karger.

Mendel, F. (1976). Postural and locomotive behavior of *Alouatta palliata* on various substrates. *Folia Primatol.* **26**:36–53.

Milton, K. (1980). Food choice and digestive strategies of two sympatric primate species. *Am. Naturalist* **177**:496–505.

———. (1980). *The Foraging Strategies of Howler Monkeys: A Study in Primate Economics*. New York: Columbia University Press.

Mittermeier, R.A., and van Roosmalen, M.G.M. (1981). Preliminary observations on habitat utilization and diet in eight Surinam monkeys. *Folia Primatol.* **36**:1–39.

Rockwood, L.L., and Glander, K.E. (1979). Howling monkeys and leaf-cutting ants: Comparative foraging in a tropical deciduous forest. *Biotropica* **11**(1):1–10.

Rudran, R. (1979). The demography and social mobility of a red howler (*Alouatta seniculus*) population in Venezuela. In *Vertebrate Ecology in the Northern Neotropics*, ed. J. Eisenberg, pp. 107–126. Washington, D.C.: Smithsonian Institution Press.

Schon, M.A. (1968). *The Muscular System of the Red Howling Monkey*. Bulletin no. 273. Washington, D.C.: Smithsonian Institution Press.

———. (1971). The anatomy of the resonating mechanism in the howling monkey. *Folia Primatol.* **19**:117–163.

Sekulic, R. (1982). Daily and seasonal patterns of roaring and spacing in four red howler (*Alouatta seniculus*) troops. *Folia Primatol.* **39**:22–48.

Southwick, C.H. (1963). Challenging aspects of the behavioral ecology of howling monkeys. In *Primate Social Behavior*, ed. C.H. Southwick, pp. 185–191. New Jersey: Van Nostrand.

Woolly Monkeys

Fooden, J. (1963). Revision of the woolly monkeys (genus *Lagothrix*). *J. Mammal.* **44**(2):213–247.

Klein, L.L., and Klein, D. (1975). Social and ecological contrasts between four taxa of neotropical primates. In *Socioecology and Psychology of Primates*, ed. R.H. Tuttle, pp. 59–86. The Hague: Mouton.

Mittermeier, R.A., de Macedo-Ruiz, H., and Luscombe, A. (1975). A woolly monkey rediscovered in Peru. *Oryx* **13**(1):41–46.

Mittermeier, R.A., de Macedo-Ruiz, H., Luscombe, A., and Cassidy, J. (1977). Rediscovery and conservation of the Peruvian yellow-tailed woolly monkey

(*Lagothrix flavicauda*). In *Primate Conservation*, ed. HSH Prince Ranier III and G.H. Bourne, pp. 95–115. New York: Academic Press.

Spider Monkeys

Cant, J.G.H. (1977). Ecology, locomotion, and social organization of spider monkeys (*Ateles geoffroyi*). Ph.D. Dissertation. University of California, Davis, Ca.
———. (1986). Locomotion and feeding postures of spider and howling monkeys: Field study and evolutionary interpretation. *Folia Primatol.* **46**:1–14.
Eisenberg, J.F., and Kuehn, R.E. (1966). The behavior of *Ateles geoffroyi* and related species. *Smithson. Misc. Coll.* **151**:1–63.
Klein, L.L., and Klein, D. (1975). Social and ecological contrasts between four taxa of neotropical primates. In *Socioecology and Psychology of Primates*, ed. R.H. Tuttle, pp. 59–86. The Hague: Mouton.
———. (1976). Neotropical primates: Aspects of habitat usage, population density and regional distribution in La Macarena, Columbia. In *Neotropical Primates: Field Studies and Conservation*, ed. R.W. Thorington and P.G. Heltne, pp. 70–78. Washington, D.C.: National Academy of Sciences.
McFarland, M.J. (1986). Ecological determinants of fission-fusion sociality in *Ateles* and *Pan*. In *Primate Ecology and Conservation*, ed. J.G. Else and P.C. Lee, pp. 181–190. Cambridge: Cambridge University Press.
Mittermeier, R.A. (1978). Locomotion and posture in *Ateles geoffroyi* and *Ateles paniscus*. *Folia Primatol.* **30**:161–193.
Mittermeier, R.A., and van Roosmalen, M.G.M. (1981). Preliminary observations on habitat utilization and diet in eight Surinam monkeys. *Folia Primatol.* **36**:1–39.
van Roosmalen, M.G.M. (1980). Habitat preference, diet, feeding strategy and social organization of the black spider monkey (*Ateles paniscus paniscus* Linnaeus 1758) in Surinam. Ph.D. Dissertation, Agricult. Univ. Wageningen.
White, F. (1986). Census and preliminary observations on ecology of the black-faced spider monkey (*Ateles paniscus chamek*) in Manu National Park, Peru. *Am. J. Primatol.* **11**:125–132.

Woolly Spider Monkeys

Milton, K. (1984). Habitat, diet and activity patterns of free-ranging woolly spider monkeys (*Brachyteles arachnoides*, E. Geoffroy, 1806). *Int. J. Primatol.* **5**(5): 491–514.

———. (1985). Mating patterns of woolly spider monkeys, *Brachyteles arachnoides:* Implications for female choice. *Behav. Ecol. Sociobiol.* **17**:53–59.
Mittermeier, R.A., Coimbra-Filho, A.F., Constable, I.D., Rylands, A.B., and Valle, C. (1982). Conservation of primates in the Atlantic forest region of east Brazil. *Int. Zoo Yrbk.* **22**:2–17.
Nishimura, A. (1979). In search of woolly spider monkeys. *Kyoto Univ. Primate Res. Inst., Reports of New World Monkeys*, (1979) pp. 21–37.
Strier, K.B. (1987). Activity budgets of woolly spider monkeys, or muriquis (*Brachyteles arachnoides*). *Am. J. Primatol.* **13**:385–395.

CALLITRICHINES

Ford, S.M. (1980). Callitrichids as phyletic dwarfs and the place of the Callitrichidae in Platyrrhini. *Primates* **21**(1):31–43.
Garber, P.A. (1984). Proposed nutritional importance of plant exudates in the diet of the Panamanian tamarin, *Saguinus oedipus geoffroyi. Int. J. Primatol.* **5**(1):1–15.
Goldizen, A.W. (1986). Tamarins and marmosets: Communal care of offspring. In *Primate Societies*, ed. B.B. Smuts, D.L. Cheney, R.M. Seyfarth, R.W. Wrangham, and T.T. Struhsaker, pp. 34–43. Chicago: University of Chicago Press.
Hershkovitz, P. (1977). *Living New World Monkeys (Platyrrhini), with an Introduction to the Primates*, vol. 1. Chicago: University of Chicago Press.
Kleiman, D.G. (1977). *The Biology and Conservation of the Callitrichidae*. Washington, D.C.: Smithsonian Institution Press.
Leutenegger, W. (1980). Monogamy in callitrichids: A consequence of phyletic dwarfism? *Int. J. Primatol.* **1**(1):95–98.
Rosenberger, A.L. (1984). Aspects of the systematics and evolution of the marmosets. In *A Primatologica No Brazil*, ed. M.T. de Mello, pp. 159–180. Angis do 1. Congresso Brasiliero de Primatologia, Sociedad de Primatologica.
Sussman, R.W., and Garber, P.A. (1987). A new interpretation of the social organization and mating system of the Callitrichidae. *Int. J. Primatol.* **8**:73–92.
Sussman, R.W., and Kinzey, W.G. (1984). The ecological role of the Callitrichidae: A review. *Am. J. Phys. Anthropol.* **64**(4):419–449.
Terborgh, J., and Goldizen, A.W. (1985). On the mating system of the cooperatively breeding saddle-backed tamarin (*Saguinus fuscicollis*). *Behav. Ecol. Sociobiol.* **16**:293–299.

Goeldi's Monkey

Heltne, P.G., Wojcik, J.F., and Pook, A.G. (1981).
   Goeldi's monkey, genus *Callimico*. In *Ecology and
   Behavior of Neotropical Primates*, ed. A.F. Coimbra-
   Filho and R.A. Mittermeier, pp. 169–209. Rio de
   Janeiro: Academia Brasiliera de Ciencias.
Hershkovitz, P. (1977). *Living New World Monkeys
   (Platyrrhini), with an Introduction to the Primates*, vol. 1.
   Chicago: University of Chicago Press.
Moynihan, M. (1976). *The New World Primates—Adaptive
   Radiation and the Evolution of Social Behavior,
   Language and Intelligence*. Princeton, N.J.: Princeton
   University Press.
Pook, A.G., and Pook, G. (1981). A field study of the
   socio-ecology of the Goeldi's monkey (*Callimico
   goeldii*) in northern Brazil. *Folia Primatol.*
   35:288–312.
———. (1982). Polyspecific associations between
   Saguinus fuscicollis, *Saguinus labiatus*, *Callimico goeldii*
   and other primates in northwestern Bolivia. *Folia
   Primatol.* **38**:196–216.

Crested Tamarin

Dawson, G.A. (1978). Composition and stability of
   social groups of the tamarin, *Saguinus oedipus
   geoffroyi* in Panama: Ecology and behavioral
   implications. In *The Biology and Conservation of the
   Callitrichidae*, ed. D.G. Kleiman, pp. 23–38.
   Washington, D.C.: Smithsonian Institution Press.
Garber, P.A. (1984a). Proposed nutritional importance
   of plant exudates in the diet of the Panamanian
   tamarin, *Saguinus oedipus geoffroyi*. *Int. J. Primatol.*
   5(1):1–15.
———. (1984b). Use of habitat and positional behavior
   in a neotropical primate, *Saguinus oedipus*. In
   Adaptations for Foraging in Non-human Primates, ed.
   P.S. Rodman and J.G.H. Cant, pp. 112–133. New
   York: Columbia University Press.
Nehman, P.F. (1978). Aspects of the ecology and social
   organization of free ranging cotton-tamarins
   (*Saguinus oedipus*) and the conservation status of the
   species. In *The Biology and Conservation of the
   Callitrichidae*, ed. D.G. Kleiman, pp. 39–72.
   Washington, D.C.: Smithsonian Institution Press.

Golden-handed Tamarin

Fleagle, J.G., and Mittermeier, R.A. (1980). Locomotor
   behavior, body size and comparative ecology of
   seven Surinam monkeys. *Am. J. Phys. Anthropol.*
   52:301–314.
Mittermeier, R.A., and van Roosmalen, M.G.M. (1981).
   Preliminary observations on habitat utilization and

diet in eight Surinam monkeys. *Folia Primatol.*
   36:1–39.
Thorington, R.W., Jr. (1968). Observations of squirrel
   monkeys in a Colombian forest. In *The Squirrel
   Monkey*, ed. L.A. Rosenblum and R.W. Cooper,
   pp. 69–85. New York: Academic Press.

Saddle-backed Tamarin

Izawa, K., and Yoneda, M. (1981). Habitat utilization of
   non-human primates in a forest of the West Pando,
   Brazil. *Kyoto Univ. Primate Res. Inst., Reports of New
   World Monkeys*, (1981) pp. 13–21.
Janson, C.H., Terborgh, J., and Emmons, L.H. (1981).
   Non-flying mammals as pollinating agents in the
   Amazonian forest. *Biotropica, Reprod. Botany
   Suppl.*:1–6.
Terborgh, J. (1983). *Five New World Primates*. Princeton,
   N.J.: Princeton University Press.
Terborgh, J., and Goldizen, A.W. (1985). On the mating
   system of the cooperatively breeding saddle-backed
   tamarin (*Saguinus fuscicollis*). *Behav. Ecol. Sociobiol.*
   16:293–299.

White-lipped Tamarin

Izawa, K., and Yoneda, M. (1981). Habitat utilization of
   non-human primates in a forest of the West Pando,
   Brazil. *Kyoto Univ. Primate Res. Inst., Reports of New
   World Monkeys*, (1981):13–21.
Terborgh, J. (1983). *Five New World Primates*. Princeton,
   N.J.: Princeton University Press.
Yoneda, M. (1981). Ecological studies of *Saguinus
   fuscicollis* and *Saguinus labiatus* with reference to
   habitat segregation and height preference. *Kyoto
   Univ. Primate Res. Inst., Reports of New World Monkeys*,
   (1981):43–50.
———. (1984). Comparative studies on vertical
   separation, foraging behavior and traveling mode of
   saddle-backed tamarins (*Saguinus fuscicollis*) and red-
   chested moustached tamarins (*Saguinus labiatus*) in
   northern Bolivia. *Primates* **25**(4):414–422.

Emperor Tamarin

Terborgh, J. (1983). *Five New World Primates*. Princeton,
   N.J.: Princeton University Press.
Terborgh, J., and Goldizen, A.W. (1985). On the mating
   system of the cooperatively breeding saddle-backed
   tamarin (*Saguinus fuscicollis*). *Behav. Ecol. Sociobiol.*
   16:293–299.

Moustached Tamarin

Garber, P.A., Moya, L., and Malaga, C. (1984). A
   preliminary field study of the moustached tamarin

monkey (*Saguinus mystax*) in northeastern Peru: Questions concerned with the evolution of a communal breeding system. *Folia Primatol.* **42**:17–32.

Stacey, Peter B., and Keonig, W.D. (1984). Cooperative breeding in the acorn woodpecker. *Sci. Am.* **251**(2):114–121.

Lion Tamarins

Coimbra-Filho, A.F., and Mittermeier, R.A. (1973). Distribution and ecology of the genus *Leontopithecus* (Lesson 1840) in Brazil. *Primates* **14**(1):47–66.

Rosenberger, A.F., and Coimbra-Filho, A.F. (1984). Morphology, taxonomic status and affinities of the lion tamarin, *Leontopithecus* (Callitrichinae, Cebidae). *Folia Primatol.* **42**:149–179.

Marmosets

Coimbra-Filho, A.F., and Mittermeier, R.A. (1978). The gouging, exudate eating and the short tusked condition in *Callithrix* and *Cebuella*. In *The Biology and Conservation of the Callitrichidae*, ed. D.G. Kleiman, pp. 105–117. Washington, D.C.: Smithsonian Institution Press.

Lacker, T.E., Jr., Bouchardet de Fonseca, G.A., Alves, C., Jr., and Magalhaes-Castro, B. (1984). Parasitism of trees by marmosets in a central Brazilian gallery forest. *Biotropica* **16**(3):202–209.

Rosenberger, A.L. (1978). Loss of incisor enamel in marmosets *J. Mammal.* **59**:207–208.

Rylands, A.B. (1984). Tree gouging and exudate feeding in marmosets (Callitrichidae, Primates). In *Tropical Rainforest: Ecology and Management, Supplemental Reports*, ed. S.L. Sulton, T. Whitmore, and A.C. Chadwick. Proceedings of Leeds Philosophical and Literature Society, Leeds, London.

Common Marmosets

Rylands, A.B. (1981). Preliminary field observations on the marmoset, *Callithrix humeralifer intermedius* (Hershkovitz, 1977) at Dardanelos, Rio Aripuana, Mato Grosso. *Primates* **22**:46–59.

———. (1984). Tree gouging and exudate feeding in marmosets (Callitrichidae, Primates). In *Tropical Rainforest: Ecology and Management Supplemental Reports*, ed. S.L. Sulton, T. Whitmore, and A.C. Chadwick. Proceedings of Leeds Philosophical and Literature Society, Leeds, London.

Pygmy Marmosets

Hernandez-Camacho, J., and Cooper, R.W. (1976). The
nonhuman primates of Colombia. In *Neotropical Primates: Field Studies and Conservation*, ed. R.W. Thorington and P.G. Heltne, pp. 35–69. Washington, D.C.: National Academy of Sciences.

Kinzey, W.G., Rosenberger, A.L., and Ramirez, M. (1975). Vertical clinging and leaping in a neotropical anthropoid. *Nature* **255**:327–328.

Moynihan, M. (1976). *The New World Primates—Adaptive Radiation and the Evolution of Social Behavior, Language and Intelligence*. Princeton, N.J.: Princeton University Press.

———. (1976). Notes on the ecology and behavior of the pygmy marmoset, *Cebuella pygmaea*, in Amazonian Colombia. In *Neotropical Primates: Field Studies and Conservation*, ed. R.W. Thorington and P.G. Heltne, pp. 79–84. Washington, D.C.: National Academy of Sciences.

Ramirez, M.F., Freese, C.H., and Revilla, C.J. (1978). Feeding ecology of the pygmy marmoset, *Cebuella pygmaea*, in northeastern Peru. In *The Biology and Conservation of the Callitrichidae*, ed. D.G. Kleiman, pp. 91–104. Washington, D.C.: Smithsonian Institution Press.

Soine, P. (1982). Ecology and population dynamics of the pygmy marmoset, *Cebuella pygmaea*. *Folia Primatol.* **39**:1–21.

Terborgh, J. (1983). *Five New World Primates*. Princeton, N.J.: Princeton University Press.

PHYLETIC RELATIONSHIPS OF PLATYRRHINES

Ciochon, R.L., and Chiarelli, A.B. (1980). *Evolutionary Biology of the New World Monkeys and Continental Drift*. New York: Plenum Press.

Ford, S.M. (1986). Systematics of the New World monkeys. In *Comparative Primate Biology*, vol. 1: *Systematics, Evolution, and Anatomy*, ed. D.R. Swindler and J. Erwin, pp. 73–135. New York: Alan R. Liss.

Rosenberger, A.L. (1981). Systematics: The higher taxa. In *Ecology and Behavior of Neotropical Primates*, ed. A.F. Coimbra-Filho and R.A. Mittermeier, pp. 9–27. Rio de Janeiro: Academia Brasiliera de Ciencias.

———. (1984). Aspects of the systematics and evolution of the marmosets. In *A Primatologica No Brasil*, ed. M.T. de Mello, pp. 159–180. Angis do 1. Congresso Brasiliero de Primatologia, Sociedad de Primatologica.

Sarich, V.M., and Cronin, J.E. (1980). South American mammal molecular systems, evolutionary clocks, and continental drift. In *Evolutionary Biology of the New World Monkeys and Continental Drift*, ed. R.L. Ciochon and A.B. Chiarelli, pp. 399–421. New York: Plenum Press.

Szalay, F.S., and Delson, E. (1979). *Evolutionary History of the Primates*. New York: Academic Press.

Thorington, R.W., Jr., and Anderson, S. (1984). Primates. In *Orders and Families of Recent Mammals of the World*, ed. S. Anderson and J. Knox Jones, Jr., pp. 187–217. New York: Wiley.

Old World Monkeys

The platyrrhine monkeys are the only primates in the neotropics, and they fill a diverse array of ecological niches there. In the Old World, on the other hand, the ecological niches are occupied by primates with more disparate phyletic backgrounds. As we discussed in Chapter 1, the nocturnal niches are the domain of the lorises and galagos in Africa and the lorises and tarsiers in Asia. The Old World diurnal niches are occupied by members of two very distinct radiations of higher primates, the Old World monkeys (Cercopithecoidea) and the hominoids (Hominoidea)—two superfamilies that make up infraorder Catarrhini. As we shall see in later chapters, the evolutionary history of these two groups is quite different, as is their current diversity. The hominoids are restricted to a few species from the tropical forests of Africa and Asia and one cosmopolitan species—humans; they are the subject of Chapter 7. There are many more species and genera of Old World monkeys than there are of hominoids, and they occupy a wide range of habitats; they are the subject of the present chapter.

Catarrhine Anatomy

The higher primates of Africa, Asia, and Europe, the catarrhines, are characterized by numerous anatomical specializations that set them apart from the more primitive New World monkeys. The name is derived from the shape of their nostrils, which are usually narrow and facing downward rather than round and facing laterally as in most New World monkeys. In their dentition, catarrhines have two rather than three premolars in each quadrant for a dental formula of $\frac{2.1.2.3.}{2.1.2.3.}$ On the external surface of the side wall of their skull, the frontal bone contacts the sphenoid bone and separates the zygomatic bone anteriorly from the parietal bone posteriorly. In the auditory region, the tympanic bone extends laterally to form a tubular external auditory meatus (see Fig. 5.3). All Old World monkeys, all gibbons, and some chimpanzees have expanded ischial tuberosities and well-developed sitting pads, and all lack an entepicondylar foramen on the humerus. In general, the living catarrhines are much larger than the living platyrrhines, and they include more folivorous and terrestrial species.

159

Old World Monkeys

Apes

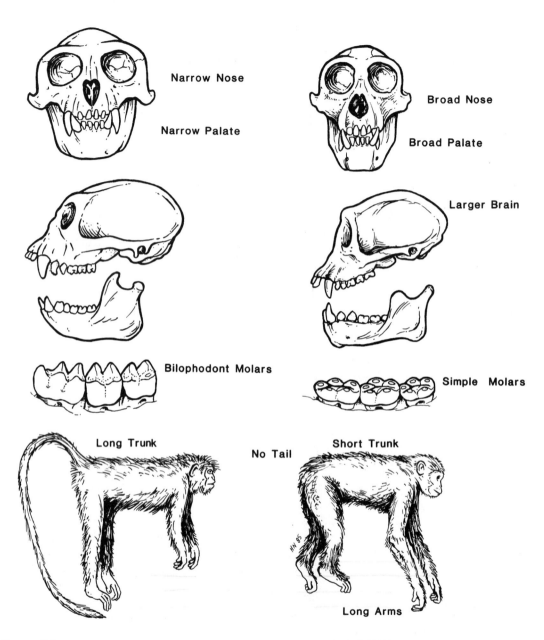

Narrow Nose

Narrow Palate

Broad Nose

Broad Palate

Larger Brain

Bilophodont Molars

Simple Molars

Long Trunk

No Tail

Short Trunk

Long Arms

FIGURE 6.1

Characteristic features that distinguish the two groups of catarrhine primates, Old World monkeys (Cercopithecoidea) and apes (Hominoidea).

Cercopithecoids

The more taxonomically diverse and numerically successful catarrhines are the cercopithecoid monkeys. It has long been the conventional wisdom that cercopithecoid monkeys are the more primitive Old World higher primates and retain many features of the common ancestors of monkeys and apes. We now know that the earliest catarrhines were quite different in many ways from either of the extant superfamilies, and that both Old World monkeys and apes are quite specialized with respect to the earliest catarrhines. We return to this issue in later chapters as we deal with catarrhine evolution.

Cercopithecoid monkeys have several anatomical features that distinguish them from apes and humans (Fig. 6.1). Most characteristic are the specialized molar teeth in which the anterior two cusps and the posterior two cusps are aligned to form two ridges, or lophs. Teeth with this structure are described as **bilophodont**. Most Old World monkeys have daggerlike canines in the males and smaller ones in the females, all of which are sharpened by a narrow anterior lower premolar. In cranial anatomy, Old World monkeys have relatively narrow nasal openings and narrow tooth rows compared with apes.

The limbs of Old World monkeys are characterized by a very narrow elbow joint with a reduced medial epicondyle and a relatively long olecranon process on the ulna. Sitting pads on the expanded ischial tuberosities are a distinctive feature of the group, and many have a long tail.

Cercopithecoid monkeys are found throughout Africa and Asia (Fig. 6.2). In Europe, they are found only on the Gibraltar

FIGURE 6.2

Geographic distribution of extant cercopithecoid monkeys.

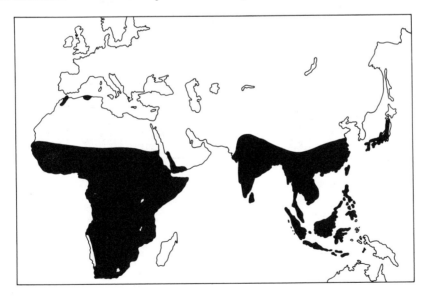

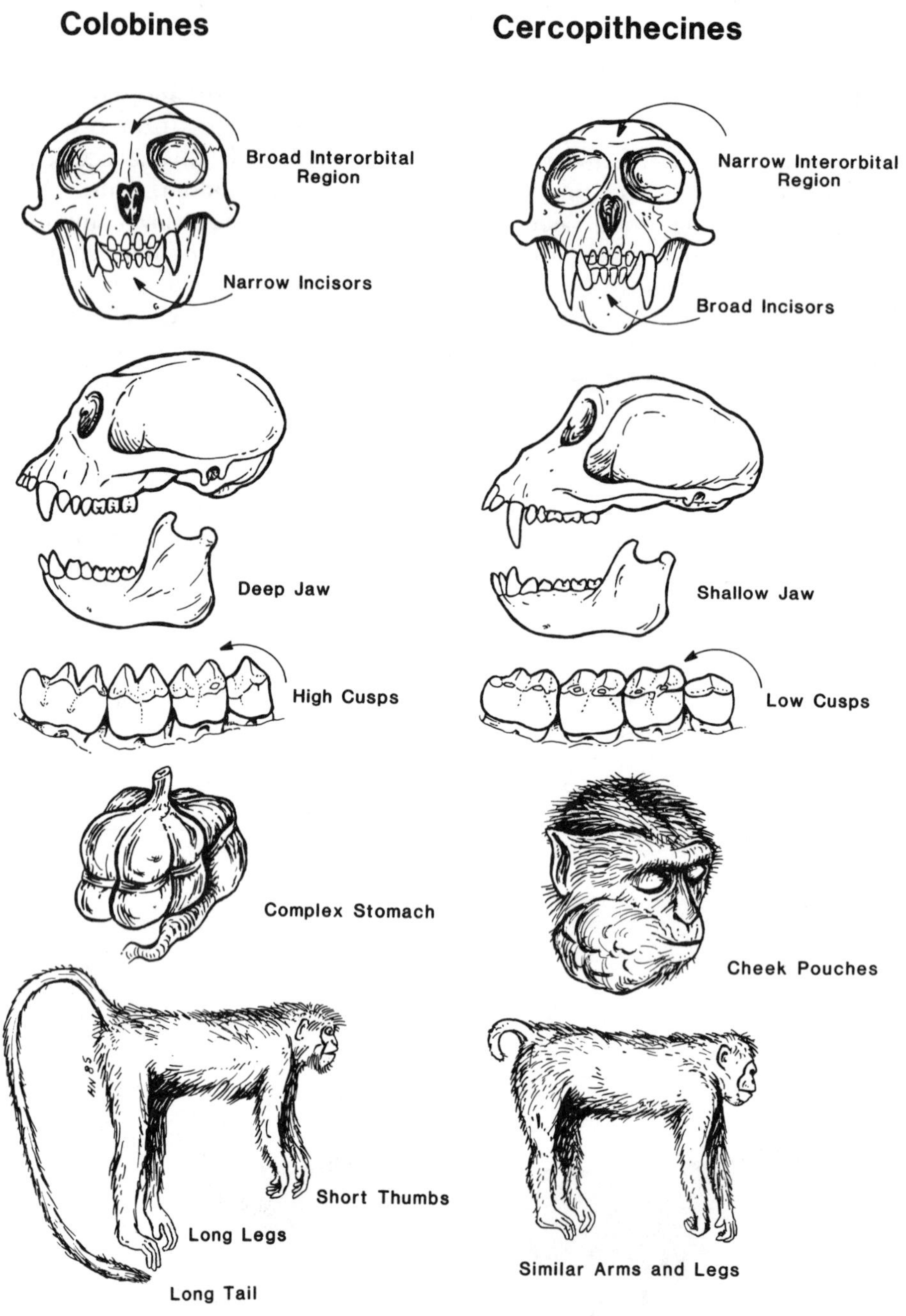

Colobines

Cercopithecines

Broad Interorbital Region

Narrow Incisors

Narrow Interorbital Region

Broad Incisors

Deep Jaw

Shallow Jaw

High Cusps

Low Cusps

Complex Stomach

Cheek Pouches

Short Thumbs

Long Legs

Long Tail

Similar Arms and Legs

FIGURE 6.3

Characteristic features of the two extant subfamilies of Old World monkeys, colobines and cercopithecines.

headland, but in the recent past they had a much more extensive distribution on that continent. Old World monkeys are found in a wider range of latitudes, climates, and vegetation types than any other group of living primates. There are two very different groups of Old World monkeys: the cercopithecines, or cheek-pouch monkeys, and the colobines, or leaf-eating monkeys. Both have undergone extensive adaptive radiations and are represented by numerous genera and species.

The two subfamilies are distinct in many aspects of their anatomy (Figs. 6.3, 6.4). Many of their differences are related to basic dietary adaptations. The colobines are predominantly leaf and seed eaters, whereas the cercopithecines are predominantly fruit eaters. Cercopithecines have cheek pouches, broader incisor teeth, and molar teeth with high crowns and relatively low cusps, whereas colobines have no cheek pouches, narrower incisors, and molar teeth with high cusps. Colobines have a large, complex stomach. In cranial anatomy, cercopithecines have a narrow interorbital region, and the lacrimal canal is formed by both the maxillary and lacrimal bones; in colobines, the interorbital region is broader and the lacrimal bone completely surrounds the canal. In general, cercopithecines have longer snouts and shallower mandibles than do colobines. Most cercopithecines have longer thumbs and shorter fingers than colobines, which often lack a thumb. Cercopithecine forelimbs and hindlimbs tend to be similar in size, whereas colobines usually have much longer hindlimbs.

FIGURE 6.4

Skulls of three cercopithecine monkeys (above) and three colobine monkeys (below).

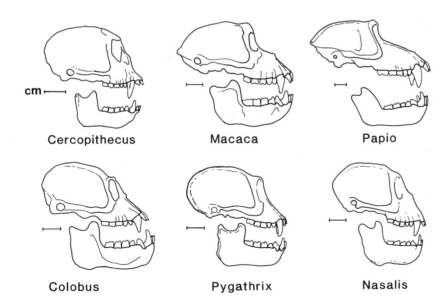

Cercopithecus   Macaca   Papio

Colobus   Pygathrix   Nasalis

FIGURE 6.5

Two macaque species that are found sympatrically throughout Southeast Asia: upper right, the crab-eating or long-tailed macaque (*Macaca fascicularis*); below, the pig-tailed macaque (*Macaca nemestrina*).

TABLE 6.1
Infraorder Catarrhini
Family Cercopithecidae
Subfamily CERCOPITHECINAE, macaques

Common Name	Species	Intermembral Index	Body Weight (g)	
Lion-tailed macaque	*Macaca silenus*			
Pig-tailed macaque	*M. nemestrina*	92	M	10,210
			F	6,350
Tonkean macaque	*M. tonkeana*	95		—
Moor macaque	*M. maura*	—		—
Ochre macaque	*M. ochreata*	100		—
Muna-Butung macaque	*M. brunescens*	99		—
Heck's macaque	*M. hecki*	93		—
Gorontalo macaque	*M. nigriscens*	—		—
Celebes black macaque	*M. nigra*	94		—
Barbary macaque	*M. sylvanus*	—		—
Toque macaque	*M. sinica*	—		—
Bonnet macaque	*M. radiata*	—	M	6,280
			F	4,530
Assamese macaque	*M. assamensis*	96	M	9,060
			F	5,800
Thibetan macaque	*M. thibetana*	95		—
Crab-eating macaque	*M. fascicularis*	93	M	4,930
			F	3,130
Taiwan macaque	*M. cyclopes*	—		—
Rhesus macaque	*M. mulatta*	93	M	7,730
			F	5,210
Japanese macaque	*M. fuscata*	—		—
Bear macaque	*M. arctoides*	98	M	9,060
			F	6,200

Cercopithecines

The cercopithecines are a predominantly African group. Only a single, very successful genus, *Macaca*, is found in Asia or Europe. Cercopithecines range in size from the tiny arboreal talapoin monkey of western Africa (just over 1 kg) to the large (as much as 50 kg), mostly terrestrial baboons found throughout the African continent (Tables 6.1–6.4).

Macaques

Macaques (*Macaca*) are medium-size cerco-pithecines (Fig. 6.5) and are relatively generalized in many aspects of their anatomy compared with other members of the subfamily. Macaques are characterized by moderately long snouts, high-crowned molar teeth with very low cusps, and long third molars (Fig. 6.4). They share several features with the African baboons and mangabeys, including long snouts, large incisors, and a chromosome number of 44. In general, their limbs are more slender than those of the African baboons and mangabeys and more robust than those of the smaller guenons.

Macaca has the widest distribution of any

nonhuman primate genus. The nineteen species of *Macaca* range from Morocco and Gibraltar in the west to Japan, Taiwan, the Philippines, Sulawesi, and Bali in the east. *Macaca sylvanus*, the Barbary macaque, is the only living nonhuman primate in Europe; *M. fuscata*, the Japanese macaque, ranges farther to the north and east than any other primate species; and *M. fascicularis*, the crab-eating macaque, from the island of Bali, extends farthest to the southeast of any nonhuman primate species. Based on the anatomy of their reproductive organs, the species of macaques fall into four major groups, with broad areas of overlap in their distribution (Table 6.1; Delson, 1980; Fooden, 1980).

Because macaques and especially the rhesus macaque, *Macaca mulatta*, are the most common laboratory primates, the anatomy, physiology, and captive behavior of this genus have been more thoroughly studied than those of any other nonhuman primate. Much less is known about the natural behavior and ecology of most macaques. The two geographically peripheral species, *M. fuscata* from Japan and *M. sylvanus* in Gibraltar and North Africa, have been most thoroughly studied, although there is increasing information on other species. Macaques' ability to coexist with humans surpasses that of all other nonhuman primates. As a result, much of the data on macaque behavior and ecology come from settings in which the monkeys derive large parts of their food either directly or indirectly from humans. This ecological relationship is an important feature of macaque biology, but such ecological data are difficult to use in comparisons with other, less adaptable species.

Macaques occupy a wider range of habitats and climates than any other nonhuman primate genus. Even more, the habitat preferences and foraging strategies for individual species vary in many ways, a factor that has contributed to the diversity of species.

The ecological differences among macaques have been most clearly documented for two Southeast Asian species, *M. nemestrina*, the pig-tailed macaque, and *M. fascicularis* (Figs. 6.5, 6.6). The smaller (3–5 kg) *M. fascicularis* prefers lowland and secondary forests, especially near rivers (Fittinghoff and Lindberg, 1980), whereas the larger (6–10 kg) *M. nemestrina* prefers upland and more hilly environments.

Macaque species vary considerably in the extent to which they are arboreal or terrestrial. All species use both settings to some extent, but they differ in frequencies. *Macaca fascicularis* is primarily an arboreal species that normally feeds and travels in the trees. These macaques are most often found in the lower levels of the main canopy, but they utilize all levels, including the ground. *Macaca nemestrina* travels more on the ground, but it feeds frequently in the trees. The locomotion of macaques is almost totally quadrupedal walking and running, with very little leaping and no suspensory behavior aside from occasional hindlimb hanging during feeding. Macaques are extremely dexterous and have short fingers and an opposable thumb.

All macaques are frugivores (Fig. 6.6), but many consume considerable amounts of leaves, flowers, and other plant materials as well as various animal prey. Japanese macaques subsist on bark during the cold winters. *Macaca fascicularis*, the crab-eating macaque, eats a variety of invertebrates—not only crabs but also termites and small vertebrates.

All macaques live in relatively large, multimale social groups, with troops of some species containing fifty or more individuals.

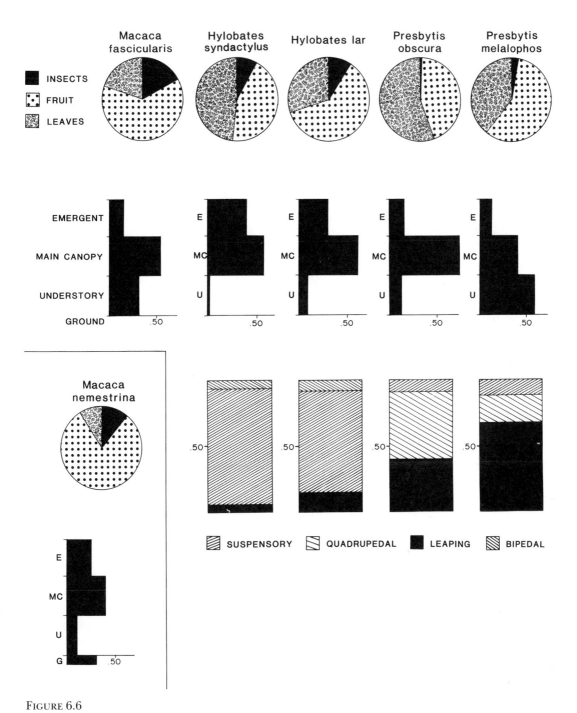

FIGURE 6.6

Diet, forest height preference, and locomotor behavior of six sympatric catarrhines from Malaysia (data from Chivers, 1980).

TABLE 6.2
Infraorder Catarrhini
Family Cercopithecidae
Subfamily CERCOPITHECINAE, mangabeys

Common Name	Species	Intermembral Index	Body Weight (g)	
Gray-cheeked mangabey	*Cercocebus albigena*	78	M	8,980
			F	6,400
Black mangabey	*C. aterrimus*	—		—
Tana River mangabey	*C. galeritus*	84		—
White-collared mangabey	*C. torquatus*	83	M	10,625

During the day these groups regularly split into smaller foraging parties. Home range size and patterns of habitat use vary considerably from species to species. Groups of about twenty *M. fascicularis* have home ranges of 40 to 100 ha and day ranges of less than a kilometer. Home ranges and day ranges for the larger groups of *M. nemestrina* are considerably larger.

Social relations within macaque groups are complex. Female hierarchies and matrilineages seem to be particularly important in interindividual relations and troop politics. Males usually migrate from troop to troop many times during their lifetime. Macaques have single births at yearly intervals.

Mangabeys

Mangabeys are large, forest-living monkeys with long molars, very large incisors, relatively long snouts, and hollow cheeks. They have relatively long limbs and long tails. Although generally placed in a single genus, *Cercocebus*, the living mangabeys include two distinct groups that differ in numerous aspects of their dentition, biochemistry, and ecology (Table 6.2). There are indications that these two types of mangabeys may even be the result of separate evolutionary branches (see Fig. 6.22; Cronin and Sarich, 1976).

Cercocebus galeritus and *C. torquatus* (each with many subspecies) form one group. These monkeys are found in a very wide range of forest types, including primary and secondary dry forests, swamp forests, and mangrove forests. They prefer the understory and frequently come to the ground for both traveling and feeding. *Cercocebus albigena* and *C. aterrimus* form the second group, sometimes placed in a separate genus, *Lophocebus*. These are strictly arboreal monkeys that prefer the main canopy levels in primary, flooded, and semideciduous forests. Both groups of mangabeys are predominantly frugivorous and specialize on figs, which are generally found in large patches. Both spend 25–30 percent of their feeding time foraging for invertebrates.

All mangabey species live in groups of ten to twenty individuals. Both single-male and multi-male groups are common.

Baboons

Baboons (Fig. 6.7; Table 6.3) are the largest and among the best known of all cercopithecines. They were important figures in the mythology of ancient Egypt and were well

FIGURE 6.7

A troop of savannah baboons (*Papio anubis*) in eastern Africa.

TABLE 6.3
Infraorder Catarrhini
Family Cercopithecidae
Subfamily CERCOPITHECINAE, baboons

Common Name	Species	Intermembral Index	Body Weight (g)	
Hamadryas baboon	*Papio hamadryas*	95	M	21,300
			F	12,000
Guinea baboon	*P. papio*	—		—
Olive baboon	*P. anubis*	97	M	25,100
			F	14,100
Yellow baboon	*P. cynocephalus*	96	M	22,800
			F	12,350
Chacma baboon	*P. ursinus*	—	M	31,200
Mandrill	*Mandrillus sphinx*	95	M	26,900
			F	11,500
Drill	*M. leucophaeus*	—	M	20,000
Gelada	*Theropithecus gelada*	100	M	19,000
			F	11,700

known to Greek and Roman scholars. As savannah-dwelling primates, they have played an important role as models for various aspects of early human evolution (Washburn and DeVore, 1961; Rose, 1976; Strum and Mitchell, 1987).

Baboons are very large monkeys and are all sexually dimorphic in body size; in many species, females are as little as half the size of males. Baboons are characterized by long molars and broad incisors. Their canines are very sexually dimorphic, and the long anterior lower premolars form a sharpening blade for the daggerlike canines. Baboons have long snouts (Fig. 6.4), a long mandible, and pronounced brow ridges. Their limbs (Fig. 6.8) are nearly equal in length, and their forearm is much longer than their humerus; they have relatively short digits on their hands and feet. Compared with other cercopithecines, baboons have relatively short tails and large ischial callosities. Females have very pronounced sexual swellings during estrus.

Baboons are found throughout the forests and savannahs of sub-Saharan Africa. They are usually divided into three genera: *Papio*, the savannah baboons; *Mandrillus*, the drill and mandrill; and *Theropithecus*, the gelada.

There are five living "species" of **savannah baboons**: *Papio papio*, *P. anubis*, *P. cynocephalus*, *P. ursinus*, and *P. hamadryas*. They are all allopatric with variable amounts of interbreeding at their boundaries and are probably best regarded as one superspecies.

The ecology and behavior of savannah baboons has been the subject of many studies over the past twenty-five years. These baboons live in woodland savannahs, grasslands, acacia scrubs, and other open areas, but also in gallery forests and some rain forest environments. They forage and travel primarily on the ground by quadrupedal walking and running, but they almost always climb trees or rocky cliffs for sleeping and often for resting. They are extremely eclectic feeders that subsist mainly on ripe fruits, roots, and tubers, as well as on grass seeds,

FIGURE 6.8

The skeleton of a baboon (*Papio*).

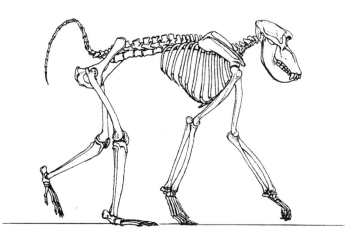

gums, and leaves. Most baboons are opportunistic faunivores (Strum, 1981) and have been reported to catch and eat numerous small mammals (hares, young gazelles, vervet monkeys) as well as many invertebrates and insects. They also eat bird eggs.

The four common savannah baboons (*P. papio*, *P. anubis*, *P. cynocephalus*, and *P. ursinus*) live in large, socially complex multimale troops ranging from forty to eighty individuals. There is often a pronounced dominance hierarchy among males and intense competition among the adult males for access to estrous females. This competition involves a whole repertoire of social maneuvers, not just simple physical prowess (see, e.g., Strum, 1987). Females frequently mate with several males during the course of their cycle. Baboon troops occupy large (4,000 ha) home ranges and travel long distances (over 5 km) every day, usually as a single group.

Social organization and foraging patterns in Hamadryas baboons are quite different from those found among other savannah species (Kummer, 1968; Nagel, 1973). These handsome silver baboons from the treeless savannahs of Ethiopia live in groups of a single adult male with one to four females plus their offspring. Males jealously guard the females in their harem and actually herd them by chasing any straying females and biting them on the neck to keep the group together. Several one-male groups, probably led by related males, regularly associate to form clans. The individual harems forage separately during the day but congregate at night on rocky cliffs in troops of up to 150 animals.

Mandrills (*Mandrillus sphinx*) and **drills** (*M. leucophaeus*) are large (up to 50 kg) forest baboons from western Africa. They are extremely sexually dimorphic in both size and coloration. Males of both species are characterized by brightly colored faces and rumps. In both mandrills and drills, males have long muzzles with pronounced maxillary ridges and long tooth rows. Females are much smaller. Like savannah baboons, mandrills and drills have forelimbs and hindlimbs of nearly equal length. Both have very short tails.

The ecology and behavior of drills and mandrills is much less known than that of the savannah species because they live in dense forest and tend to be shy as a result of human hunting. Within the forest they are primarily terrestrial, but females and young regularly climb into trees. The diet of drills and mandrills has been described as fruit, leaves, pith, and insects.

Mandrills and drills have been seen in one-male groups of thirteen to nineteen individuals, multi-male groups of thirty-seven to sixty-eight animals, and large congregations numbering up to 250 individuals. The larger groups, which are most common in the dry season, seem to be aggregations of individual one-male groups. Mandrill groups have a high ratio of adult females to adult males; solitary males are common.

The most distinctive of the living baboons in both morphology and ecology is the **gelada** (*Theropithecus gelada*) from the highlands of Ethiopia (Fig. 6.9). *Theropithecus gelada* is the only surviving species of a more successful and widespread radiation during the Pliocene and Pleistocene. Like other baboons, geladas are extremely sexually dimorphic in both size and appearance. Males have long, shaggy manes and pronounced facial whiskers, whereas the female pelage is much shorter. Both sexes have striking red hourglass patches of skin on their chests, and in females these are outlined with white vesicles. The molar teeth of *Theropithecus* are characterized by complex enamel foldings.

FIGURE 6.9

The gelada (*Theropithecus gelada*) from the highlands of Ethiopia.

Male canines are very large, even by baboon standards. The snout and mandible are relatively short and deep. The hands of geladas are characterized by a relatively long thumb compared with the length of the other digits, an adaptation for manipulation.

Geladas live in the treeless montane grasslands of the Ethiopian highlands, where they forage on the ground all day and sleep on rocky cliffs at night. They are the most terrestrial nonhuman primates and always move by quadrupedal walking and running. They are exclusively herbivorous, eating grass, seeds, and roots throughout the year, and, only occasionally, fruit. They feed by sitting upright, plucking grass blades and seeds by hand.

Geladas live in one-male groups of three to twenty individuals, which may gather in bands. This relationship between bands is different from that found among the clans of Hamadryas baboons (Stammbach, 1986). Unattached males converge into all-male groups. Groups occasionally join together in temporary herds of up to 400 individuals. Several groups usually share a single home range, and day ranges are relatively small for baboons (1–2 km), reflecting both the small foraging units and sedentary feeding habits of geladas.

Guenons

Guenons (*Cercopithecus*) are the small forest monkeys of sub-Saharan Africa. There are at

TABLE 6.4
Infraorder Catarrhini
Family Cercopithecidae
Subfamily CERCOPITHECINAE, guenons

Common Name	Species	Intermembral Index	Body Weight (g)	
Blue monkey	*Cercopithecus mitis*	82		—
Spot-nosed monkey	*C. nictitans*	82	M	6,500
			F	4,000
Red-tailed monkey	*C. ascanius*	79	M	4,170
			F	3,000
Lesser spot-nosed monkey	*C. petaurista*	—		—
Red-bellied monkey	*C. erythrogaster*	—		—
Moustached monkey	*C. cephus*	81	M	4,000
			F	2,880
Red-eared monkey	*C. erythrotis*	—		—
Mona monkey	*C. mona*	86		—
Campbell's monkey	*C. campbelli*	—		—
Crowned monkey	*C. pogonias*	—	M	4,500
			F	3,000
Wolf's monkey	*C. wolfi*	—		—
Diana monkey	*C. diana*	79		—
Zaire diana monkey	*C. salongo*	—		—
Preuss's monkey	*C. preussi*	—		—
L'Hoest's monkey	*C. lhoesti*	—		—
De Brazza's monkey	*C. neglectus*	82	M	6,320
			F	3,960
Dryas monkey	*C. dryas*	—		—
Hamlyn's monkey	*C. hamlyni*	—		—
Vervet monkey	*C. aethiops*	83	M	5,370
			F	3,360
Allen's swamp monkey	*Allenopithecus nigroviridis*	84	M	6,900
			F	3,100
Talapoin monkey	*Miopithecus talapoin*	83	M	1,380
			F	1,120
Patas monkey	*Erythrocebus patas*	92	M	11,100
			F	5,900

least nineteen guenon species (Table 6.4), which are remarkably diverse in color and appearance (Fig. 6.10) but relatively uniform in size and body proportions (Schultz, 1970). Guenons are of medium size and have a moderate amount of sexual dimorphism in most species. All have sexually dimorphic canines, relatively narrow molar teeth, and short third molars. They have relatively short snouts compared with the larger cercopithecines (see Fig. 6.4). Guenons have longer hindlimbs than forelimbs and long tails. They are all forest dwellers and show considerable variation from species to species in their forest preference and use of canopy levels. All species are basically arbo-

Cercopithecus mitis   *Cercopithecus hamlyni*   *Cercopithecus diana*

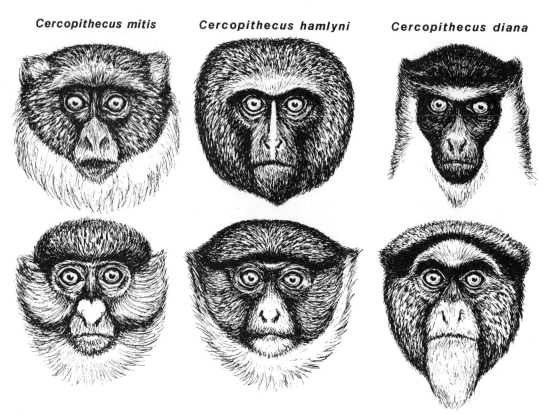

Cercopithecus ascanius *Allenopithecus nigroviridis* *Cercopithecus neglectus*

FIGURE 6.10

The faces of six guenons.

real quadrupeds, but some (*C. aethiops*) frequently come to the ground, and some are quite good leapers.

In general, guenons are both frugivorous and insectivorous; many species spend large parts of their day foraging for insects. A few species eat moderate amounts of leaves in some parts of the year. Most species live in single-male groups.

Guenons often feed and travel in mixed-species groups. Despite the overall structural uniformity of guenons, individual species have evolved unique foraging strategies that distinguish them from sympatric species

(Gautier-Hion, 1978; Galat and Galat-Luong, 1985). Three frequently associated guenon species that have been well studied are *C. cephus*, the **moustached monkey**; *C. pogonias*, the **crowned monkey**; and *C. nictitans*, the **spot-nosed monkey**, all from Gabon (Figs. 6.11, 6.12). *Cercopithecus cephus* is the smallest species, *C. pogonias* is slightly larger, and *C. nictitans* is much larger and more sexually dimorphic. All are found in primary rain forests, but *C. cephus* is also very common in secondary forests. All are arboreal quadrupeds. *Cercopithecus pogonias* and *C. nictitans* are found primarily in the middle and

FIGURE 6.11

Three guenon species from Gabon that frequently forage together in mixed-species groups: above, the crowned guenon (*Cercopithecus pogonias*); middle, the spot-nosed guenon (*C. nictitans*); below, the moustached monkey (*C. cephus*).

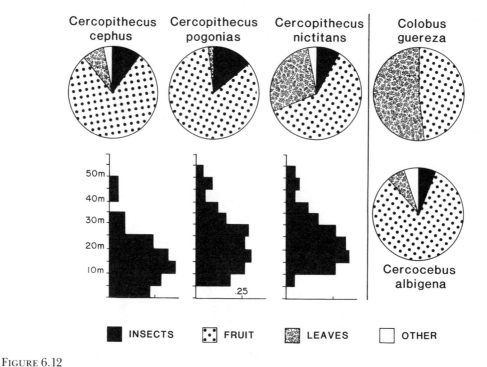

FIGURE 6.12

Diet and forest height preference of five monkeys from Gabon.

upper levels of the main canopy, whereas *C. cephus* prefers the lower levels of the canopy and the understory and occasionally comes to the ground.

The diets of the three monkeys have been documented through examination of stomach contents (Fig. 6.12). *Cercopithecus pogonias* is the most frugivorous, the most insectivorous, and the least folivorous, *C. cephus* is intermediate, and *C. nictitans* is the most folivorous, with leaves, flowers, and other vegetable materials accounting for almost 30 percent of its diet. The two species that overlap most in canopy use overlap least in diet. The types of insects eaten by the species also differ considerably: *C. pogonias* appears to specialize on mobile prey, whereas *C. nictitans* almost exclusively eats cryptic

immobile prey. Again, *C. cephus* is intermediate, but it occupies a different level of the forest. All three species live in moderately small home ranges with mean group size increasing from *C. cephus* (ten individuals) to *C. pogonias* to *C. nictitans*. In each species there is usually a single adult male.

Two other guenon species that have been particularly well studied are *C. mitis*, the **blue monkey**, and *C. ascanius*, the **red-tailed monkey**; both are found in the Kibale forest of Uganda. *Cercopithecus ascanius*, a close relative of *C. cephus*, is smaller and less dimorphic than *C. mitis*, a relative of *C. nictitans*.

One of the most handsome guenons is *Cercopithecus neglectus*, **De Brazza's monkey** (Fig. 6.13), from western and central Africa.

Cercopithecus neglectus is one of the largest and most sexually dimorphic guenons, with males averaging over 7 kg and females less than 4 kg. De Brazza's monkeys prefer flooded forests, including islands in rivers, where they move primarily in the understory and on the ground. They are slow quadrupedal monkeys. Their diet is predominantly frugivorous and includes lesser amounts of leaves, animal matter, and mushrooms.

The most unusual feature of De Brazza's monkey is its social organization. In some parts of their range, they live in polygynous groups of eight to twelve individuals, but in Gabon, where they have been quite well studied (Gautier-Hion and Gautier, 1978), most live in monogamous families. The two sexes seem to forage somewhat independently and have different strategies for dealing with predators. The small females hide in the undergrowth while the large males climb trees and call at the predator, apparently to distract it from the female and young. Both the day ranges and the home ranges of these family groups are very small.

Cercopithecus aethiops, the **vervet monkey**

FIGURE 6.13

A family of De Brazza's monkeys (*Cercopithecus neglectus*).

FIGURE 6.14

A troop of vervet monkeys (*Cercopithecus aethiops*) in woodland savannah habitat.

or **green monkey** (Fig. 6.14), is the most widespread guenon species and ranges throughout sub-Saharan Africa. Its large range is undoubtedly associated with its preference for woodland savannah and gallery forests. Vervet monkeys are the most terrestrial guenons and frequently cross open areas between feeding trees and forage on the ground. Terrestrial movements account for approximately 20 percent of their locomotion, and the remainder is arboreal. Vervets are predominantly quadrupedal both on the ground and in the trees; leaping accounts for only 10 percent of their locomotor activity. Their diverse diet consists of fruits, gums, shoots, and a variety of invertebrates. One population in Senegal catches

crabs, a behavior reminiscent of the crab-eating macaque of Southeast Asia.

Vervet monkeys differ from most forest guenons in that they live in relatively large troops with several adult males. As in baboons, there is a clear dominance hierarchy among males, especially during the mating season.

The **talapoin monkey** (*Miopithecus talapoin*) is closely related to the guenons. It is the smallest Old World monkey, weighing just over a kilogram, and also the smallest living catarrhine. Many of the distinguishing features of talapoins—a relatively large head, large eyes, and short snout—are characteristics of young animals, suggesting that talapoins are a neotenic guenon.

Talapoins live in the riverine forests of western and central Africa. They are found most frequently in the dense undergrowth, where they move by leaping and quadrupedal walking and running. They are among the best leapers of all cercopithecines, and they also seem to be good swimmers. Their diet contains large amounts of insects and fruit. They are among the most insectivorous of Old World monkeys (Gautier-Hion, 1978).

Talapoins live in large social groups averaging between seventy and one hundred individuals (the larger groups studied were parasitic on manioc farms), with roughly one adult male for every two adult females. Talapoin troops are always on the move, flushing insects. They always sleep by the water, which they leap into when frightened.

Talapoins have an unusual social organization characterized by little interaction between males and females during most of the year (Rowell and Dixon, 1975). They usually forage in subgroups composed of individuals of the same sex. During the three-month breeding season, males join the female groups and some females join all-male subgroups. After breeding, the individuals return to their single-sex groups.

The **swamp monkey** (*Allenopithecus nigroviridis*) is a medium-size guenon that lives in flooded forests in western and central Africa. Compared with other small cercopithecines, this species has broader macaquelike molar teeth suggestive of a frugivorous diet. On the basis of its dental and skeletal anatomy, several authorities have suggested that this species is the most primitive guenon. Virtually nothing is known of its behavior in a natural environment.

The **patas** or **hussier monkey** (*Erythrocebus patas*) is a close relative of the guenons which has developed extreme specializations for its life on the open grasslands. Patas are medium-size, very sexually dimorphic monkeys with slender bodies, long limbs, and long tails. They have narrow hands and feet with short digits and a reduced pollex and hallux.

Patas monkeys live on the grassy plains and savannahs of western and north central Africa. Although they usually sleep in trees at the edge of the forest, most of their foraging takes place in the open grass, where they move by quadrupedal walking and running. They are extremely agile, fast runners (55 km/h, according to Kingdon, 1971), and they frequently stand bipedally to look over the tall grass for potential predators or conspecifics. The bulk of their diet seems to be grass seeds, new shoots, and acacia gums. They also eat the beans of tamarind trees and a variety of other tough savannah fruits, seeds, and berries. They supplement their herbivorous diet with insects and various other prey items. They seem normally to eat on the move, picking up bits of food items as they walk.

Groups of patas monkeys average about a dozen individuals, with usually a single adult male. Nongroup males often live together in all-male bands, but there seems to be considerable turnover of males associated with patas groups. Patas' day ranges are extraordinarily variable, ranging from 700 to nearly 12,000 m, with a central tendency of between 2,000 and 2,500 m. Sometimes the group forages coherently and other times members of a group are separated by as much as 800 m. The estimated home ranges are over 5,000 ha, the largest known for any nonhuman primate species. It is not surprising that they can run so fast.

Colobines

The second subfamily of Old World monkeys are the colobines, or leaf-eating monkeys, of Africa and Asia. Colobines (Tables 6.5–6.7) are easily distinguished from cercopithecines by their sharp-cusped cheek teeth and relatively narrow incisors. Their skulls have relatively short snouts (see Fig. 6.4), narrow nasal openings, broad interorbital areas, and deep mandibles (see Fig. 6.3). They have complex, sacculated stomachs (similar to cattle) that enable them to maintain bacterial colonies for digesting cellulose (Bauchop, 1978). Their skeletons are characterized by relatively long legs, long tails, and thumbs that are usually short or absent.

In general, the living colobine monkeys are more arboreal and folivorous than are cercopithecines, and they are also better leapers. Socially, they are characterized by single-male groups. Aunting behavior, in which infants are frequently passed around among females within a troop, is common in many colobines.

There are two major groups of colobine monkeys, the colobus monkeys of Africa and the langurs of Asia.

Colobus Monkeys

The African colobus monkeys come in three color schemes—black and white, red, and olive. The three groups are quite distinct behaviorally and ecologically (Table 6.5).

The **black-and-white colobus monkey**, or **guereza** (Fig. 6.15), is the largest and most spectacular of the African colobine monkeys. It is a quite robust monkey, with considerable sexual dimorphism in size. There are three black-and-white species: *Colobus guereza*, *C. polykomos*, and *C. angolensis*. Guerezas live in a wide range of forest types throughout sub-Saharan Africa, in both primary rain forests and patchy dry forests. They are extremely hardy and can survive in a variety of habitats. They are arboreal and prefer the main canopy levels. All colobines are good leapers, but guerezas move more frequently by quadrupedal walking and bounding. They virtually never engage in suspensory behavior. Guerezas usually feed by sitting on branches and pulling food toward themselves.

The diet of guerezas is predominantly leaves (see Fig. 6.12), often of only a few tree species. They rely extensively on mature leaves for much of the year. Unripe fruits and young leaves are important only in some seasons.

Black-and-white colobus usually live in very small groups of a single adult male, one or two females, and offspring. They have

TABLE 6.5
Infraorder Catarrhini
Family Cercopithecidae
Subfamily COLOBINAE, African colobus monkeys

Common Name	Species	Intermembral Index	Body Weight (g)	
Guereza	*Colobus guereza*	79	M	10,100
			F	8,040
King colobus	*C. polykomos*	78	M	10,000
			F	7,700
Angolan colobus	*C. angolensis*	—		—
Black colobus	*C. satanas*	—	F	9,000
Red colobus	*Piliocolobus badius*	87	M	8,250
			F	8,240
Olive colobus	*Procolobus verus*	80	M	4,280
			F	3,600

FIGURE 6.15

Two sympatric colobines from Africa: above and below, the red colobus (*Piliocolobus badius*); center, the black-and-white colobus (*C. guereza*).

extremely small home ranges, which they advertise with loud vocal calls, as well as very small day ranges. In some seasons they are virtually sedentary and many spend days feeding in a single tree.

The **black colobus** (*Colobus satanas*), from western and central Africa, is quite distinct in both anatomy and ecology from the guerezas. This species has larger, flatter teeth and a more robust skull than guerezas. In contrast with the black-and-whites, black colobus feed predominantly on hard seeds; leaves and fruits are less important parts of their diet. As a result of this specialization on seeds as a protein source, *C. satanas* is able to thrive in areas that are uninhabitable for more folivorous colobines because the leaves have high levels of poisonous tannins.

Black colobus live in multi-male groups that are larger than the single-male groups of guerezas, and they also have larger home ranges. Day ranges are, on average, quite small but vary dramatically from season to season, depending on the availability of food. When food is abundant, black colobus often feed in a few trees for several days; when food is scarce, they move long distances and feed briefly from many different trees.

Red colobus (*Piliocolobus badius*) are found sympatric with guerezas throughout most of sub-Saharan Africa. They are smaller than guerezas, and their relatively short pelage gives them a much more slender appearance. They have broader incisors, more gracile skulls, longer legs, and less sexual size dimorphism than guerezas, and they are strikingly different from guerezas in many aspects of their behavior and ecology as well. They seem to reach higher densities in primary forests than black-and-whites but are not found in drier forests. They range through all levels of the main canopy and emergents. In comparison with guerezas, red colobus eat fewer mature leaves; they prefer fruit, young leaves, and shoots.

Red colobus live in large troops of forty to eighty animals with numerous adults of both sexes. The troops have overlapping home ranges of 100 ha or more, and there seems to be direct competition and a dominance hierarchy between troops over access to specific feeding areas. Day ranges of red colobus are larger than those of black-and-white colobus. All of these ranging differences can be related to the distribution of the major food sources preferred by the two species. The guerezas' more spatially and temporally homogenous diet of mature leaves can be exploited effectively in small groups. The fruit, shoots, and new leaves preferred by *Piliocolobus badius* are less evenly distributed, but when available they are quite abundant; thus the size of food resources probably does not limit their group size. Furthermore, larger groups are better able to obtain and defend large food sources.

Many aspects of the social organization of red colobus are unusual for colobines. Females have a large estrus swelling, like the cercopithecines that live in multi-male groups, and it is females rather than males which generally emigrate from their natal group.

The **olive colobus** (*Procolobus verus*) is the smallest colobine and the most poorly known of the African species. It is sexually dimorphic in size and has the smallest thumb and the largest feet of all African colobines.

Olive colobus live in swamp forest and range almost exclusively in the understory. They are reported to show an extraordinary diversity of locomotor abilities, depending on what other species they are associated with—an ability that has earned them the appellation "magic monkey." They have been frequently reported to travel on the ground. Olive colobus forage as a very dispersed group in thick vegetation and regularly associate with groups of *Cercopithecus*. They eat mainly new leaves.

The social organization of olive colobus seems to be similar to that of red colobus. They live in small multi-male groups of ten to fifteen animals. Like red colobus, female olive colobus show sexual swellings. Females carry young infants in their mouth, like many prosimians. Little is known of their ranging behavior. Olive colobus have only soft vocalizations and emit a strong odor, suggesting that olfactory communication may be especially important in this species.

Langurs

It is in Asia that the colobine monkeys have reached their greatest diversity and abundance. Two or three sympatric species are common, and the density of leaf-eating monkeys in southern and eastern Asia exceeds that of any other forest vertebrates.

TABLE 6.6
Infraorder Catarrhini
Family Cercopithecidae
Subfamily COLOBINAE, langurs

Common Name	Species	Intermembral Index	Body Weight (g)	
Hanuman langur	*Presbytis (Semnopithecus) entellus*	83	M	19,100
			F	15,200
Purple-faced leaf monkey	*P. (S.) senex*	—	M	6,860
			F	5,110
John's or Nilgiri leaf monkey	*P. (S.) johnii*	80	M	11,500
Banded leaf monkey	*P. (Presbytis) melalophos*	78	M	6,543
			F	6,753
Sunda Island leaf monkey	*P. (P.) aygula*	76	M	6,396
			F	6,671
Maroon leaf monkey	*P. (P.) rubicunda*	76	M	6,339
			F	6,056
Thomas's leaf monkey	*P. (P.) thomasi*	—		—
Mentawai leaf monkey	*P. (P.) potenziani*	—		—
White-fronted leaf monkey	*P. (P.) frontata*	76	M	5,570
	P. (P.) hosei	75	M	6,200
			F	5,570
Dusky leaf monkey	*P. (Trachypithecus) obscura*	83	M	7,540
			F	6,080
Phayre's leaf monkey	*P. (T.) phayrei*	—	M	7,300
			F	6,250
Silvered leaf monkey	*P. (T.) cristata*	82	M	6,930
			F	5,950
Capped leaf monkey	*P. (T.) pileata*	82		—
Golden leaf monkey	*P. (T.) geei*	—		—
Francois' leaf monkey	*P. (T.) francoisi*	—		—

The most common, most diverse, and most abundant are the langurs, members of the genus *Presbytis*. In addition, there are a number of genera commonly grouped as the "odd-nosed" monkeys.

The genus *Presbytis* can be divided into three distinct species groups—one from India and Sri Lanka (*Semnopithecus*) and two from Southeast Asia (*Presbytis*) and (*Trachypithecus*) (Table 6.6).

The **sacred** or **Hanuman langur** (*Presbytis entellus*) of India (Fig. 6.16), is one of the most adaptable of all higher primates. This species is found from Sri Lanka in the south, to the Rajastan Desert in the west, and well into the Himalaya mountains of Nepal (Bishop, 1979). Over this broad geographic range the species shows considerable diversity in body weight and skeletal morphology. In general, they are long-limbed, gray to brown monkeys with a very long tail, short thumbs, and long feet. They thrive in virtually all imaginable habitats found on the Indian subcontinent, including tropical rain forests, deciduous forests, temperate dry forests, and conifer forests, as well as deserts and cities. Hanuman langurs are the most terrestrial of the colobines. In the trees they use both quadrupedal gaits and leaps, but mainly the former. They rely primarily on seated feeding postures.

Considering the diversity of habitats they occupy, it is not surprising that the diet of Hanuman langurs is quite eclectic. In general, however, they eat fruit, flowers, and new leaves and seem unable to subsist on a diet of mature leaves.

Social groups of Hanuman langurs vary considerably in size, from ten to one hundred individuals with a mean of about twenty. There is typically one adult male per ten individuals. Home ranges vary, according to group size and habitat, from 24 to 200 ha.

Langur troops seem to be centered around groups of related adult females that aid one another and care for each other's offspring. By contrast, male residence in a troop is usually relatively short-lived. Most langurs live in one-male troops in which the single adult male fathers all of the offspring and drives out rival males. At fairly regular intervals, bands of roving males attack a group, drive out the resident male, and kill dependent infants. One of these intruders then establishes himself as the dominant male, drives away the others, and starts the cycle anew (Hrdy, 1977a). Hanuman langurs have single births at sixteen- to eighteen-month intervals. Infants are regularly passed around among the adult females and are often cared for by females other than their mother.

The other langurs from the Indian subcontinent are *Presbytis vetulus* (= *senex)*, the **purple-faced langur**, from Ceylon (Fig. 6.16), and a closely related species, *P. johnii*, **John's langur**, from western India. Both species are characterized by purple faces, white sideburns, and relatively small size. Compared with Hanuman langurs, both are restricted to forested areas. They are almost totally arboreal and are excellent leapers. Both are more folivorous than *P. entellus* and can subsist almost exclusively on mature leaves.

Like the black-and-white colobus of Africa, purple-faced langurs live in very small, one-male groups or families, with home ranges of a hectare or less. Their day ranges are small and, like black-and-white colobus, they often spend an entire day in one tree. Male takeovers and infanticide have also been reported in purple-faced langurs (Rudran, 1973).

The langurs east of India are spendidly diverse in their coloration but can be divided into two distinct groups on the basis of body

Figure 6.16

Two colobine species from India and Sri Lanka: above, the purple-faced monkeys (*Presbytis vetulus* (= *senex*)); below, the Hanuman langurs (*P. entellus*).

FIGURE 6.17

Two sympatric leaf monkey species from Malaysia: above, the spectacled langur, or dusky leaf monkey (*Presbytis obscura*); below, the banded leaf monkey (*P. melalophos*).

proportions, vocalizations, infant color patterns, and ecology (Figs. 6.6, 6.17). The most diverse of these is the **Presbytis melalophos group**, which contains seven species from Thailand and west Malaysia (*P. melalophos*), Sumatra (*P. melalophos* and *P. thomasi*), Borneo (*P. aygula, P. frontata, P. hosei,* and *P. rubicunda*), the Mentawai Islands (*P. potenziani*), Java (*P. comata*), and various other islands of the Sunda Shelf. There are numerous subspecies with a dazzling variability in adult coat color. By contrast, all infants show a distinct banded or cruciform pattern with a dark head, dark body and tail, and dark arms.

The best-known species, *P. melalophos*, the banded leaf monkey, is a relatively small colobine (6 kg) that shows little sexual size dimorphism. Like other members of its species group, the banded leaf monkey has a relatively short face. Its postcranial skeleton is characterized by relatively long legs, a long trunk, and slender arms that are associated with its leaping locomotion (Fig. 6.18).

Leaf monkeys of this group seem to be found in a variety of inland forests, but not in swamps or montane forests. In west Malaysia the banded leaf monkey is more common in secondary forests than upland primary forests and moves and feeds more frequently in the understory and lower canopy levels than other sympatric langurs (see Fig. 6.6). Banded leaf monkeys are extraordinary leapers and move less frequently by arboreal quadrupedalism. They occasionally use forelimb suspension as well (Figs. 6.6, 6.18).

Banded leaf monkeys eat primarily young leaves, seeds, and fruit and rarely, if ever, partake of mature leaves. They live in small groups of a dozen or fewer animals, frequently with a single adult male. There is suggestive evidence that they defend their relatively small home ranges (21 ha) at the boundaries. Troops of banded leaf monkeys forage as a group over their day range of 750 to 1,140 m. They are active throughout the day and often well into early evening. Although many males show wounds, scars, and other evidence of conflicts, there are no records of infanticide in this species group.

Presbytis potenziani, the **Mentawai Island leaf monkey**, is a colobine with an unusual social organization for an Old World monkey. Very little is known about the ecology and diet of these monkeys except that they live in monogamous family groups. Like many monogamous primates, the males and females sing together in daily vocal duets (Tilson and Tenaza, 1976).

The other widespread group of Southeast Asian langurs is the **Presbytis cristata–Presbytis obscura** (or **Trachypithecus**) group. These monkeys are similar in size to those of the *P. melalophos* group, but they are more sexually dimorphic. They are mostly silver or gray monkeys as adults, but they have bright yellow or orange infants. *Presbytis obscura*, the spectacled langur (Fig. 6.17), has striking white eye rings. This species group has a more robust skull, and arms and legs that are more similar in length. Some species have clear habitat preferences, but others do not. *Presbytis obscura* is more abundant in upland primary forests or secondary forests; *P. cristata* is commonly found in mangrove swamps and in flooded land along long rivers in west Malaysia and Borneo, but some populations are also found inland. In west Malaysia, where it has been most thoroughly studied, *P. obscura* prefers the main canopy levels and, to a lesser degree, the emergents (see Fig. 6.6). These monkeys are primarily arboreal quadrupeds that leap much less frequently than *P. melalophos* (Fig. 6.18). Both *P. obscura* and *P. cristata* are primarily folivores, but they also eat unripe fruit.

The social organization and foraging pat-

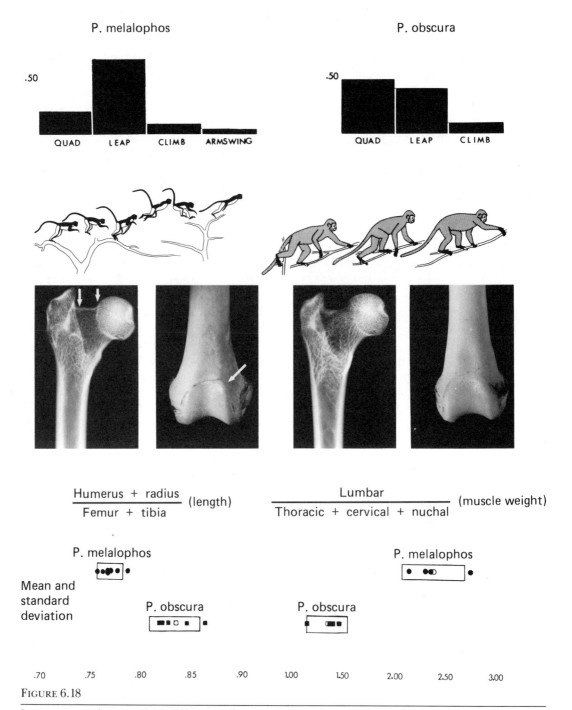

FIGURE 6.18

Locomotor and anatomical differences between *Presbytis melalophos* and *P. obscura*. *P. melalophos* leaps more frequently and its femur is characterized by a short, straight neck and a prominent lateral ridge on the patellar groove (arrows). This species also has relatively longer hindlimbs and larger back muscles.

FIGURE 6.19

The proboscis monkey (*Nasalis larvatus*) in a nipa mangrove swamp.

terns of *P. obscura* are quite different from those of *P. melalophos*. They live in slightly larger multi-male groups and have larger home ranges. More significant, however, they often do not forage as a group. The group, which may forage together in the early morning or late afternoon, splits up during the day into small foraging units, each of which travels a relatively short distance and may spend hours in one place. Like Hanuman langurs, these langurs pass their brightly colored infants around among the females. *Presbytis cristata* seem to live more frequently in single-male groups.

Odd-nosed Monkeys

The remaining colobine monkeys from Asia are all characterized by odd-shaped noses.

The best known of these is *Nasalis larvatus*, the **proboscis monkey** of Borneo (Fig. 6.19, Table 6.7). This large red monkey is the most sexually dimorphic of all colobines. Males are almost twice the size of females and have an enormous pendulous nose; females have a smaller, turned-up proboscis.

Proboscis monkeys live primarily in the nipa mangrove swamps, where they are reported to come regularly to the ground to feed. Little is known about their locomotion except that they are good leapers and excellent swimmers. Their diet consists mainly of leaves. Proboscis monkeys live in multi-male groups averaging about twenty animals. Virtually nothing is known about their ranging behavior.

Simias concolor, the **Simakobu monkey**, is a

TABLE 6.7
Infraorder Catarrhini
Family Cercopithecidae
Subfamily COLOBINAE, odd-nosed monkeys

Common Name	Species	Intermembral Index	Body Weight (g)	
Proboscis monkey	*Nasalis larvatus*	94	M	20,370
			F	9,820
Simakobu	*Simias concolor*	—	M	8,750
			F	7,100
Douc langur	*Pygathrix nemaeus*	94	M	10,900
			F	8,200
Golden snub-nosed monkey	*Rhinopithecus roxellana*	—	M	15,000–39,000
			F	6,500–10,000
Tonkin snub-nosed monkey	*R. avunculus*	—		—
Brelich's snub-nosed monkey	*R. brelichi*	—		—

close relative of *Nasalis* from the Mentawai Islands off the west coast of Sumatra. Both males and females of this species have a short, turned-up nose, much like that of a female proboscis monkey. This species has very unusual body proportions for a colobine, with similar-size forelimbs and hindlimbs and a short tail. These monkeys' daily activities, including feeding and much of their travel, are predominately in trees, but when frightened they descend to the ground and flee terrestrially. They generally feed on leaves and also eat fruits and berries.

Simakobu monkeys live in monogamous family groups that average three to four individuals. They have very few vocalizations.

Pygathrix nemaeus, the **Douc langurs** from Vietnam, Cambodia, and Laos, are colorful monkeys with little or no sexual dimorphism. Little is known about their natural behavior. They have been reported to live in small groups of three to eleven individuals and to feed on buds and new leaves.

Rhinopithecus, the **golden monkey** (Fig.

6.20), is probably the largest of all colobines, weighing 30 kg or more. One species is from the monsoon forests of Vietnam (*R. avunculus*) and another (*R. roxellana*, with three subspecies) is from China. All have a short, turned-up nose, a bluish face, and long shaggy hair that forms a spectacular cape in males. They have a long tail, relatively short limbs, and short digits that are covered with long fur on the dorsal surface.

The behavior and ecology of *R. roxellana* are only just becoming known through the work of Chinese zoologists. These monkeys are restricted to the mountainous forests of southern China. Little has been reported on their locomotion, but their limb anatomy, foot structure, and some anecdotal field observations suggest that they are probably among the most terrestrial of all Asian colobines. They are largely folivorous and eat the foliage of the Chinese fir tree as well as lichens during the cold winter months.

Golden monkeys live in multi-male groups of twenty to thirty animals in the winter. These groups form much larger congrega-

FIGURE 6.20

The golden monkey (*Rhinopithecus roxellana*) from China.

tions (up to 300 individuals) in the summer months. Births seem to be strictly seasonal.

ADAPTIVE RADIATION OF OLD WORLD MONKEYS

Compared with either prosimians or neotropical anthropoids, Old World monkeys are a remarkably uniform group, both morphologically (Schultz, 1970) and behaviorally (Fig. 6.21). This uniformity may be due partly to the recency of their adaptive radiation. In size, they barely range over one order of magnitude and have no very small species and no extremely large species. Sexual dimorphism in canine teeth is the rule and body size dimorphism is common. Although their bilophodont molars vary according to dietary habits in details of cusp height, length, and breadth and in the length of shearing crests, they are remarkably similar overall (Kay, 1978).

In their locomotor habits, cercopithecoid monkeys are all primarily quadrupedal walkers and runners, and some (mostly colobines) are also good leapers. When feeding in trees, they always sit above branches rather than use suspensory postures. No other primate radiation includes so many terrestrial species or is so limited in gait patterns and locomotor repertoire. Old World monkeys are primarily frugivorous or folivorous; a few are seed specialists. Al-

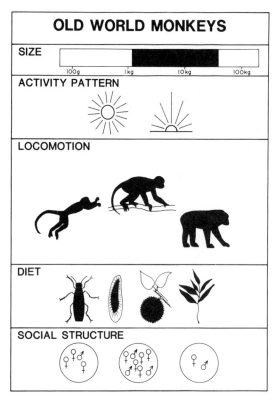

OLD WORLD MONKEYS

SIZE

| 100g | 1kg | 10kg | 100kg |

ACTIVITY PATTERN

LOCOMOTION

DIET

SOCIAL STRUCTURE

FIGURE 6.21

The adaptive radiation of Old World monkeys.

though many species feed to varying degrees on insects (most cercopithecines) and gums (baboons, patas monkeys), there are no species that specialize on these foods to the degree found in many smaller prosimians and platyrrhines.

Old World monkeys generally live in either single-male or multi-male polygynous groups (Wrangham, 1980). In most species, females spend their entire lives in their natal troop, and the troop's social organization centers around matrilines and female hierarchies. Males regularly emigrate from their natal group and may move through several troops during their lifetime (see, e.g., Strum,

1987; van Noordwijk and van Schaik, 1985). Monogamy is rare. All have single births.

Despite this relative uniformity, Old World monkeys have successfully colonized more vegetative zones and climates than any other group of living primates. In numbers of individuals, numbers of species, and biomass density they are probably the most successful nonhuman primates. This success results from a variety of factors. For the colobines, it is almost certainly their ability to digest cellulose and to exploit folivorous food sources that are unavailable to other animals. For the cercopithecines, their terrestrial locomotor potential, manipulative abilities, and intelligence enable them to exploit a wide range of foods and environments which less flexible species cannot endure. Cercopithecines are also relatively prolific breeders for higher primates, with many species giving birth on an annual basis.

PHYLETIC RELATIONSHIPS OF OLD WORLD MONKEYS

There are relatively few problems concerning the phyletic relationships among cercopithecoids. The differences among authorities concern low-level taxonomic questions such as whether the African and Asian colobines should be in one or several genera. The most interesting phyletic issue concerns *Cercocebus*. On morphological and behavioral grounds, this group seems to be a coherent genus, but on immunological evidence it seems that members of one species group are more closely related to some baboons than to other mangabeys. Above the genus level, cercopithecoid phylogenies based on dental and cranial studies and those based on immunology are in almost total agreement (Fig. 6.22).

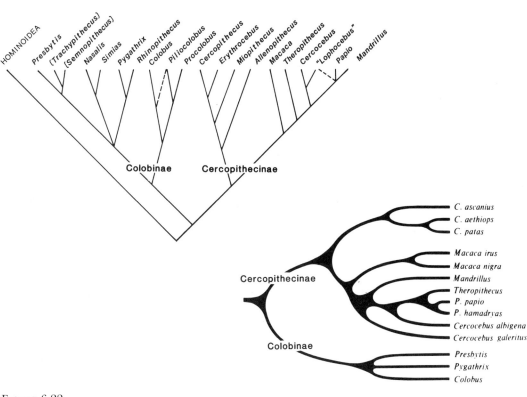

FIGURE 6.22

Two phylogenies of Old World monkeys—one based on dental and skeletal features (above, Strasser and Delson, 1987) and one based on immunology (below, Cronin and Sarich, 1976).

BIBLIOGRAPHY

GENERAL

Andrews, P. (1985). Family group systematics and evolution among catarrhine primates. In *Ancestors: The Hard Evidence*, ed. E. Delson, pp. 14–22. New York: Alan R. Liss.

Delson, E. (1985). Catarrhine evolution. In *Ancestors: The Hard Evidence*, ed. E. Delson, pp. 9–13. New York: Alan R. Liss.

Huxley, T.H. (1863). *Evidence as to Man's Place in Nature.* London: Williams and Norgate.

Napier, J.R., and Napier, P.H., eds. (1970). *Old World Monkeys.* New York: Academic Press.

Rose, M.D. (1974). Ischial tuberosities and ischial callosities. *Am. J. Phys. Anthropol.* **40**(3):375–384.

Strasser, E., and Delson, E. (1987). Cladistic analysis of cercopithecid relationships. *J. Hum. Ev.* **16**:81–100.

CERCOPITHECINES

Aldrich-Blake, F.P.G. (1970). Problems of social structure in forest monkeys. In *Social Behavior in Birds and Mammals*, ed. J.H. Crook, pp. 79–101. London: Academic Press.

Andelman, S.J. (1986). Ecological and social determinants of cercopithecine mating systems. In *Ecological Aspects of Social Evolution*, ed. D.I. Rubenstein and R.W. Wrangham, pp. 201–216. Princeton, N.J.: Princeton University Press.

Murray, P. (1975). The role of cheek pouches in cercopithecine monkey adaptive strategy. In *Primate Functional Morphology and Evolution*, ed. R.H. Tuttle, pp. 151–194. The Hague: Mouton.

Struhsaker, T.T. (1969). Correlates of ecology and social organization among African cercopithecines. *Folia Primatol.* **11**:80–118.

Struhsaker, T.T., and Leland, L. (1979). Socioecology of five sympatric monkey species in the Kibale Forest, Uganda. In *Advances in the Study of Behavior*, vol. 9, pp. 159–228. New York: Academic Press.

Macaques

Aldrich-Blake, F.P.G. (1980). Long-tailed macaques. In *Malayan Forest Primates: Ten Years' Study in the Tropical Forest*, ed. D.J. Chivers, pp. 147–166. New York: Plenum Press.

Caldicott, J.O. (1981). Findings on the behavioral ecology of the pig-tailed macaque. *Malays. Appl. Biol.* **10**(2):213–220.

———. (1986). An ecological and behavioral study of the pig-tailed macaque. *Contrib. Primatol.* **21**:1–259.

Chivers, D.J., ed. (1980). *Malayan Forest Primates: Ten Years' Study in the Tropical Forest*. New York: Plenum Press.

Crockett, C.M., and Wilson, M.L. (1980). The ecological separation of *Macaca nemestrina* and *Macaca fascicularis* in Sumatra. In *The Macaques: Studies in Ecology, Behavior and Evolution*, ed. D.G. Lindberg, pp. 148–181. New York: Van Nostrand Reinhold.

Deag, J.M., and Crook, J.H. (1971). Social behavior and "agonistic buffering" in wild Barbary macaque, *Macaca sylvanus* L. *Folia Primatol.* **15**:183–200.

Delson, E. (1980). Fossil macaques, phyletic relationships and a scenario of deployment. In *The Macaques: Studies in Ecology, Behavior and Evolution*, ed. D.G. Lindberg, pp. 10–30. New York: Van Nostrand Reinhold.

Fittinghoff, N.A., Jr., and Lindberg, D.G. (1980). Riverine refuging in eastern Bornean *Macaca fascicularis*. In *The Macaques: Studies in Ecology, Behavior and Evolution*, ed. D.G. Lindberg, pp. 182–214. New York: Van Nostrand Reinhold.

Fleagle, J.G. (1980). Locomotion and posture. In *Malayan Forest Primates: Ten Years' Study in the Tropical Forest*, ed. D.J. Chivers, pp. 191–207. New York: Plenum Press.

Fooden, J. (1980). Classification and distribution of living macaques (*Macaca* Lacepede, 1799). In *The Macaques: Studies in Ecology, Behavior and Evolution*, ed. D.G. Lindberg, pp. 1–9. New York: Van Nostrand Reinhold.

Furuya, Y. (1965). Social organization of the crab-eating monkey. *Primates* **6**:285–337.

Goldstein, S.J. (1981). Disturbed-site feeding by rhesus monkeys in Pakistan. *Am. J. Phys. Anthropol.* **54**(2):225–226.

Hartman, C.G., and Straus, W.L., eds. (1933). *The Anatomy of the Rhesus Monkey*. New York: Haffner.

Itani, J. (1983). Sociological studies of Japanese monkeys. In *Recent Progress of Natural Sciences in Japan*, vol. 8, pp. 89–94. Tokyo: Sci. Council of Japan.

Kawai, M., and Ohsawa, H. (1983). Ecology of Japanese monkeys, 1950–1982. In *Recent Progress of Natural Sciences in Japan*, vol. 8, pp. 95–108. Tokyo: Sci. Council of Japan.

Kurland, J.A. (1973). A natural history of Kra macaques (*M. fascicularis* Raffles, 1821) at the Kutai Reserve, Kalimantan, Timur, Indonesia. *Primates* **14**:245–262.

Lindberg, D.G. (1971). The rhesus monkey in northern India: An ecological and behavioral study. In *Primate Behavior: Developments in Field and Lab Research*, vol. 2, ed. A. Rosenblum, pp. 2–106. New York: Academic Press.

Lindberg, D.G., ed. (1980). *The Macaques: Studies in Ecology, Behavior and Evolution*. New York: Van Nostrand Reinhold.

MacKinnon, J.R., and MacKinnon, K.S. (1980). Niche differentiation in a primate community. In *Malayan Forest Primates: Ten Years' Study in the Tropical Forest*, ed. D.J. Chivers, pp. 167–190. New York: Plenum Press.

Neville, M.K. (1968). Ecology and activity of Himalayan foothill rhesus monkeys (*M. mulatta*). *Ecol.* **49**:110–123.

Poirier, F.E., and Smith, E.O. (1974). The crab-eating macaques (*M. fascicularis*) of Angaur Island, Pelau, Micronesia. *Folia Primatol.* **22**:258–306.

Richard, A.F., and Goldstein, S.J. (1981). Primates as weeds: The implications for macaque evolution. *Am. J. Phys. Anthropol.* **54**(2):267.

Roonwal, M.L., and Mohnot, S.M. (1977). *Primates of South Asia*. Cambridge, Mass.: Harvard University Press.

Simonds, P. (1965). The bonnet macaques of south India. In *Primate Behavior*, ed. I. DeVore, pp. 175–196. New York: Holt, Rinehart and Winston.

Suzuki, A. (1965). An ecological study of wild Japanese monkeys in snowy areas: Focused on their food habits. *Primates* **6**:31–72.

Taub, D.M. (1977). Geographic distribution and habitat diversity of the Barbary macaque, *Macaca sylvanus* L. *Folia Primatol.* **27**:108–133.

van Noordwijk, M.A. (1985). *The Socio-ecology of Sumatran Long-tailed Macaques* (Macaca fascicularis), II: *The Behavior of Individuals*. Ph.D. Dissertation. Utrecht: Rijksuniversiteit.

van Schaik, C.D. (1985). *The Socio-ecology of Sumatran Long-tailed Macaques* (Macaca fascicularis), I: *Costs and Benefits of Group Living*. Ph.D. Dissertation. Utrecht: Rijksuniversiteit.

Wheatley, B.P. (1980). Feeding and ranging of East Bornean *Macaca fascicularis*. In *The Macaques: Studies in Ecology, Behavior and Evolution*, ed. D.G. Lindberg, pp. 215–246. New York: Van Nostrand Reinhold.

Mangabeys

Chalmers, N.R. (1968). Group composition, ecology and daily activity of free living mangabeys in Uganda. *Folia Primatol.* **8**:247–262.

Cronin, J.E., and Sarich, V.M. (1976). Molecular evidence for dual origin of mangabeys among Old World monkeys. *Nature* **260**:700–702.

Dunbar, R.I.M. (1974). Observations on the ecology and social organization of the green monkey, *Cercocebus sabaeus*, in Senegal. *Primates* **15**:341–350.

Freeland, W.J. (1979). Mangabey (*C. albigena*) social organization and population density in relation to food use and availability. *Folia Primatol.* **32**:108–124.

Homewood, K.M. (1978). Feeding strategy of the Tana mangabey (*C. galeritus galeritus*). *J. Zool (London)* **186**:375–392.

Quiris, R. (1975). Ecologie et organisation sociale de *Cercocebus galeritus agilis* dans le Nord-Est du Gabon. *Terre Vie* **29**:337–398.

Waser, P. (1977). Feeding, ranging and group size in the mangabey *C. albigena*. In *Primate Ecology: Studies of Feeding and Ranging Behavior in Lemurs, Monkeys and Apes*, ed. T.H. Clutton-Brock, pp. 183–222. New York: Academic Press.

———. (1984). Ecological differences and behavioral contrasts between two mangabey species. In *Adaptations for Foraging in Non-human Primates*, ed. P.S. Rodman and J.G.H. Cant, pp. 195–216. New York: Columbia University Press.

Baboons

Aldrich-Blake, F.P.G., Bunn, T.K., Dunbar, R.I.M., and Headley, P.M. (1971). Observations on baboons, *P. anubis*, in an arid region in Ethiopia. *Folia Primatol.* **15**:1–35.

Altmann, J. (1980). *Baboon mothers and infants*. Cambridge, Mass.: Harvard University Press.

Altmann, S.A. (1974). Baboons, space, time and energy. *Am. Zool.* **14**:221–248.

Altmann, S.A., and Altmann, J. (1970). *Baboon Ecology*. Chicago: University of Chicago Press.

———. (1977). Life history of yellow baboons: Physical development, reproductive parameters and infant mortality. *Primates* **18**(2):315–330.

DeVore, I., and Hall, K.R.L. (1965). Baboon ecology. In *Primate Behavior*, ed. I. DeVore, pp. 20–52. New York: Holt, Rinehart and Winston.

DeVore, I., and Washburn, S. (1962). Baboon ecology and human evolution. In *African Ecology and Human Evolution*. Viking Fund. Publ. in Anthropology, no. 36, pp. 335–367.

Dunbar, R.I.M., and Nathan, M.F. (1972). Social organization of the Guinea baboon, *Papio papio*. *Folia Primatol.* **17**:321–334.

Hall, K.R.L. (1962). Numerical data, maintenance activities and locomotion of the wild Chacma baboon, *Papio ursinus*. *Proc. Zool. Soc. London* **139**:181–220.

Harding, R.S.O. (1975). Meat eating and hunting in baboons. In *Paleoanthropology, Morphology and Paleoecology*, ed. R.H. Tuttle, pp. 245–258. The Hague: Mouton.

———. (1976). Ranging patterns of a troop of baboons (*Papio anubis*) in Kenya. *Folia Primatol.* **25**:143–185.

Hausfater, G. (1975). Dominance and reproduction in baboons (*Papio cynocephalus*): A quantitative analysis. *Contributions to Primatology*, vol. 7. Basel: S. Karger.

Jolly, C.J. (1970). The large African monkeys as an adaptive array. In *Old World Monkeys*, ed. J.R. Napier and P.H. Napier, pp. 139–174. New York: Academic Press.

Kummer, H. (1968). *Social Organization of Hamadryas Baboons*. Chicago: University of Chicago Press.

Moreno-Black, G., and Maples, W.R. (1977). Differential habitat utilization of four Cercopithecidae in a Kenyan forest. *Folia Primatol.* **27**:85–107.

Nagel, U. (1973). A comparison of anubis baboons, Hamadryas baboons and their hybrids at a species border in Ethiopia. *Folia Primatol.* **19**:104–165.

Popp, J. (1983). Ecological determinism in the life histories of baboons. *Primates* **24**(2):198–210.

Ransom, T. (1981). *Beach Troop of the Gombe*. London: Bucknell University Press.

Rose, M.D. (1974). Ischial tuberosities and ischial callosities. *Am. J. Phys. Anthropol.* **40**(3):375–384.

———. (1976). Bipedal behavior of olive baboons

(*Papio anubis*) and its relevence to an understanding of the evolution of human bipedalism. *Am. J. Phys. Anthropol.* **44**:247–261.

Rowell, T. (1966). Forest living baboons in Uganda. *J. Zool. (London)* **149**:344–364.

Smuts, B.B. (1985). *Sex and friendship in baboons.* Hawthorne, N.Y.: Aldine.

Stammbach, E. (1986). Desert, forest, and montane baboons: Multilevel societies. In *Primate Societies*, ed. B.B. Smuts, D.L. Cheney, R.M. Seyfarth, R.W. Wrangham, and T.T. Struhsaker, pp. 112–120. Chicago: University of Chicago Press.

Strum, S.C. (1981). Processes and products of change: Baboon predatory behavior at Gilgil, Kenya. In *Omnivorous Primates: Gathering and Hunting in Human Evolution*, ed. R.S.O. Harding and G. Teleki, pp. 255–302. New York: Columbia University Press.

Strum, S.C. (1987). *Almost Human.* New York: Random House.

Strum, S.C., and Mitchell, W. (1987). Baboons: Baboon models and muddles. In *The Evolution of Human Behavior: Primate Models*, ed. W.G. Kinzey, pp. 87–104. Albany: SUNY Press.

Washburn, S.L., and DeVore, I. (1961). The social life of baboons. *Sci. Am.* **204**(6):62–71.

Mandrills and Drills

Gartlan, J.S. (1970). Preliminary notes on the ecology and behavior of the drill (*Mandrillus leucophaeus* Ritgen). In *Old World Monkeys*, ed. J.R. Napier and P.H. Napier, pp. 445–480. New York: Academic Press.

Gartlan, J.S., and Struhsaker, T.T. (1972). Polyspecific association and niche separation of rain forest anthropoids in Cameroon, West Africa. *J. Zool. (London)* **168**:221–266.

Grubb, P. (1973). Distribution, divergence and speciation of the drill and mandrill. *Folia Primatol.* **20**:161–177.

Hosino, J., Mori, A., Kudo, H., and Kawai, M. (1984). Preliminary report on the grouping of mandrills (*Mandrillus sphinx*) in Cameroon. *Primates* **25**:295–307.

Jouventin, P. (1975). Observations sur la socio-ecologie du mandrill. *Terre Vie* **29**:493–532.

Lahm, S. (1985). Mandrill ecology and the status of Gabon's rainforests. *Primate Conservation* **6**:32–33.

——. (1986). Diet and habitat preference of *Mandrillus sphinx* in Gabon: Implications of foraging strategy. *Am. J. Primatol.* **11**:9–26.

Stammbach, E. (1986). Desert, forest, and montane baboons: Multilevel societies. In *Primate Societies*, ed. B.B. Smuts, D.L. Cheney, R.M. Seyfarth, R.W.

Wrangham, and T.T. Struhsaker, pp. 112–120. Chicago: University of Chicago Press.

Geladas

Dunbar, R.I.M. (1980). Demographic and life history variables of a population of gelada baboons (*Theropithecus gelada*). *J. Anim. Ecol.* **49**:485–506.

——. (1984). *Reproductive Decisions: An Economic Analysis of Gelada Baboon Social Strategies.* Princeton, N.J.: Princeton University Press.

——. (1986). The social ecology of gelada baboons. In *Ecological Aspects of Social Evolution*, ed. D. Rubenstein and R.W. Wrangham, pp. 332–351. Princeton, N.J.: Princeton University Press.

Kawai, M., Dunbar, R.I.M., Ohsawa, H., and Mori, U. (1983). Social organization of gelada baboons: Social units and definitions. *Primates* **24**(1):13–24.

Stammbach, E. (1986). Desert, forest, and montane baboons: Multilevel societies. In *Primate Societies*, ed. B.B. Smuts, D.L. Cheney, R.M. Seyfarth, R.W. Wrangham, and T.T. Struhsaker, pp. 112–120. Chicago: University of Chicago Press.

Guenons

Brennan, E.J. (1985). DeBrazza's monkeys (*Cercopithecus neglectus*) in Kenya: Census, distribution and conservation. *Am. J. Primatol.* **8**(4):269–278.

Cords, M. (1986). Forest guenons and patas monkeys: Male-male competition in one-male groups. In *Primate Societies*, ed. B.B. Smuts, D.L. Cheney, R.M. Seyfarth, R.W. Wrangham, and T.T. Struhsaker, pp. 98–111. Chicago: University of Chicago Press.

Devos, A., and Omar, A. (1971). Territories and movements of Sykes monkeys (*Cercopithecus mitis kolbi* Neuman) in Kenya. *Folia Primatol.* **16**:196–205.

Galat, G., and Galat-Luong, A. (1985). La communaute de primates diurnis de la foret de Tai, Cote d'Ivoire. *Terre Vie* **40**:3–32.

Gautier, J., and Gautier-Hion, A. (1969). Associations polyspecifiques chez les Cercopitheques du Gabon. *Terre Vie* **23**:164–201.

Gautier-Hion, A. (1975). Dimorphisme sexual et organisation sociale chez les cercopithecines forestiers africains. *Mammalia* **39**(3):365–374.

——. (1978). Food niches and coexistence in sympatric primates in Gabon. In *Recent Advances in Primatology*, vol. 1, ed. D.J. Chivers and J. Herbert, pp. 269–286. London: Academic Press.

Gautier-Hion, A., and Gautier, J.P. (1974). Les associations polyspecifiques de Cercopitheques du Plateau du M'passa (Gabon). *Folia Primatol.* **22**:134–177.

——. (1976). Croissance, maturite sexuelle et sociale, reproduction chez les cercopithecines forestiers africains. *Folia Primatol.* **26**:165–184.

——. (1978). Le singe de Brazza: Une strategie originale. *Z. Tierpsychol.* **46**:84–104.

——. (1979). Niche ecologique et diversite des especes sympatriques dans le genre *Cercopithecus*. *Terre Vie* **33**:493–507.

Haddow, A.J. (1952). Field and laboratory studies of an African monkey, *Cercopithecus ascanius schmidti*. *Proc. Zool. Soc. London* **122**:297–394.

Hylander, W.L. (1975). Incisor size and diet in anthropoids with special reference to Cercopithecoidea. *Science* **189**:1095–1098.

Kano, T. (1971). Distribution of the primates in the eastern shore of Lake Tanganyika. *Primates* **12**:281–304.

Kay, R.F. (1978). Molar structure and diet in extant Cercopithecoidae. In *Development, Function and Evolution of Teeth*, ed. P.M. Butler and K.A. Joysey, pp. 309–339. New York: Academic Press.

Kay, R.F., and Hylander, W.F. (1978). The dental structure of mammalian folivores with special reference to primates and phalangeroids (Marsupialia). In *The Ecology of Arboreal Folivores*, ed. G.G. Montgomery, pp. 173–192. Washington, D.C.: Smithsonian Institution Press.

Kingdon, J. (1971). *East African Mammals*, vol. 1. New York: Academic Press.

Rudran, R. (1978). Socioecology of the blue monkeys (*Cercopithecus mitis stuhlmanni*) of the Kibale Forest, Uganda. *Smithson. Contrib. Zool.* **249**:1–88.

Schultz, A.H. (1970). The comparative uniformity of the Cercopithecoidea. In *Old World Monkeys*, ed. J.R. Napier and P.H. Napier, pp. 39–52. New York: Academic Press.

Struhsaker, T.T. (1978). Food habits of five monkey species in the Kibale Forest, Uganda. In *Recent Advances in Primatology*, vol. 1, ed. D.J. Chivers and J. Herbert, pp. 225–248. London: Academic Press.

Vervets

Galat, G., and Galat-Loung, A. (1976). La colonisation de la mangrove par *Cercopithecus aethiops sabaeus* au Senegal. *Terre Vie* **30**:3–30.

Hall, K.R.L., and Gartlan, J.S. (1965). Ecology and behavior of the vervet monkeys, *Cercopithecus aethiops*, Lolui Island, Lake Victoria. *Proc. Zool. Soc. London* **145**:37–56.

Kavanagh, M. (1978). The diet and feeding behavior of *Cercopithecus aethiops tantalus*. *Folia Primatol.* **30**:30–63.

Struhsaker, T.T. (1967). Ecology of vervet monkeys (*Cercopithecus aethiops*) in the Masai-Amboseli Game Reserve, Kenya. *Ecology* **48**:891–904.

Talapoins

Gautier-Hion, A. (1966). L'ecologie et l'ethologie du talapoin (*Miopithecus talapoin talapoin*). *Revue Biol. Gabon* **2**:311–329.

——. (1970). L'organisation sociale d'une bande de talapoins dan le Nord-Est du Gabon. *Folia Primatol.* **12**:116–141.

——. (1971). L'ecologie de talapoin du Gabon. *Terre Vie* **25**:427–490.

——. (1973). Social and ecological features of talapoin monkeys—comparisons with sympatric cercopithecines. In *Comparative Ecology and Behavior of Primates*, ed. R.P. Michael and J.H. Crook, pp. 147–170. New York: Academic Press.

——. (1978). Food niches and coexistence in sympatric primates in Gabon. In *Recent Advances in Primatology*, vol. 1, ed. D.J. Chivers and J. Herbert, pp. 269–286. London: Academic Press.

Rowell, T.E. (1972). Toward a natural history of the talapoin monkey in Cameroon. *Ann. Fac. Sci. Cameroun* **10**:121–134.

——. (1973). Social organization of wild talapoin monkeys. *Am. J. Phys. Anthropol.* **38**(2):593–598.

Rowell, T.E., and Dixon, A.F. (1975). Changes in social organization during the breeding season of wild talapoin monkeys. *J. Reprod. Fert.* **43**:419–434.

Swamp Monkeys

Zeeve, S.R. (1985). Swamp monkeys of the Lomako Forest, Central Zaire. *Primate Conservation* **5**:32–33.

Patas Monkeys

Chism, J., Rowell, T.E., and Olson, D. (1984). Life history patterns of female patas monkeys. In *Female Primates: Studies by Women Primatologists*, ed. M.F. Small, pp. 175–190. New York: Alan R. Liss.

Gartlan, S.J. (1974). Adaptive aspects of social structure in *Erythrocebus patas*. *Symp. Fifth Cong. Int. Primatol. Soc.*, pp. 161–171.

Hall, K.R.L. (1965). Behavior and ecology of the wild patas monkey, *Erythrocebus patas*, in Uganda. *J. Zool.* **148**:15–87.

COLOBINES

Bauchop, T. (1978). Digestion of leaves in vertebrate arboreal folivores. In *The Ecology of Arboreal Folivores*, ed. G.G. Montgomery, pp. 193–204. Washington, D.C.: Smithsonian Institution Press.

Hylander, W.L. (1979). The functional significance of primate mandibular form. *J. Morphol.* **160**(2):223–240.

Struhsaker, T.T. (1986). Colobines: Infanticide by adult males. In *Primate Societies*, ed. B.B. Smuts, D.L. Cheney, R.M. Seyfarth, R.W. Wrangham, and T.T. Struhsaker, pp. 83–97. Chicago: University of Chicago Press.

Vogel, C. (1966). Morphologische Studien am Gesichtsschadel catarriner Primaten. *Bibl. Primatol. fasc.* **4**, Basel.

———. (1968). The phylogenetical evolution of some characteristics and some morphological trends in the evolution of the skull in catarrhine primates. In *Taxonomy and Phylogeny of Old World Primates with Reference to the Origin of Man*, ed. A.B. Chiarelli, pp. 21–55. Turin: Rosenberg and Sellier.

Yamada, H., and Sakai, T. (1983). Tooth size and its sexual dimorphism in colobus monkeys. *J. Anthropol. Soc. Nippon* **91**(1):79–98.

Guerezas

Clutton-Brock, T.H. (1975a). Feeding behavior of red colobus and black and white colobus in East Africa. *Folia Primatol.* **23**:165–208.

———. (1975b). Ranging behavior of red colobus monkeys. *Anim. Behav.* **23**:706–722.

Dunbar, R.I.M., and Dunbar, E.P. (1974). Ecology and population dynamics of *Colobus guereza* in Ethiopia. *Folia Primatol.* **21**:188–208.

Hull, D.B. (1979). A craniometric study of the black-and-white colobus Illiger 18ll (Primates, Cercopithecoidea). *Am. J. Phys. Anthropol.* **51**:163–182.

Mittermeier, R.A., and Fleagle, J.G. (1976). The locomotor and postural repertoires of *Ateles geoffroyi* and *Colobus guereza*, and a re-evaluation of the locomotor category of semibrachiation. *Am. J. Phys. Anthropol.* **45**:235–256.

Morbeck, M.E. (1977). Positional behavior, selective use of habitat substrate and associated non-positional behavior in free ranging *Colobus guereza* (Ruppel 1835). *Primates* **18**(1):35–58.

Oates, J.F. (1977a). The guereza and its food. In *Primate Ecology: Studies of Feeding and Ranging Behavior in Lemurs, Monkeys and Apes*, ed. C.H. Clutton-Brock, pp. 275–321. New York: Academic Press.

———. (1977b). The social life of a black-and-white colobus monkey, *C. guereza. Z. Tierpsychol.* **45**:1–60.

Oates, J.F., and Trocco, T.F. (1983). Taxonomy and phylogeny of black-and-white colobus monkeys. *Folia Primatol.* **40**:83–113.

Rose, M.D. (1978). Feeding and associated positional behavior of black-and-white colobus monkeys (*Colobus guereza*). In *The Ecology of Arboreal Folivores*, ed. G.G. Montgomery, pp. 253–264. Washington, D.C.: Smithsonian Institution Press.

Struhsaker, T.T., and Oates, J.F. (1975). Comparison of the behavior and ecology of red colobus and black-and-white colobus monkeys in Uganda: A summary. In *Socioecology and Psychology of Primates*, ed. R.H. Tuttle, pp. 103–124. The Hague: Mouton.

Black Colobus

Harrison, M.J.S. (1986). Feeding ecology of black colobus (*Colobus satanas*) in central Gabon. In *Primate Ecology and Conservation*, ed. J.G. Else and P.C. Lee, pp. 31–38. Cambridge: Cambridge University Press.

McKey, D.B. (1978). Soils, vegetation and seed eating by black colobus monkeys. In *The Ecology of Arboreal Folivores*, ed. G.G. Montgomery, pp. 423–437. Washington, D.C.: Smithsonian Institution Press.

McKey, D.B., and Gartlan, J.S. (1981). Food selection by black colobus monkeys (*C. satanas*) in relation to plant chemistry. *Biol. J. Linn. Soc.* **16**:115–146.

McKey, D.B., Waterman, P.G., Mbi, C.N., Gartlan, J.S., and Struhsaker, T.T. (1978). Phenolic content of vegetation in two African rain forests: Ecological implications. *Science* **202**:61–64.

Red Colobus

Clutton-Brock, T.H. (1973). Feeding levels and feeding sites of red colobus (*Colobus badius tephrosceles*) in the Gombe National Park. *Folia Primatol.* **219**:368–379.

———. (1975a). Feeding behavior of red colobus and black and white colobus in East Africa. *Folia Primatol.* **23**:165–208.

———. (1975b). Ranging behavior of red colobus monkeys. *Anim. Behav.* **23**:706–722.

Gunderson-Coolen, V. (1977). Some observations on the ecology of *Colobus badius temmincki*, Abuko Nature Reserve, the Gambia, West Africa. *Primates* **18**:305–314.

Marsh, C.W. (1981). Time budget of Tana River red colobus. *Folia Primatol.* **35**:30–50.

Starin, E.D. (1981). Monkey moves. *Nat. Hist.* **90**(9):36–43.

Struhsaker, T.T. (1975). *The Red Colobus Monkey.* Chicago: University of Chicago Press.

———. (1978). Food habits of five monkey species in the Kibale Forest, Uganda. In *Recent Advances in Primatology*, vol. 1, ed. D.J. Chivers and J. Herbert, pp. 229–248. London: Academic Press.

Struhsaker, T.T., and Oates, J.F. (1975). Comparison of the behavior and ecology of red colobus and black-and-white colobus monkeys in Uganda: A summary. In *Socioecology and Psychology of Primates*, ed. R.H. Tuttle, pp. 103–124. The Hague: Mouton.

Olive Colobus

Booth, A.H. (1957). Observations on the natural history of the olive colobus monkey, *Procolobus verus* (Van Beneden). *Proc. Zool. Soc. London* **129**:421–430.

Galat, G., and Galat-Luong, A. (1985). La communaute de primates diurnis de la foret de Tai, Cote d'Ivoire. *Terre Vie* **40**:3–32.

Langurs

Brandon-Jones, D. (1978). The evolution of the recent Asian Colobinae. In *Recent Advances in Primatology*, vol. 1, ed. D.J. Chivers and J. Herbert, pp. 323–325. London: Academic Press.

Eisenberg, J.F., Muckenhirn, N.A., and Rudran, R. (1972). The relation between ecology and social structure in primates. *Science* **176**:863–874.

Pocock, R.I. (1935). On monkeys of the genera *Pithecus* (or *Presbytis*) and *Pygathrix* found to the east of Bengal. *Proc. Zool. Soc. London* (1935):895–961.

Hanuman Langurs

Bishop, N.H. (1979). Himalayan langurs: Temperate colobines. *J. Hum. Evol.* **8**:251–281.

Hladik, C.M. (1977). A comparative study of the feeding strategies of two sympatric species of leaf monkeys: *Presbytis senex* and *Presbytis entellus*. In *Primate Ecology: Studies of Feeding and Ranging Behaviour in Lemurs, Monkeys and Apes*, ed. C.H. Clutton-Brock, pp. 324–353. New York: Academic Press.

Hrdy, H.S. (1974). Male-male competition and infanticide among the langurs (*Presbytis entellis*) of Abu, Rajastan. *Folia Primatol.* **22**:19–58.

———. (1977a). Infanticide as a primate reproductive strategy. *Am. Sci.* **65**:40–49.

———. (1977b). *The Langurs of Abu*. Cambridge, Mass.: Harvard University Press.

Jay, P.C. (1965). The common langur of northern India. In *Primate Behavior: Field Studies of Monkeys and Apes*, ed. I. DeVore, pp. 197–249. New York: Holt, Rinehart and Winston.

Ripley, S. (1967). The leaping of langurs: A problem in the study of locomotor behavior. *Am. J. Phys. Anthropol.* **26**:149–170.

———. (1970). Leaves and leaf-monkeys: Social organization of foraging. In *Old World Monkeys*, ed. J.R. Napier and P.H. Napier, pp. 483–509. New York: Academic Press.

Sommer, V., and Mohnot, S.M. (1984). New observations on infanticide among Hanuman langurs (*Presbytis entellus*) near Jodhpur (Rajasthan/India). *Behav. Ecol. Sociobiol.* **16**:245–248.

Sugiyama, Y. (1964). Group composition, population density and some sociological observations of Hanuman langurs (*Presbytis entellus*). *Primates* **5**(3–4):7–48.

———. (1965). On the social change of Hanuman langurs (*Presbytis entellus*) in their natural condition. *Primates* **6**(3–4):381–418.

———. (1967). Social organization of Hanuman langurs. In *Social Communication among Primates*, ed. S.A. Altmann. Chicago: University of Chicago Press.

———. (1976). Characteristics of the ecology of the Himalayan langurs. *J. Hum. Evol.* **5**:249–277.

Yoshiba, K. (1968). Local and intertroop variability in ecology and social behavior of common Indian langurs. In *Primates: Studies in Adaptation and Variability*, ed. P. Jay, pp. 217–242. New York: Holt, Rinehart and Winston.

Purple-faced and John's Langurs

Eisenberg, J.F., Muckenhirn, N.A., and Rudran, R. (1972). The relation between ecology and social structure in primates. *Science* **176**:863–874.

Hladik, C.M. (1977). A comparative study of the feeding strategies of two sympatric species of leaf monkeys: *Presbytis senex* and *Presbytis entellus*. In *Primate Ecology: Studies of Feeding and Ranging Behaviour in Lemurs, Monkeys and Apes*, ed. C.H. Clutton-Brock, pp. 324–353. New York: Academic Press.

Hladik, C.M., and Hladik, A. (1972). Disponsibilites alimentaire et domaines vitaux des primates a Ceylan. *Terre Vie* **26**:149–215.

Poirier, F.E. (1970). The Nilgiri langur (*P. johnii*) of South India. In *Primate Behavior: Developments in Field and Lab Research*, vol. 1, ed. L.A. Rosenblum, pp. 251–383. New York: Academic Press.

Rudran, R. (1973). Adult male replacement in one-male troops of purple-faced langurs (*Presbytis senex senex*) and its effect on population structure. *Folia Primatol.* **19**:166–192.

Leaf Monkeys

Bernstein, I.S. (1968). The lutong of Kuala Selangor. *Behaviour* **32**:1–16.

Curtin, P., and Dolhinow, P. (1978). Primate social behavior in a changing world. *Am. Sci.* **66**:468–475.

Curtin, S.H. (1977). Niche separation in sympatric Malaysian leaf-monkeys (*Presbytis obscura* and *Presbytis melalophos*). *Yrbk. Phys. Anthropol.* **20**:421–439.

———. (1980). Dusky and banded leaf-monkeys. In *Malayan Forest Primates: Ten Years' Study in the Tropical Forest*, ed. D.J. Chivers, pp. 107–146. New York: Plenum Press.

Curtin, S.H., and Chivers, D.J. (1978). Leaf-eating primates of peninsula Malaysia: The siamang and the dusky leaf-monkey. In *The Ecology of Arboreal Folivores*, ed. G.G. Montgomery, pp. 441–464. Washington, D.C.: Smithsonian Institution Press.

Fleagle, J.G. (1977a). Locomotor behavior and muscular anatomy of sympatric Malaysian leaf-monkeys (*Presbytis obscura* and *Presbytis melalophos*). *Am. J. Phys. Anthropol.* **46**:297–308.

———. (1977b). Locomotor behavior and skeletal anatomy of sympatric Malaysian leaf-monkeys. *Yrbk. Phys. Anthropol.* **20**:440–453.

———. (1978). Locomotion, posture and habitat use of two sympatric leaf-monkeys in West Malaysia. In *Recent Advances in Primatology*, vol. 1, ed. D.J. Chivers and J. Herbert, pp. 331–336. London: Academic Press.

Fooden, J. (1971). Report on the primates collected in western Thailand, January–April, 1967. *Fieldiana Ser. Zool.* **59**:62.

———. (1976). Primates obtained in peninsular Thailand June–July, 1973, with notes on the distribution of continental Southeast Asian leaf-monkeys (*Presbytis*). *Primates* **17**:95–118.

Furuya, Y. (1962). The social life of silvered leaf-monkeys (*Trachypithecus cristatus*). *Primates* **3**:41–60.

Hausfater, G., and Vogel, C. (1982). Infanticide in langur monkeys (genus *Presbytis*): Recent research and a review of hypotheses. In *Advanced Views in Primate Biology*, ed. A.B. Chiarelli and R.S. Corruccini, pp. 160–176. Berlin: Springer-Verlag.

MacKinnon, J.R., and MacKinnon, K.S. (1978). Comparative feeding ecology of six sympatric primates in western Malaysia. In *Recent Advances in Primatology*, vol. 1, ed. D.J. Chivers and J. Herbert, pp. 309–321. London: Academic Press.

———. (1980). Niche differentiation in a primate community. In *Malayan Forest Primates*, ed. D.J. Chivers, pp. 167–190. New York: Plenum Press.

Mukherjee, R.P., and Saha, S.S. (1974). The golden langurs (*Presbytis geei* Khajuria 1956) of Assam. *Primates* **15**(4):327–340.

Stott, K., Jr., and Selsor, C.J. (1961). Observations of the maroon leaf monkey in North Borneo. *Mammalia* **25**:184–189.

Tilson, R.L. (1976). Infant coloration and taxonomic affinity of the Mentawai Islands leaf monkey, *Presbytis potenziani*. *J. Mammal.* **57**(4):766–769.

Tilson, R.L., and Tenaza, R.R. (1976). Monogamy and duetting in an Old World monkey. *Nature* **263**:230–231.

Washburn, S.L. (1944). The genera of Malaysian langurs. *J. Mammal.* **25**:289–294.

Wilson, C.C., and Wilson, W.L. (1977). Behavioral and morphological variation among primate populations in Sumatra. *Yrbk. Phys. Anthropol.* **20**:207–233.

Wilson, W.L., and Wilson, C.C. (1975). Species-specific vocalizations and the determination of phylogenetic affinities of the *Presbytis aygula-melalophos* group in Sumatra. In *Contemporary Primatology*, ed. S. Kondo, M.Kawai, and A. Ehara, pp. 459–463. Basel: S. Karger.

Wolf, K.E., and Fleagle, J.G. (1977). Adult male replacement in a group of silvered leaf-monkeys (*Presbytis cristata*) at Kuala Selangor, Malaysia. *Primates* **18**:949–955.

Proboscis Monkeys

Kawabe, M., and Mano, T. (1972). Ecology and behavior of the wild proboscis monkey (*Nasalis larvatus*, Wurmb) in Sabah, Malaysia. *Primates* **13**:213–228.

Kern, J.A. (1964). Observations on the habits of proboscis monkey, *Nasalis larvatus* (Wurmb), made in the Brunei Bay area, Borneo. *Zoologica* **49**:183–192.

———. (1965). The proboscis monkey. *Animal Kingdom* **68**:67–73.

Simakobu Monkeys

Tilson, R.L. (1977). Social organization of Simakobu monkeys (*Nasalis concolor*) in Siberut Island, Indonesia. *J. Mammal.* **58**(2):202–212.

———. (1979). Der uberkannte Affe: Die ersten Bilder der Pageh-Stumpfnasse. *Tier* **5**:20–23.

Watanabe, K. (1981). Variations in group composition and population density of the two sympatric Metawaian leaf monkeys. *Primates* **22**:145–160.

Douc Langurs

Gochfield, M. (1974). Douc langurs. *Nature* **247**:167.

Kavanagh, M. (1972). Food sharing behavior within a group of douc monkeys (*Pygathrix nemaeus*). *Nature* **239**:406–407.

Lippold, L.K., and Brockman, D.K. (1974). San Diego's douc langurs. *Zoonooz* **47**:4–11, San Diego, Ca.

Golden Monkeys

Bangjie, T. (1985). The status of primates in China. *Primate Conservation* **5**:63–81.

Davison, G.W.H. (1982). Convergence with terrestrial cercopithecines by the monkey *Rhinopithecus roxellanae*. *Folia Primatol.* **37**:209–215.

Poirier, F.E. (1983). The golden monkey in the People's Republic of China. *IUCN/SSC, Primate Specialist Group Newsletter*, no. 3, March 1983.

Schaller, G.B. (1982). Zhen-zhen, rare treasure of Sichuan. *Animal Kingdom* **85**(6):5–14.

———. (1985). First published photos of China's golden treasure. *International Wildlife* **15**(1):29–31.

Zhang, Y.Z., Wang, S., and Quan, C.Q. (1981). On the geographical distribution of primates in China. *J. Hum. Evol.* **10**:215–226.

Zhixiang, L., Shilai, M., Chenghui, H., and Yingxiang, W. (1982). The distribution and habitat of the Yunnan golden monkey, *Rhinopithecus bieti. J. Hum. Evol.* **11**:633–638.

ADAPTIVE RADIATION OF OLD
WORLD MONKEYS

Kay, R.F. (1978). Molar structure and diet in extant Cercopithecoidae. In *Development, Function and Evolution of Teeth*, ed. P.M. Butler and K.A. Joysey, pp. 309–339. New York: Academic Press.

Ripley, S. (1970). Leaves and leaf-monkeys: Social organization of foraging. In *Old World Monkeys*, ed. J.R. Napier and P.H. Napier, pp. 483–509. New York: Academic Press.

Schultz, A.H. (1970). The comparative uniformity of the Cercopithecoidea. In *Old World Monkeys*, ed. J.R. Napier and P.H. Napier, pp. 39–52. New York: Academic Press.

Strum, S.C. (1987). *Almost Human*. New York: Random House.

van Noordwijk, M.A., and van Schaik, C.D. (1985). Male migration and rank acquisition in wild long-tailed macaques (*Macaca fascicularis*). *Anim. Behav.* **33**:849–861.

Wrangham, R.W. (1980). An ecological model of female-bonded primate groups. *Behaviour* **75**:262–300.

SEVEN

Apes and Humans

SUPERFAMILY HOMINOIDEA

In this chapter we review the less diverse group of catarrhine primates, the hominoids. There are only five genera of living hominoids. They range in size from about 4 kg for the smallest gibbons to over 200 kg for male gorillas. With the notable exception of our own species, hominoids have a rather restricted distribution—the tropical forests of Africa and Southeast Asia (Fig. 7.1). As we see in later chapters, hominoids were much more diverse and abundant in Europe at earlier times. Humans are the only hominoids that occur "naturally" in the New World, despite persistent rumors of Sasquatch or other apelike animals from the western parts of North America. However, both humans and chimpanzees have been sighted from time to time in extraterrestrial environments.

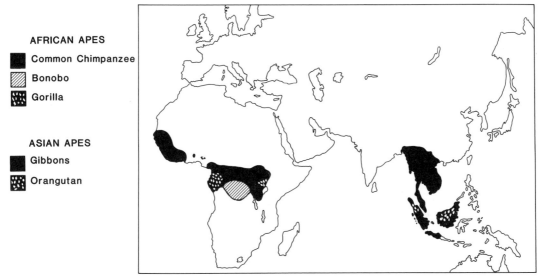

AFRICAN APES
- ◼ Common Chimpanzee
- ▨ Bonobo
- ▦ Gorilla

ASIAN APES
- ◼ Gibbons
- ▦ Orangutan

FIGURE 7.1

Geographic distribution of extant apes.

Hominoids

Hominoids are distinguished from Old World monkeys by a variety of both primitive catarrhine features and unique specializations (see Fig. 6.1). Like cercopithecoids, all living hominoids have a tubular tympanic bone and a dental formula of $\frac{2.1.2.3}{2.1.2.3}$. Compared to Old World monkeys, hominoids have relatively primitive molar teeth—with rounded cusps rather than the bilophodont crests of monkeys. The lower molars are characterized by an expanded talonid basin surrounded by five main cusps. The upper molars are quadrate and have a distinct trigon anteriorly and a large hypocone posteriorly. The anterior lower premolar varies in shape from an elongate shearing blade in gibbons to a bicuspid tooth in humans. Most hominoids have relatively broad incisors. Hominoid canines are much more variable than those of cercopithecoids in both shape and degree of sexual dimorphism.

Hominoids are characterized by relatively broad palates, broad nasal regions, and large brains. Hominoid skeletons show a variety of distinctive features (Fig. 7.2). The axial skeleton is characterized by a reduced lumbar region, an expanded sacrum, and the absence of a tail. All hominoids have a relatively broad thorax with a dorsally positioned scapula. Hominoids have relatively long up-

FIGURE 7.2

Characteristic skeletal features of extant apes, illustrated by a siamang.

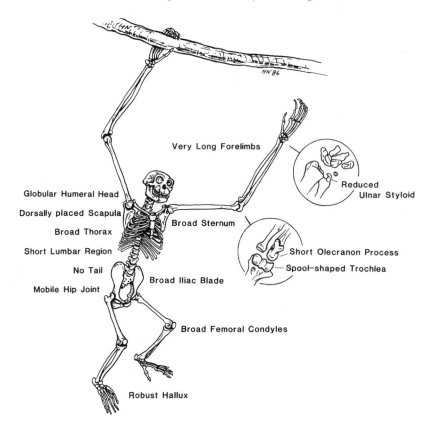

Very Long Forelimbs

Reduced Ulnar Styloid

Globular Humeral Head

Dorsally placed Scapula

Broad Thorax

Short Lumbar Region

No Tail

Mobile Hip Joint

Broad Sternum

Short Olecranon Process

Spool-shaped Trochlea

Broad Iliac Blade

Broad Femoral Condyles

Robust Hallux

per limbs, and their elbow joint is characterized by a spool-shaped trochlea on the humerus and a short olecranon process on the ulna. Their wrist lacks an articulation between the ulna and the carpal bones; instead, a fibrous meniscus separates the two bones. The hindlimbs of hominoids are characterized by a broad ilium, broad femoral condyles, and usually a large, robust hallux.

The five hominoid genera are traditionally placed in three separate families: hylobatids, pongids, and hominids. Although in this chapter we follow this classification, it is important to realize that this is unquestionably a gradistic rather than a phyletic arrangement. The phyletic relationships among hominoids are discussed later in the chapter (see also Chapter 1).

Hylobatids

The gibbons, *Hylobates*, from Southeast Asia (Fig. 7.3) are the smallest, the most specifically diverse, and the most numerically suc-

cessful of living apes (Table 7.1). These lesser apes are anatomically the most primitive of living apes and retain many monkeylike features, but in some aspects, such as their limb proportions, they are the most specialized of the living hominoids. The numerous gibbon species are relatively uniform in morphology. They are all relatively small (4–11 kg), with no sexual size dimorphism. They have simple molar teeth characterized by low rounded cusps and broad basins (Fig. 7.4). Their incisors are relatively short, but broad. Both sexes have long daggerlike canines and bladelike lower anterior premolars for sharpening the upper canine.

Gibbons have short snouts and shallow faces, large orbits with protruding rims, and a wide interorbital distance. Their braincase is globular and has no nuchal cresting, and only occasionally do they develop a sagittal crest. The mandible is shallow and has a broad ascending ramus.

Gibbons are outstanding among living primates in their limb proportions (Fig. 7.5). They have the longest forelimbs relative to

TABLE 7.1
Infraorder Catarrhini
Family HYLOBATIDAE

Common Name	Species	Intermembral Index	Body Weight (g)	
Siamang	*Hylobates syndactylus*	147	M	10,900
			F	10,600
Crested gibbon	*H. concolor*	140	MF	6,300
Hoolock gibbon	*H. hoolock*	129	M	6,930
			F	6,480
Kloss's gibbon	*H. klossii*	126	M	5,670
			F	5,920
White-handed gibbon	*H. lar*	130	M	5,700
			F	5,300
Agile gibbon	*H. agilis*	129	M	5,830
			F	5,410
Pileated gibbon	*H. pileatus*	—		—
Silvery gibbon	*H. moloch*	127		5,700
Mueller's gibbon	*H. muelleri*	129	M	5,760
			F	5,700

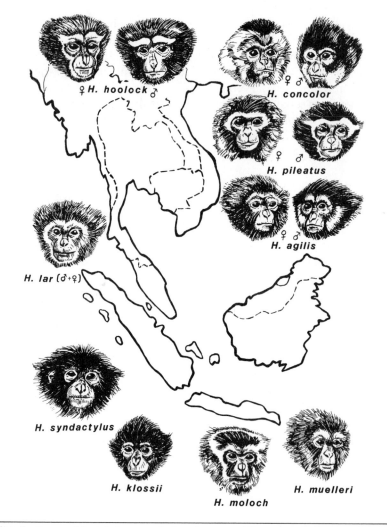

FIGURE 7.3

Geographic distribution and facial characteristics of extant gibbon populations.

body size of any primates, and they also have very long legs. They have long, curved, slender digits on both hands and feet as well as a long muscular pollex and a large hallux. Gibbons are the only apes that consistently have ischial callosities.

Gibbons are found throughout the evergreen forests of Southeast Asia from eastern India to southern China on the mainland, as well as on Borneo, Java, Sumatra, and nearby islands of the Sunda Shelf. There is general agreement on the number of distinct morphological groups of gibbons but debate on how many of these should be considered separate species. The three major groups of lesser apes are the **siamang** (*H. syndactylus*), the **concolor gibbon** (*H. concolor*), and the others. This last group contains numerous

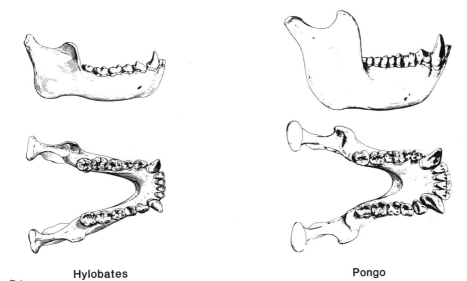

Hylobates   **Pongo**

FIGURE 7.4

Lower jaws of a siamang (left) and an orangutan (right).

FIGURE 7.5

The skeleton of a gibbon (*Hylobates*).

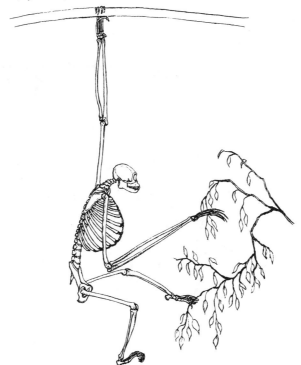

FIGURE 7.6

Two sympatric gibbons from west Malaysia: upper left, the white-handed gibbon (*Hylobates lar*); below, the siamang (*Hylobates syndactylus*).

allopatric gibbon populations that some authorities recognize as distinct species but that others consider subspecies. The hoolock gibbon (*H. hoolock*) and Kloss's gibbon (*H. klossii*) are the most distinctive species in this group. The remaining populations, the white-handed, or lar gibbon (*H. lar*), the agile gibbon (*H. agilis*), the pileated gibbon (*H. pileatus*), the silvery gibbon (*H. moloch*), and Mueller's gibbon (*H. muelleri*), are considered by many authorities to be subspecies of *H. lar*. Like the savannah baboons of Africa, the "lar" gibbons form what is called a superspecies. The populations are all allopatric and have distinct vocalizations. There is, however, evidence of interbreeding at species borders, and the level of morphological differences in cranial morphology between the populations is less than that found between other so-called distinct species of primates.

The behavior and ecology of the different species of gibbons is remarkably uniform considering their broad geographic range (Chivers, 1984). The greatest differences in gibbon behavior and ecology are between the two sympatric species, the siamang and the white-handed gibbon, which are found together in the forests of west Malaysia (Fig. 7.6; see Fig. 6.6) and Sumatra. All gibbon species show a preference for moist primary forests rather than secondary or riverine forests, but siamang are found at higher elevations and more commonly in mountain regions than are the sympatric white-handed gibbons. Gibbons move and feed mainly in the middle and upper levels of the canopy and virtually never descend to the ground. They are the most suspensory of all primates and are aerialists par excellence, moving almost exclusively by two-armed brachiation and by slower quadrumanous climbing (Fig. 7.7). The larger siamang travel mainly by slow, pendulum-like arm-over-arm brachia-tion, whereas the smaller gibbon species use more rapid ricocheting brachiation in which they throw themselves from one tree to the next over gaps of 10 m or more. During feeding, all gibbons use more deliberate quadrumanous climbing when moving among small terminal branches. They use a wide variety of both seated and suspensory feeding postures.

Gibbons specialize on a diet of ripe fruit, part of which is found in small, widely scattered clumps throughout the forest and part, such as many figs, which occurs in large bonanzas. Gibbon species also eat varying amounts of new leaves and invertebrates such as termites and arachnids. The proportions of these foods in their diet vary from season to season and from species to species. The larger siamang rely more on new leaves than do the smaller lar gibbons. The Kloss's gibbon from the island of Siberut is unusual in that it does not eat leaves, only fruit and invertebrates.

Most gibbons live in small monogamous families composed of a mated pair and up to four dependent offspring. Families of the more folivorous siamang usually forage as a unit and have smaller day ranges (1 km or less) and home ranges (18–50 ha) than the lar gibbons. The latter often forage individually, with the members of a single family separated by as much as several hundred meters, and they have larger day ranges (2 km) and home ranges. These differences in ranging pattern have been related to the different distribution of preferred foods—leaves for the siamang and scattered fruits for the gibbons. A clumped resource such as leaves can be readily exploited by a group, while widely scattered fruits are more easily sought out by individuals. Moreover, when feeding on clumped fruit resources, such as figs, siamang are dominant over the smaller lar gibbons and are able to exclude them

Brachiation   **Bipedalism**

Climbing   **Feeding postures**

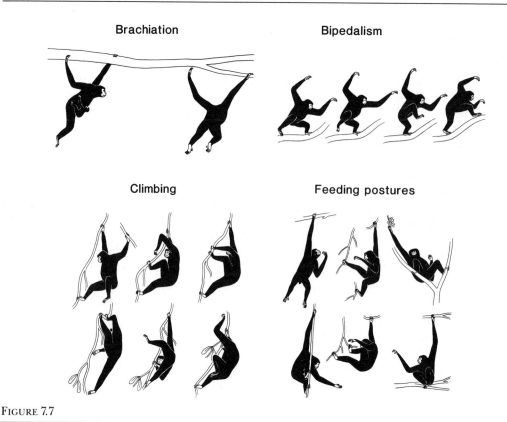

FIGURE 7.7

Locomotor behavior and feeding postures of the Malayan siamang (*Hylobates syndactylus*).

from food trees until the siamang have had their fill.

The one possible exception to the monogamous social structure of gibbons is the black-crested gibbon (*H. concolor concolor*) from China, which has been reported to live in larger polygynous groups. The ecological reasons for this difference are under study (Haimoff *et al.*, 1986).

All gibbons are fiercely territorial and defend their core areas with daily calling bouts and occasional intergroup conflict. Social interactions within a gibbon group are limited and consist primarily of occasional grooming bouts. Gibbons have single births every two or three years. In siamang, the

male carries the offspring during its second year of life. In other species, male investment is not so extensive. Young gibbons spend up to ten years in their family group before leaving, usually after harrassment by their parents.

Pongids

There are four living species of pongids, or great apes (Table 7.2). The orangutan is the only great ape from Asia. The gorilla, the common chimpanzee, and the bonobo or pygmy chimpanzee from Africa are a closely related group of apes that some authors regard as size variants of a single type. Great

Table 7.2
Infraorder Catarrhini
Family PONGIDAE

Common Name	Species	Intermembral Index	Body Weight (g)	
Orangutan	*Pongo pygmaeus*	139	M	81,000
			F	37,000
Common chimpanzee	*Pan troglodytes schweinfurthii*	103	M	43,000
			F	33,200
	P. troglodytes troglodytes	106	M	60,000
			F	47,400
Pygmy chimpanzee, Bonobo	*P. paniscus*	102	M	45,000
			F	33,200
Gorilla	*Gorilla gorilla gorilla*	116	M	169,500
			F	71,500
	G. gorilla beringei	116	M	159,200
			F	97,700
	G. gorilla graueri	—	M	175,200
			F	80,000

apes share many skeletal features with the lesser apes, such as relatively short trunks, the absence of a tail, a broad chest, long arms, and long hands and feet, but they are distinguished by their large size and many more detailed anatomical characteristics, such as more robust canine teeth, broader premolars, a very broad ilium, and a robust fibula (see Martin, 1986).

The great apes are primarily forest primates. All great apes are herbivorous and eat varying proportions of fruit and leaves. These large primates are less committed to a suspensory life than are the lesser apes, and all are to varying degrees terrestrial in their habits. They are, however, more suspensory than are living cercopithecoid monkeys, and all are characterized by some use of quadrumanous climbing. Even large male gorillas climb trees for feeding on particular foods in some seasons of the year.

All great apes build nests for sleeping and resting, but they share few unifying features in their social behavior. Each species seems to show a different pattern of social organization, sexual dimorphism, and individual grouping tendencies. In keeping with their large size, all have a relatively long life span and a long ontogeny. All have single births and gestation periods similar to those of humans.

Orangutans

The shaggy red orangutan (*Pongo pygmaeus*) is the largest living Asian ape (Fig. 7.8). There are two living subspecies, one on Borneo and one on Sumatra, but in prehistoric times orangutans had a much larger range, which included Java and parts of southern China. Orangutans are extremely sexually dimorphic in size, with females weighing approximately 60 kg and males roughly twice that. They are quite distinct morphologically from the African apes. Dentally, they are characterized by cheek teeth with thick enamel, low, flat cusps, and

FIGURE 7.8

The orangutan (*Pongo pygmaeus*).

crenulated occlusal surfaces (Fig. 7.4; see also Fig. 13.21). They have large upper central incisors and small peglike upper laterals. The canines are large and sexually dimorphic. Orangutan crania are characterized by a relatively high, rounded braincase, poorly developed brow ridges, a deep face with small orbits set close together, and a uniquely prognathic snout with a large convex premaxilla. The mandible is deep and has a high ascending ramus (Fig. 7.4).

The limbs of orangutans show extreme specializations for suspensory behavior. They have very long forelimbs and long, hooklike hands with long curved fingers and a short pollex. Their extremely mobile hindlimbs are relatively short, and they have handlike feet with long curved digits and a reduced hallux.

Orangutans seem to prefer upland rather than lowland forest areas. Females and immature individuals are almost totally arboreal, whereas adult (especially old) males on Borneo frequently descend to the ground to travel. In the trees, orangutans move almost exclusively by slow quadrumanous climbing in which they use their hands and feet interchangeably as they move within tree crowns and transfer themselves from tree to tree. On the ground they move quadrupedally with their hands held in a fist (Tuttle, 1969b). When feeding they use both seated and suspensory postures. They frequently use their strong arms to bend or break branches to bring food to their mouth rather than change positions. Orangutans eat primarily fruits (many of which contain hard seeds that they crush with their flat molars), and they also consume considerable amounts of new leaves, shoots, and bark. There are differences in the diets of male and female orangutans on Borneo (Galdikas and Teleki, 1981). Male orangutans eat more termites, presumably in conjunction with their terrestrial forays.

Adult orangutans are usually solitary and have a noyau social organization similar to that found among nocturnal prosimians. The only consistent social groups among orangutans consist of females with their immature offspring. Individual females (with young) live in relatively small home ranges of approximately 70 ha and have day ranges of only 300 m. Mature adult males occupy much larger home ranges that overlap the ranges of several adult females. They also move much farther each day, partly in search of more food to supply their greater bulk, and also, presumably, to monitor the whereabouts of their female consorts and male competitors.

Young adult males that have not acquired their own territory seem to have a very different ranging behavior that is probably more appropriately described as a reproductive strategy (Galdikas, 1985; Mitani, 1985a,b). They forage with adult females for weeks or months at a time and forceably mate with the usually uncooperative female. Interactions between adult male orangutans are usually aggressive, occasionally involving fierce battles but more often only vocal exchanges. Male–female sexual encounters vary dramatically from violent interactions

between young males and adult females, which can best be described as rape, to occasional, long erotic treetop trysts, which usually occur between older adult males and females.

The care and upbringing of young orangutans is totally the responsibility of the females. Female orangutans become sexually mature at about seven years. Sexual maturation in male orangutans is a more variable and interesting phenomenon. They become sexually competent sub-adults somewhere between the ages of eight and fifteen years, but they may not become fully mature with cheek pouches for many years after that. Final maturation seems to occur very rapidly and to be influenced by social factors rather than by age alone.

Gorillas

The gorilla (*Gorilla gorilla*) is the largest living primate and shares with the chimpanzee the distinction of being our closest primate relative (Fig. 7.9). The single species of gorilla is normally divided into three geographically isolated subspecies, the western lowland gorilla, the eastern lowland gorilla, and the mountain gorilla.

Gorillas have extreme sexual size dimorphism; females weigh on average 90 kg and males over 200 kg. This dimorphism is also evident in many aspects of their skeletal anatomy, where it manifests itself in the greater general robustness of the males. The molar teeth of gorillas have a greater development of crests than those of any other hominoid, a feature associated with their more folivorous diet. They have large, tusk-like canines and relatively small incisors. Gorillas have relatively long snouts, pronounced brow ridges, and in males well-developed sagittal and nuchal crests.

Gorillas have relatively long forelimbs (Fig. 7.10). Their hands are very broad and have a

FIGURE 7.9

The mountain gorilla (*Gorilla gorilla*) from the Virunga volcanoes of Rwanda and Zaire.

large pollex and (like all African apes) dermal ridges on the dorsal surface of their digits. Their trunk is relatively short and broad and has a wide thorax and a broad basinlike pelvis. Their hindlimbs are relatively short and their feet are broad. In mountain gorillas the hallux is somewhat adducted and connected to the other digits by webbing, giving them a very humanlike footprint.

Gorillas have a limited distribution in the tropical forests of sub-Saharan Africa. They show a definite preference for secondary, herbaceous forests and are among the most terrestrial of all primates. Adult mountain gorillas rarely climb trees; more frequently they destroy them. Like male orangutans, male gorillas use their strength to great advantage when foraging. The lowland forms are more frequently arboreal, especially females and youngsters, but they normally feed, rest, and build their sleeping nests on the ground, where they move by quadrupedal walking and running. Like chimpanzees, gorillas have an unusual hand posture for quadrupedal standing and moving, called knuckle-walking (Fig. 7.10). Rather than support their forelimb on the palm of their hand (like most primates) or on the palmar surface of their fingers (like many baboons), they support it on the dorsal surface of the third and fourth digits of their curled hands. In the trees they are relatively good climbers, but they do not use suspensory feeding postures.

Gorillas have the most herbaceous diet of any living ape. They eat mainly leaves and pith in great quantities. Lowland gorillas seem to eat more fruit than the mountain subspecies. Reports of gorillas eating meat of any sort are extremely rare.

Gorillas live in groups of about a dozen animals which usually contain one mature (silverback) adult male, one or more younger

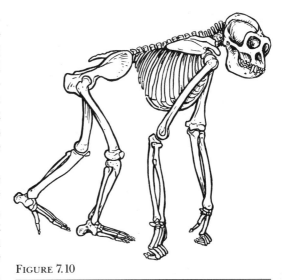

FIGURE 7.10

The skeleton of a gorilla (*Gorilla gorilla*).

males, and several adult females with offspring. Mountain gorillas have home ranges of approximately 160 ha and travel as a coherent group through day ranges of about 500 m. In their feeding activities, gorillas are extremely destructive of the vegetation, and their ranging patterns seem to involve harvesting and destroying favorite patches of rapidly regenerating vegetation on a systematic and regular basis.

Although the composition of a gorilla group is similar to that of many primate groups, with a single adult male and several adult females, its formation and maintenance seem to be based on different demographic and social relationships. Most primate groups (among prosimians and Old World monkeys) seem to be organized around groups of related females who grow up and remain in their natal group while males transfer from one troop to another. In gorilla society, females migrate between groups (Harcourt, 1978). As a result, a normal gorilla group is composed of un-related females that generally do not inter-act with each other very much. Most in-teractions are between the adult male and individual females. Gorillas do, however, resemble other single-male groups of pri-mates in their intense competition between males for control of a troop, and takeovers are often accompanied by infanticide.

Common Chimpanzee

Pan troglodytes, the common chimpanzee (Fig. 7.11), is the more widespread of the two chimpanzee species. The three subspecies extend in a broad belt across much of central Africa, from Senegal in the west to Tanzania in the east, and differ considerably in body weight. All have moderate levels of sexual dimorphism.

Compared with gorillas, chimpanzees have broader incisors and cheek teeth with broader basins and lower, more rounded cusps. Their skulls are very similar in overall shape to those of gorillas, but they have shallower faces and mandibles and do not show such extensive development of sagittal and nuchal crests. The limbs of chimpanzees are more similar in length than those of gorillas and are also less robust. They have narrower hands and feet with more slender, curved digits.

Chimpanzees occupy a variety of habitats, from rain forests to dry savannah areas with very few trees. Most of our knowledge of chimpanzee behavior comes from woodland, open forested environments rather than jun-gles. In these habitats chimpanzees generally feed in trees (57–88 percent of each day, depending on the season) and travel on the ground between feeding sites using a quad-rupedal, knuckle-walking locomotion. In an arboreal setting they use both quadrupedal and suspensory locomotion to move about within a feeding source, and they also use a variety of seated and suspensory feeding postures. Chimpanzees are more suspensory than gorillas but considerably less suspen-sory than either gibbons or orangutans.

Chimpanzees eat primarily fruit (60 per-cent) and leaves (21 percent), but they are also opportunistic faunivores that catch and eat social insects and various smaller mam-mals, including other primates (colobus monkeys and baboons). Predation seems to be almost exclusively an activity of adult males, who subsequently share the food with other members of the group.

Social groups of chimpanzees are more fluid than those of many higher primates and have a fission–fusion system like that of platyrrhines such as *Ateles*. Adults of both sexes spend large portions of their time foraging alone, but they join from time to time with other individuals in temporary associations or parties. Females are more solitary than males. They spend more time alone (65 percent vs. 14–29 percent for males), have shorter day ranges, and have smaller individual home ranges. Female–female social interactions are relatively rare except among close relatives such as mothers and daughters. Adult male chimpanzees are more gregarious. They often groom each other, more frequently join in feeding par-ties, and also seem to form patrol groups that monitor the boundaries of the commu-nity home range. Chimpanzee mating oc-curs in a number of contexts, including promiscuous mating within the group, pos-sessive behavior on the part of an individual male within the group toward the fertile female, and consortship in which an individ-ual male and a receptive female travel to-gether for several days at a time, apart from other individuals.

In view of the largely solitary nature of adult chimpanzees, it has been difficult to determine the extent to which groups of

FIGURE 7.11

The common chimpanzee (*Pan troglodytes*).

Figure 7.12

The bonobo, or pygmy chimpanzee (*Pan paniscus*), from central Zaire.

individuals form larger social groups or communities with any permanence. Temporary foraging and feeding parties do seem regularly to contain the same individuals—from a relatively closed social network of fifteen to eighty chimpanzees sharing a composite range of 800 ha or more. Interactions between adults (except estrous females) of neighboring communities are usually aggressive. As with gorillas, it is primarily young females that migrate between communities; males tend to remain in their natal groups. As a result, the adult males of a community are probably all closely related, a fact that would account for their gregariousness and cooperation. Any long-term community structure in chimpanzees is more likely to be based on continuity of the male members than of the more mobile female members.

Bonobos

Pan paniscus, the pygmy chimpanzee or bonobo (Fig. 7.12), is similar in adult body weight to the smaller subspecies of the common chimpanzee (*Pan troglodytes schweinfurthii*) but has a darker face, a more gracile skull, more slender limbs, and longer hands and feet. Although there seems to be sexual dimorphism in the body weight of bonobos, there is virtually no sexual dimorphism in either the dentition (only the canines) or the limb skeleton. Bonobos have a relatively restricted distribution in central Africa south of the Zaire River, where they live in a more forested environment than do most common chimpanzees.

Like common chimpanzees, bonobos travel mainly on the ground by knuckle-walking and feed both on the ground and in trees. Their arboreal locomotion involves quadrupedal, suspensory, and bipedal activities. They also use a variety of seated, standing, and suspensory feeding postures.

Bonobos seem to be the most suspensory of the African apes (Susman, 1984).

Like common chimpanzees, bonobos eat primarily fruit, pith, and leaves, as well as occasional prey items including small ungulates, insects, snakes, and fish. There are no long-term data on the daily ranging patterns of bonobos, but studies to date suggest that a community of about fifty bonobos ranges over an area of 2,200 ha. Like common chimpanzees, bonobos are normally seen in small groups of four to five individuals, but their fission-fusion society differs in the more frequent occurrence of feeding groups containing both males and females rather than of the single-sex groups seen in common chimpanzees. Observations on grooming provide further evidence that affiliations both within and between the sexes in this species are different from those described for common chimpanzees and gorillas. In bonobos, male–female grooming is the most frequent kind, with female–female next in frequency. This contrasts with the patterns found in common chimpanzees, in which male–male grooming is most common, and in gorillas, in which male–female grooming is common but female–female grooming virtually never occurs. The greater cohesiveness of the sexes in bonobos compared with common chimpanzees seems to be related to their use of larger food patches (White and Wrangham, 1988).

Size and Evolution of the African Apes

Despite their differences in size, diet, and social organization, the three living African apes are quite similar in many aspects of cranial and skeletal anatomy and virtually identical in many biochemical assays. Many authors have suggested that most of their differences are a corollary of their differences in body size. A simplistic summary of

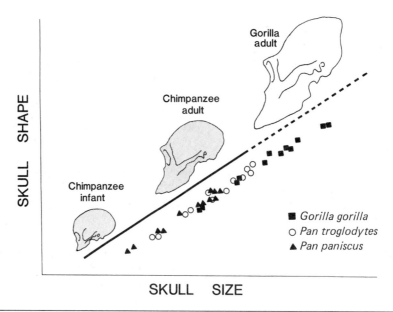

FIGURE 7.13

A schematic illustration of the hypothesis that the African apes occupy different positions on the same ontogenetic trajectory. The shape of a gorilla skull is just an extension of the chimpanzee growth curve to a larger size (courtesy of Brian Shea).

this view is that gorillas are overgrown chimpanzees and that bonobos are morphologically similar to immature chimpanzees; that is, that adult bonobos have a morphology that resembles the juvenile morphology of common chimpanzees, and that adult chimpanzees have a morphology that corresponds to that of a juvenile gorilla. The adults of each species, in this view, represent different points on the same growth curve.

Closer examination of this hypothesis has shown that the actual relationships between size and shape in the three African apes are somewhat more complex than this simple model suggests, but for many features of the skull and many limb proportions it accurately describes the species differences in shape (Fig. 7.13). In other features, growth patterns of individual species are clearly distinct.

The relationships between relative growth, absolute growth rates, and sexual dimor-phism among the African apes are a particularly intriguing and promising area of research (see, e.g., Shea, 1985). For example, despite their size difference, the temporal length of ontogeny in gorillas and chimps appears to be virtually the same. Gorillas attain their larger size by having a rate of absolute growth that is twice that of the common chimpanzee. Each species of African ape has a unique pattern of growth differences between the sexes, resulting in species-specific amounts of sexual dimorphism in the adults. Compared with common chimpanzees, gorillas are characterized by earlier female sexual maturity and thus greater differences in the timing of sexual maturity and in size dimorphism. In bonobos, however, both males and females retain more juvenile-like cranial features than common chimpanzees, and, like immature chimpanzees, they show less sexual dimorphism. These developmental and social differences

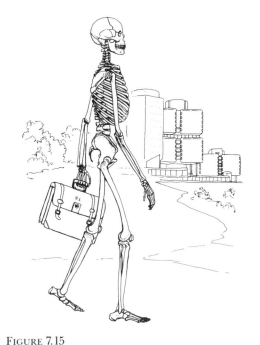

FIGURE 7.14

A member of the human race (*Homo sapiens*).

FIGURE 7.15

The skeleton of a human.

are probably related to the differing ecological adaptations of each ape.

Hominids

Like the aye-aye, *Homo sapiens*, the only living hominid, is a very odd primate species—and a very unusual creature by any standard (Figs. 7.14, 7.15, Table 7.3). We are more similar to other hominoids in our dental and skeletal anatomy than our striking external and postural features would suggest, and we also have many outstanding specializations that set us apart. We are distinguished dentally by our relatively small canines, broad premolars, and reduced or absent third molars in many populations; otherwise our teeth are similar to those of many chimpanzees. Our mandible, with its protruding chin, is very unusual. Our skull has a small, short face and a large balloonlike cranium with poorly developed crests and a

TABLE 7.3
Infraorder Catarrhini
Family HOMINIDAE

Common Name	Species	Intermembral Index	Body Weight (g)	
Human	*Homo sapiens*	72	M	68,000
			F	55,000

foramen magnum that lies well beneath the skull base. Relative to our body size, we have extremely large brains.

Like all hominoids, we have a relatively short trunk and long arms. Our hand is characterized by short slender fingers and an extremely opposable thumb. The most distinctive features of our skeleton are those of our long lower extremities associated with our upright, bipedal locomotion. Our pelvic bone is extremely short and broad, and the femur, tibia, and fibula are extremely long. Unlike all other primates, we have a hallux that is not opposable but aligned with the other toes. We have a long heel, long metatarsals, and very short pedal phalanges. The bones of our foot form two arches, one longitudinal and one transverse. These give us our characteristic footprint, in which we land on the heel, pass our weight along the lateral border of the foot to the front, then push off with the ball of the foot and ultimately the great toe.

One of our most striking features is the apparent lack of hair over most of our body, which contrasts with the noticeable concentration of hair on our heads, under our arms, and in the genital region. The development of human facial hair is an extremely variable feature that differs not only between sexes but also among different human populations. In fact, the density of hairs on the human body is not that different from that in large apes such as chimpanzees and gorillas; human hairs are just very short and often lightly pigmented. Another striking feature that distinguishes humans from other primates is the distribution of subcutaneous body fat. Like body hair, this is not only quite different between sexes of many human populations—men tend to store fat in their abdomen, women store it in their breasts and on their hips and buttocks—but

is also quite variable among major human population groups.

As primates, we humans are uniquely cosmopolitan. Only Antarctica has steadfastly resisted permanent colonization by our species. All other habitats, including tropical rain forests, woodlands, savannahs, plains, deserts, mountains, and arctic coastlines, have supported human populations for many thousands of years. We are also the most terrestrial of all primates. Only humans (and possibly gelada baboons) regularly live their entire lives without ever climbing a tree for food or sleep. Our bipedal gait is unique among mammals, but it does not seem to endow us with particularly striking speed or locomotor efficiency compared with other mammals, including nonhuman primates. There are indications, however, that it permits more endurance at slow speeds.

The "natural" human diet is probably something that exists only in television commercials and on billboards. Humans are opportunistic and probably omnivorous—we eat virtually anything. We lack the notable digestive specializations that characterize the more vegetarian primate species and show greater similarities to the faunivores.

There is also no obviously "natural" pattern of social organization among humans. There is more variability in the social organization of human societies than is found in any other primate species. Monogamous families and single–male groups of one male with several females are the most common arrangements throughout the world, but there are human populations in which polyandry (one female with several mates) or even more promiscuous multi-male and multi-female groups are the norm.

Culture complicates comparisons with nonhuman primate social organizations, since human social mores are often mandated by religion or law. Comparing human sexual dimorphism with that found in other primates does not offer any more convincing evidence of a natural social structure for our species. Our body size dimorphism allies us with polygynous mammal species (Alexander *et al.*, 1979). Our lack of canine dimorphism is more similar to that found among monogamous primates, or those living in fission-fusion societies. However, humans of both sexes have small canines, whereas in many monogamous primates (except *Callicebus*) both sexes have large canines.

It is becoming increasingly obvious that human sexuality is not as dramatically different from that of other primates as had once been thought (see, e.g., Morris, 1967). All of the features of human sexual behavior that have long been held to be unique to our species seem to be found to various degrees and in various combinations among our primate relatives, from lovers' stares to female orgasms. Indeed, from a primate perspective, the most striking aspect of human reproduction is the size of newborn infants; they are extremely large compared with adult body size, despite being born at a relatively immature stage of development.

ADAPTIVE RADIATION OF HOMINOIDS

Compared with other major radiations of primates, living hominoids (except humans) show a striking lack of taxonomic, morphological, and ecological diversity but a range in body size exceeding that found among any other extant primates (Fig. 7.16). Only the gibbons could be considered "normal-size" primates.

Ecologically, all apes are diurnal and all

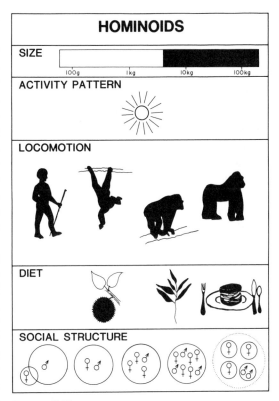

FIGURE 7.16

The adaptive radiation of living hominoids.

are more or less restricted to forested areas. Although all apes seem to utilize some suspensory locomotion, the African apes more commonly travel using arboreal and terrestrial quadrupedal (knuckle-walking) gaits. Human bipedalism is unique among primates. Living apes are all frugivorous and folivorous; there are no seed, insect, or gum specialists. Again, human dietary diversity and our regular use of agriculture and animal domestication are unique.

The most diverse aspect of the behavior of hominoids is their social organization. It is different for every genus, between species of chimpanzees, and among populations of humans. Certainly this diversity belies most attempts to identify an ancestral social system for humans by extrapolation from our nearest relatives (Wrangham, 1987; but see Ghiglieri, 1987). This diversity in hominoid social interaction is difficult to account for, but several factors probably contribute. One is the large size. Since hominoids are probably less subject to predation than other primates, the antipredator advantages of group living are lessened. Associated with our large size is a considerable longevity compared with that of most primates. Our extended life spans permit a considerable versatility in reproductive strategies which is not available to animals with shorter life spans. Finally, the larger brain size and increased intelligence of apes probably permits a greater flexibility of social interactions based on memory and unique interindividual relationships.

PHYLETIC RELATIONSHIPS OF HOMINOIDS

The proper systematic grouping of hominoids is a difficult and very subjective prob-

FIGURE 7.17

The phyletic relationships among hominoids.

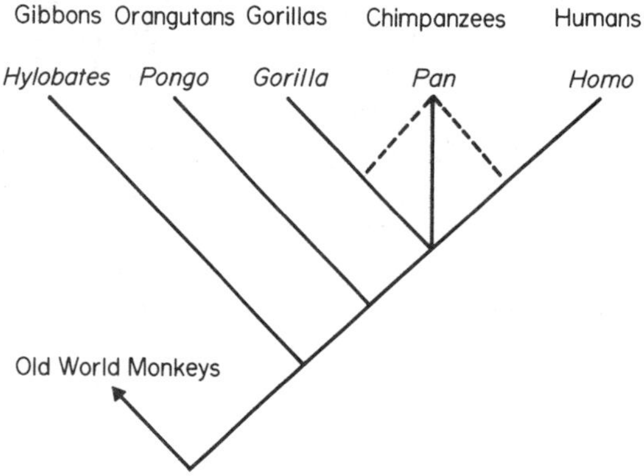

lem. As with the classification of prosimians and anthropoids, the issue is one of deciding between a gradistic and a phylogenetic classification (Fig. 1.3). Morphologically and behaviorally, there seem to be three groups of hominoids: hylobatids (the lesser apes), pongids (the great apes), and hominids (humans). Gibbons are undoubtedly the hominoids closest to cercopithecoid monkeys, and they seem to retain many primitive catarrhine features in their skeleton and visceral anatomy. For the great apes, I have adopted the gradistic approach in this book—grouping the great apes in Pongidae and separating out our own species (and our fossil relations) in Hominidae. But as noted in Chapter 1, there is considerable biomolecular evidence, as well as evidence from overall gross anatomy, to indicate that humans share a more recent heritage with chimpanzees and gorillas than with orangutans (e.g., Goodman, 1963). The unique morphological features distinguishing hominids are relatively recent specializations, whereas the similarities linking the great apes with each other are older, ancestral features that have been lost in the human lineage (Fig. 7.17).

The more difficult question concerns the phyletic relationships among humans and the African apes. Several types of biomolecular data indicate that humans and chimpanzees are more closely related to one another than either is to the gorilla, a relationship that seems in conflict with the fact that chimpanzees and gorillas share a number of unique dental and skeletal specializations, including thin dental enamel and the many musculoskeletal features of the forelimb associated with knuckle-walking (see, e.g., Miyamoto *et al.*, 1987; Andrews and Martin, 1987). Indeed, it seems unlikely that the exact relationships among chimpanzees, gorillas, and humans can ever be clearly resolved, since there is already abundant evidence to support alternative views.

BIBLIOGRAPHY

GENERAL

Cronin, J.E. (1983). Apes, humans and molecular clocks: A reappraisal. In *New Interpretations of Ape and Human Ancestry*, ed. R.L. Ciochon and R.S. Corruccini, pp. 115–150. New York: Plenum Press.

Fleagle, J.G., and Jungers, W.L. (1982). Fifty years of higher primate phylogeny. In *A History of American Physical Anthropology (1930–1980)*, ed. F. Spencer, pp. 187–230. New York: Academic Press.

Ghiglieri, M.D. (1987). Sociobiology of the great apes and the hominid ancestor. *J. Hum. Evol.* **16**:319–357.

Goodman, M. (1963). Man's place in the phylogeny of the primates as reflected in serum proteins. In *Classification and Human Evolution*, ed. S.L. Washburn, pp. 204–234. Chicago: Aldine.

Hamburg, D.A., and McCown, E.R., eds. (1979). *Perspectives on Human Evolution*, vol. 5: *The Great Apes*. Menlo Park, Ca.: Benjamin-Cummings.

Preuschoft, H., Chivers, D.J., Brockelman, W.Y., and Creel, N., eds. (1984). *The Lesser Apes—Evolutionary and Behavioral Biology*. Edinburgh: Edinburgh University Press.

Reynolds, V. (1967). *The Apes*. New York: E.P. Dutton.

Schultz, A.H. (1968). The recent hominoid primates. In *Perspectives on Human Evolution*, ed. S.L. Washburn and P.C. Jay, pp. 122–195. New York: Holt, Rinehart and Winston.

Simpson, G.G. (1966). The biological nature of man. *Science* **152**(3721):472–478.

Tuttle, R.H. (1987). *Apes of the World*. Park Ridge, N.J.: Noyes.

Wrangham, R.W. (1979). On the evolution of ape social systems. *Social Science Information* **18**:335–368.

———. (1987). The significance of African apes for reconstructing human social evolution. In *The Evolution of Human Behavior: Primate Models*, ed. W.G. Kinzey, pp. 51–71. Albany: SUNY Press.

HYLOBATIDS

Carpenter, C.R. (1940). A field study in Siam of the behavior and social relations of the gibbon. *Comp. Psychol. Monogr.* **16**(5):1–212.

Chivers, D.J. (1974). *The Siamang in Malaya. A Field Study of a Primate in a Tropical Rain Forest*. Contributions to Primatology, vol. 4. Basel: S. Karger.

———. (1977). The lesser apes. In *Primate Conservation*, ed. HSH Prince Ranier III and G.H. Bourne, pp. 539–598. New York: Academic Press.

———. (1984). Feeding and ranging in gibbons: A summary. In *The Lesser Apes—Evolutionary and Behavioral Biology*, ed. H. Preuschoft, D.J. Chivers, W.Y. Brockelman, and N. Creel, pp. 267–281. Edinburgh: Edinburgh University Press.

Creel, N., and Preuschoft, H. (1984). Systematics of the lesser apes: A quantitative taxonomic analysis of craniometric and other variables. In *The Lesser Apes—Evolutionary and Behavioral Biology*, ed. H. Preuschoft, D.J. Chivers, W.Y. Brockelman, and N. Creel, pp. 562–613. Edinburgh: Edinburgh University Press.

Fleagle, J.G. (1976). Locomotion and posture of the Malayan siamang and implications for hominoid evolution. *Folia Primatol.* **26**:245–269.

Gittens, G.P., and Raemaekers, J.J. (1980). Siamang, lar and agile gibbons. In *Malayan Forest Primates*, ed. D.J. Chivers, pp. 63–105. New York: Plenum Press.

Haimoff, E.H., Yang, X.-J., He, S.-J., and Chen, H. (1986). Census and survey of wild black-crested gibbons (*Hylobates concolor concolor*) in Yunnan Province, People's Republic of China. *Folia Primatol.* **46**:205–214.

Leighton, D.R. (1986). Gibbons: Territoriality and monogamy. In *Primate Societies*, ed. B.B. Smuts, D.L. Cheney, R.M. Seyfarth, R.W. Wrangham, and T.T. Struhsaker, pp. 135–145. Chicago: University of Chicago Press.

Marshall, J., and Sugardjito, J. (1986). Gibbon systematics. In *Comparative Primate Biology*, vol. 1: *Systematics, Evolution, and Anatomy*, ed. D.R. Swindler and J. Erwin, pp. 137–186. New York: Alan R. Liss.

Mitani, J.C. (1984). The behavioral regulation of monogamy in gibbons (*Hylobates muelleri*). *Behav. Ecol. Sociobiol.* **15**:225–229.

Raemaekers, J.J. (1979). Ecology of sympatric gibbons. *Folia Primatol.* **31**:227–245.

Raemaekers, J.J., and Raemaekers, P.M. (1985). Field playback of loud calls to gibbons (*Hylobates lar*): Territorial, sex specific and species-specific responses. *Anim. Behav.* **33**:481–493.

Rumbaugh, D.M., ed. (1972–1976). *Gibbons and Siamang*, vols. 1–4. Basel: S. Karger.

Schultz, A.H. (1973). The skeleton of the Hylobatidae and other observations on their morphology. In *Gibbon and Siamang*, vol. 2, ed. D.M. Rumbaugh, pp. 1–54. Basel: S. Karger.

Tuttle, R.H. (1972). Functional and evolutionary biology of hylobatid hands and feet. In *Gibbon and*

Siamang, vol. 1, ed. D.M. Rumbaugh, pp. 136–206. Basel: S. Karger.

PONGIDS

Orangutans

Galdikas, B.M.F. (1979). Orang-utan adaptations at Tanjung Puting Preserve: Mating and ecology. In *The Great Apes*, ed. D.A. Hamburg and E.R. McCown, pp. 195–233. Menlo Park, Ca.: Benjamin-Cummings.

———. (1985). Subadult male orangutan sociality and reproductive behavior at Tanjung Puting. *Am. J. Primatol.* **8**:87–99.

Galdikas, B.M.F., and Teleki, G. (1981). Variations in subsistence activities of female and male pongids: New perspectives on the origins of hominid labor division. *Curr. Anthropol.* **22**(3):241–256.

MacKinnon, J. (1974a). The behavior and ecology of wild orang-utans (*Pongo pygmaeus*). *Anim. Behav.* **22**:3–74.

———. (1974b). *In Search of the Red Ape*. New York: Ballantine Books.

———. (1977). A comparative ecology of Asian apes. *Primates* **18**(4):747–772.

———. (1978). *The Ape within Us*. New York: Holt, Rinehart and Winston.

Maple, T.L. (1980). *Orang-utan Behavior*. New York: Van Nostrand Reinhold.

Martin, L. (1986). Relationships among great apes and humans. In *Major Topics in Primate and Human Evolution*, ed. B. Wood, L. Martin, and P. Andrews, pp. 161–187. Cambridge: Cambridge University Press.

Mitani, J. (1985a). Mating behavior of male orangutans in the Kutai Game Reserve, Indonesia. *Anim. Behav.* **33**:392–402.

———. (1985b). Sexual selection and adult male orangutan loud calls. *Anim. Behav.* **33**:272–283.

Rijksen, H.D. (1978). *A Field Study on Sumatran Orangutans (Pongo pygmaeus abelii Lesson 1827)*. Wageninger: H. Veenman and Zonen.

Rodman, P.S. (1973). Population composition and adaptive organization among orang-utans of the Kutai Reserve. In *Comparative Ecology and Behavior of Primates*, ed. R.P. Michael and J.H. Crook, pp. 171–209. New York: Academic Press.

———. (1977). Feeding behavior of orang-utans of the Kutai Nature Reserve, East Kalimantan. In *Primate Ecology: Studies of Feeding and Ranging in Lemurs,*

Monkeys and Apes, ed. T.H. Clutton-Brock, pp. 383–413. New York: Academic Press.

———. (1979). Individual activity patterns and the solitary nature of orangutans. In *The Great Apes*, ed. D.A. Hamburg and E.R. McCown, pp. 235–256. Menlo Park, Ca.: Benjamin-Cummings.

———. (1984). Foraging and social systems of orangutans and chimpanzees. In *Adaptations for Foraging in Nonhuman Primates*, ed. P.S. Rodman and J.G.H. Cant, pp. 134–160. New York: Columbia University Press.

Rodman, P.S., and Mitani, J. (1986). Orangutans: Sexual dimorphism in a solitary species. In *Primate Societies*, ed. B.B. Smuts, D.L. Cheney, R.M. Seyfarth, R.W. Wrangham, and T.T. Struhsaker, pp. 146–154. Chicago: University of Chicago Press.

Schultz, A.H. (1941). Growth and development of the orang-utan. In *Contributions to Embryology*. Carnegie Institute Washington Pub. no. 525, vol. 29, pp. 57–110.

Schwartz, J.H. (1988). *Aspects of the Biology of the Orang-utan*. Oxford: Oxford University Press.

Sugardjito, J. (1982). Locomotor behavior of the Sumatran orang-utan (*Pongo pygmaeus* Abelii) at Ketambe, Gunung Leuser National Park. *Malay Nat. J.* **35**:57–64.

Sugardjito, J., te Boekhorst, I.J.A., and van Hooff, J.A.R.A.M. (1987). Ecological constraints on the grouping of wild orang-utans (*Pongo pygmaeus*) in the Gunung Leuser National Park, Sumatra, Indonesia. *Int. J. Primatol.* **8**:17–42.

Tuttle, R.H. (1969a). Knuckle-walking and the problem of human origins. *Science* **166**:953–961.

———. (1969b). Quantitative and functional studies on the hands of the Anthropoidea—I. The Hominoidea. *J. Morphol.* **128**(3):309–364.

Gorillas

Casimer, M.J. (1975). Feeding ecology and nutrition of an eastern gorilla group in the Mt. Kuhuzi region (Republic de Zaire). *Folia Primatol.* **24**:1–136.

Dixon, F. (1981). *The Natural History of the Gorilla*. New York: Columbia University Press.

Fossey, D. (1983). *Gorillas in the Mist*. Boston: Houghton Mifflin.

Groves, C.P. (1970). *The World of Animals: Gorillas*. New York: Arco.

Harcourt, A.H. (1978). Strategies of emigration and transfer by primates, with particular reference to gorillas. *Z. Tierpsychol.* **48**:401–420.

———. (1979). The social relations and group structure

of wild mountain gorillas. In *The Great Apes*, ed. D.A. Hamburg and E.R. McCown, pp. 187–192. Menlo Park, Ca.: Benjamin-Cummings.

Maple, T.L., and Hoff, M.P. (1982). *Gorilla Behavior*. New York: Van Nostrand Reinhold.

Merfield, F.G., and Miller, H. (1956). *Gorilla Hunter: The African Adventures of a Hunter Extraordinary*. New York: Farrar, Straus and Cadahy.

Reynolds, V. (1979). Some behavioral comparisons between chimpanzees and gorillas in the wild. In *Primate Ecology*, ed., R.W. Sussman, pp. 323–349. New York: Wiley.

Schaller, G.B. (1963). *The Mountain Gorilla*. Chicago: University of Chicago Press.

Stewart, K.J., and Harcourt, A.H. (1986). Gorillas: Variation in female relationships. In *Primate Societies*, ed. B.B. Smuts, D.L. Cheney, R.M. Seyfarth, R.W. Wrangham, and T.T. Struhsaker, pp. 155–164. Chicago: University of Chicago Press.

Tutin, C.E.G., and Fernandez, M. (1985). Foods consumed by sympatric populations of *Gorilla g. gorilla* and *Pan t. troglodytes* in Gabon: Some preliminary data. *Int. J. Primatol.* **6**:27–43.

Tuttle, R.H., and Watts, D.P. (1985). The positional behavior and adaptive complexes of *Pan gorilla*. In *Primate Morphophysiology, Locomotor Analyses and Human Bipedalism*, ed. S. Kondo, pp. 261–288. Tokyo: University of Tokyo Press.

Vedder, Amy L. (1984). Movement patterns of a group of free-ranging mountain gorillas (*Gorilla gorilla beringei*) and their relation to food availability. *Am. J. Primatol.* **7**(2):73–88.

Watts, D. (1984). Composition and variability of mountain gorilla diets in the central Virungas. *Am. J. Primatol.* **7**:323–356.

———. (1985). Relations between group size and composition and feeding competition in mountain gorilla groups. *Anim. Behav.* **33**:72–85.

Common Chimpanzees

Baldwin, P.J., McGrew, W.C., and Tutin, C.E.G. (1982). Wide ranging chimpanzees at Mt. Asserik Senegal. *Int. J. Primatol.* **3**(4):367–385.

Boesch, C., and Boesch, H. (1981). Sex differences in the use of natural hammers by wild chimpanzees: A preliminary report. *J. Hum. Evol.* **10**:265–286.

Bourne, G.H., ed. (1969–1970). *The Chimpanzee*, vols. 1–2. Basel: S. Karger.

deWaal, F. (1982). *Chimpanzee Politics: Power and Sex among Apes*. New York: Harper and Row.

Ghiglieri, M.P. (1984). *The Chimpanzees of Kibale Forest*.

New York: Columbia University Press.

———. (1985). The social ecology of chimpanzees. *Sci. Amer.* **252**:36–40.

———. (1988). *East of the Mountains of the Moon*. New York: The Free Press.

Goodall, J. van L. (1965). Chimpanzees of the Gombe Stream Reserve. In *Primate Behavior*, ed. I. Devore, pp. 425–473. New York: Holt, Rinehart and Winston.

———. (1968). The behavior of free-living chimpanzees of the Gombe Stream Reserve. *Anim. Behav. Monographs* **1**:161–311.

———. (1971). *In the Shadow of Man*. New York: Dell.

———. (1983). Population dynamics during a fifteen year period in one community of free-living chimpanzees in the Gombe National Park, Tanzania. *Z. Tierpsychol.* **64**:1–60.

———. (1986). *Chimpanzees of Gombe*. Cambridge, Mass.: Harvard University Press.

McBeath, N.M., and McGrew, W.C. (1982). Tools used by wild chimpanzees to obtain termites at Mt. Asserik, Senegal: The influence of habitat. *J. Hum. Evol.* **11**:65–72.

McGrew, W.C. (1983). Animal foods in the diets of wild chimpanzees (*Pan troglodytes*): Why cross-cultural variation? *J. Ethol.* **1**:46–61.

McGrew, W.C., Baldwin, C.J., and Tutin, C.E.G. (1981). Chimpanzees in a hot, dry and open habitat: Mt. Asserik, Senegal, West Africa. *J. Hum. Evol.* **10**:227–244.

Nishida, T. (1979). The social structure of chimpanzees of the Mahale Mountains. In *The Great Apes*, ed. D.A. Hamburg and E.R. McCown, pp. 73–121. Menlo Park, Ca.: Benjamin-Cummings.

Nishida, T., and Hiraiwa-Hasegawa, M. (1986). Chimpanzees and bonobos: Cooperative relationships among males. In *Primate Societies*, ed. B.B. Smuts, D.L. Cheney, R.M. Seyfarth, R.W. Wrangham, and T.T. Struhsaker, pp. 165–177. Chicago: University of Chicago Press.

Reynolds, V. (1965). *Budongo: An African Forest and Its Chimpanzees*. New York: Natural History Press.

Rohles, F.H., ed. (1972). *The Chimpanzee: A Topical Bibliography*. Seattle, Wa.: Primate Information Center, Regional Primate Research Center.

Sugiyama, Y. (1984). Population dynamics of wild chimpanzees at Bossou, Guinea, between 1976 and 1983. *Primates* **25**:391–400.

Teleki, G. (1981). The omnivorous diet and eclectic feeding habits of chimpanzees in Gombe National Park, Tanzania. In *Omnivorous Primates: Gathering and Hunting in Human Evolution*, ed. R.S.O. Harding and G. Teleki, pp. 303–343. New York: Columbia University Press.

Tutin, C.E.G. (1980). Reproductive behaviour of wild chimpanzees in the Gombe National Park, Tanzania. *J. Reprod. Fert., Suppl.* **28**:43–57.

Tutin, C.E.G., and McGinnis, P.R. (1981). Chimpanzee reproduction in the wild. In *Reproductive Biology of the Great Apes*, ed. C.E. Graham, pp. 239–264. New York: Academic Press.

Wrangham, R.W. (1979). On the evolution of ape social systems. *Social Science Information* **18**(3):335–368.

Wrangham, R.W., and Smuts, B.B. (1980). Sexual differences in the behavioral ecology of chimpanzees in the Gombe National Park, Tanzania. *J. Reprod. Fert., Suppl.* **28**:13–31.

Bonobos

Badrian, A., and Badrian, N. (1980). The other chimpanzee. *Animal Kingdom* (Aug.–Sept. 1980):173–181.

Badrian, N., Badrian, A., and Susman, R.L. (1981). Preliminary observations on the feeding behavior of *Pan paniscus* in the Lomako Forest of Central Zaire. *Primates* **22**(2):173–181.

Kano, T. (1979). A pilot study on the ecology of pygmy chimpanzees, *Pan paniscus*. In *The Great Apes*, ed. D.A. Hamburg and E.R. McCown, pp. 123–136. Menlo Park, Ca.: Benjamin-Cummings.

MacKinnon, J. (1978). *The Ape within Us*. New York: Holt, Rinehart and Winston.

Susman, R.L., ed. (1984). *The Pygmy Chimpanzee: Evolutionary Biology and Behavior*. New York: Plenum Press.

Susman, R.L., Badrian, N.L., and Badrian, A.J. (1980). Locomotor behavior of *Pan paniscus* in Zaire. *Am. J. Phys. Anthropol.* **53**:69–80.

Susman, R.L., and Jungers, W.L. (1981). Comments on: Bonobos—general hominid prototype or special insular dwarfs? *Curr. Anthropol.* **22**(4):369–370.

White, F.J., and Wrangham, R.W. (1988). Feeding competition and patch size in the chimpanzee species *Pan paniscus* and *Pan troglodytes*. *Behavior* **105**(2):148–164.

Wrangham, R.W. (1986). Ecology and social relationships in two species of chimpanzee. In *Ecological Aspects of Social Evolution*, ed. D.I. Rubenstein and R.W. Wrangham, pp. 352–378. Princeton, N.J.: Princeton University Press.

Size and Evolution of African Apes

Shea, B.T. (1985). Ontogenetic allometry and scaling: A discussion based on the growth and form of the skull in African apes. In *Size and Scaling in Primate Biology*, ed. W.L. Jungers. New York: Plenum Press.

HOMINIDS

Alexander, R.D., Hoogland, J.L., Howard, R.D., Noonan, K.M., and Sherman, P.W. (1979). Sexual dimorphisms and breeding systems in pinnipeds, ungulates, primates and humans. In *Evolutionary Biology and Human Social Organization*, ed. N.A. Chagnon and W. Irons, pp. 402–435. Boston, Mass.: Duxbury Press.

Harrison, G.A., Tanner, J.M., Pilbeam, D.R., and Baker, P.T. (1988). *Human Biology*. Oxford: Oxford University Press.

Hrdy, S.B. (1981). *The Woman That Never Evolved*. Cambridge, Mass.: Harvard University Press.

Morris, D. (1967). *The Naked Ape*. New York: McGraw-Hill.

ADAPTIVE RADIATION OF HOMINOIDS

Fleagle, J.G. (1976). Locomotion and posture of the Malayan Siamang and implications for hominoid evolution. *Folia Primatol.* **26**:245–269.

Ghiglieri, M.P. (1987). Sociobiology of the great apes and the hominid ancestor. *J. Hum. Evol.* **16**:319–357.

Schultz, A.H. (1968). The recent hominoid primates. In *Perspectives on Human Evolution*, ed. S.L. Washburn and P.C. Jay, pp. 122–195. New York: Holt, Rinehart, and Winston.

Wrangham, R.W. (1987). The significance of African apes for reconstructing human social evolution. In *The Evolution of Human Behavior: Primate Models*, ed. W.G. Kinzey. Albany, N.Y.: SUNY Press.

PHYLETIC RELATIONSHIPS OF HOMINOIDS

Andrews, P., and Martin, L. (1987). Cladistic relationships of extant and fossil hominoids. *J. Hum. Evol.* **16**:101–118.

Cronin, J.E. (1983). Apes, humans and molecular clocks: A reappraisal. In *New Interpretations in Ape and Human Ancestry*, ed. R.L. Ciochon and R.S. Corruccini, pp. 115–150. New York: Plenum Press.

Fleagle, J.G., and Jungers, W.L. (1982). Fifty years of higher primate phylogeny. In *A History of American Physical Anthropology (1930–1980)*, ed. F. Spencer, pp. 187–230. New York: Academic Press.

Goodman, M. (1963). Man's place in the phylogeny of the primates as reflected in serum proteins. In *Classification and Human Evolution*, ed. S.L. Washburn, pp. 204–234. Chicago: Aldine.

Miyamoto, M.M., Slightom, J.L., and Goodman, M. (1987). Phylogenetic relations of humans and African apes from DNA sequences in the $\psi\eta$-globin region. *Science* **238**:369–373.

Sibley, C.G., and Ahlquist, J.E. (1987). DNA hybridization evidence of hominoid phylogeny: Results from an expanded data set. *J. Mol. Evol.* **26**:99–121.

Simpson, G.G. (1966). The biological nature of man. *Science* **152**(3721):472–478.

Primate Adaptations

ANATOMY AND BEHAVIOR

In the preceding chapters we have discussed the anatomical and behavioral features of the major radiations of living primates. In this chapter we examine size, diet, locomotor behavior, and other aspects of comparative ecology for consistent associations between anatomical and behavioral characteristics as well as for correlations among the different aspects of behavior and ecology. By investigating the functional relationships between morphological features such as size, tooth shape, bone shape, and behavioral habits, we can understand why primate species have evolved many of their anatomical differ-

ences. We can also use this information to reconstruct some aspects of the behavior and ecology of extinct species.

Effects of Size

Body size is a basic aspect of the adaptive strategy of any primate species. An animal's size puts considerable restrictions on its ecological options, and many of the differences between species in structure, behavior, and ecology are correlated with absolute body size. Much of this size-dependent variation in morphology, physiology, and ecology can be explained by the impact of simple mathematical considerations on basic physiological and mechanical phenomena.

As linear dimensions of any object—including an animal—increase, so too do its areal (e.g., cross-sectional) dimensions and its volume. But they do not increase at the

same rates. Area increases as a function of the square of linear dimensions (L^2) and volume increases as a function of the cube of linear dimensions (L^3) (Fig. 8.1). If an animal were to double in length, breadth, and width, for example, its cross-sectional dimensions would increase fourfold, and its volume would be eight times as great. When these simple mathematical considerations are applied to animal bodies, the consequences are great. An animal's weight is a function of its volume; the strength of any of its bones is a function of the cross-sectional area of the bone. Thus an animal whose linear dimensions double would weigh eight

231

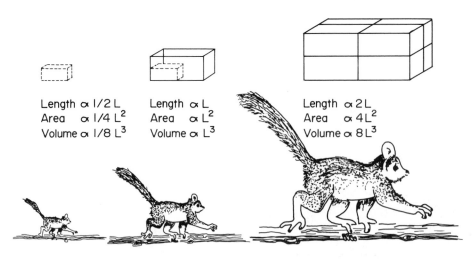

Length ∝ 1/2 L   Length ∝ L   Length ∝ 2L
Area   ∝ 1/4 L²   Area   ∝ L²   Area   ∝ 4L²
Volume ∝ 1/8 L³   Volume ∝ L³   Volume ∝ 8L³

FIGURE 8.1

Differences in linear dimension involve much greater differences in area and volume which affect many physiological and structural aspects of an animal's life.

times as much, but its structural supports would be only four times as strong. We expect, then, that animals of greatly different size will not be similarly proportioned—and this is just what we find. Figure 8.2 shows the femur of a pygmy marmoset expanded to the same length as a gorilla femur. The gorilla femur is far thicker—an adaptation to support the gorilla's much greater volume and weight. Such size-related scaling, first discussed by Galileo, is apparent throughout the animal kingdom. It explains why there are no 1-ton birds and why humans cannot jump like grasshoppers or carry weight like ants. It also explains why cinematic fantasies of incredible shrinking or growing men and women are indeed fantasies.

Similar considerations affect the scaling of other physiological functions, such as absorption in the digestive system. If the surface area of the digestive system increased in proportion to L^2 while the mass that must be fed increased in proportion to L^3, larger animals would have a relatively small digestive tract with which to process foods for much larger bodies (see Chivers and Hladik, 1980). If primate brains were all the same shape, larger species would have a relatively smaller brain surface area for any particular brain weight. Primates have evolved in different ways to accommodate these scaling difficulties. Some primates have relatively longer intestines; some have different kinds of digestive organs; some have different diets. Larger species tend to have more convoluted brains, so the ratio of surface area to brain weight remains approximately the same from species to species.

Metabolism is another physiological function that does not scale linearly with changes in body size—and thus it is another factor in explanations of size-related differences in ecology and behavior. A primate's metabolic rate or the amount of energy an individual

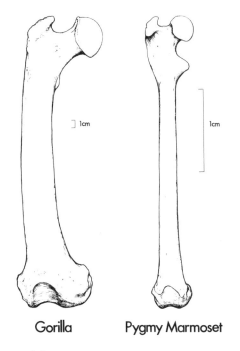

Gorilla   Pygmy Marmoset

FIGURE 8.2

The right femur of a gorilla (*Gorilla gorilla*) and of a pygmy marmoset (*Cebuella pygmaea*) drawn to the same length. The bone of the larger gorilla is relatively much thicker than the bone of the small marmoset because it must support a relatively greater body mass for its length.

requires for either basic body functions (basal metabolism) or daily activities (daily metabolic rate) scales in proportion to body mass raised to the power .75, not simply in direct proportion to body mass—or to surface area of the body, as had long been believed. Thus larger animals expend relatively less energy and consequently need less food than small animals. Put more simply, two 5-kg monkeys require more food than a single 10-kg monkey.

Primates have evolved to accommodate the constraints of scaling in two ways: by evolving different physical proportions (as in increasing brain convolutions) and by adapting life style to capitalize on the scaling consequences. We can see each of these kinds of adaptation in the evolution of primates. There is an increasingly large and sophisticated literature on size-related differences in many aspects of primate biology, including limb length, brain size, reproductive physiology, tooth size, and locomotor behavior (see Jungers, 1985). Altogether these studies are termed **allometry**, of which there are three general types (Fig. 8.3): (a) **growth allometry** is the study of shape changes associated with size changes in ontogeny; (b) **intraspecific allometry** is the study of size-related differences in adults of the same species; and (c) **interspecific allometry** examines size-related differences across a wide range of different species for broader principles of scaling.

Considerations of size are critical to our understanding of both primate adaptation and evolution, but a detailed discussion of this topic is beyond the scope of this book. In this discussion we concentrate instead on the role of size in ecological adaptation—the way primates of different size tend to have different ways of life. In considering these adaptive differences, it is often impossible to determine whether size-related differences in behavior and ecology are behavioral adaptations a species has adopted to "accommodate its size" or whether they are better viewed as gross morphological adaptations that enable a species to better exploit a particular ecological niche. Size and adaptation are so intertwined that determining which precedes the other in evolution is akin to determining whether chickens come before eggs. Furthermore, any particular body size comes with both advantages and disadvantages. The best approach is to look for

GROWTH ALLOMETRY

INTRASPECIFIC ALLOMETRY

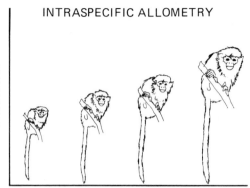

INTERSPECIFIC ALLOMETRY

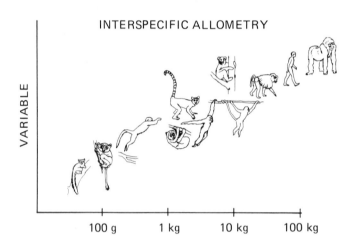

100 g   1 kg   10 kg   100 kg

FIGURE 8.3

Representative graphs of three ways of examining the association of shape changes with size changes: growth allometry examines the shape changes associated with ontogenetic size increases; intraspecific allometry examines the shape changes associated with size differences among adults of a single species; interspecific allometry examines shape changes associated with size differences across a wide sample of different species.

consistent associations between size and behavioral ecology, associations that may provide us with insight into the structure of primate communities, both living and fossil.

Size and Diet

Primate diets are closely linked with body size (Fig. 8.4). Species that eat insects tend to be relatively small, whereas those that eat leaves tend to be relatively large. Fruit eaters tend to supplement their diets with either insects or leaves, depending on their size. These patterns result from the interaction of several independent, size-related phenomena. First, primates need a balanced diet that not only meets their caloric (energy) needs but also their other nutritional re-

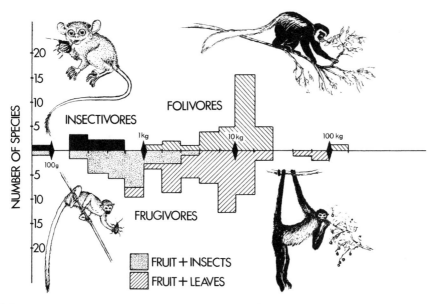

FIGURE 8.4

Primate dietary habits are correlated with body size. Insectivorous primates are relatively smaller than folivorous primates. Smaller frugivorous species tend to supplement their diet with insects, and larger frugivorous species supplement their diet with leaves (redrawn from Kay, 1984).

quirements, such as protein and a variety of trace elements and vitamins. Although fruits are high in calories, they are very low in protein content; most primates must therefore turn to other sources for their protein. The two most abundant sources of dietary protein for primates are other animals (such as insects) and folivorous materials such as leaves, shoots, and buds. Why, then, do small primates tend to eat insects and large ones folivorous material? Although these two protein strategies are in a nutritional sense complementary, the physiological and behavioral problems faced by a primate that feeds on these two dietary items are quite different.

Insects (and animal material in general) are an excellent source of nutrients, fulfilling nearly all a primate's requirements. Furthermore, insects are relatively high in calories per unit weight. This is particularly important for small animals, which have relatively higher energy requirements than large ones (the shrew must eat several times its body weight in food every day). Insects are so good a food source that the real question is not why small primates eat them but why large ones do not. The answer seems to lie in the time normally involved in catching and handling insects. No primates have evolved the specialized abilities of ant-eaters to prey on large colonies of social insects; rather, they depend largely on locating and catching isolated individuals. It has been suggested, and it seems quite reasonable, that the number of insects that a primate can find and catch in a given day (or night) is likely to be relatively similar from species to species—regardless of size—assuming, of course, that they look in the appropriate places and have the right grasping abilities, eyesight, and so on. In an eight-hour active period, any two primates might be able to ingest forty insects of one type or another. For a small prosimian, this catch could supply all the energy requirements for the day; for a larger monkey, however, this much food might supply all its protein needs but not enough energy. Thus, although larger primates might supplement their fruity (high-energy) diet with insects, they cannot rely solely on insects in the way a small primate might.

Unlike insects, leaves are neither cryptic nor hard to catch, but they do pose their own problems. Although relatively high in protein (particularly young leaves, buds, and shoots), leaves also contain large amounts of less palatable components such as cellulose or even toxins (a strategy plants have evolved to prevent the loss of their own energy factories). Compared with insects or fruits, leaves are generally low in energy yield for their weight. Large body size helps a primate overcome some of these problems inherent in a leafy diet. First, large animals need less energy per kilogram of mass than do small animals. Thus they can more easily afford to have a diet that is relatively low in energy sources. Second, although primates do not have the enzymes needed to break down the cellulose in leaves, many are able to maintain colonies of microorganisms in part of their digestive tract to perform this task for them. This kind of digestion takes time, but the time food travels through an animal's digestive tract is roughly proportional to the length of the gut and thus to the animal's size. For this reason, a small primate with a short gut has less opportunity to digest plant fibers than does a larger animal with a longer gut. Furthermore, these longer, slower guts with special chambers for fermenting cellulose also seem to help detoxify some of the poisons. Thus, whereas the

upper size limit of insect eaters seems to be imposed by the time required to locate and catch their prey, the lower size limit of folivores seems to be determined by metabolic and digestive parameters. In general, folivorous primates have body weights of no less than 500 g, whereas insectivores tend to weigh less than this limit. This natural physiological break at 500 g, known as **Kay's threshold**, applies throughout the order Primates.

Size and Locomotion

Like diet, locomotion shows general patterns of size-related scaling in primates. Terrestrial primates are usually larger than arboreal ones, both within taxonomic groups and for the order as a whole (Kay and Simons, 1980). Presumably this difference reflects both the limited capability of arboreal supports to sustain large animals and perhaps also some amount of selection for large size among terrestrial species as a means of deterring potential predators.

Within arboreal primates, there are size-related trends in the use of different types of locomotion. Although we lack the extensive quantitative data on primate locomotion that we have for diet, the allometry of locomotor behavior has been quantitatively assessed for South American monkeys, and similar patterns seem to hold for the rest of the order (with some notable exceptions). In general, we find that leaping is more common among small primates (Fig. 8.5a), whereas suspensory behavior is more common in larger species (Fig. 8.5b). Like fruit eating, quadrupedal walking and running does not seem to show any pattern with respect to body size.

The trends we find in leaping and suspensory behavior seem to be primarily the result of simple mechanical phenomena (Fig. 8.6). Two primates, one small and one large, traveling through the forest canopy will each

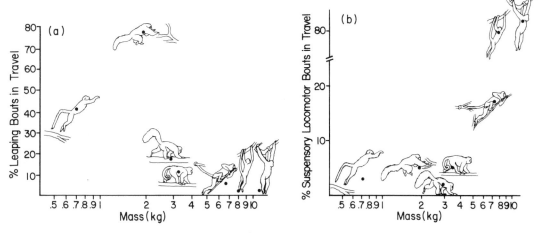

FIGURE 8.5

Primate locomotor behavior is correlated with body size. Among platyrrhine monkeys, (a) leaping is more common for smaller than for larger species and (b) suspensory behavior is more common for larger than for smaller species.

FIGURE 8.6

A small primate and a large primate traveling through the same forest are confronted with different locomotor problems because of their difference in size. The small primate encounters relatively more gaps that can only be crossed by leaping, while the larger species encounters relatively more gaps that can be crossed by suspensory behavior or bridging.

encounter gaps between trees that they somehow must cross to continue their journey. In the same forest, the smaller one will more frequently encounter gaps that it can cross only by leaping; the larger one will more frequently encounter gaps that can be crossed by bridging or by suspending itself between the terminal supports. Leaping, of course, involves the generation of high propulsive forces from the hindlimbs—and larger animals must generate greater forces to leap. Smaller animals will find more supports that can sustain their leaps than will larger animals. On the other hand, during both locomotion and feeding, larger animals will more frequently encounter supports too narrow and too weak to support their larger bodies and will more often

need to suspend themselves below multiple branches for both support and balance (Fig. 8.7). Another relevant factor is the amount of energy a tree climber must absorb when it falls from a tree to the ground. Those animals with greater weight are likely to adopt the more cautious form of locomotion.

All of these arguments support the scaling patterns seen in New World primates and suggested for the order as a whole. As with diet, there are notable exceptions, such as the small suspensory lorises or some of the larger saltatory colobines, but within taxonomic groups these broad patterns seem to hold.

Quadrupedal behavior seems to show no major size restrictions; there are both large

FIGURE 8.7

During feeding, small primates encounter more supports that can easily support their weight, while larger primates have to spread their weight over a large number of supports to feed at the same place.

and small quadrupeds. Larger quadrupeds tend to move on larger supports, however, and the largest support is the ground. The interesting exceptions to this pattern are animals that show other special adaptations, such as marmosets, which have claws for clinging to large tree trunks, or very suspensory animals, such as spider monkeys, which spread their weight over several relatively small branches.

Size and Reproduction

In addition to diet and locomotion, primate reproduction seems particularly closely linked to size. Both gestation period and life span (and duration of infancy, childhood, and so forth) increase with body size; larger primates generally take longer to grow up and live longer. In contrast, litter weight, or the size of the newborn, scales negatively with body size; that is, smaller primates have relatively larger babies. These factors may

well be linked to the scaling of metabolism. A relatively higher metabolism may, for example, enable small primates to produce relatively larger babies in a shorter time. Regardless of the underlying causal relationships, these patterns clearly affect the ecology of primate species (see Harvey *et al.*, 1986).

Size and Ecology

Various other aspects of primate ecology show likely, but less clear cut, relationships with size. Within any habitat, smaller primates are certainly more susceptible to predation than are larger species (Terborgh, 1986). In some cases, size-related features of ecology may be just alternate expressions of the factors discussed above. For example, home range size for primate species increases with body size (Fig. 8.8), presumably reflecting the need for larger animals to cover a wider area to support themselves. It

FIGURE 8.8

The home range occupied by a group of living primates is linearly correlated with the weight of the species. Folivorous species (solid symbols) have relatively smaller home ranges than do

frugivorous or omnivorous species (open symbols). Triangles represent nocturnal species; squares, diurnal terrestrial; circles, diurnal arboreal.

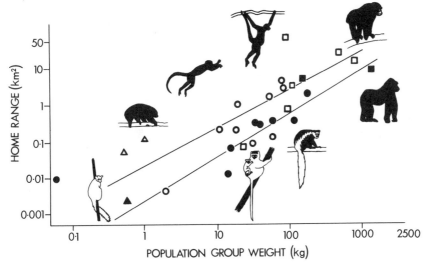

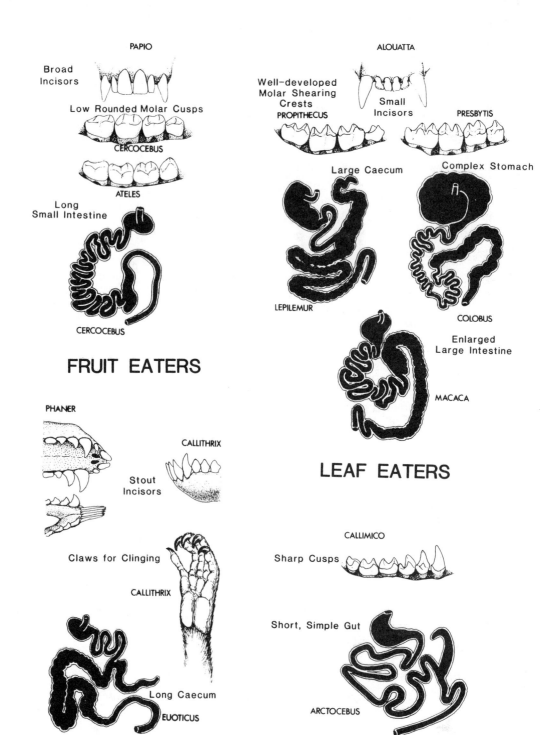

FRUIT EATERS

PAPIO

Broad Incisors

Low Rounded Molar Cusps

CERCOCEBUS

ATELES

Long Small Intestine

CERCOCEBUS

LEAF EATERS

ALOUATTA

Well-developed Molar Shearing Crests

Small Incisors

PROPITHECUS

PRESBYTIS

Large Caecum

Complex Stomach

LEPILEMUR

COLOBUS

Enlarged Large Intestine

MACACA

GUM EATERS

PHANER

CALLITHRIX

Stout Incisors

Claws for Clinging

CALLITHRIX

Long Caecum

EUOTICUS

INSECT EATERS

CALLIMICO

Sharp Cusps

Short, Simple Gut

ARCTOCEBUS

has also been shown that primate group sizes increase with body size—larger species live in larger groups—but this relationship is more suspect and difficult to explain. In part it seems to reflect aggregation for predator protection, since many of the species which aggregate into large groups are terrestrial species and often actually forage in smaller groups.

Adaptations to Diet

Diet is generally recognized as the single most important parameter underlying the behavioral and ecological differences among living primates, and primate diets have been more thoroughly documented than any other aspect of behavior. Food provides the energy that primates need for reproduction and seems to be the main objective of most of their daily activities. The use of hands to obtain and prepare food is a distinctive feature unifying the feeding habits of all primates, but, as the previous chapters emphasize, primate species show a wide range of behavioral and morphological adaptations for obtaining and processing different types of food (Fig. 8.9).

Dental Adaptations

The best-documented morphological adaptations to diet are those found in primate teeth, the organs primarily responsible for initial processing of food once it has been located. Fortunately, since teeth are also the parts most commonly preserved in the fossil record, they also provide us with an opportunity for reconstructing the diets of extinct species.

The anterior part of the tooth row, the incisors and canines, is related to ingestion, and it also serves a wide range of nondietary functions such as grooming and fighting. The role of canines and incisors in procuring or ingesting food is often not as food-specific as that of other parts of the tooth row; strong procumbent incisors for removing the bark from trees may be used by primates that subsequently eat the bark itself, or insects in the underlying wood, or exudates that flow from the hole. Nevertheless, there are some general patterns linking incisor form with diet. Relative to the size of their molars, folivores tend to have smaller incisors than do frugivores, because leaves require less incisive preparation. Primates that feed extensively on exudates frequently have large procumbent incisors for digging

FIGURE 8.9 (facing page)

Morphological adaptations to diet among living primates. Fruit eaters tend to have relatively large incisors for ingesting fruits, simple molar teeth with low cusps for crushing and pulping soft fruits, and relatively simple digestive tracts without any elaboration of either the stomach or the large intestine. Leaf eaters have relatively small incisors, molar teeth with well-developed shearing crests, and an enlargement of part of the digestive tract for the housing of bacteria for the breakdown of cellulose. Gum (exudate) eaters usually have specialized incisor teeth for digging holes in bark and scraping exudates out of the holes, and claws or clawlike nails for clinging to the vertical trunks of trees. Many also have an enlarged caecum, suggesting that they may use bacteria in the gut to breakdown the structural carbohydrates in gums or resins. Insect eaters are characterized by molar and premolar teeth with sharp cusps and well-developed shearing crests and a digestive tract with a simple stomach and a short large intestine.

holes in the bark of trees to elicit the flow of these fluids.

The cheek teeth—the premolars and particularly the molars—break up food mechanically and prepare it for additional chemical processing further along the digestive system. Thus the particular adaptations we see in molar teeth are generally not for specific foods but for food items with particular structural properties or consistencies.

There are major functional differences among primate molar teeth in the development of shearing crests or dental blades for cutting food items into small particles. Physiological experiments have demonstrated that the digestion of both insect skeletons and leaves is enhanced by chopping these food items into small pieces and thereby increasing the surface area. Thus we find that insect eaters and folivores are characterized by molars with extensive development of these shearing crests. In folivores, this development of shearing crests is also associated with thin enamel on the tooth crown, an adaptation that creates even more shearing edges on the border between the superficial enamel and the underlying dentine once the teeth are slightly worn.

Although both insect eaters and folivores are characterized by well-developed shearing crests, other criteria distinguish these two groups. As we discussed above, one difference is size: insect eaters are usually smaller. Moreover, insect eaters generally have higher, more pointed cusps for puncture crushing in addition to the well-developed crests for mastication. In contrast, fruit eaters are characterized by molar teeth with lower, more rounded cusps, fewer crests, and broad, flat basins for crushing and pulping rather than cutting. Those primates that specialize on hard nuts or seeds also have low, rounded (often barely distinguishable) cusps that, in addition, are characterized by

extremely thick enamel for withstanding high chewing forces.

Attempts to link differences in mandible shape and skull form with dietary differences have been considerably less successful than dental studies, probably because cranial morphology serves so many diverse and often conflicting functions. The most successful attempts have been Hylander's (1979) demonstration that more folivorous Old World monkeys have deeper mandibles than less folivorous monkeys and Jolly's (1970) demonstration that, in baboons, subfossil lemurs, and presumably hominids, those species that eat tougher foods have reduced prognathism and a deeper ascending ramus on the mandible than those species with a softer diet.

Digestive Tract Adaptations

Although of little use to the paleontologist, the soft anatomy of the primate digestive system shows dietary adaptations as distinctive as those seen in the dentition (Fig. 8.9). Whereas dentition shows adaptations to the size and mechanical characteristics of particular foods, the remainder of the digestive system shows adaptations to the chemical or nutritive properties of dietary items. Leaves and gums, for instance, which are very different in their consistency and require different dental adaptations, present similar problems for the remainder of the digestive system, since both are composed of long chains of structural carbohydrates that require extra processing chambers.

In general, primate digestive systems show three different patterns of dietary adaptation. Faunivorous primates (mainly insect eaters but also some omnivorous species) have a relatively short, simple digestive system with a small, simple stomach, usually a small caecum, and a very small colon relative to the size of the small intestine. In essence, the digestive system of a faunivore is devoted to absorption, the function of the small intestine. Frugivores also have relatively simple digestive systems, although large frugivores tend to have relatively large stomachs.

Folivores show the most elaborate adaptations in the visceral part of their digestive system because they must process foods containing large amounts of structural carbohydrates and also must overcome various toxins. Because primates have no natural ability to digest the cellulose contained in the cell walls of plants, these elaborations of the visceral digestive system involve forming an enlarged pouch somewhere in the digestive tract to maintain a colony of microorganisms that can digest cellulose or other structural carbohydrates. There are several possible solutions to this ranching situation, and different primate folivores seem to grow their bacteria and break down cellulose in at least three different places.

Some folivorous prosimians have an enlarged caecum, a feature also seen in rabbits and horses. Colobine monkeys have an enlarged stomach with numerous sections, similar to but much less elaborate than that of cows. Most other partly folivorous species, including indriids, apes (siamang and gorillas), New World monkeys (*Alouatta*), and some cercopithecine monkeys (*Macaca sylvanus*), accommodate the leafy portion of their diet by enlarging their colon. In addition to their role in breaking down the cellulose, it seems likely that the "fermenting" areas in the digestive systems of primate folivores help them overcome the various toxins found in many plant parts. This detoxification seems to be facilitated both directly, through actual chemical breakdown, and indirectly, by slowing down the rate at which food is processed to allow the liver more time to detoxify.

Although the role of the visceral modifications in the digestion of plant materials has been well studied in primates, there is less evidence about how and where primates break down other structural carbohydrates such as those in gums (see Nash, 1986) and the chitinous exoskeleton of invertebrates. There are anatomical indications, and a few physiological studies, suggesting that the process may be similar to that involved in cellulose digestion, since primates with specialized diets of gums (*Galago*), insects (*Tarsius*), or both (*Cebuella*) are also characterized by a large caecum.

Diet and Ranging

Adaptations to diet extend well beyond the digestive system. The many foods primates eat are found in various places, and many of the differences we see in ranging patterns seem to be adaptations for harvesting foods with unique distributions in both time and space. Primate ranging behavior is clearly correlated with diet (see Oates, 1986). Folivores tend to have relatively smaller home ranges for their size than do frugivores, reflecting the fact that foliage is more uniformly distributed and more common than fruits (see Fig. 8.8). Folivores tend to have shorter day ranges for the same reason. Because of their smaller ranges, folivores also are found in higher population densities and biomass densities than are frugivores.

In conjunction with the different distributions of primate foods, it has been shown that fruit-eating primates have relatively larger brains for their body weight than do leaf-eating primates, and several authors have suggested that the need to remember the location and fruiting cycles of trees may have been the most important factor leading to the relatively large brain size and intelligence that characterize higher primates.

Diet and Social Groups

Despite many determined efforts, broad correlations between diet and social structure have proven difficult to identify. Although it is fairly easy to explain the social organization of any single species or pair of species in terms of dietary differences and the distribution of preferred foods, broader predictive patterns are more elusive (see, e.g., Rubenstein and Wrangham, 1986). Monogamous species include frugivores like gibbons, folivores like the indriids, and insectivores like *Tarsius*. For any dietary group such as folivores, we can find species that live in monogamous families (indriids or siamang), single-male groups (*Colobus guereza, Alouatta, Gorilla*), or large, multimale groups (*Piliocolobus badius, Papio*), not to mention *Lepilemur*, which lives in a noyau arrangement.

Why does group organization seem to have so little to do with the single activity that occupies most of a primate's time? The most likely reason is that a broad categorization of diets into insects, fruits, and leaves does not reflect the patterns of food distribution that are likely to be important for determining foraging group size. For example, while mature leaves may be ubiquitous and easy to harvest, new leaves and shoots are more like fruits in their seasonal abundance and restricted availability. Thus folivores specializing on these two types of foliage show dramatic differences in both ranging patterns and social organization (Clutton-Brock, 1974). Among frugivores, some primates specialize on fruits that are found in large numbers at a given time but may be widely distributed in time and space. Such foods may be exploited best by a large, wide-ranging group. Other primates specialize on fruits that are found in small numbers but on a more regular temporal basis. Unfortunately, such details about spatial distribution of primate food items have not been as well documented as other aspects of primate diet (Oates, 1986). Moreover, it is difficult to map out possible food items in advance of a field study. Thus, while the foraging pattern of a species makes good sense in retrospect, it has proved very difficult to actually evaluate how their diet compares with the available foods.

In addition, although food is certainly the major determinant of molar shape, it is only one of the factors likely to influence the grouping behavior of primates. Other factors, such as predation and reproductive considerations (access to mates, parental care), influence social groupings (see Chapter 3). There is also probably a large amount of phylogenetic inertia involved. Closely related species with different diets often seem to show subtle modifications of a basically similar social system rather than a dramatically different arrangement. Nevertheless, increasingly sophisticated field studies are succeeding in isolating the important variables and identifying aspects of diet, such as patch size (whether food items are widely scattered or clumped), that seem to have the greatest influence on group structure.

Locomotor Adaptations

Primate locomotor adaptations are found in many parts of the body. Most of the differences we see in the anatomy of the limbs and trunk of living primates are clearly related to differences in their locomotor and postural abilities—the way they move, hang, and sit. Locomotion and posture also affect the orientation of the head on the trunk, the shape of the thorax, and the positioning of abdominal viscera.

Like many other adaptations, the modifications of the musculoskeletal system related to locomotor differences are influenced by the ancestry of the group being considered, and primates often have evolved different solutions to the same problem. Evolution by natural selection has worked with the available material. Thus quadrupedal lemurs, quadrupedal monkeys, and quadrupedal apes all show similarities related to their quadrupedal habits, but they show affinities to other lemurs, monkeys, and apes as well. For the paleontologist, this is a real advantage; it means that bones can provide information about both phylogeny and adaptation (see, e.g., Szalay, 1978)—if the two can be accurately distinguished.

Because locomotor adaptations may have different expressions in different species, our best approach is to examine the mechanical problems that different types of locomotion present. Then we can consider how living primate species have evolved musculoskeletal differences to meet these mechanical demands. We will concentrate on features of the skeleton (Figs. 8.10–8.14) that can be related to different postures and methods of progression, because these are the best-documented aspects of primate locomotor anatomy and those that are most useful in reconstructing the locomotor habits of fossils. It is important to realize that such correlations between bony morphology and locomotor behavior are constantly being tested and refined by experimental studies that permit a clearer understanding of the biomechanical and physiological mechanisms of primate locomotion (see, e.g., Fleagle, 1979).

Arboreal Quadrupeds

Arboreal quadrupedalism is the most common locomotor behavior among primates, and most groups of primates include arbo-

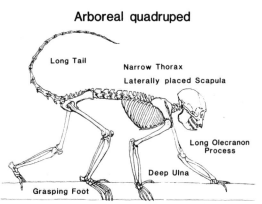

Arboreal quadruped

Long Tail

Narrow Thorax

Laterally placed Scapula

Long Olecranon Process

Deep Ulna

Grasping Foot

Short, Similar-length Forelimb and Hindlimb

FIGURE 8.10

The skeleton of a primate arboreal quadruped illustrating some of the distinctive anatomical features associated with that type of locomotion.

real quadrupeds. In many respects, arboreal quadrupeds show a generalized skeletal morphology that can easily be modified into any of the more specialized locomotor types, and it is likely that this type of locomotor behavior characterized both the earliest mammals and the earliest primates (Fig. 8.10).

Quadrupeds, by definition, use four limbs in locomotion. Experimental evidence suggests that in primates, compared with other mammals, the hindlimbs play a heavy role in support and propulsion, while the forelimbs are more important in "steering" (Kimura *et al.*, 1979). The major problem arboreal quadrupeds face in their locomotion is providing propulsion on an inherently unstable, uneven support that is usually very small compared with the size of the animal. Stability and balance are their major concerns.

The overall body proportions of arboreal quadrupeds are adapted to meeting these problems of balance and stability in several ways. These primates have forelimbs and hindlimbs that are more similar in length than are those of either leapers, which have relatively long hindlimbs, or climbers, which have relatively long forelimbs. In addition, arboreal quadrupeds' forelimbs and hindlimbs are both usually short, to bring the center of gravity closer to the arboreal support. Many arboreal quadrupeds also bring the center of gravity closer to the support by using abducted, flexed limbs when they walk. Finally, many have long tails, which aid balance. The grasping hands and feet of most arboreal quadrupedal primates provide both a firm base for propulsion and a guard against falling.

The forelimbs of arboreal quadrupeds show a number of distinctive osteological features related to their typical postures and method of progression. The shoulder joint is characterized by an elliptically shaped glenoid fossa on the scapula and a broad humeral head surrounded by relatively large tubercles for the attachment of the scapular muscles that control the position of the head of the humerus. The humeral shaft is usually moderately robust, since the forelimb plays a major role in both support and propulsion.

The elbow region of an arboreal quadruped is particularly diagnostic. On the distal end of the humerus, the medial epicondyle is large and directed medially. This process, where the major flexors of the wrist and some of the finger flexors originate, provides leverage for these muscles when the hand and wrist are in different degrees of pronation and supination. The olecranon process of the ulna is long to provide leverage for the triceps muscles when the elbow is in the flexed position characteristic of arboreal quadrupeds. Because the elbow rarely reaches full extension, the olecranon fossa of the humerus is shallow. The ulna shaft is relatively robust and often bowed and deep, suggesting that it plays a more important role in support of the body in

arboreal quadrupeds than in many other locomotor types. At the wrist, arboreal quadrupeds are characterized by a relatively broad hamate, presumably for weight bearing, and a midcarpal joint that seems to permit extensive pronation.

As a group, primates are characterized by relatively long digits and grasping hands. Among primates, however, arboreal quadrupeds usually have digits of moderate length—longer than those of terrestrial quadrupeds but shorter than those of suspensory species. They show a wide range of grasps.

The most distinctive features of the hindlimb joints of an arboreal quadruped reflect the characteristic abducted posture of that limb. The femoral neck is set at a high angle relative to the shaft, enhancing abduction at the hip. At the knee, the abduction of the hindlimb is expressed in the asymmetrical size of the femoral condyles and their articulating facets on the top of the tibia. At the ankle, the tibio-talar joint is also asymmetrical. The lateral margin of the proximal talar surface is higher than the medial margin, reflecting the normally inverted posture of the grasping foot. Arboreal quadrupeds all have a large hallux and moderately long digits.

Terrestrial Quadrupeds

There are relatively few terrestrial quadrupeds among primates compared with their abundance among other orders of mammals, and none show the striking morphological adaptations found in such runners as cheetahs or antelopes. The main group of primate terrestrial quadrupeds alive today are the larger Old World monkeys—baboons, some macaques, and the patas monkey. These species show a number of distinctive anatomical features that sepa-

Terrestrial quadruped

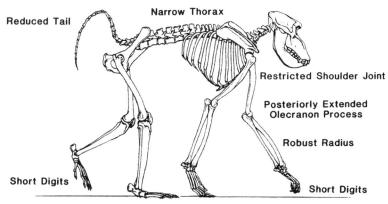

Narrow Thorax

Reduced Tail

Restricted Shoulder Joint

**Posteriorly Extended
Olecranon Process**

Robust Radius

Short Digits

Short Digits

Long, Similar–length Forelimb and Hindlimb

FIGURE 8.11

The skeleton of a primate terrestrial quadruped illustrating some of the distinctive anatomical features associated with that type of locomotion.

rate them from more arboreal species. Most of these features relate to use of more extended, adducted limb postures on a broad flat surface. Since balance is not a problem, these primates have a narrow, deep trunk and relatively long limbs, designed for long strides and speed, and their tails are often short or absent (Fig. 8.11).

The limbs of terrestrial quadrupeds seem designed for speed and simple fore-aft motions rather than for power and more complex rotational movements at the joints. At the shoulder joint, the articulating surfaces of the scapula and head of the humerus provide only a limited anterior-posterior motion, and the greater tuberosity of the humerus is high and positioned in front of the shoulder joint to facilitate rapid forward movement of the limb during running (Jolly, 1966).

Terrestrial quadrupeds have an elbow joint that reflects their more extended limb postures. Instead of being long and extending proximally, as in arboreal quadrupeds, the olecranon process extends dorsally to

the long axis of the ulna, an orientation that maximizes the leverage of the elbow-extending muscles when the elbow is nearly straight rather than flexed. A related feature is that the olecranon fossa on the posterior surface of the humerus is deep. The articulation of the ulna with the humerus is relatively narrow, whereas the head of the radius is relatively large, suggesting that the latter bone plays a more important role in transmitting weight from the elbow to the wrist in terrestrial quadrupeds than in other primates. The medial epicondyle on the humerus is short and directed posteriorly, an orientation that facilitates the use of the wrist and hand flexors when the forearm is pronated, the normal position for terrestrial species.

The carpal bones of terrestrial quadrupeds are relatively short and broad, more suitable for weight bearing and less adapted for rotational movements. Their hands are characterized by robust metacarpals and short phalanges.

The hindlimbs of terrestrial quadrupeds,

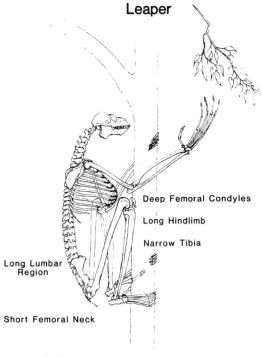

Leaper

Deep Femoral Condyles

Long Hindlimb

Narrow Tibia

Long Lumbar Region

Short Femoral Neck

FIGURE 8.12

The skeleton of a primate leaper illustrating some of the distinctive anatomical features associated with that type of locomotion.

like their forelimbs, are long. Their feet have robust tarsal elements, robust metatarsals, and short phalanges.

Leapers

Many primates are excellent leapers, and leaping adaptations have almost certainly evolved independently in many primate groups. Although there are many differences among primate leapers, there are also a number of similarities resulting from the mechanical demands of such movement (Fig. 8.12). In leaping, most of the propulsive force comes from a single rapid extension of the hindlimbs with little or no

contribution from the forelimbs. The leaper's takeoff speed, and hence the distance the animal can travel during a leap, is proportional to the distance over which the propulsive force is applied—the length of its hindlimbs. Longer legs thus enable a leaper to obtain a longer leap from the same locomotor force. Although the forelimbs are certainly used in landing after leaps, in clinging between leaps, and for various other tasks including feeding, they have a minor role in locomotion. Thus leapers are characterized by relatively long powerful legs and relatively short, slender forelimbs. Since they also gain an extra increase in propulsion by flexing and then rapidly extending their back, they also have relatively long trunks, particularly in the lumbar region, the site of most flexion and extension in the spine.

There are many skeletal adaptations for leaping to be found in the hindlimb. Because hip extension is a major source of propulsive force in leaping, primate leapers usually have a long ischium, which increases the leverage of the hamstring muscles. The direction in which the ischium is extended depends on the postural habits of the species. In primates that leap from a quadrupedal position, the ischium extends distally in line with the blade of the ilium, enhancing hip extension when the hindlimb is at a right angle to the trunk. In prosimians that normally leap from a vertical clinging posture, the ischium is usually extended posteriorly rather than distally, increasing the moment arm of the hamstrings when the limb is near full extension, a common situation for vertical clingers.

Whereas arboreal quadrupeds use abducted limbs for balancing on small supports, leapers restrict their limb excursions to simple hingelike flexion and extension movements, both for greater mechanical efficiency and to avoid twisting and damaging joints during the powerful takeoff. In this regard, leapers resemble swift quadrupedal mammals. Many features of the hindlimbs of leapers seem related to this alignment of movement and to increasing the range of flexion and extension. For example, the neck of the femur is very short and thick in leapers, and in many species the head of the femur has a cylindrical shape for simple flexion-extension movements rather than the ball-and-socket joint found at the hips of most primates. At the knee joint, the femoral condyles are very deep to permit an extensive range of flexion and extension, and they are symmetrical because of the adducted limb postures. The patellar groove has a pronounced lateral lip to prevent displacement of the patella during powerful knee extension. The tibia is usually very long and laterally compressed, reflecting the emphasis on movement in an anterior-posterior plane, and the attachments for the hamstring muscles on the tibial shaft are relatively near the proximal end so that when these muscles extend the hip they do not flex the knee as well. In many leapers, the fibula is very slender and bound to the tibia distally so that the ankle joint becomes a simple hinge joint for flexion and extension. The morphology of the tarsal region varies considerably among leapers. In many small leapers, the calcaneus and navicular are extremely long, providing a long load arm for rapid leaping. The digits of leapers, like their forelimbs, reflect postural habits rather than adaptations directly related to leaping.

Suspensory Primates

Many living primates hang below arboreal supports by various combinations of arms and legs. Because of the acrobatic nature of such behavior, the skeletons of suspensory

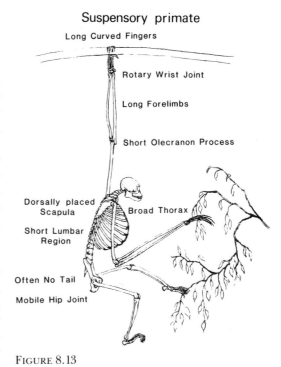

Suspensory primate

Long Curved Fingers

Rotary Wrist Joint

Long Forelimbs

Short Olecranon Process

Dorsally placed Scapula

Broad Thorax

Short Lumbar Region

Often No Tail

Mobile Hip Joint

FIGURE 8.13

The skeleton of a suspensory primate illustrating some of the distinctive anatomical features associated with that type of locomotion.

primates show features that enhance their abilities to reach supports in many directions (Fig. 8.13). In their body proportions, suspensory primates have long limbs, especially forelimbs. Their trunks are relatively short and have a broad thorax, a broad fused sternum, and a very short lumbar region to reduce bending of the trunk during hanging and reaching.

The relatively deep narrow scapula of climbers is positioned on the dorsal rather than the lateral side of the broad thorax, enhancing their reach in all directions. The shoulder joint, which faces upward to aid reaching above the head, is composed of a relatively small round glenoid fossa and a very large, globular humeral head with low tubercles—a combination that permits a wide range of movement. Because elbow extension is important but does not need to be powerful, the olecranon process on the ulna is short. The medial epicondyle of the humerus is large and medially oriented to enhance the action of the wrist flexors at all ranges of pronation and supination. Both the ulna and the radius are usually relatively long and slender, since they play no role in support.

Suspensory primates show numerous features of the wrist which seem to increase the mobility of that joint. In many species the ulna does not articulate with the carpals, and the distal and proximal rows of carpal bones form a ball-and-socket joint with increased rotational ability. Suspensory species have long fingers with curved phalanges for grasping a wide range of arboreal supports.

Like the forelimb, the hindlimb of suspensory primates is characterized by very mobile joints. Mobility at the hip joint is increased by a spherical head of the femur set on a highly angled femoral neck to permit extreme degrees of abduction. The knee joint is characterized by broad, shallow femoral condyles and a shallow patellar groove. There is very little bony relief on the talus at the ankle joint, a condition that allows movement in many directions rather than restricting it in one direction. In most species, the calcaneus has a short lever arm for the calf muscles that extend the ankle; there is, however, an additional process for the origin of the short flexor muscle of the toes to enhance grasping. The feet of suspensory primates, like their hands, have long curved phalanges for grasping branches.

Bipeds

One of the most distinctive types of primate locomotion is the bipedalism that characterizes humans. The mechanics and dynamics of human locomotion have been more thoroughly studied than those of any other type of animal movement, but many aspects of human locomotion are still poorly understood. Compared with other types of primate locomotion, bipedalism is unusual in that there is only one living species that moves in this way.

The major bony features associated with bipedalism are found in the trunk and lower extremity. The upper extremity of humans, like that of leapers, does not normally play a role in locomotion and is adapted for other functions. The major mechanical problems faced by a bipedal primate are balance, particularly from side to side, and the difficulty of supporting all of the body weight on a single pair of limbs. One of the most striking correlates of our upright posture is the dual curvature of our spine, with a dorsal convexity (kyphosis) in the thoracic region and a ventral convexity (lordosis) in the lumbar region. In most other primates the kyphosis extends the entire length of the spine; the unique human lumbar curvature moves the center of mass of the trunk forward and also brings the center of mass closer to the hip joint. In keeping with our vertical posture, the size of each vertebra increases dramatically from the cervical region to the lumbar region, for each successive vertebra must support a greater part of the body mass (Fig. 8.14).

The human pelvis is the most unusual in the entire primate order. It has a very short, broad iliac blade that serves to lower the center of gravity and to provide better balance and stability. This arrangement also places many of the large hip muscles on the side of the lower limb rather than behind it; in this position, they can act to balance the trunk over the lower limbs during walking and running. The human ischium, where the hip extensors originate, is extended posteriorly (as in vertical leapers) rather than inferiorly (as in most other primates). This position provides greater leverage for the major hip extensors to move the lower limbs behind the trunk.

The human femur is characterized by a very large head, which must support the weight of the entire body during much of the locomotor cycle. Unlike most other primates, humans are naturally knock-kneed; our femur is normally aligned obliquely, with the proximal ends much further apart than the distal ends. The alignment of the femur (called a valgus position) has the effect of placing the knees—as well as the legs, ankles, and feet—directly beneath the body rather than at the sides. As a result, successive footsteps involve less lurching from side to side, and, during those parts of the walking cycle when only one limb is on the ground, that limb is always near the midline of the body (its center of gravity). This oblique orientation of the femur is reflected in many of its bony details, such as the long oblique neck and the angle between the distal condyles and the shaft. A disadvantage of this oblique alignment of our femur is that it predisposes us to a dislocation of the patella, because the muscles extending the knee are now located lateral to the knee itself. To keep the small patella in place we have developed a very large bony lip on the lateral side of the patellar groove.

In contrast with the grasping, handlike foot of most primates, our foot has been transformed into a rather rigid lever for propulsion. The long tuberosity on the calcaneus forms the lever arm, while the stout metatarsals and the large hallux aligned with the other digits provide a firm load arm. The phalanges on our toes are extremely small because they are not used for grasping, only for pushing off. The strong ligaments on the sole of the foot bind the tarsals and metatarsals together to form two bony arches that act to some degree as springlike

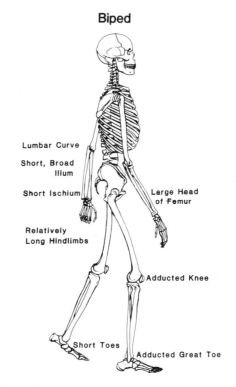

Biped

Lumbar Curve
Short, Broad Ilium
Short Ischium
Large Head of Femur
Relatively Long Hindlimbs
Adducted Knee
Short Toes
Adducted Great Toe

FIGURE 8.14

The skeleton of a bipedal primate illustrating some of the distinctive anatomical features associated with that type of locomotion.

shock absorbers. In addition, they direct the body weight through the outside of the foot during each stride, providing us with our characteristic human footprint.

Locomotor Compromises

In the previous sections we have portrayed primates that are somewhat hypothetical and idealistic—primates adapted for a single type of locomotion. But, as we discussed in earlier chapters, most primates habitually use many types of locomotion, just as they eat many types of food. Many arboreal quadrupeds often leap, some leapers are also suspensory, and so on. Nevertheless, it is reassuring that the features discussed above seem to distinguish not only primates that always leap from primates that always move quadrupedally, but also those species that leap more and are less quadrupedal from those that leap less and are more quadrupedal (see Fig. 6.18). We can therefore have confidence that these features are likely to be useful in reconstructing the habits of extinct primates known only from bones.

There are other factors to consider in trying to understand how primate skeletons are related to locomotor habits. The same parts of the body that are used in locomotion play other roles as well in the animal's life. Hands are used both in locomotion and in obtaining food, perhaps catching insects, picking leaves, or opening seed pods. The bony pelvis is an anchor for the hindlimb and also a site for the origin of many hip muscles, but it also supports the abdominal viscera and serves as the birth canal in females. These multiple functional demands, which are placed on almost every part of an animal's body, often complicate attempts to identify features that are uniquely related to one type of movement or to reconstruct the locomotor abilities of an extinct primate from bits of the skeleton. Still, many of the bony features discussed above, as well as numerous others (which can be found in more technical articles), have proved to be generally characteristic of animals with particular locomotor habits and should provide useful evidence for reconstructing fossils.

Locomotion and Ecology

Why do primates show such diverse locomotor abilities and all the morphological specializations that accompany them? One factor is certainly size. As we discussed earlier in this chapter, within the same

habitat large and small primates are likely to face very different problems in terms of balance and the availability of strong enough supports. Thus larger species are more likely to be suspensory or terrestrial.

Apart from size, the major adaptive significance of different locomotor habits seems to be the access they provide to different parts of a forest habitat. In different types of forests and at different vertical levels within a forest, the density and the arrangement of available supports for a primate to move on are often quite different. Primates that live in open areas are best adapted to terrestrial walking and running. Even within a tropical rain forest, the available supports in the understory are different from those higher in the canopy, and species that travel and feed in different levels have different methods of moving. The lowest levels of most forests are characterized by many vertical supports such as tree trunks and lianas, but there are few pathways that are continuous in a horizontal direction (see Fig. 3.3). Primates that feed and travel in the understory are often leapers that can best move between discontinuous vertical supports. Higher, in the main canopy levels, the forest is usually more continuous horizontally and suitable for other methods of progression, such as quadrupedal walking and running or suspensory behavior.

In addition to this general relationship between forest level and locomotion, there are particular types of locomotion or posture that seem related to specific habitats or food sources. Primates that live in bamboo forests (*Hapalemur, Callimico*) are almost always leapers because of the predominance of vertical supports. Primates that regularly eat gums or other tree exudates often have claws or clawlike nails so that they can cling to large tree trunks (Fig. 8.9).

Oddly enough, except for special cases such as gum eaters, there are very few general associations between the patterns of locomotion used by primates and their dietary habits. It is more frequently the case that, among sympatric species, those with the most similar diets show the greatest locomotor differences; at the same time, those with the most similar locomotion show the greatest dietary differences. This suggests that primates have often evolved locomotor differences for exploiting similar foods in different parts of their environment, and vice versa (Fleagle and Mittermeier, 1980).

It is also likely that many of the ways locomotion contributes to a species' foraging habits have not been properly studied. As noted in Chapter 3, we normally categorize foods into fruits, leaves, and insects, a classification that accords well with the mechanical and nutritional properties of dietary items. But for understanding locomotion, we should perhaps classify foods according to their distribution in the forest, the shapes of the trees in which they are found, or the size of the branches from which they can best be harvested. We also know very little about the way postural abilities may enable different species to forage in different parts of the same tree. In any case, it is clear that locomotor habits are an integral part of primate feeding strategies, and the subtle nature of this relationship deserves more study.

Anatomical Correlates of Social Organization

As we have discussed in previous chapters, primates live in many different types of social groups, and the reproductive strategies of individuals of different ages and sexes vary dramatically from species to species. There are a number of general anatomical and physiological features that seem to characterize species that live in particular types of social groups (Harvey and Harcourt, 1984). Among higher primates, the degree of canine dimorphism is closely associated with the amount of direct competition among males for mating access to females (Kay *et al.*, 1987). Those species in which there is intense male–male competition, compared with the amount of female–female competition, are characterized by greater dimorphism than those in which competition is less, or equal, among the sexes (Figs. 8.15, 8.16). Body size dimorphism shows a similar pattern: monogamous and polyandrous species show virtually no dimorphism, species living in single-male groups show very high dimorphism, and species living in multi-male groups show intermediate levels of dimorphism. Testes size shows a different distribution. In monogamous and single-male species, there is little mating competition within a group, so testes are relatively small. In multi-male and polyandrous groups there is considerable competition for mating success, and males have relatively large testes. Concomitantly, female higher primates that live in multi-male groups usually have sexual swellings that advertise their reproductive status throughout the menstrual cycle. There are, however, several very different hypotheses as to why these swellings evolved (Hrdy and Whitten, 1986).

These associations between social organization and anatomy are, of course, rough generalizations. As we learn more about the nature of interindividual interactions among primates, many of these correlations become less absolute—and more intricate. Nevertheless, in the case of dental correlations we are provided with some clues to reconstructing aspects of the social behavior of extinct primate species, the topic of the remaining chapters of this book (see, e.g., Fleagle *et al.*, 1980).

Polygynous Social System
Dimorphic Canines

Monogamous Social System
Monomorphic Canines

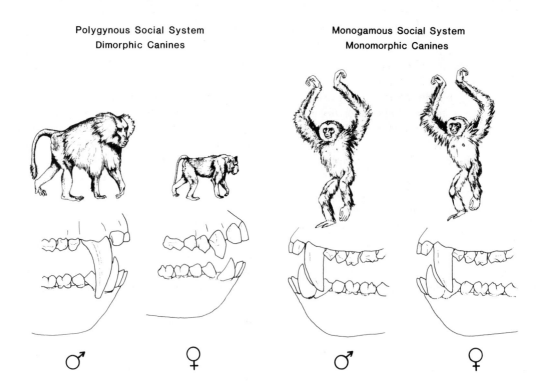

♂   ♀   ♂   ♀

FIGURE 8.15

Canine differences between monogamous gibbons (*Hylobates*) and polygynous baboons (*Papio*).

FIGURE 8.16

Morphological features associated with differences in social organization. Relative canine size (male canine length ÷ female canine length) and body size dimorphism (male weight ÷ female weight) separate monogamous and polyandrous species from primates living in other types of social groups. Relative testes size separates primate species living in multi-male groups from other types of social groups (adapted from Harvey and Harcourt, 1987).

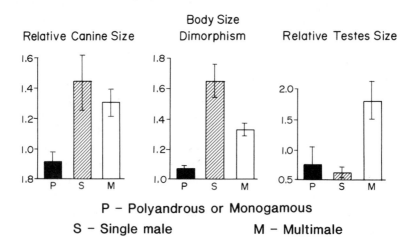

P – Polyandrous or Monogamous
S – Single male
M – Multimale

BIBLIOGRAPHY

GENERAL

Bock, W., and von Wahlert, G. (1965). Adaptation and the form-function complex. *Evolution* **19**:269–299.

Fisher, D.C. (1985). Evolutionary morphology: Beyond the analogous, the anecdotal, and the ad hoc. *Paleobiology* **11**(1):120–138.

Hildebrand, M., Bramble, D.M., Liem, K.F., and Wake, D.B., eds. (1985). *Functional Vertebrate Morphology.* Cambridge, Mass.: Harvard University Press.

Kay, R.F., and Cartmill, M. (1977). Cranial morphology and adaptations of *Palaechthon nacimienti* and other Paromomyidae (Plesiadapoidea, ?Primates), with a description of a new genus and species. *J. Hum. Evol.* **6**:19–35.

Morbeck, M.E., Preuschoft, H., and Gomberg, N. (1979). *Environment, Behavior, and Morphology: Dynamic Interactions in Primates.* New York: Gustav Fischer.

Tuttle, R.H. (1972). *The Functional and Evolutionary Biology of Primates.* Chicago: Aldine.

EFFECTS OF SIZE

Chivers, D.J., and Hladik, C.M. (1980). Morphology of the gastrointestinal tract in primates: Comparisons with other mammals in relation to diet. *J. Morphol.* **166**:337–386.

Clutton-Brock, T.H., and Harvey, P.H. (1983). The functional significance of variation in body size among mammals. In *Recent Advances in the Study of Mammalian Behavior*, ed. J.F. Eisenberg and D.G. Kleiman. Special Publication of the American Society of Mammalogists, no. 7. Shippensburg, Pa.

Harvey, P.H., Martin, R.D., and Clutton-Brock, T.H. (1986). Life histories in comparative perspective. In *Primate Societies*, ed. B.B. Smuts, D.L. Cheney, R.M. Seyfarth, R.W. Wrangham, and T.T. Struhsaker, pp. 181–196. Chicago: University of Chicago Press.

Jungers, W.L. (1985). *Size and Scaling in Primate Biology.* New York: Plenum Press.

Kay, R.F., and Simons, E.L. (1980). The ecology of Oligocene African Anthropoidea. *Int. J. Primatol.* **1**:21–38.

Leutenegger, W. (1973). Maternal-fetal weight relationships in primates. *Folia Primatol.* **20**:280–293.

———. (1979). Evolution of litter size in primates. *Am. Naturalist* **114**:525–531.

ADAPTATIONS TO DIET

Chivers, D.J., and Hladik, C.M. (1980). Morphology of the gastrointestinal tract in primates: Comparisons with other mammals in relation to diet. *J. Morphol.* **166**:337–386.

Chivers, D.J., Wood, B.A., and Bilsborough, A. (1984). *Food Acquisition and Processing in Primates.* New York: Plenum Press.

Clutton-Brock, T.H. (1974). Primate social organization and ecology. *Nature* **250**:539–542.

Clutton-Brock, T.H., and Harvey, P.H. (1977). Primate ecology and social organization. *J. Zool.* **183**:1–39.

Glander, K.E. (1982). The impact of plant secondary compounds on primate feeding behavior. *Yrbk. Phys. Anthropol.* **25**:1–18.

Hylander, W.L. (1979). The functional significance of primate mandibular form. *J. Morphol.* **160**:223–240.

Jolly, C.F. (1970). The seed-eaters: A new model of hominid differentiation based on a baboon analogy. *Man* **5**:5–28.

Kay, R.F. (1984). On the use of anatomical features to infer foraging behavior in extinct primates. In *Adaptations for Foraging in Nonhuman Primates: Contributions to an Organismal Biology of Prosimians, Monkeys and Apes*, ed. P.S. Rodman and J.G.H. Cant, pp. 21–53. New York: Columbia University Press.

Kay, R.F., and Hylander, W.L. (1978). The dental structure of mammalian folivores with special reference to primates and phalangeroidea (Marsupialia). In *The Ecology of Arboreal Folivores*, ed. G.G. Montgomery, pp. 173–191. Washington, D.C.: Smithsonian Institution Press.

Lucas, P.W., Corlett, R.T., and Luke, D.A. (1986). A new approach to postcanine tooth size applied to Plio-Pleistocene hominids. In *Primate Evolution*, ed. J.G. Else and P.C. Lee, pp. 191–201. Cambridge: Cambridge University Press.

Milton, K. (1978). The quality of diet as a possible limiting factor on the Barro Colorado Island howler monkey population. In *Recent Advances in Primatology*, vol. 1, *Behaviour*, ed. D.J. Chivers and K.A. Joysey, pp. 387–389. London: Academic Press.

Milton, K., and May, M.L. (1976). Body weight, diet, and home range size in primates. *Nature* **259**:459–462.

Nash, L.T. (1986). Dietary, behavioral, and morphological aspects of gumnivory in primates. *Yrbk. Phys. Anthropol.* **29**:113–138.

Oates, J.F. (1986). Food distribution and foraging behavior. In *Primate Societies*, ed. B.B. Smuts, D.L. Cheney, R.M. Seyfarth, R.W. Wrangham, and T.T.

Struhsaker, pp. 197–209. Chicago: University of Chicago Press.

Parra, R. (1978). Comparison of foregut and hindgut fermentation in herbivores. In *The Ecology of Arboreal Folivores*, ed. G.G. Montgomery, pp. 205–230. Washington, D.C.: Smithsonian Institution Press.

Rodman, P.S., and Cant, J.G.H., eds. (1984). *Adaptations for Foraging in Nonhuman Primates: Contributions to an Organismal Biology of Prosimians, Monkeys and Apes*. New York: Columbia University Press.

Rubenstein, D.I., and Wrangham, R.W. (1986). *Ecological Aspects of Social Evolution*. Princeton, N.J.: Princeton University Press.

Terborgh, J. (1983). *Five New World Primates: A Study in Comparative Ecology*. Princeton, N.J.: Princeton University Press.

———. (1986). The social systems of New World primates. An adaptationist view. In *Primate Ecology and Conservation*, ed. J.G. Else and P.C. Lee, pp. 199–212. Cambridge: Cambridge University Press.

Wright, P.C. (1984). Biparental care in *Aotus trivirgatus* and *Callicebus moloch*. In *Female Primates: Studies by Women Primatologists*, ed. M.F. Small, pp. 59–75. New York: Alan R. Liss.

LOCOMOTOR ADAPTATIONS

Cant, J.G.H., and Temerin, L.A. (1984). A conceptual approach to foraging adaptations of primates. In *Adaptations for Foraging in Nonhuman Primates: Contributions to an Organismal Biology of Prosimians, Monkeys and Apes*, ed. P.C. Rodman and J.G.H. Cant, pp. 304–342. New York: Columbia University Press.

Fleagle, J.G. (1977a). Locomotor behavior and muscular anatomy of sympatric Malaysian leaf monkeys (*Presbytis obscura* and *melalophos*). *Am. J. Phys. Anthropol.* **46**:297–308.

———. (1977b). Locomotor behavior and skeletal anatomy of sympatric Malaysian leaf monkeys (*Presbytis obscura* and *melalophos*). *Yrbk. Phys. Anthropol.* **20**:440–453.

———. (1979). Primate positional behavior and anatomy: Naturalistic and experimental approaches. In *Environment, Behavior and Morphology: Dynamic Interactions in Primates*, ed. M.E. Morbeck, H. Preuschoft, and N. Gomberg, pp. 313–325. New York: Gustav Fischer.

———. (1984). Primate locomotion and diet. In *Food Acquisition and Processing in Primates*, ed. D.J. Chivers,

B.A. Wood, and A.L. Bilsborough, pp. 105–117. New York: Plenum Press.

Fleagle, J.G., and Mittermeier, R.A. (1980). Locomotor behavior, body size and comparative ecology of seven Surinam monkeys. *Amer. J. Phys. Anthropol.* **52**:301–322.

Grand, T.I. (1972). A mechanical interpretation of terminal branch feeding. *J. Mammal.* **53**:198–201.

Jenkins, F.A., Jr. (1974). *Primate Locomotion*. London: Academic Press.

Jolly, C.F. (1966). The evolution of the baboon. In *The Baboon in Medical Research*, vol. 2, ed. H. Vogtborg. Austin: University of Texas Press.

Kay, R.F., and Covert, H.H. (1984). Anatomy and behavior of extinct primates. In *Food Acquisition and Processing in Primates*, ed. D.J. Chivers, B.A. Wood, and A. Bilsborough, pp. 467–508. New York: Plenum Press.

Kimura, T., Okada, M., and Ishida, H. (1979). Kinesiological characteristics of primate walking: Its significance for human walking. In *Environment, Behavior and Morphology: Dynamic Interactions in Primates*, ed. M.E. Morbeck, H. Preuschoft, and N. Gomberg, pp. 297–311. New York: Gustav Fischer.

Morbeck, M.E., Preuschoft, H., and Gomberg, N. (1979). *Environment, Behavior and Morphology: Dynamic Interactions in Primates*. New York: Gustav Fischer.

SOCIAL ORGANIZATION

Fleagle, J.G., Kay, R.F., and Simons, E.L. (1980). Sexual dimorphism in early anthropoids. *Nature* **287**:328–330.

Harvey, P.H., and Harcourt, A.H. (1984). Sperm competition, testes size and breeding systems in primates. In *Sperm Competition and the Evolution of Animal Mating Systems*, ed. R.L. Smith, pp. 589–600. London: Academic Press.

Hrdy, S.B., and Whitten, P.L. (1986). Patterning of sexual activity. In *Primate Societies*, ed. B.B. Smuts, D.L. Cheney, R.M. Seyfarth, R.W. Wrangham, and T.T. Struhsaker, pp. 370–384. Chicago: University of Chicago Press.

Kay, R.F., Plavkin, M., Wright, P.C., Glander, K., and Albrecht, G.H. (1987). Behavioral and size correlates of canine dimorphism in platyrrhine primates. *Am. J. Phys. Anthropol.* **87**:218.

The Fossil Record

PALEONTOLOGICAL RESEARCH

In the previous chapters we have discussed the anatomy, behavior, and ecology of extant primates, with only a passing mention of their evolutionary history. In the following chapters we discuss primate adaptation and evolution from a paleontological perspective. Although most of our understanding of the relationships among living organisms is based on the study of living species themselves, the fossil record provides us with many types of information about the biology of primates which we could never know from the living species alone.

The unique aspect of the fossil record is that it establishes a temporal framework for evolution. It provides a crude dating for individual events, such as the first appearance of particular taxonomic groups or particular anatomical features. It also provides evidence for the patterns and rates of evolutionary change—whether it was gradual or occurred in fits and starts.

We can extract several kinds of information valuable for understanding the phylogeny of living primate species from the fossil record. It often shows us intermediate or primitive forms that link more distinct living groups, and it demonstrates how the living species came to be the way they are by documenting the sequence of evolutionary changes that led to their present differences. In addition, the increased diversity in the fossil record is important when we attempt to reconstruct the morphology of ancestral species and evolutionary pathways from a theoretical perspective.

The fossil record also enables us to examine adaptive changes through time. Knowledge of past adaptations can help us understand how the adaptive characteristics of extant radiations came to be the way they are and can also suggest tests for examining causal changes between morphology and environment.

Most important, the fossil record provides us with a record of life in the past. It is our only evidence of extinct primates—in most cases, animals whose existence we could never have predicted or even imagined had we not been confronted with their bones. As we shall see, some groups of primates were far more diverse in morphology, ecology, and biogeography during the very recent past than they are today, and other successful radiations from previous epochs have no living representatives.

The information available from the fossil record is quite different from that we can obtain about living species, most noticeably in its incompleteness. Time extracts its price, and our insights into the past are, alas, more often glimpses than panoramas. If we hurry, we can still observe living primates in the forest as they go about their daily activities and we can record their behavior in scientific papers, books, photographs, and films. We can examine their pelage, measure and dissect their bodies, and study their physiology, communication, and learning abilities, in addition to measuring their bones and teeth. For fossils we have only bones and teeth— mainly the latter. The occasional impression of the bushy tail of an archaic primate, or footprint of an early hominid, is, unfortunately, a rare and remarkable occurrence. As a result, our discussions of the behavior of extinct primates, and even the identification of different sexes and species, require a much larger dose of guesswork than our descriptions of living species.

Our greatest tool is, of course, our ability to extrapolate from the consistent patterns we see among living species to these more poorly known animals in the fossil record. We must keep an open mind, however; the fossil record is likely to be full of unique events. Thus, before we discuss primate evolution, we must consider briefly the special attributes of the fossil record and the types of information that are available for understanding primate history.

Geological Time

The evolution of primates has taken place on a time scale that is virtually impossible to comprehend in anything but a comparative sense. As individuals, one hundred years is the most we are likely to ever experience, yet few events in primate evolution can be dated to within one or even five million years. The scale of events is more commonly on the order of tens of millions of years.

The evolution of primates, like that of most other groups of modern mammals, has occurred almost totally within the Cenozoic era—the Age of Mammals—roughly the last 65 million years. Paleontologists have traditionally divided this period into smaller units (epochs and land mammal ages) on the basis of animals commonly contained in the sediments. Through **faunal correlation**, sediments from different places and the fossils in them can be placed in a relative time scale.

In recent decades, this relative time framework has been calibrated; absolute dates have been determined for events through the use of **radiometric dating** techniques.

Many elements on earth are naturally unstable and change to more stable elements at a characteristic rate. By examining the percentages of an unstable element and its more stable form in a rock, it is possible to calculate how long ago the rock was formed. Radiometric dating of geological sediments in absolute numbers of years is only possible for certain types of rocks, however—usually relatively pure volcanic ashes or lava flows. Determining the age of particular events in primate evolution therefore usually requires a combination of both relative and absolute dating methods. Figure 9.1 summarizes determinations of the age of geological epochs and faunal ages relevant to primate evolution together with representative events in the history of primates.

Paleomagnetism

One of the many startling geological discoveries of the last few decades has been that the earth's magnetic field has reversed polarity frequently during the past—approximately once every 700,000 years, but not

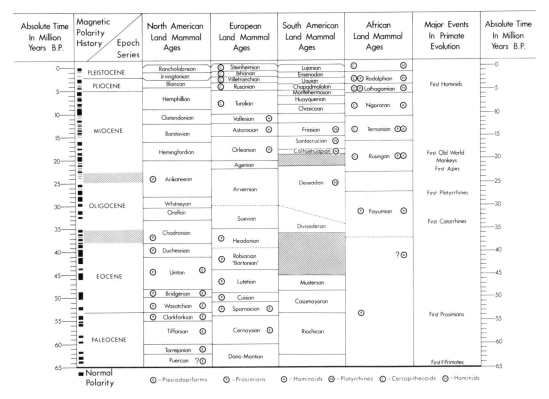

FIGURE 9.1

A geological time scale for the Cenozoic era, showing the epoch series, major land mammal ages, paleomagnetic changes, distribution of fossil primates on different continents, and first appearance of major phyletic groups.

at regular intervals. How and why these changes have occurred is not well understood, but geologists have compiled a history of them over the past 500 million years through combined studies of paleomagnetism and radiometric dating techniques (Fig. 9.1). Paleomagnetism provides another method of relative dating of sediments and fossils. Thus, at many fossil sites that lack the appropriate rocks for absolute dating, geologists can use the sequence of magnetic reversals in conjunction with faunal correlation to determine the position of the rocks in the geological timetable and to estimate the absolute age.

Continental Drift

Even more exciting than the discovery that the earth's magnetic poles have frequently changed polarity in the geological past has been the realization that the continents of the earth are constantly in motion with respect to one another (Fig. 9.2). Thus the sizes, orientations, and connections of the continents and the positions of the oceans

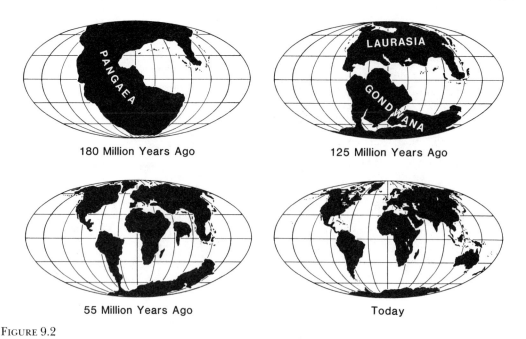

180 Million Years Ago   125 Million Years Ago

55 Million Years Ago   Today

FIGURE 9.2

Positions of the continents at various times during the past 180 million years.

surrounding them have been quite different in the past than they are today. Needless to say, these geographic arrangements have greatly influenced the routes of migration and dispersal available to plants and animals. In addition, the relative positions of land masses have had major effects on ocean currents and climate—effects with global consequences for primates and all other living things. Many of the most dramatic changes in the earth's surface took place well before the first appearance of primates and so have little bearing on the subject of this book. Nevertheless, during the past 65 million years a number of changes in continental positions and connections have influenced primate evolution.

Paleoclimate

Through studies of fossil land plants and various marine organisms, geologists have been able to reconstruct the major changes in the earth's climate during the Cenozoic era. These studies show several general trends over the past 65 million years (Fig. 9.3) which have undoubtedly been important in primate evolution. It is important to remember, however, that climatic changes in a restricted area are quite likely to show different patterns than those that may characterize the earth as a whole, and our knowledge of climatic changes in any one place is usually quite crude.

The formation of glaciers at polar latitudes is one of the most far-reaching global climatic events. In addition to dramatically altering regional climates and landforms, glaciers profoundly affect sea levels by changing the distribution of water on the earth's surface. In turn, these changes in sea level can affect the erosional and depositional rates of rivers, streams, and beaches. Over the past 65 million years there have been dramatic changes in global sea levels, generally associated with the development of glaciers at the poles. Like the positions of continents, changes in sea level can have important effects on plant and animal dispersal. In the following chapters we attempt to relate these changes in continental position, climate, and sea level to the major events in primate evolution.

Fossils and Fossilization

Fossils are any remains of life preserved in rocks. We most commonly think of fossils as petrified bones and teeth, but fossils also include such things as impressions, natural molds of brains or even bodies, and traces of life such as footprints, worm burrows, or termite nests (Fig. 9.4).

Although fossils often preserve shapes of bones or teeth very accurately, most fossils are usually formed by replacement of the original biological materials with minerals derived from the sedimentary environments in which they are buried. In many cases, however, this replacement takes place at a molecular level, so even microscopic details of morphology, such as muscle attachments, fine tooth-wear scratches, dental enamel prisms, or delicate bone structures are preserved and can be analyzed with many of the same tools we apply to the study of living primate skeletons. One worker has even

FIGURE 9.3

Temperatures during the Cenozoic era: above, global climate changes (based on North Sea foraminifera); below, relative sea levels (based on seismic reflections).

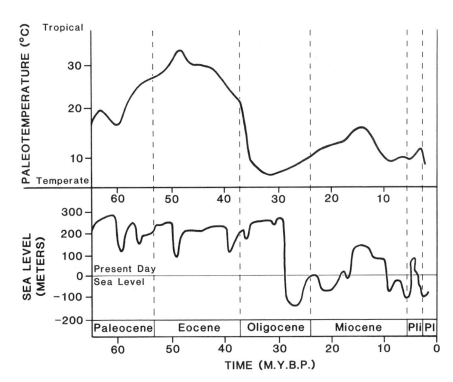

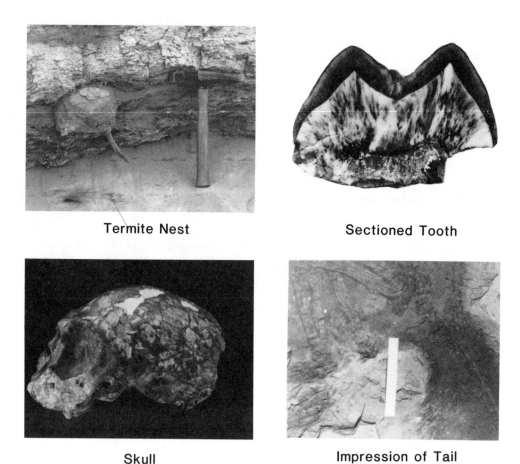

FIGURE 9.4

Different kinds of fossils.

succeeded in extracting collagen molecules from fossil bones and studying them immunologically as if they were living tissues (Lowenstein, 1985).

The type of remains available to a scientist today from an animal that lived some time in the geological past is determined by many events and processes. The study of the factors that determine which animals become fossils, what parts of their bodies are preserved, how they are preserved, and how they appear to scientists many millions of years later is **taphonomy**. Taphonomists seek to reconstruct as well as possible everything that has happened to a bone between the time it was climbing a tree 35 million years ago in the body of an early fossil monkey until the time it was discovered along with other fossils in Egyptian sandstone (Fig. 9.5). They want to know such things as why teeth and ankle bones are commonly found as fossils but other parts may not be, or why

FIGURE 9.5

The history of a fossil (adapted from Shipman, 1981).

some fossils are found as whole bodies and others as fragments. In pursuit of answers to such questions, taphonomists engage in many unusual activities—such as staking out dead antelopes on the Serengeti Plains to see what happens to them, or placing bones in cement mixers to simulate the effects of rolling down a rocky stream. Taphonomy is a young science, barely in its infancy, but it is providing many new insights into primate evolution as it progresses.

Taphonomic studies enable paleontologists to determine if the remains of animals found at a particular locality or site have been transported to the site by the action of streams or perhaps predatory birds, or whether the animals are more likely to have lived and died where their bones are recovered. Studies of the proportions of different skeletal elements recovered, the absence of abrasion, and the abundance of bite marks on bones of fossil primates and other mammals from the early Eocene of Wyoming, for example, indicate that the fossils are the result of long-term accumulations on the surface and were not transported long distances and concentrated by stream action. Thus the proportions of species in the fossil record at this site probably represent a relatively accurate estimate of the proportions of different species living in this area 50 million years ago (Bown and Kraus, 1981a,b).

In contrast, studies of the proportions of bony elements, their positioning, and the presence of cuts, breaks, and burned surfaces have shown that remains of Pleistocene elephants from Michigan are not the result of natural death and dismemberment by carnivores. Fisher's (1984) work shows that the elephants (or mastodons) were butchered by humans, who used tools fashioned from the bones of the animals previously killed.

Paleoenvironments

A primate fossil is usually found along with other fossils, both plant and animal, and within a particular geological setting—all of which can yield useful information about the environment in which the animal lived and died. Whether a fossil primate is found associated with forest rodents or savannah rodents, for example, can provide clues to its habitat preferences. Land snails seem to have narrow habitat preferences, and fossil snails have proved very useful in determining the extent to which a particular fossil locality represents a forested or an open habitat. Similarly, fossil plants can yield information about both local habitat and climate.

The sediments containing fossils can provide many kinds of information about the fossils' origin. They can tell us if a fossil deposit was preserved on a floodplain, on a river delta, in a stream channel, or on the shores of a lake. This information about where an animal's bones were preserved provides clues to where it lived. In addition, sediments can provide detailed information about the climatic regime during which they were formed. Was it hot, cold, wet, or dry? Was the weather relatively uniform or was it seasonal? In addition to the information they convey about a particular fossil site, sedimentary deposits can tell us about climatic trends in a particular region.

In reconstructing fossil environments, there are obvious limits to the amount of detailed information we can infer about the life of an extinct primate. There are also potential pitfalls in extrapolating from the events surrounding fossilization to the habits of an animal. For example, most fossil primates are found in sediments that were originally deposited by streams, rivers, or lakes, often channels within a stream or along floodplains of rivers that overflowed their banks during floods. One early worker, finding fossil lemur bones mixed with the bones of turtles and crocodiles, argued that the lemurs must have been aquatic. Obviously, finding lemur bones in deposits formed by water need not imply aquatic lemurs; more probably, the bones of many different animals were just buried together in stream or lake deposits. While the crocodiles may have lived in the river, the lemurs probably lived in trees overhanging the water, or perhaps their bodies were washed into the river during a rainstorm.

Reconstructing Behavior

Generally, the best and most reliable information about the habits of an extinct primate is obtained by comparing details of its dental and skeletal anatomy with those of living primates. Sediments may tell us where it died, and taphonomy may tell us how and why it was preserved, but its teeth and bones can tell us how it lived—what it ate, how it moved, and possibly in what kind of social group it lived. In the previous chapter we discussed many of the associations between behavior and anatomy among living primates that form the basis for our interpretations of fossil behavior. Our ability to reconstruct the habits of an extinct primate from its bones is intimately linked to our understanding of how the shape of bones in living primates varies with their behavior. Associations between bony morphology and behavior that are true only "some of the time" among living primates cannot be expected to yield reliable reconstructions when applied to fossils (see Kay and Covert, 1984).

Furthermore, we have to remain always aware that uniformitarianism has its limits; the present is our best key to the past, but the past was not necessarily just like the present. We know, for example, that tooth size and many aspects of behavior are highly correlated with body size among living primates, but we cannot necessarily extrapolate these relationships based on a finite sample of living species to a fossil primate whose teeth are considerably larger or smaller than those of any living species. Likewise, many fossil primates had anatomical features that were quite different from anything we find among living species. We are sure to have problems interpreting such structures and may need to compare the fossil primates with another type of mammal for an analogy.

We commonly find that fossil primates differ from living species in the combinations of anatomical features they exhibit. A fossil ape may have a humerus that resembles those of a howling monkey in some features, a variegated lemur in others, and a macaque in still others. In such a case we must examine closely the mechanical implications of the individual features rather than simply look for a living species that matches the fossil in all aspects. Our reconstructions of the behavior of extinct primates from their bones and teeth must not be based on simple analogy, but on an understanding of the physiological and mechanical principles underlying the associations between bony structure and behavior (Fleagle, 1979).

Paleobiogeography

It is a common tale that primate fossils are rare because primates typically live in jungles—which have acid soils that destroy their bones before they can be preserved—while animals such as horses live on savannahs—where bones are more easily saved for posterity. Although different soils may well affect the chances of fossilization in different environments, there are many examples of tropical environments in the Cenozoic fossil record indicating that the tree-dwelling habits of primates are not primarily responsible for the gaps in our knowledge of primate evolution. In fact, the primate fossil record is, overall, probably more complete than that of almost any other group of mammals.

The large gaps in the primate fossil record are more directly the result of a remarkably meager geological record from those parts of the world in which primates have almost certainly been most successful for tens of millions of years—the Amazon Basin in South America, the Zaire Basin in central Africa, and the tropical forests of Southeast Asia. For huge amounts of time and space we lack not just fossils but even rocks from critical places and ages. Thus the seemingly poor fossil record of primates compared with, for example, that of horses is likely due to the fact that primates have evolved in places with virtually no fossil record or one that is still covered with forests and recent sediments, while horses were evolving in temperate areas of Europe and western North America which have an excellent fossil record and miles of well-exposed sediments resulting from recent climatic events. For the most part, our knowledge of extinct primates comes from places that today are too dry and poorly vegetated to support living primates—Wyoming, Egypt, and northern Kenya. This terrain is excellent for geological and paleontological research; however, all of the paleoenvironmental evidence tells us that, during the earlier epochs when primates were abundant, these places were lush forests.

Because so much of our understanding of major events in primate evolution is based on an absence of evidence, no aspect of primate evolution is open to more surprises than biogeography. As new fossils are discovered from parts of the world that were previously poorly known, such as China, many of our notions about the evolution, diversity, and biogeography of primates will be dramatically revised. For example, it now seems most likely that platyrrhines arrived in South America about 30 million years ago, because we have no record of earlier primates on that continent. But an unsuspected discovery of fossil prosimians from Brazil would dramatically change our view of the evolutionary history and biogeography of higher primates. Similarly, our current view that hominids originated in Africa is based on a lack of early hominids from other continents. We must keep in mind that our current understanding of primate evolution will continue to change with new finds and new interpretations. In the following chapters we try to evaluate the nature of the evidence for our present understanding of primate evolution with an eye toward particular issues that are presently unresolved.

BIBLIOGRAPHY

PRIMATE EVOLUTION

Ciochon, R.L., and Fleagle, J.G. (1987). *Primate Evolution and Human Origins*. New York: Aldine.

Piveteau, J. (1957). *Primates. Traite de Paleontologie*, vol. 7. Paris: Masson et Cie.

Simons, E.L. (1972). *Primate Evolution*. New York: Macmillan.

Szalay, F.S., and Delson, E. (1979). *Evolutionary History of the Primates*. New York: Académic Press.

GEOLOGICAL TIME

Berggren, W.A., Kent, D.V., Flynn, J.J., and Van Couvering, J.A. (1985). Cenozoic geochronology. *G.S.A. Bulletin* **96**:1407–1418.

Cocks, L.R.M., ed. (1981). *The Evolving Earth*. Cambridge: Cambridge University Press.

Haq, B.U., Hardenbol, J., and Vail, P.R. (1987). Chronology of fluctuating sea levels since the Triassic. *Science* **235**:1156–1167.

MacFadden, B.J., Campbell, K.E., Cifelli, R.L., Stiles, O., Johnson, N.M., Naeser, C.W., and Zeitler, P.K. (1985). Magnetic polarity stratigraphy and mammalian biostratigraphy of the Deseaden (middle Oligocene–early Oligocene) Salla beds of northern Bolivia. *J. Geol.* **93**:223–250.

Marshall, L.G., Hoffstetter, R., and Pascual, R. (1983). Mammals and stratigraphy: Geochronology of the continental mammal-bearing Tertiary of South America. *Paleovertebrata, Mem. Extraordinaire*. **1983**:1–93.

Pickford, M. (1986). The geochronology of Miocene higher primate faunas of East Africa. In *Primate Evolution*, ed. J.G. Else and P.C. Lee, pp. 19–33. Cambridge: Cambridge University Press.

Savage, D.E., and Russell, D.E. (1983). *Mammalian Paleofaunas of the World*. Reading, Mass.: Addison-Wesley.

Vail, P.R., and Hardenbol, J. (1979). Sea-level changes during the Tertiary. *Oceanus* **22**:71–79.

VanAndel, T. (1985). *New Views of an Old Planet*. Cambridge: Cambridge University Press.

Wolfe, J.A. (1978). A paleobotanical interpretation of Tertiary climates in the Northern Hemisphere. *Am. Sci.* **66**:694–703.

FOSSILS, ENVIRONMENT, BEHAVIOR

Behrensmeyer, A.K., and Hill, A., eds. (1980). *Fossils in the Making*. Chicago: University of Chicago Press.

Behrensmeyer, A.K., and Kidwell, S.M. (1985). Taphonomy's contributions to paleobiology. *Paleobiology* **11**(1):105–119.

Bown, T.M., and Kraus, M.J. (1981a). Lower Eocene alluvial paleosols (Willwood Formation, northwest Wyoming, U.S.A.) and their significance for paleoecology, paleoclimatology, and basin analysis. *Palaeogeog., Palaeoclimatol., Palaeoecol.* **34**:1–30.

———. (1981b). Vertebrate fossil-bearing paleosol units (Willwood Formation, Lower Eocene, northwest Wyoming, U.S.A.): Implications for taphonomy, biostratigraphy, and assemblage analysis. *Palaeogeogr., Palaeoclimatol., Palaeoecol.* **34**:31–56.

Bown, T.M., Kraus, M.J., Wing, S.L., Fleagle, J.G., Tiffany, B., Simons, E.L., and Vondra, C.F. (1982). The Fayum forest revisited. *J. Hum. Evol.* **11**(7):603–632.

Fisher, D.C. (1984). Taphonomic analysis of late Pleistocene mastodon occurrences: Evidence of butchery by North American Paleo-Indians. *Paleobiology* **10**(3):338–357.

Fleagle, J.G. (1979). Primate postural behavior and anatomy: Naturalistic and experimental approaches. In *Environment, Behavior, and Morphology*, ed. M.E. Morbeck, H. Preuschoft, and N. Gomberg, pp. 313–325. New York: Gustav Fischer.

Kay, R.F., and Covert, H.H. (1984). Anatomy and behavior of extinct primates. In *Food Acquisition and Processing in Primates*, ed. D.J. Chivers, B.A. Wood, and A. Bilsborough, pp. 467–508. New York: Plenum Press.

Lowenstein, J.M. (1985). Radioimmunoassay of extinct and extant species. In *Hominid Evolution—Past, Present and Future*, ed. P.V. Tobias, pp. 401–410. New York: Alan R. Liss.

Shipman, P. (1981). *Life History of a Fossil*. Cambridge, Mass.: Harvard University Press.

Archaic Primates

PALEOCENE EPOCH

The Paleocene, the first epoch in the Age of Mammals, is a poorly known part of earth history, but it provides our only record of the first major radiation of placental mammals, including possible primates. At the end of the Cretaceous, the dinosaurs that had dominated terrestrial faunas for the past 120 million years had all disappeared. No one is quite sure why they disappeared or whether their departure was abrupt or gradual. All we know is that, beginning about 65 million years ago, the fossil record contains no more than an occasional tooth or claw to suggest a last lingering dinosaur; instead, the most abundant vertebrates are mammals of various sorts.

Geologically, the late Cretaceous and Paleocene were relatively active times in earth history and were marked by the rise of several major mountain groups, including the American Rockies. Geographically, the world looked somewhat different than it does today (Fig. 10.1). The North Atlantic was considerably narrower than it is today,

FIGURE 10.1

Map of the world during the middle Paleocene (60 million years ago), with locations (★) of archaic primate fossil sites.

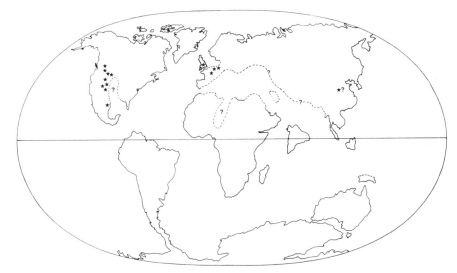

particularly in the vicinity of Greenland. The intermittent occurrence of land connections between North America and Europe is indicated by the similarity of the Paleocene faunas of the two continents. There is also faunal evidence of occasional connections between North America and Asia, presumably across the Bering Strait.

South America, Africa, and India were all island continents as far as we know (although South America and Antarctica were apparently connected until the Oligocene). The South Atlantic was an open ocean, although somewhat narrower than it is today, and perhaps there were land surfaces lying between South America and Africa. The Panama land bridge, which currently connects North and South America, would not come into being for another 50 million years. Africa was separated from Europe by the great Tethys Seaway extending from China on the east to southern France on the west. India was adrift in the Indian Ocean and had not yet collided with the Asian mainland.

Paleocene climates were relatively cooler than those of either the preceding late Cretaceous or the succeeding Eocene epochs, but temperatures fluctuated throughout the epoch (see Fig. 9.3). The flora of western North America, which has been carefully studied, was characterized by deciduous trees and conifers rather than the more tropical plants characteristic of immediately earlier and later epochs.

Primate Origins: *Purgatorius*

The earliest primates evolved from some insectivore-like mammal some time in the latest part of the Cretaceous period, the last period in the Age of Reptiles. It is impossible to determine exactly how primates are related to other orders of mammals, but there are indications from paleontology, comparative anatomy, and biomolecular studies that primates, tree shrews, flying lemurs, and bats are more closely related to one another than to other mammals (see, e.g., Wible and Covert, 1987). It is, however, difficult to identify more than a few common features linking the living members of these four orders amid the many subsequent specializations each has evolved over the past 60 million years.

Among the fossil mammals from the earliest part of the Cenozoic era, the one that most closely resembles later primates is the tiny *Purgatorius* (Table 10.1) from the earliest part of the Paleocene epoch and possibly the latest part of the Cretaceous (Van Valen and Sloan, 1965; Clemens, 1974). Known from several jaws and many isolated teeth, *Purgatorius* had a primitive dental formula of $\frac{3.1.4.3.}{3.1.4.3.}$. The features that set it apart from its contemporaries and suggest a relationship with early primates are the enlarged central incisor, the molarlike development of the last premolar, the relatively low trigonid and broad talonid basin, and the elongate last lower molar. From the relative height of its molar cusps, as well as its small size, there seems little doubt that *Purgatorius* was primarily insectivorous, but many of the molar features it shares with later primates indicate a functional shift from an emphasis on vertical shearing (for insect eating) toward more transverse shearing and crushing, which would have permitted a more omnivorous diet.

Purgatorius is known from several sites in North America. Its presence there is not necessarily indicative of a North American origin for primates; it may simply be an artifact of our poor record of this period in the evolution of placental mammals. The features *Purgatorius* shares with later primates are so general, and this species is so primitive compared with almost all later mammals of all orders, that there are no features linking *Purgatorius* with any one group of later primates. Indeed, several authorities have questioned whether there is any justification for classifying such a primitive mammal among the primates at all. Nevertheless, this genus is a placental mammal that could easily have given rise to all later primates, and it is the only primatelike mammal from the earliest part of the Paleocene. In the middle and late Paleocene, remains of primatelike mammals are more abundant. It is in sediments of this age that we find copious documentation of the first major radiation of primatelike mammals, the plesiadapiforms, or archaic primates.

Plesiadapiforms

The plesiadapiforms, usually ranked as a distinct suborder of primates, were an extremely successful group of primatelike mammals that flourished in the Paleocene and early Eocene of North America and Europe (Fig. 10.2). They are the most common mammals in many Paleocene faunas (Fig. 10.3). Their known taxonomic diversity (more than twenty-five genera and seventy-five species) is approximately twice that of living prosimians, and their diversity in size is comparable to that of either living prosimians or New World anthropoids.

Table 10.1
Suborder Plesiadapiformes
Family *incertae sedis*

Species	Body Weight (g)
Purgatorius (early Paleocene, W. North America)	
P. unio	153
?P. ceratops	—

Plesiadapiforms have long been known mainly from fragmentary jaws and teeth. Thus, affinities between these archaic primates and later primates, as well as relationships within the suborder, are based mainly on dental characteristics (Fig. 10.4). Several dental features of plesiadapiforms, including molar teeth with relatively low cusps (compared with contemporary or extant insectivores), lower molars with low trigonids and basin-shaped talonids, and unreduced lower third molars with an extended talonid, seem to link them with later primates. Their upper molars have prominent conules, a poorly developed or absent stylar shelf, and a well-developed postprotocingulum (nannopithex fold) or comparable wear facet distal to the protocone. The primitive dental formula for plesiadapiforms (excluding *Purgatorius*, which is sometimes arbitrarily added to this suborder) is $\frac{2.1.3.3}{2.1.3.3}$. Most later members of all lineages show reduction and loss of teeth, most frequently the lateral incisor and the anterior premolar. Since all known species of the suborder have a dental formula with three or fewer premolars, they are too specialized to have given rise to the earliest prosimians, many of which have four premolars. In addition, many plesiadapiforms have an extremely large and procumbent lower central incisor, which separates

FIGURE 10.2

Reconstruction of a scene from the late Paleocene of North America showing several plesiadapiforms. A small group of *Plesiadapis rex* feeds in a tree, and *Ignacius frugivorus* feeds on exudates from the trunk. A small *Picrodus silberlingi* feeds on nectar in a bush. On the ground, *Chiromyoides minor* chews on a seed, and a small microsyopid grasps its insect prey.

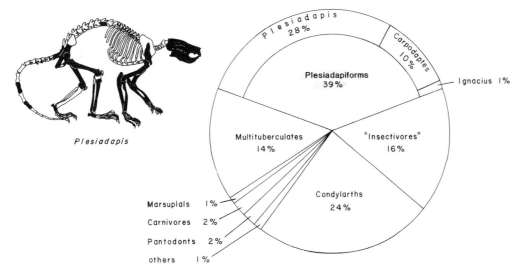

Plesiadapis

FIGURE 10.3

The abundance of different mammalian orders in a late Paleocene fossil site in western North America.

FIGURE 10.4

Mandibles of several plesiadapiforms, showing the diversity in shape and size of the dentition. a, *Plesiolestes problematicus*, lateral and occlusal views; b, *Elwynella oreas*, lateral and occlusal views; c, *Plesiadapis rex*, lateral view; d, *Elphidotarsius florencae*, lateral view; e, *Chiromyoides campanicus*, lateral view; f, *Saxonella crepaturae*, lateral view; g, *Picrodus silberlingi*, lateral view.

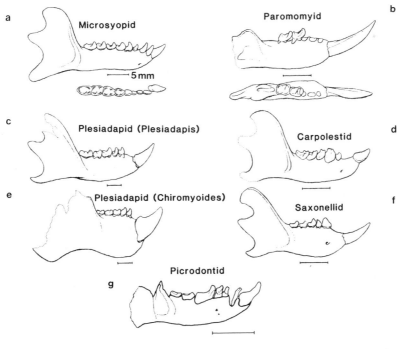

273

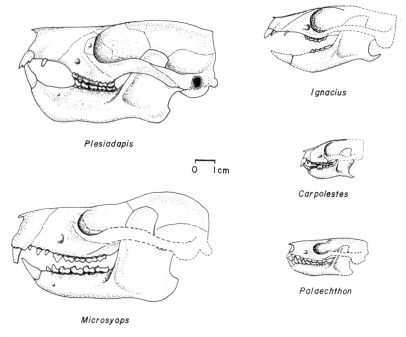

Ignacius

Plesiadapis

0   1 cm

Carpolestes

Palaechthon

Microsyops

FIGURE 10.5

Skulls of five plesiadapiforms.

them from later primates (Fig. 10.4; see Fig. 11.2).

The sharp cusps on the teeth of many species, as well as their small size, suggest that many of the archaic primate species were largely insectivorous. Nevertheless, many of the features linking plesiadapiforms with later primates suggest a general shift toward more crushing and grinding in the cheek teeth in conjunction with more omnivory and herbivory compared with contemporary insectivores.

Most plesiadapiforms have a low, flat skull with a long snout, a small brain, large zygomatic arches, and no bony ring surrounding the orbits (Fig. 10.5). In these features they are more primitive than all later primates. The arterial circulation to the brain also seems to be distinctly different from that found among later primates. Many authors have suggested that plesi-

adapiforms are linked with living primates by having an auditory bulla composed of the petrosal bone. Unfortunately, the composition of the auditory bulla is impossible to identify accurately without embryological material (MacPhee *et al.*, 1983). As a result, there are no definite cranial features linking plesiadapiforms with extant primates.

Analyses of the limb and trunk skeleton, particularly the foot and elbow, have indicated several features that unite the archaic primates and others that link them with later primates. Such analyses are, however, severely limited by the paucity of material. The one genus for which ample skeletal remains are known, *Plesiadapis*, has relatively short robust limbs, a nonopposable hallux, and clawed digits—features that are clearly more primitive than those found in the limbs of later primates (see Fig. 11.2).

There is considerable diversity among current authorities in their assignment of individual species and genera to different families of archaic primates. The taxonomic scheme adopted here is based largely on the shape of incisors and premolars because these structures are known for most genera. On these criteria, the plesiadapiforms can be divided into six families: microsyopids, plesiadapids, carpolestids, saxonellids, paromomyids, and picrodontids.

Microsyopids

The Microsyopidae (Table 10.2) are the most primitive plesiadapiforms, and there is considerable debate among authorities over

TABLE 10.2
Suborder Plesiadapiformes
Family MICROSYOPIDAE

Species	Body Weight (g)
Palaechthon (m.–l. Paleocene, North America)	
P. alticuspis	99
P. nacimienti	150
P. woodi	60
Plesiolestes (m.–l. Paleocene, North America)	
P. problematicus	300
P. sirokyi	1,000
Talpohenach (m. Paleocene, North America)	
T. torrejonia	300
Torrejonia (m. Paleocene, North America)	
T. wilsoni	740
Palenochtha (m.–l. Paleocene, North America)	
P. minor	44
Berruvius (l. Paleocene–e. Eocene, Europe)	
B. lesseroni	20
B. gingerichi	20
Navajovius (l. Paleocene–e. Eocene, North America)	
N. kohlhaasae	88
Micromomys (l. Paleocene–e. Eocene, North America)	
M. silvercouleei	30
M. willwoodensis	—
M. vossae	20
M. fremdi	20

Species	Body Weight (g)
Tinimomys (e. Eocene, North America)	
T. graybulliensis	35
Niptomomys (e. Eocene, North America)	
N. doreenae	25
N. thelmae	40
Uintasorex (m.–l. Eocene, North America)	
U. parvulus	20
Microsyops (Eocene, North America)	
M. angustidens	745
M. latidens	760
M. scottianus	1,432
M. lundeliusi	3,841
M. elegans	1,185
M. annectens	2,320
M. kratos	3,362
Arctodontomys (e. Eocene, North America)	
A. wilsoni	450
A. simplicidens	500
A. nuptus	750
Craseops (l. Eocene, North America)	
C. sylvestris	3,300
Alveojunctus (m. Eocene, North America)	
A. minutus	—

whether some of the genera are actually primatelike insectivores rather than plesiadapiforms. It was a very successful family, with species in both North America and Europe, ranging from at least the middle Paleocene through the late Eocene. Microsyopids are relatively diverse in appearance and include both the smallest of all known primates, an animal the size of a shrew, and the largest plesiadapiform, an animal the size of a raccoon. Microsyopids share with other plesiadapiforms the primatelike dental features listed above and are relatively conservative in that they retain a primitive dental morphology throughout their evolutionary history. All microsyopids have a narrow, lanceolate (spearhead-shaped), specialized lower central incisor, the feature that distinguishes the family (Fig. 10.4).

Cranially, microsyopids share most primitive mammalian features described above for plesiadapiforms (Fig. 10.5), as well as a confusing mix of primatelike and nonprimatelike features. The best-known genus, *Microsyops*, has a cranial arterial pattern that more closely resembles the cranial arterial pattern of living primates than does that of

any other plesiadapiform primate. But this taxon also has an auditory structure that is more primitive than that of other plesiadapiforms in that it lacks a bony bulla (MacPhee *et al.*, 1983).

The earliest microsyopids were five very primitive, closely related genera from the middle Paleocene of North America: ***Talpohenach, Palenochtha, Palaechthon, Plesiolestes***, and ***Torrejonia***. In size, they are comparable to the smallest living primates (60–200 g). Most species have a dental formula of $\frac{2.1.3.3.}{2.1.3.3.}$, but the canine and anteriormost premolar are very small in many species and probably lacking in some. The enlarged, lanceolate lower first incisors form a scooplike apparatus for cutting, an adaptation that suggests a partly herbivorous diet (Fig. 10.6; Szalay, 1981). Yet the molars have relatively acute cusps compared with those of many living primate species, suggesting that insects were a major part of the diet as well.

For one of these small middle Paleocene species, *Palaechthon nacimienti*, there is a relatively complete but crushed skull. Like most plesiadapiforms, *Palaechthon* has a

FIGURE 10.6

The anterior dentitions of *Schoinobates volans*, a folivorous marsupial, *Plesiolestes problematicus*, a microsyopid primate, *Plesiadapis simonsi*, a plesiadapid primate, and *Erinaceus europaeus*, an insectivorous hedgehog. Note that the incisors of the plesiadapiform primates form a bladelike structure similar to that found in the marsupial rather than independent prongs as in the hedgehog (after Szalay, 1981).

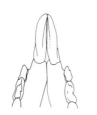 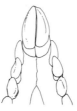  

Schoinobates   *Plesiolestes*   *Plesiadapis*   *Erinaceus*
volans   *problematicus*   *simonsi*   *europaeus*

small braincase attached to a relatively large set of teeth. Compared with extant primates, *Palaechthon* has relatively small, laterally directed orbits, suggesting limited stereoscopic abilities; a broad interorbital region, suggesting a large olfactory fossa and greater reliance on a sense of smell; and a large infraorbital foramen, suggesting the presence of a richly innervated snout bearing sensitive facial vibrissae.

Kay and Cartmill (1977) suggest that the small size and the cranial features of *Palaechthon* indicate that it was probably a terrestrial forager that hunted for concealed insects and other animal prey by "nosing around the ground," guided more by hearing, smell, and its sensitive snout than by vision (Fig. 10.2)—and that it was probably nocturnal. Unfortunately, there are no associated skeletal elements for any microsyopids to test these theories about their locomotor habits.

Palaechthon and its close relative *Plesiolestes* are very similar dentally to two Eocene genera, **Microsyops** and **Craseops**. In these later forms, the dental formula is reduced but the overall skull shape is similar. *Microsyops* lacks a petrosal bulla, but it is not clear whether it has a bulla made of cartilage or one formed by a separate entotympanic bone.

A more specialized group of microsyopids, which many authorities feel does not belong among the primates, is the Uintasoricinae. This subfamily contains four tiny shrew- to mouse-size genera: **Navajovius** and **Berruvius**, from the Paleocene and Eocene of North America and France, respectively, and **Niptomomys** and **Uintasorex**, from the Eocene of western North America. These four share the lanceolate incisor with other microsyopids, but their cheek teeth have small third molars and relatively broader talonids and smaller trigonids on the lower

molars. Their dentition suggests a soft diet of fruits, gums, or nectar.

Two other small microsyopids from western North America are **Micromomys**, from the late Paleocene and early Eocene, and **Tinimomys**, from the early Eocene. As their names indicate, these are the smallest primates known; they probably weighed no more than 30 g. Their molars resemble those of the more conservative microsyopid genera, but they are distinguished by their large posterior lower premolar (P_4). From its small size and its molars with very acute cusps, *Micromomys* appears to have been almost totally insectivorous. *Tinimomys* has more rounded cusps, suggesting more omnivorous habits.

Plesiadapids

The best known of the archaic primates, plesiadapids were very diverse and abundant in the middle Paleocene of North America and the late Paleocene and early Eocene of both North America and Europe (Fig. 10.7, Table 10.3). They are usually divided into five genera. The smallest species were comparable in size to a tamarin; the largest were the size of a guenon (5 kg).

Compared with many other archaic primates, plesiadapids have relatively generalized teeth (Figs. 10.4, 10.7). The earliest genus, **Pronothodectes**, has a dental formula of $\frac{2.1.3.3}{2.1.3.3}$, but later genera show considerable reduction and loss of incisors, canines, and premolars. All plesiadapids have relatively broad, procumbent lower incisors that occlude in a pincer fashion with the mitten-shaped upper central incisors (Figs. 10.5, 10.6). Compared with microsyopids, plesiadapids have premolars and molars with low bulbous cusps. The posterior two upper premolars are short and broad. The lower molars have a relatively low trigonid and

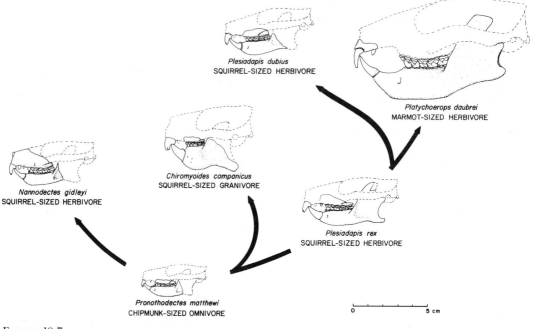

Plesiadapis dubius
SQUIRREL-SIZED HERBIVORE

Platychoerops daubrei
MARMOT-SIZED HERBIVORE

Chiromyoides campanicus
SQUIRREL-SIZED GRANIVORE

Plesiadapis rex
SQUIRREL-SIZED HERBIVORE

Nannodectes gidleyi
SQUIRREL-SIZED HERBIVORE

Pronothodectes matthewi
CHIPMUNK-SIZED OMNIVORE

0   5 cm

FIGURE 10.7

A phylogeny of plesiadapids, showing diversity in dental, mandibular, and cranial form probably associated with dietary diversity (from Gingerich, 1976).

broad talonid. These low-crowned cheek teeth are very similar to the molars of many later primates, and, together with the relatively large size of most species, they suggest that plesiadapids were probably more strictly herbivorous than other archaic primates (Figs. 10.2, 10.7).

There are several skulls of **Plesiadapis** from the late Paleocene of France (Russell, 1964), and cranial fragments of two genera are known from the western United States. The cranium of *Plesiadapis* (Fig. 10.5) has a long snout with a large premaxillary bone and a diastema between the large incisors and the cheek teeth in both the upper and lower jaws. The auditory bulla in adult individuals is continuous with the petrosal bone, and the tympanic ring (tympanic

bone) is fused to the bulla and extended laterally to form a bony tube (Szalay, 1975).

Although considerable skeletal material is known for *Plesiadapis*, there is disagreement over the probable locomotor habits of this archaic primate. Much of this uncertainty is due to the fact that *Plesiadapis* is very different in its limb structure from any living primate, and it is not immediately clear what type of living mammal is the most appropriate analogue. From the robustness of the skeleton, from the limb proportions, from similarities to living groundhogs, and from its abundance and wide geographic distribution, several authors have suggested that *Plesiadapis* must have been a gregarious terrestrial form. The short robust limbs, the long, laterally compressed claws, and the long bushy tail (known from a delicate limestone impression; see Fig. 9.5) suggest instead that it was an arboreal quadruped.

TABLE 10.3
Suborder Plesiadapiformes
Family PLESIADAPIDAE

Species	Body Weight (g)
Pronothodectes (m. Paleocene, North America)	
P. matthewi	306
P. jepi	406
Nannodectes (l. Paleocene, North America)	
N. intermedius	429
N. gazini	376
N. simpsoni	619
N. gidleyi	729
Plesiadapis (l. Paleocene–e. Eocene, North America, Europe)	
P. praecursor	592
P. anceps	786
P. rex	919
P. gingerichi	2,900
P. churchilli	1,290
P. fodinatus	981
P. dubius	706
P. simonsi	2,086
P. cookei	4,879
P. walbeckensis	714
P. remensis	1,342
P. tricuspidens	2,166
P. russelli	—
Chiromyoides (l. Paleocene–e. Eocene, North America, Europe)	
C. campanicus	—
C. caesor	—
C. minor	256
C. potior	—
C. major	—
Platychoerops (e. Eocene, Europe)	
P. daubrei	3,111
P. richardsoni	—

The phylogenetic relationships among plesiadapids have been thoroughly studied, and there is excellent documentation of the patterns of evolutionary change in this family (Gingerich, 1976; Fig. 10.7). The primitive *Pronothodectes* from the middle Paleocene of Wyoming and Montana apparently gave rise to two lineages, each placed in a separate genus. **Nannodectes**, from the late Paleocene (Tiffanian), has relatively narrow cheek teeth but shows very little dental reduction except in the latest, most advanced species. The other genus, *Plesiadapis*, from the late Paleocene and early Eocene of both North America and Europe, shows greater dental reduction between the large central incisor and the broad cheek teeth. There are two distinct lineages of species within the genus *Plesiadapis*, one decreasing in size and one increasing. In North America, the latter culminates in *Plesiadapis cookei*, one of the largest of the archaic primates. The latest European plesiadapid is another large form, **Platychoerops**.

One of the most distinctive plesiadapids was **Chiromyoides**, from both North America and Europe. This relatively short-faced form with robust incisors, a deep mandible, and relatively flat cheek teeth probably ate some type of seeds (Figs. 10.2, 10.4, 10.7).

Carpolestids

The carpolestids (Table 10.4) are a North American family characterized by the enlargement of their last lower premolar and last two upper premolars. The family contains three genera that follow one another in time: **Elphidotarsius**, from the middle and late Paleocene; **Carpodaptes**, from the late Paleocene; and **Carpolestes**, from the latest Paleocene and earliest Eocene (Fig. 10.8). There are numerous species known for each

TABLE 10.4
Suborder Pleisadapiformes
Family CARPOLESTIDAE

Species	Body Weight (g)
Elphidotarsius (m.–l. Paleocene, North America)	
E. florencae	50
E. shotgunensis	40
E. russelli	40
E. wightoni	20
Carpodaptes (l. Paleocene, North America)	
C. aulacodon	50
C. hazelae	50
C. hobackensis	40
C. cygneus	50
C. jepseni	100
Carpolestes (l. Paleocene–e. Eocene, North America)	
C. nigridens	90
C. dubius	100

genus. Because of their distinctive morphological specializations and short species durations, carpolestids are useful as biostratigraphic indicators in early Tertiary sediments.

Carpolestids were small, mouse-size primates (20–50 g). Their anterior dentition resembles plesiadapids in the shape of the large procumbent lower central incisor and the mitten-shaped upper central incisor. The characteristic hypertrophy of the posterior lower premolar to form a large serrated blade and the expansion of the occluding upper teeth into a broad, multicusped "chopping block" increases in both size and morphological uniqueness with time from the early *Elphidotarsius* through *Carpodaptes* to *Carpolestes*. Carpolestids have a first molar with an elongate trigonid and posterior

molars with relatively high narrow trigonids and compressed talonids.

Exactly how carpolestids used their teeth for processing food and what types of food they ate are difficult to reconstruct. The few

FIGURE 10.8

Differences in premolar shape among the three genera of carpolestids. Note the increasing size of the last premolar and first molar (from Rose, 1975).

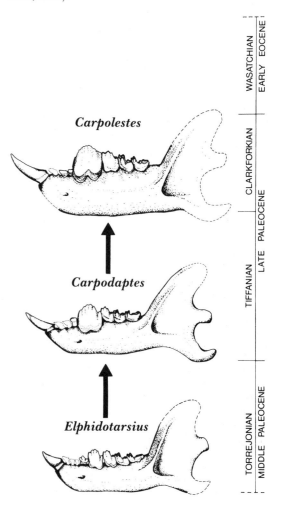

living species of mammals that show similar but not identical enlarged premolars are in some cases predominantly insectivorous (caenolestid marsupials) and in other cases partly herbivorous (rat kangeroos). An alternate approach to this problem has been to look directly at the enlarged carpolestid premolars for microscopic evidence of wear that might indicate how the teeth were used. The results (Biknevicius, 1986) suggest that they used the blades to split open some type of object with a hard outside and a soft inside, perhaps nuts, hard-shelled fruits, even insects with hard shells. Similarities between the molars of carpolestids and those of the insectivorous *Tarsius* support the latter possibility.

Carpolestes had a relatively short face compared with that of many other archaic primates (Fig. 10.5). No postcranial remains are known for members of this family.

Saxonellids

Saxonella (Fig. 10.4, Table 10.5) is a relative of the plesiadapids from the late Paleocene of North America and Germany. Like carpolestids, *Saxonella* seems to be a derivative of the plesiadapids that evolved a very large lower premolar. In contrast with carpolestids, however, which enlarged the last premolar, *Saxonella* enlarged P_3. Thus,

although probably adaptively similar, the two groups are not closely related.

Paromomyids

The paromomyids (Table 10.6) are another specialized group of archaic primates with affinities to plesiadapids. They appear to have been among the most long-lived and geographically widespread families of archaic primates. The five genera of paromomyids ranged from the middle Paleocene through the latest Eocene, and they have been found in North America as far north as the Arctic Circle, as well as in Europe. The

TABLE 10.5
Suborder Plesiadapiformes
Family SAXONELLIDAE

Species	Body Weight (g)
Saxonella (l. Paleocene, North America, Europe)	
S. crepaturae	80
Saxonella sp.	70

TABLE 10.6
Suborder Plesiadapiformes
Family PAROMOMYIDAE

Species	Body Weight (g)
Paromomys (m. Paleocene, North America)	
P. maturus	200
P. depressidens	590
Ignacius (m. Paleocene–l. Eocene, North America)	
I. graybullianus	306
I. frugivorus	216
I. fremontensis	157
I. mcgrewi	240
Phenacolemur (l. Paleocene– m. Eocene, North America, Europe)	
P. praecox	480
P. simonsi	159
P. pagei	380
P. jepseni	245
Elwynella (m. Eocene, North America)	
E. oreas	360
Arcius (e. Eocene, Europe)	
A. rougieri	160
A. fuscus	—
A. lapparenti	—

most primitive member of the family is **Paromomys**, from the middle Paleocene. The four other genera, **Ignacius, Phenacolemur, Elwynella**, and **Arcius**, appear to be the results of independent parallel lineages. Paromomyids were small to medium-size primates with a long, slender lower central incisor (Fig. 10.4). In *Phenacolemur, Ignacius,* and *Arcius*, the canine and the anterior premolars are reduced or lost, leaving a diastema between the procumbent incisor and the cheek teeth. The posterior lower premolar is usually tall and pointed. Paromomyids have relatively flat, low-crowned lower molars with short, squared trigonids and broad, shallow talonid basins; the upper molars are square with expanded basins. Both upper and lower posterior molars are conspicuously elongated. (Some authors, e.g., Szalay and Delson, 1979, include many additional genera in Paromomyidae. In this chapter, these additional genera are placed in Microsyopidae; see Bown and Rose, 1976).

The function of the large lower incisor, which occludes with the lobate upper incisors, is uncertain. Some authors note similarities to the incisors of shrews and suggest that they functioned in procuring insects; others suggest that it was used to gnaw holes in trees to elicit the flow of exudates (Fig. 10.2). The pointed P_4 seems to be adapted for puncturing food during initial preparation, and the broad, flat, lower molars suggest a herbivorous rather than an insectivorous diet for most paromomyids.

Partial skulls are known for both *Ignacius* (Fig. 10.5) and *Phenacolemur*. In both genera, the face is long and narrow and has a large infraorbital foramen, suggesting a richly innervated snout with tactile vibrissae. In both there is a bony auditory bulla continuous with the petrosal bone that extends laterally to form a bony external auditory meatus. In *Ignacius*, for which the vascular canals of the auditory region are well preserved, there is no evidence of an intrabullar cerebral blood supply (internal carotid, stapedial, or promontory) as found among most primates; instead, the main cerebral blood supply seems to have been more anterior and shows greatest similarities to the ascending pharyngeal pattern found in lorises and cheirogaleids (MacPhee *et al.*, 1983). Few postcranial elements have been attributed to paromomyids.

Paromomyids are the only nonhuman primates with a geographic range that extends above the Arctic Circle (McKenna, 1980; Hickey *et al.*, 1983). During the early Eocene, a paromomyid similar to *Ignacius* (but much larger) thrived on Ellesmere Island, at 78° north latitude. Because there are several months of total darkness at that latitude today, it seems likely that the fauna there was composed of cathemeral or crepuscular mammals.

Picrodontids

Family Picrodontidae (Table 10.7) is known only by dental remains from the middle and late Paleocene of western North America. There are three genera, **Picrodus, Zanycteris**, and **Draconodus**, all from the middle to late Paleocene.

Picrodontids resemble other plesiadapiforms in their incisor morphology (Fig. 10.4), but their cheek teeth are quite unusual. The first upper and lower molars are enlarged and oddly shaped. The lower molars have very small trigonids and large, shallow talonids with crenulated enamel. Unlike other plesiadapiforms, the last molar in picrodonts is reduced. Because of notable similarities between the molars of picrodontids and those of bats, Szalay (1972) has suggested a diet of fruit and nectar (Fig. 10.2).

Table 10.7
Suborder Plesiadapiformes
Family PICRODONTIDAE

Species	Body Weight (g)
Picrodus (m.–l. Paleocene, North America)	
P. silberlingi	88
Zanycteris (l. Paleocene, North America)	
Z. paleocenus	—
Draconodus (m. Paleocene, North America)	
D. apertus	370

ADAPTIVE RADIATION OF PLESIADAPIFORMS

The plesiadapiforms were a very successful group of early primatelike mammals that evolved a wide range of body sizes and dental adaptations. They include several species that were almost as large as the largest living prosimians or New World monkeys, as well as several species that were much smaller than any living primate. Their cranial structure is so different from that of any living primate that we have no real evidence of whether they were diurnal or nocturnal. Their great diversity in dental morphology suggests considerable diversity in dietary adaptations. It seems likely that many species specialized on insects, and many on fruit, leaves, seeds, and other herbivorous materials, and it has been suggested that some relied on nectar or gums. The size and shape differences in their incisor and mandible structure indicate that plesiadapiform feeding habits were probably quite different from anything found among living primates. Among these archaic primates are animals like the picrodontids

and carpolestids, which had very odd dental specializations by any standards.

The limb skeletons of the archaic primates are so poorly known and so different from living species that it is difficult to reach any firm conclusions regarding their locomotor adaptation. Their size range suggests considerable locomotor diversity. *Plesiadapis* was probably quadrupedal and partly arboreal, but the smallest species may well have been terrestrial.

The social habits of these archaic primates are certainly beyond our ken, but we can speculate. If they were nocturnal, they probably lived in a noyau arrangement like many primitive mammals. Diurnal species may have lived in larger groups.

The radiation of plesiadapiforms was largely during the Paleocene. Only a few microsyopids and paromomyids survived past the early Eocene. There are several explanations commonly offered for the rapid decline and disappearance of this once very successful group in the beginning of the Eocene. The most common view has been that plesiadapiform decline and extinction resulted from competition with rodents (Van Valen and Sloan, 1966). Others have suggested that early prosimians (Szalay, 1972) and possibly bats (Sussman and Raven, 1978) also played a role in their decline. In addition, Gingerich (e.g., 1976) has suggested that the diversity of some plesiadapiforms is closely linked with climatic changes (see Fig. 16.9) and that their decline and extinction at the beginning of the Eocene related to the more tropical environments of that epoch (see Fig. 9.3) as well as to competition from new groups of mammals. In a recent review of this problem, Maas *et al.* (1987, 1988) found that the changes in climate during the late Paleocene and early Eocene do not correlate well with changes in the diversity of plesiadapiforms, and that the radiation of early prosimians (see Chapter 11) came after the extinction of most plesiadapiforms. The increasing diversity of early rodents is, however, inversely correlated with the decline of the plesiadapiforms (see Fig. 16.10). Moreover, functional comparisons show that plesiadapiforms and rodents were likely to have been similar in many aspects of their ecological adaptations.

PLESIADAPIFORMS AND LATER PRIMATES

Although plesiadapiforms are the most primatelike mammals from the Paleocene, all have unique specializations that preclude them from the ancestry of the early prosimi-

ans that immediately succeeded them in the beginning of the Eocene epoch, as well as any relationship with other later primates (Fig. 10.9). Only *Purgatorius* is generalized enough in its dental formula to be a suitable ancestor for all later primates, but it is so generalized and poorly known that its primate status is marginal.

Because plesiadapiforms are undoubtedly a separate lineage from all later primates, many authors have rightly questioned their identification as primates and have suggested that they be regarded as insectivores or even as a separate order of mammals. The systematic position of plesiadapiforms remains a lively topic for debate (MacPhee *et al.*, 1983; Gingerich, 1986; Martin, 1986;

FIGURE 10.9

The phyletic position of plesiadapiforms.

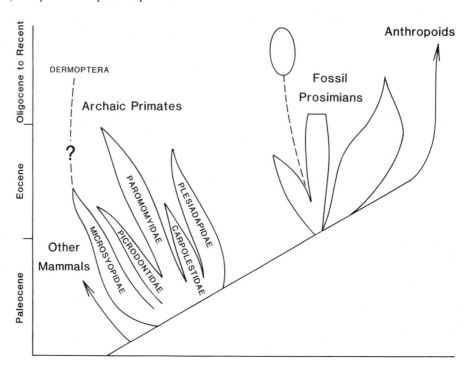

Wible and Covert, 1987), but there is increasing consensus that identification of the primate–nonprimate boundary is, to a large extent, a matter of taste or systematic philosophy. Despite their reduced dental formula and cranial specializations, the cheek teeth of plesiadapiforms are more similar to those of later primates than to any other group of mammals. The tarsal bones and other skeletal elements of plesiadapiforms show greater affinities to later primates than to members of any other order of mammals. The early evolution of placental mammals is replete with systematic problems of this nature for which there are no clean taxonomic boundaries or easy solutions (Luckett, 1980).

Any taxonomic scheme has its flaws. Cartmill's (1974) suggestion of placing all plesiadapiforms in the Insectivora has the elegance of making the order Primates more easily definable on the basis of cranial and skeletal features that link living prosimians and anthropoids. It ignores, however, the primatelike features of plesiadapiforms and places them with the insectivores on the basis of absolutely no derived similarities. Szalay (Szalay and Delson, 1979) advocates a more intermediate position, based on his view that some groups of plesiadapiforms are linked with later primates by their petrosal bulla and cranial arterial supply, but much of the evidence for such a classification is either unavailable or equivocal (MacPhee *et al.*, 1983). This leaves only the generalized dental similarities and skeletal affinities as evidence for linking plesiadapiforms and later primates. Until we can more clearly document the details of plesiadapiform morphology and reconstruct the phyletic relationships of other orders of mammals relative to later primates, the exact position of the plesiadapiforms in the early radiation of placental mammals will remain unresolved.

BIBLIOGRAPHY

PALEOCENE EPOCH

Adams, C.G. (1981). An outline of Tertiary paleogeography. In *The Evolving Earth*, ed. L.R.M. Cocks, pp. 221–235. Cambridge, Mass.: Cambridge University Press.

Hickey, L.J. (1980). Paleocene stratigraphy and flora of the Clark's Fork Basin. In *Early Cenozoic Paleontology and Stratigraphy of the Bighorn Basin, Wyoming 1880–1980*, ed. P.D. Gingerich, pp. 33–50. University of Michigan, Museum of Paleontology, Papers on Paleontology no. 24.

Krause, D. (1984). Mammalian evolution in the Paleocene: The beginning of an era. In *Mammals: Notes for a Short Course*, ed. T.D. Broadhead, pp. 87–109. Knoxville: University of Tennessee, Dept. of Geological Sciences.

PRIMATE ORIGINS

Archibald, J.D. (1977). Ectotympanic bone and internal carotid circulation of eutherians in reference to anthropoid origins. *J. Hum. Evol.* **6**:609–622.

Clemens, W.A. (1974). *Purgatorius*, an early paromomyid primate (Mammalia). *Science* **184**:903–906.

Gingerich, P.D. (1986). *Plesiadapis* and the delineation of the order Primates. In *Major Topics in Primate and Human Evolution*, ed. B.Wood, L. Martin, and P. Andrews, pp. 32–46. Cambridge: Cambridge University Press.

Rose, K.D., and Fleagle, J.G. (1981). The fossil history of nonhuman primates in the Americas. In *Ecology and Behavior of Neotropical Primates*, vol. 1, ed. A.F. Coimbra-Filho and R.A. Mittermeier, pp. 111–167. Rio de Janeiro: Academeia Brasileria de Ciencias.

Szalay, F.S. (1975). Where to draw the nonprimate-primate taxonomic boundary. *Folia Primatol.* **23**:158–163.

———. (1972). Paleobiology of the earliest primates. In *The Functional and Evolutionary Biology of Primates*, ed. R. Tuttle, pp. 3–35. Chicago: Aldine-Atherton.

Szalay, F.S., and Drawhorn, J. (1980). Evolution and diversification of the Archonta in an arboreal milieu. In *Comparative Biology and Evolutionary Relationships of Tree Shrews*, ed. W.P. Luckett, pp. 133–169. New York: Plenum Press.

Van Valen, L., and Sloan, R.E. (1965). The earliest primates. *Science* **150**:743–745.

Wible, J.R., and Covert, H.H. (1987). Primates: Cladistic diagnosis and relationships. *J. Hum. Evol.* **16**:1–20.

PLESIADAPIFORMS

Covert, H.H. (1986). Biology of early Cenozoic primates. In *Comparative Primate Biology*, vol. 1: *Systematics, Evolution, and Anatomy*, ed. D.R. Swindler and J. Erwin, pp. 335–359. New York: Alan R. Liss.

Gidley, J.W. (1923). Paleocene primates of the Fort Union, with discussion of relationships of Eocene primates. *Proc. U.S. Nat. Mus.* **63**:1–38.

Gingerich, P.D. (1986). *Plesiadapis* and the delineation of the order Primates. In *Major Topics in Primate and Human Evolution*, ed. B. Wood, L. Martin, and P. Andrews, pp. 32–46. Cambridge: Cambridge University Press.

Krause, D.W. (1978). Paleocene primates from Western Canada. *Canadian J. Earth Sci.* **15**:1250–1271.

MacPhee, R.D.E., Cartmill, M., and Gingerich, P.D. (1983). New Palaeogene primate basicrania and the definition of the order Primates. *Nature* **301**:509–511.

Simons, E.L. (1967). Fossil primates and the evolution of some primate locomotor systems. *Am. J. Phys. Anthropol.* **26**:241–253.

Simpson, G.G. (1937). The Fort Union of the Crazy Mountain Field, Montana and its mammalian faunas. *Bull. U.S. Nat. Mus.* **169**:1–287.

Szalay, F.S. (1972). Paleobiology of the earliest primates. In *The Functional and Evolutionary Biology of Primates*, ed. R. Tuttle, pp. 3–35. Chicago: Aldine-Atherton.

Szalay, F.S., and Dagosto, M. (1980). Locomotor adaptations as reflected in the humerus of Paleogene primates. *Folia Primatol.* **34**:1–45.

Szalay, F.S., Tattersall, I., and Decker, R. (1975). Phylogenetic relationships of *Plesiadapis*—postcranial evidence. *Contrib. Primatol.* **5**:136–166.

Teilhard de Chardin, P. (1922). Les mammiferes de l'Eocene inferieur francais et leurs gisements. *Ann. Paleontol.* **11**:9–116.

Microsyopids

Bown, T.M., and Gingerich, P.D. (1973). The Paleocene primate *Plesiolestes* and the origin of Microsyopidae. *Folia Primatol.* **19**:1–18.

Bown, T.M., and Rose, K.D. (1976). New early Tertiary primates and a reappraisal of some Plesiadapiformes. *Folia Primatol.* **26**:109–138.

Fox, R.C. (1984). The dentition and relationships of the Paleocene primate *Micromomys* Szalay, with description of a new species. *Can. J. Earth Sci.* **21**(11):1262–1267.

Gunnell, G.F. (1985). Systematics of early Eocene Microsyopidae (Mammalia, Primates) in the Clark's Fork Basin, Wyoming. *Contr. Mus. Paleontol. Univ. Michigan* **27**:51–71.

Hoffstetter, R. (1986). Paleontologie. Limite entre primates et non-primates; position des Plesiadapiformes et des Microsyopidae. *C.R. Acad. Sci. (Paris)*, t. 302, serie II, no. 1, pp. 43–45.

Kay, R.F., and Cartmill, M. (1977). Cranial morphology and adaptations of *Palaechthon nacimienti* and other Paromomyidae (Plesiadapoidea, Primates), with a description of a new genus and species. *J. Hum. Evol.* **6**:19–53.

MacPhee, R.D.E., Cartmill, M., and Gingerich, P.D. (1983). New Palaeogene primate basicrania and the definition of the order Primates. *Nature* **301**:509–511.

Rose, K.D., and Bown, T.M. (1982). New plesiadapiform primates from the Eocene of Wyoming and Montana. *J. Vert. Paleontol.* **2**(1):63–69.

Szalay, F.S. (1968). The beginnings of primates. *Evolution* **22**:19–36.

———. (1969). Mixodectidae, Microsyopidae, and the insectivore-primate transition. *Bull. Am. Mus. Nat. Hist.* **140**:195–330.

———. (1973). New Paleocene primates and a diagnosis of the new suborder Paramomyiformes. *Folia Primatol.* **19**:73–87.

———. (1981). Phylogeny and the problems of adaptive significance: The case of the earliest primates. *Folia Primatol.* **36**:157–182.

Plesiadapids

Gingerich, P.D. (1973). First record of the Paleocene primate *Chiromyoides* from North America. *Nature* **244**:517–518.

———. (1974). Dental function in the Paleocene primate *Plesiadapis*. In *Prosimian Biology*, ed. R.D. Martin, G.A. Doyle, and A.C. Walker, pp. 531–541. London: Duckworth.

———. (1975). New North American Plesiadapidae (Mammalia, Primates) and a biostratigraphic zonation of the middle and upper Paleocene. *Contr. Mus. Paleontol. Univ. Michigan* **24**:135–148.

———. (1976). Cranial anatomy and evolution of early Tertiary Plesiadapidae (Mammalia, Primates). University of Michigan, Museum of Paleontology, Papers on Paleontology no. 15.

Russell, D. (1964). Les mammiferes paleocenes d'Europe. *Mem. Mus. Nat. d'Hist. Natur., ser. C.* **13**:1–324.

Szalay, F.S. (1975). Phylogeny of primate higher taxa: The basicranial evidence. In *Phylogeny of the Primates: A Multidisciplinary Approach*, ed. W.P. Luckett and F.S. Szalay, pp. 91–125. New York: Plenum Press.

Szalay, F.S., Tattersall, I., and Decker, R. (1975). Phylogenetic relationships of *Plesiadapis*—postcranial evidence. *Contrib. Primatol.* **5**:136–166.

Carpolestids

Biknevicius, A. (1986). Dental function and diet in the Carpolestidae (Primates: Plesiadapiformes) *Am. J. Phys. Anthropol.* **71**:157–172.

Fox, R.C. (1984). A new species of the Paleocene primate *Elphidotarsius* Gidley: Its stratigraphic position and evolutionary relationships. *Can. J. Earth Sci.* **21**(11):1268–1277.

Rose, K.D. (1975). The Carpolestidae, early Tertiary primates from North America. *Bull. Mus. Comp. Zool.* **147**:1–74.

———. (1977). Evolution of carpolestid primates and chronology of the North American middle and late Paleocene. *J. Paleontol.* **51**(3):536–542.

Saxonellids

Fox, R.C. (1984). First North American record of the Paleocene primate *Saxonella. J. Paleontol.* **58**(3):892–894.

Russell, D. (1964). Les mammiferes paleocenes d' Europe. *Mem. Mus. Nat. d'Hist. Natur., ser. C.* **13**:1–321.

Paromomyids

Gingerich, P.D. (1974). Function of pointed premolars in *Phenacolemur* and other mammals. *J. Dent. Res.* **53**:497.

Godinot, M. (1984). Un noveau genre de Paromomyidae (Primates) de l'Eocene Inferieur d'Europe. *Folia Primatol.* **43**:84–96.

Hickey, L.J., West, R.M., Dawson, M.R., and Choi, D.K. (1983). Arctic terrestrial biota: Paleomagnetic with mid-northern latitudes during the late Cretaceous and early Tertiary. *Science* **221**:1153–1156.

MacPhee, R.D.E., Cartmill, M., and Gingerich, P.D. (1983). New Palaeogene primate basicrania and the definition of the order Primates. *Nature* **301**:509–511.

McKenna, M.C. (1980). Eocene paleolatitude, climate and mammals of Ellesmere Island. *Palaeogeogr., Palaeoclimatol., Palaeoecol.* **30**:349–362.

Rose, K.D., and Bown, T.M. (1982). New plesiadapiform primates from the Eocene of Wyoming and Montana. *J. Vert. Paleontol.* **2**(1):63–69.

Rose, K.D., and Gingerich, P.D. (1976). Partial skull of the plesiadapiform primate *Ignacius* from the early Eocene of Wyoming. *Contrib. Mus. Paleontol., Univ. Michigan* **24**:181–189.

Russell, D.E., Louis, P., and Savage, D.E. (1967). Primates of the French early Eocene. *Univ. California Geol. Sci.* **73**:1–46.

Simpson, G.G. (1955). The Phenacolemuridae, a new family of early primates. *Bull. Am. Mus. Nat. Hist.* **105**:415–441.

Szalay, F.S. (1972). Cranial morphology of the early Tertiary *Phenacolemur* and its bearing on primate phylogeny. *Am. J. Phys. Anthropol.* **36**:59–76.

Picrodontids

Gingerich, P.D., Houde, P., and Krause, D.W. (1983). A new earliest Tiffanian (Late Paleocene) mammalian fauna from Bangtail Plateau, Western Crazy Mountain Basin, Montana. *J. Paleontol.* **57**:957–970.

Szalay, F.S. (1968). The Picrodontidae, a family of early primates. *Am. Mus. Nov*, no. 2329, pp. 1–55.

———. (1972). Paleobiology of the earliest primates. In *The Functional and Evolutionary Biology of Primates*, ed. R. Tuttle, pp. 3–35. Chicago: Aldine-Atherton.

Tomida, Y. (1982). A new genus of picrodontid primate from the Paleocene of Utah. *Folia Primatol.* **37**:37–43.

ADAPTIVE RADIATION OF PLESIADAPIFORMS

Gingerich, P.D. (1976). Cranial anatomy and evolution of early Tertiary Plesiadapidae (Mammalia, Primates). University of Michigan, Papers on Paleontology no. 15, pp. 1–40.

Mass, M.C., Krause, D.W., and Strait, S.G. (1988). Decline and extinction of plesiadapiforms in North America: Displacement or replacement? *Paleobiology*, in press.

Maas, M.C., Strait, S.G., and Krause, D.W. (1987). Decline and extinction of Plesiadapiformes (?Primates: Mammalia) in North America. *Am. J. Phys. Anthropol.* **72**:228.

Sussman, R.W., and Raven, P.H. (1978). Pollination by lemurs and marsupials: An archaic coevolutionary system. *Science* **200**:731–736.

Szalay, F.S. (1972). Paleobiology of the earliest primates. In *The Functional and Evolutionary Biology of Primates*, ed. R.L. Tuttle, pp. 3–35. Chicago: Aldine-Atherton.

Van Valen, L., and Sloan, R.E. (1966). The extinction of the multituberculates. *Syst. Zool.* **15**:261–278.

PLESIADAPIFORMS AND LATER PRIMATES

Cartmill, M. (1972). Arboreal adaptations and the origin of the order Primates. In *The Functional and Evolutionary Biology of Primates*, ed. R. Tuttle, pp. 97–122. Chicago: Aldine-Atherton.

———. (1974). Rethinking primate origins. *Science* **184**:436–443.

Gingerich, P.D. (1981). Why study fossils? *Am. J. Primatol.* **1**:293–295.

———. (1986). *Plesiadapis* and the delineation of the order Primates. In *Major Topics in Primate and Human Evolution*, ed. B. Wood, L. Martin, and P. Andrews, pp. 32–46. Cambridge: Cambridge University Press.

Luckett, W.P., ed. (1980). *Comparative Biology and Evolutionary Relationships of Tree Shrews*. New York: Plenum Press.

MacPhee, R.D.E., Cartmill, M., and Gingerich, P.D. (1983). New Palaeogene primate basicrania and the definition of the order Primates. *Nature* **301**:509–511.

Martin, R.D. (1968). Towards a new definition of primates. *Man* **3**(3):377–401.

———. (1986). Primates: A definition. In *Major Topics in Primate and Human Evolution*, ed. B. Wood, L. Martin, and P. Andrews, pp. 1–31. Cambridge: Cambridge University Press.

Simpson, G.G. (1940). Studies on the earliest primates. *Bull. Am. Mus. Nat. Hist.* **77**:185–212.

Szalay, F.S. (1975a). Phylogeny of primate higher taxa: The basicranial evidence. In *Phylogeny of the Primates: A Multidisciplinary Approach*, ed. W.P. Luckett and F.S. Szalay, pp. 91–125. New York: Plenum Press.

———. (1975b). Where to draw the nonprimate-primate taxonomic boundary. *Folia Primatol.* **23**:158–163.

Szalay, F.S., and Decker, R.L. (1974). Origins, evolution and function of the tarsus in late Cretaceous eutherians and Paleocene primates. In *Primate Locomotion*, ed. F.A. Jenkins, pp. 239–259. New York: Academic Press.

Szalay, F.S., and Delson, E. (1979). *Evolutionary History of the Primates*. New York: Academic Press.

Szalay, F.S., Rosenberger, A.L., and Dagosto, M. (1987). Diagnosis and differentiation of the order Primates. *Yrbk. Phys. Anthropol.* **30**:75–105.

Wible, J.R., and Covert, H.H. (1987). Primates: Cladistic diagnosis and relationships. *J. Hum. Evol.* **16**:1–20.

Fossil Prosimians

EOCENE EPOCH

In North America and Europe, the beginning of the Eocene epoch (57–37 million years ago) was marked by a major change in faunas. Many modern types of mammals, including the earliest artiodactyls, perissodactyls, and rodents, replaced more archaic types of mammals. In primate evolution, this epoch is marked by the disappearance of most plesiadapiforms and the first appearance of primates that resemble living prosimians (Fig. 11.1). These faunal changes took place in a series of waves rather than in a single broad sweep (Rose, 1981; Gingerich and Rose, 1977); they seem to be the result of both climatic changes and new connections between continents or major continental areas.

The paleogeography in the Eocene was not strikingly different from that at the beginning of the Paleocene (see Fig. 10.1).

FIGURE 11.1

Geographic distribution of fossil prosimian sites.

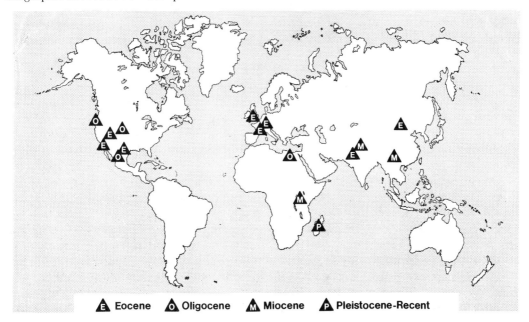

▲E Eocene ▲O Oligocene ▲M Miocene ▲P Pleistocene-Recent

North America and Europe became increasingly separated and distinct in their mammalian faunas throughout the epoch, and there is faunal evidence for intermittent connections between North America and Asia. While the Tethys Seaway remained open across most of the Mediterranean region and southern Asia, farther east, India was coming in contact with Asia. South America remained isolated from other continents except Antarctica. Little is known of Africa during this time, but there are indications that a large seaway separated the northwest corner from the rest of the continent.

Eocene climates in Europe and North America were warmer and more equable than those of the preceding epoch (see Fig. 9.3). Both the sediments and the flora indicate tropical climates for North America. Climates were so warm during the early part of the Eocene that there was a relatively diverse fauna of mammals, including a paromomyid, living well within the Arctic Circle. It has been suggested that some of the early Eocene immigrants to northwestern North America came from farther south along with the increasingly warmer climate.

The First Modern Primates

The primates that made their debut in the early Eocene were quite different from the plesiadapiforms of the preceding Paleocene epoch. They were considerably more advanced and had all the anatomical features characteristic of living primates. They had shorter snouts, smaller infraorbital foramina, and a postorbital bar completing the bony ring around their orbits (Fig. 11.2). They had larger, more rounded braincases, and their auditory region and cerebral blood supply were like those of living prosimians. Their skeletons had more slender limbs with a divergent, grasping hallux, and they possessed nails rather than claws on most digits (Dagosto, 1988).

All of these morphological differences indicate that the Eocene primates practiced a very different way of life from the archaic primates they succeeded. Many of the cranial differences indicate an increased reliance on vision rather than smell and tactile vibrissae. The postcranial changes suggest increased importance of manipulative abilities, with the replacement of claws by nails,

and the locomotor skeletons of many species suggest leaping abilities and more primate-like, acrobatic locomotion. In one species there are even suggestions of a social organization similar to that seen in living platyrrhines. As Simons (1972) has aptly dubbed them, they were the "first primates of modern aspect."

Like the plesiadapiforms they replaced, the early prosimians were among the most abundant mammals of their day, but they were not equally successful on all continents. They are common elements in the mammalian faunas throughout North America and Europe, are less well known from Asia or Africa, and are unknown from South America or Antarctica (Fig. 11.1). There are few indications of the geographic or phyletic origin of early prosimians. Their first appearances in Europe and North America seem to be at approximately the same time; indeed, early Eocene faunas of those two continents are virtually identical. The earliest Asian prosimians may be slightly older than the earliest in Europe and North Amer-

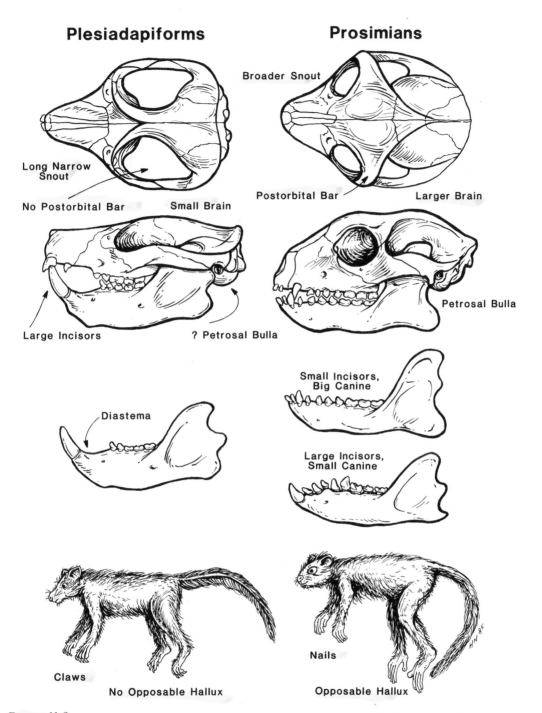

Plesiadapiforms

Long Narrow Snout

No Postorbital Bar   Small Brain

Large Incisors   ? Petrosal Bulla

Diastema

Claws

No Opposable Hallux

Prosimians

Broader Snout

Postorbital Bar   Larger Brain

Petrosal Bulla

Small Incisors, Big Canine

Large Incisors, Small Canine

Nails

Opposable Hallux

FIGURE 11.2

Comparison of fossil prosimians and more archaic plesiadapiforms, showing major anatomical contrasts.

ica, but the Asian species are poorly known and even their prosimian status has been questioned. As we noted in Chapter 10, there are no good phyletic ancestors for early prosimians among the plesiadapiforms; all the latter are too specialized in their dentition.

From their first appearance in the early Eocene, these prosimians can be readily divided into two distinct groups—the lemur-like adapids and the tarsier- or galago-like omomyids. The earliest members of the two families (*Donrussellia, Cantius,* and *Teilhardina*) are very similar, suggesting a divergence just prior to the earliest Eocene. Both families subsequently produced adaptive radiations of species that flourished throughout the epoch, and their collateral relatives are thriving today in the forests of Africa, Madagascar, and Asia.

Adapids

In many aspects of their anatomy, adapids are the most primitive of all known primates, fossil or living. Most of the specializations

FIGURE 11.3

Mandibles of representative adapids.

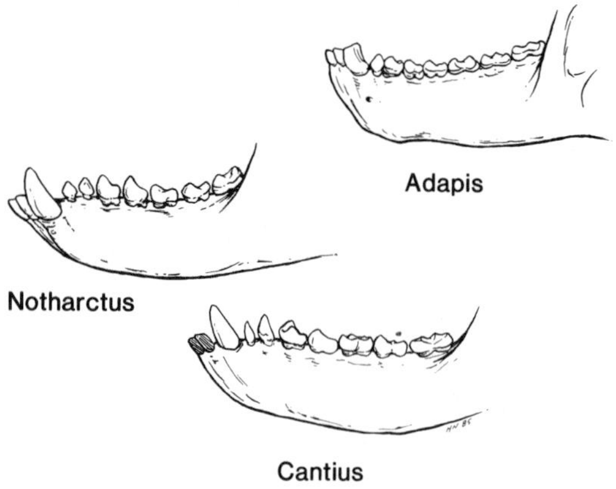

found among later primates could easily be derived from an early adapid morphology. As we discuss later, such a basically primitive morphology poses interesting difficulties and virtually unlimited possibilities in ascertaining the phyletic relationships of adapids with later primate groups.

Compared with the earlier plesiadapiforms and the contemporaneous omomyids, most adapids were rather large primates, comparable in size to living lemurids. The primitive adapid dental formula (Figs. 11.2, 11.3), retained by many relatively late members of the family, is $\frac{2.1.4.3}{2.1.4.3}$. Adapids differ from plesiadapiforms and omomyids and superficially resemble living anthropoids in their anterior dentition. The lower incisors are small and positioned vertically in the mandible, and the uppers are relatively broad, but short, and are separated by a median gap. Both upper and lower canines are larger than the incisors and, in some taxa, are sexually dimorphic. The anterior premolars are often caniniform and the posterior ones are often molariform. The upper molars are broad, and the two major lineages independently evolved a hypocone. Lower molars are relatively long and narrow in most taxa. Development of numerous shearing crests, presumably as an adaptation to folivory, appears to have evolved independently in many adapid lineages along with fusion of the two halves of the mandible.

Adapids have relatively long but broad snouts with a small infraorbital foramen (Figs. 11.2, 11.4). Their orbits, like those of living prosimians, are encircled by a complete bony ring. They have a large ethmoid recess with numerous ethmoturbinates, as in lemurs and primitive mammals generally. The braincase is larger than that of the archaic primates but smaller than in extant lemurs or anthropoids. The tympanic ring is suspended within the inflated bony bulla, much as in extant lemurs. The bony canals for stapedial and promontory branches of the internal carotid artery are apparently quite variable. Even within a single species, some individuals apparently have a larger canal for the stapedial, some have a larger promontory canal, and still others have similar-size canals for the two (Gingerich and Martin, 1981).

The skeletal anatomy, which is well known for several North American genera, shows that adapid limbs are similar to those of living strepsirhines but more robust. These Eocene prosimians have relatively long legs, a long trunk, and a long tail. Their extremities have a divergent pollex and a grasping foot (Fig. 11.5).

The systematics of adapids has been studied by many workers—and not without disagreement. The biostratigraphy of species from the western United States is particularly well documented (Gingerich, 1984). Eocene adapids are divided into two subfamilies that are largely (but not completely) distinct biogeographically. The notharctines are a predominantly North American group whose earliest genus is also found in western

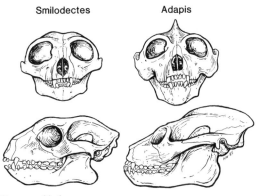

FIGURE 11.4

Reconstructed skulls of two adapids (approximately one half actual size).

Europe; the adapines are a typically European group with one relatively late genus from North America. By the early Oligocene, both subfamilies seem to have become extinct in North America and Europe. The third subfamily of adapids, the sivaladapines, is from the late Miocene of Asia. In addition, there are various African and Asian fossil primates whose adapid affinities are dubious or unsettled.

FIGURE 11.5

Reconstructed skeleton of *Smilodectes gracilis* (redrawn from Simons, 1964).

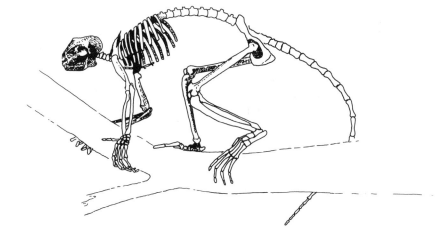

TABLE 11.1
Infraorder Lemuriformes
Family Adapidae
Subfamily NOTHARCTINAE

Species	Body Weight (g)
Cantius (e. Eocene, North America, Europe)	
C. torresi	1,100
C. ralstoni	1,300
C. mckennai	1,600
C. trigonodus	2,000
C. abditus	3,000
C. frugivorus	2,800
C. venticolis	3,000
C. eppsi	—
C. savagei	—
Copelemur (e. Eocene, North America)	
C. tutus	3,600
C. feretutus	2,000
C. consortutus	1,600
C. praetutus	1,300
Notharctus (m. Eocene, North America)	
N. robinsoni	4,700
N. tenebrosus	4,200
N. pugnax	5,500
N. robustior	6,900
Smilodectes (m. Eocene, North America)	
S. mcgrewi	3,000
S. gracilis	2,100
Pelycodus (e. Eocene, North America)	
P. jarrovii	4,500

Notharctines

The notharctines (Table 11.1) were among the most common mammals in the early and middle Eocene faunas of western North America, but they had only limited diversity (Fig. 11.6). There were never more than two or three synchronic species and only a total of five genera in the 5 million years from which the group is known. The earliest notharctine, and the earliest adapid, is **Can-**

tius, with numerous species from North America (Gingerich, 1986) and two from Europe (Simons, 1962). *Cantius* was a small- to medium-size adapid ranging from about 1.5 kg in the earliest and smallest species to over 4 kg in the latest.

Cantius (Fig. 11.3) has a dental formula of $\frac{2.1.4.3}{2.1.4.3}$. The lower molars have a simple trigonid with three cusps and a broad-basined talonid; the upper molars are simple tritubercular teeth in the early species, but later species (in North America) developed a hypocone from the postprotocingulum (or nannopithex fold). All species have four premolars, prominent canines, and two small vertical incisors. The mandibular symphysis is unfused in this early adapid genus. *Cantius* was probably largely frugivorous.

The partial skulls and few skeletal remains of *Cantius* resemble those of the better-known, later genera in most aspects. They indicate a diurnal species that moved primarily by arboreal quadrupedal running and leaping (Rose and Walker, 1985). There is inconclusive evidence as to whether *Cantius* was sexually dimorphic in canine size like most higher primates or resembled most living prosimians in lacking any obvious dental sexual dimorphism.

In North America, *Cantius* gave rise to **Pelycodus** in the early Eocene and **Notharctus** in the middle Eocene (Fig. 11.6). **Copelemur**, from the early Eocene, and **Smilodectes**, from the middle Eocene, seem to represent a separate lineage that was more common at southern latitudes (Beard, 1988).

Notharctus is larger (up to 8 kg) than *Cantius* and has larger hypocones and meso-styles on the upper molars, reduced paraconids on the lower molars, and a fused mandibular symphysis. Because the transition from *Cantius* to *Notharctus* was a gradual and essentially continuous one, this last feature is arbitrarily used to delineate the

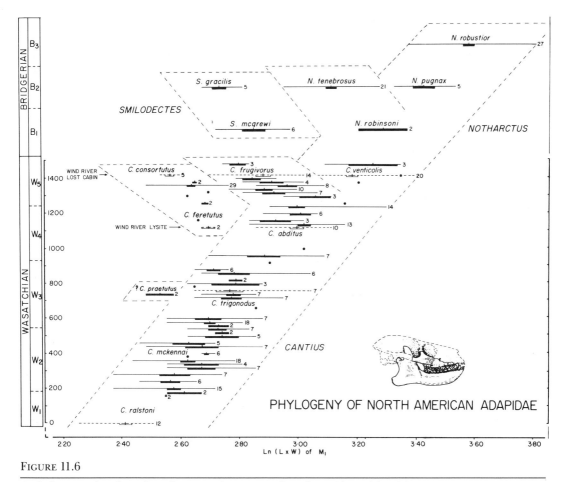

A phylogeny of notharctines from northern Wyoming (from Gingerich, 1984).

two genera (Fig. 11.6). The cheek teeth of *Notharctus* have well-developed shearing crests and the genus was certainly folivorous (Covert, 1986).

Notharctus is similar to *Lemur* in both overall cranial proportions and in details of its basicranial anatomy. The Eocene genus is more robustly built and has a smaller braincase and more pronounced sagittal and nuchal crests. There is a moderately long snout with a large premaxillary bone. The lacrimal bone is at the edge of the orbit rather than anterior to it, as in extant lemurs. The auditory region has a free tympanic ring lying within the bulla and stapedial and promontory arterial canals of similar size. Although the size and position of these canals are widely used to reconstruct patterns of cranial circulation in fossil mammals, it is important to keep in mind that there is not necessarily a one-to-one correspondence between bony canals and arteries in living primates (see Conroy and Wible, 1978).

Several virtually complete skeletons are known for *Notharctus*. Gregory (1920) found that the Eocene genus is most similar (but not identical) in skeletal proportions and details of limb architecture to the extant genera *Lemur*, *Varecia*, *Lepilemur*, and *Propithecus* but has relatively more robust bones. *Notharctus* has extremely long hindlimbs (intermembral index = 60), a long flexible trunk, and a long tail. The ilium is sickle-shaped, as in extant lemurs, and the ischium is rather long. The pollex and hallux are large and opposable; the digits are long and tipped with nails. In most but not all details of muscle attachment that could be reconstructed, it is similar to living prosimian leapers. The calcaneus is rather short, as in *Varecia* (Martin, 1972). There is little doubt that *Notharctus* was an adept leaper and quadrupedal runner, but it probably did not cling to vertical supports as do living indriids (Gebo, 1985).

Smilodectes (Figs. 11.4, 11.5) was a smaller (2 kg) middle Eocene contemporary of *Notharctus* characterized by narrower teeth, a shorter snout, and a more rounded frontal bone. Like *Notharctus*, it was diurnal and folivorous. Its external brain morphology is known from several endocasts. Compared with other Eocene mammals, *Smilodectes* had an expanded visual cortex and reduced olfactory bulbs; its brain was larger than that of most contemporaneous mammals but smaller than that of extant prosimians (Radinsky, 1975, 1977). Both *Notharctus* and *Smilodectes* apparently became extinct in the middle Eocene (Fig. 11.6).

Adapines

The adapines (Table 11.2) are known from the Eocene of Europe and also from the late Eocene of North America. Adapines had a much more diverse evolutionary radiation than the relatively uniform notharctines (Fig. 11.7). They ranged in size from tiny, presumably insectivorous species the size of a pygmy marmoset (100 g) to large folivorous species similar in dental size to a howling monkey. There is abundant cranial material of several species, but associated skeletal material is extremely rare (Filhol, 1883; Dagosto, 1983; Godinot and Jouffroy, 1984).

Adapines seem to have evolved from the genus *Donrussellia*, a genus similar to *Cantius* with a full dental formula of $\frac{2.1.4.3}{2.1.4.3}$, simple tritubercular upper molars, and lower molars with a simple trigonid and a broad talonid. In contrast with the North American notharctines, the European adapines developed a hypocone from the lingual cingulum rather than from the protocone as in the notharctines. Because of the diversity of the group, the relative difficulty of placing many isolated localities in a reliable stratigraphic framework, and, not least of all, the radically different systematic philosophies of recent students of European adapines, the systematics of this group is more complicated than that of their North American relatives (cf. Gingerich, 1977b; Szalay and Delson, 1979; Schwartz and Tattersall, 1982a, b, 1983). There are at least ten genera and twenty species of adapines from western Europe (Fig. 11.7).

In dental morphology, most genera and species of adapines retained a full complement of teeth (Figs. 11.3, 11.4); a few reduced the number of premolars to three. Most taxa have a small, cingulum-derived hypocone on the upper molars, but they vary in many details of molar structure, such as cusp height and crest development. The adapines evolved numerous dental adaptations, indicative of considerable dietary diversity within the subfamily. The small (120 g) *Anchomomys gaillardi*, for example,

TABLE 11.2
Infraorder Lemuriformes
Family Adapidae
Subfamily ADAPINAE

Species	Body Weight (g)	Species	Body Weight (g)
Donrussellia (e. Eocene, Europe)		*Cryptadapis* (l. Eocene, Europe)	
D. gallica	—	C. tertius	—
D. provincialis	—	*Microadapis* (l. Eocene, Europe)	
D. magna	—	M. sciureus	600
Protoadapis (e.–m. Eocene, Europe)		*Anchomomys* (l. Eocene, Europe)	
P. curvicuspidens	2,500	A. gaillardi	110
P. filholi	—	A. pygmaea	250
P. lemoinei	—	A. quercyi	—
P. recticuspidens	1,600	*Agerinia* (m. Eocene, Europe, Asia)	
P. russelli	700	A. roselli	—
P. louisi	1,100	A. sp.	
P. weigelti	3,000	*Adapis* (l. Eocene–e. Oligocene, Europe)	
P. ulmensis	1,400	A. betillei	—
Europolemur (l. Eocene, Europe)		A. parisiensis	1,300
E. klatti	1,700	A. sudrei	1,400
E. koenigswaldi	—	A. laharpei	1,700
E. dunaefi	—	*Leptadapis* (l. Eocene–e. Oligocene, Europe)	
Periconodon (m. Eocene, Europe)			
P. helveticus	250	L. magnus	4,000
P. huerzleri	570	L. assolicus	3,000
P. roselli	650	L. capellae	—
P. lemoinei	—	L. priscus	1,300
Caenopithecus (l. Eocene, Europe)		L. ruetimeyeri	2,500
C. lemuroides	3,500	*Mahgarita* (l. Eocene, North America)	
Pronycticebus (l. Eocene, Europe)		M. stevensi	700
P. gaudryi	1,100		
Cercamonius (l. Eocene, Europe)			
C. brachyrhynchus	4,000		

has extremely simple upper molars, not unlike those of marmosets (Gingerich, 1977a). Judging from its sharp molar cusps and tiny size, this species was almost certainly insectivorous. The larger *Pronycticebus gaudryi* has relatively simple molar teeth with sharp cusps, a robust, tusklike upper canine, and a long row of sharp premolars, suggesting a

carnivorous diet (Szalay and Delson, 1979). *Periconodon* and *Microadapis* have molars with broader, more bulbous cusps suggestive of fruit eating. Others, such as *Adapis*, *Leptadapis*, and *Caenopithecus*, have large but linear hypocones, extensive buccal styles on the upper molars, and extreme development of crests rather than individual cusps on all

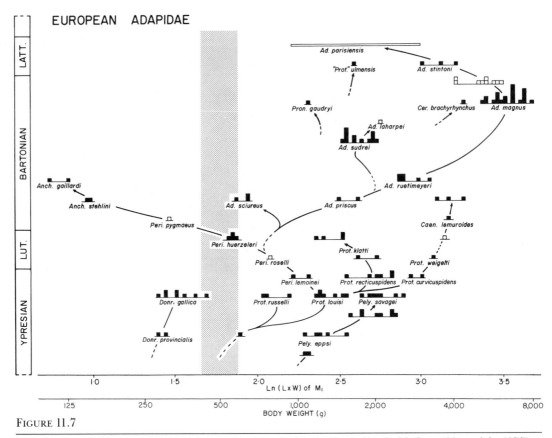

FIGURE 11.7

A phylogeny of European adapines; cross-hatching indicates Kay's threshold (from Gingerich, 1977).

teeth, suggesting that they were predominantly folivorous.

The best known of the European adapines is **Adapis parisiensis** (Fig. 11.4), a medium-size (2 kg) species from several late Eocene deposits in France. This species was first described by Cuvier in 1822, well before the discovery of any other fossil primates, but its primate affinities were not recognized until some fifty years later. *Adapis parisiensis*, the latest adapine in Europe, disappeared during the major European faunal turnover known as the Grand Coupure, which coin-

cided with a major drop in temperature near the Eocene–Oligocene boundary.

Adapis parisiensis has a full complement of teeth (Figs. 11.3, 11.4). Like most adapids and living lemurs, it has upper central incisors that are relatively broad and spatulate with a gap between their bases, presumably for an organ of Jacobson. The upper lateral incisors are smaller and positioned behind the upper centrals. There is no indication of sexual dimorphism in size of the upper canines.

The lower anterior dentition of *A. parisien-*

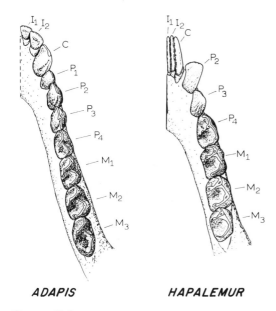

ADAPIS   HAPALEMUR

FIGURE 11.8

The lower dentition of *Adapis parisiensis* compared with that of *Hapalemur griseus*, a Malagasy prosimian. Note the similarities in the molar and premolar morphology and the contrasts in the anterior dentition.

sis is unusual in that the lower incisors and canines form a single cutting edge (Fig. 11.8). Gingerich has suggested that this morphology represents incipient development of a tooth comb as seen in extant strepsirhines. Like the tooth combs of many mammals, the lower incisors of *Adapis* have fine parallel striations on the enamel, indicating that they were used in grooming. *Adapis* has long narrow molars and premolars with well-developed shearing crests. They are strikingly similar to the molars of *Hapalemur*, suggesting a folivorous diet for *Adapis*.

Adapis has a very low, broad skull (Fig. 11.4) with flaring zygomatic arches, a small braincase, and prominent sagittal and nuchal crests in the larger individuals (males?).

The orbits are relatively small, suggesting diurnal habits, and are oriented slightly upward rather than directly forward. The snout is moderately short. From the robust zygomatic arches and the extremely large temporal fossa, it is clear that *Adapis* had extremely large chewing muscles, concordant with the folivorous nature of its dentition.

The auditory region of *Adapis*'s skull has an inflated bulla with a free tympanic, as in extant strepsirhines. There is always a canal for the stapedial artery and a groove for the promontory artery, but the relative sizes of these canals vary from one specimen to another. The brain is relatively small compared with that of extant prosimians and has a large olfactory bulb.

There are several relatively complete limb bones of *A. parisiensis*. Analyses (Dagosto, 1983) of these bones suggest that in its locomotor abilities *Adapis* was most similar to the living lorises *Nycticebus* and *Perodicticus*—slow arboreal quadrupeds (Fig. 11.9). However, a nearly complete hand of this species shows no indication of special grasping abilities comparable to that of lorises (Godinot and Jouffroy, 1984). The joint between the ulna and the wrist in *Adapis* shows features linking it with extant lemurs and lorises (Beard *et al.*, 1988).

Leptadapis magnus was a large (8.5 kg), earlier relative of *Adapis* that is often placed in the same genus. Like *Adapis*, it was probably a large diurnal folivore that moved by quadrupedal climbing. *Leptadapis* is one of the only fossil or living prosimians for which there is good evidence of sexual dimorphism in both cranial size and canine size, and it seems likely that this large Eocene adapine lived in polygynous social groups.

One of the most unusual adapid fossils is a half-skeleton of a small primate from the

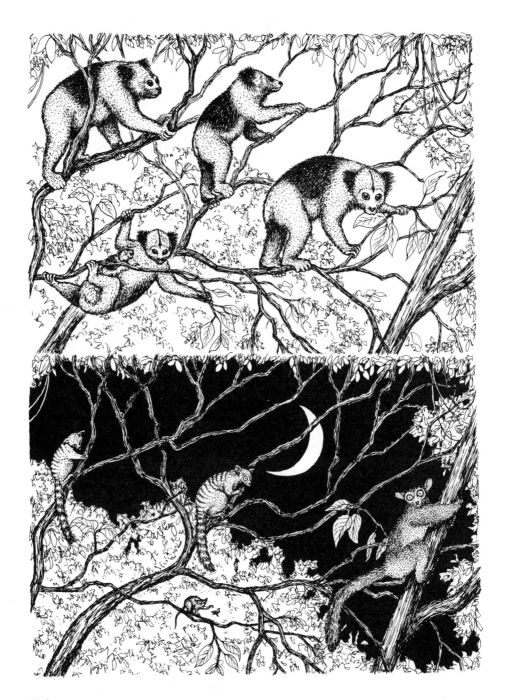

FIGURE 11.9

Scene from the late Eocene of the Paris Basin. Above, the diurnal *Adapis parisiensis* feed on leaves. Below are several nocturnal microchoer- ines; the tiny *Pseudoloris* attempts to catch an insect while *Necrolemur* (left) and *Microchoerus* (right) cling to branches.

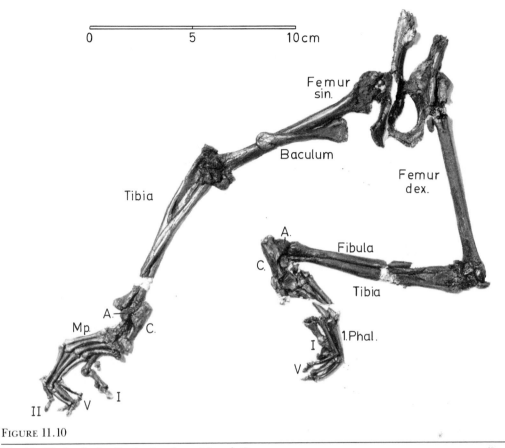

FIGURE 11.10

The hindlimb of a fossil adapine from Messel, Germany.

oil shales of Messel, Germany (Fig. 11.10). Because only the lower half of the skeleton has been found, it cannot be confidently assigned to any genus or species. The limbs are different from those of either *Adapis* or *Leptadapis* and suggest a leaping form. Like living strepsirhines, this species has a "grooming claw" on the second digit of its foot. It also has a very large baculum (penis bone) for an animal of its size.

Mahgarita stevensi, from the late Eocene of Texas, is the only North American adapine. It is of interest that it occurs after the apparent extinction of the notharctines at the end of the middle Eocene (Wilson and Szalay, 1976, 1977). *Mahgarita* has relatively small premolars and, like the European adapines, the hypocone on the upper molars is derived from the lingual cingulum. The mandibular symphysis is fused. The strong development of crests on the molar teeth, as well as its moderate size (1200 g), suggest that it was probably folivorous.

Sivaladapines

Several primates from the Eocene of China have from time to time been identified as

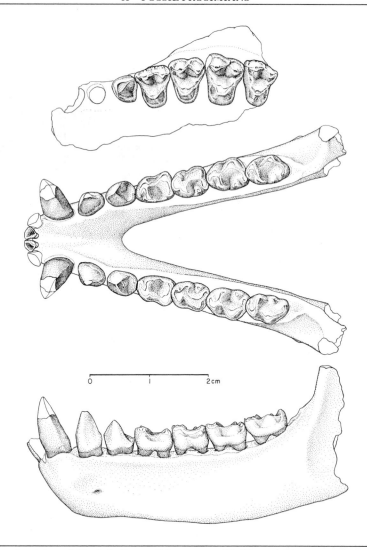

FIGURE 11.11

Upper and lower dentition of *Sivaladapis nagrii* (courtesy of P. Gingerich).

either adapids or omomyids. Russell and Gingerich (1987) have described several Eocene adapids from Pakistan, including a new genus, **Panobius**. However, well after the notharctines and adapines disappeared from North America and Europe, there were a number of adapid-like primates thriving alongside fossil apes in the late Miocene of India, Pakistan, and China. The best known

of these, **Sivaladapis nagrii** (Table 11.3), from the late Miocene of India, was fairly large (5–6 kg) with a dental formula of $\frac{2.1.3.3.}{2.1.3.3.}$ (Fig. 11.11). The sharp crests on its molars and premolars suggest a folivorous diet for *Sivaladapis*. Unlike the latest members of either the European adapines or the North American notharctines, *Sivaladapis* has simple upper molars with no hypocone. There

TABLE 11.3
Infraorder Lemuriformes
Family Adapidae
Subfamily SIVALADAPINAE

Species	Body Weight (g)
Indraloris (l. Miocene, Asia)	
I. himalayensis	2,500
Sivaladapis (l. Miocene, Asia)	
S. nagrii	2,700
S. palaeindicus	—
Sinoadapis (l. Miocene, Asia)	
S. carnosus	—

is a similar, large adapid, **Sinoadapis**, from the latest Miocene site of Lufeng in China (Wu and Pan, 1985). The relationship of sivaladapines to either of the Eocene subfamilies is unclear, and there are as yet no skulls or limbs of the Asian adapids.

Possible African Adapids

There are two poorly known genera from the Eocene and Oligocene of North Africa that have been identified by some authorities as adapids; other authorities place them in different families (Table 11.4). *Azibius trerki*

TABLE 11.4
Infraorder Lemuriformes
Family Adapidae
Subfamily *incertae sedis*

Species	Body Weight (g)
Azibius (Eocene, Africa)	
A. trerki	120
Panobius (?e.–m. Eocene, Asia)	
P. afridi	130
Hoanghonius (Eocene, Asia)	
H. stehlini	700
Lushius (l. Eocene, Asia)	
L. qinlinensis	—

(Sudre, 1975) is a tiny mammal from the Eocene of Algeria known from a single jaw with three teeth. The present material is insufficient to either deny or confirm adapid affinities.

Oligopithecus savagei (see Fig. 12.15) is a moderate-size (1.5 kg) early Oligocene primate from Egypt that is regarded by many authors as an early anthropoid (Simons, 1972; Szalay and Delson, 1979) and by others as an adapid (Gingerich, 1980). The limited material (one jaw and a few isolated teeth) and numerous differences from other adapids (or anthropoids) preclude a reliable identification of its affinities, but it seems more likely that *Oligopithecus* is an early anthropoid (see Chapter 12).

ARE ADAPIDS STREPSIRHINES?

Since adapids were first identified as primates, virtually all authors have noted their many anatomical similarities to living strepsirhines, and particularly to lemurs. Adapids are lemurlike in their cheek teeth, in the overall configuration of their skull with its simple postorbital bar and moderately long snout, and in the morphology of the nasal region. The auditory region is also lemurlike, with an inflated bulla and a free ectotympanic ring, and the carotid circulation is more similar to that of lemurs than to that of either haplorhines or lorises in that most individuals have a stapedial canal of moderate size. In virtually all of these features, both adapids and strepsirhines probably retain the primitive primate condition found in many other mammals rather than share unique specializations. Furthermore, adapids lack a tooth comb, the derived feature that most clearly distinguishes living strepsirhines from other primates, and they also seem to have retained more primitive hands and feet than many Malagasy species.

Adapids and living strepsirhines share only a few anatomical features that may be unique specializations linking the two but also precluding ancestral relations to other primates. One is the grooming claw of the second toe, which is present in the Messel adapid and in all extant strepsirhines. Eocene adapids and strepsirhines also share two unusual features of the ankle, a flaring fibular surface on the talus and the arrangement of the cuneiform facets of the navicular (Dagosto, 1988), and adapines (but not notharctines) have a strepsirhine-like articulation between the ulna and the carpus (Beard *et al.*, 1988). Finally, adapids and extant strepsirhines are characterized by tiny, spatulate upper incisors that are quite different from the incisors of either extant anthropoids or other prosimians (Rosenberger *et al.*, 1985).

TABLE 11.5
Infraorder Lemuriformes
Superfamily LORISOIDEA

Species	Body Weight (g)
Family GALAGIDAE	
Progalago (e. Miocene, Africa)	
P. dorae	1,200
P. songhorensis	800
Komba (e. Miocene, Africa)	
K. robusta	300
K. minor	125
Galago (Pliocene–Recent, Africa)	
G. howelli	700
G. sadimanensis	200
Family LORISIDAE	
Mioeuoticus (e. Miocene, Africa)	
M. bishopi	?300
M. spp.	
Nycticeboides (l. Miocene, Asia)	
N. simpsoni	500

The overall anatomical similarity between adapids and strepsirhines clearly demonstrates that living strepsirhines have retained many aspects of an adapid-like morphology for nearly 60 million years, but at present there is very little evidence to indicate how the later, tooth-combed strepsirhines are related to the radiation of Eocene (and Miocene) adapids.

In addition to their traditional link with strepsirhines, the adapids have also been proposed as the ancestors of higher primates (Gingerich, 1980; Rasmussen, 1986). This suggestion has been based largely on their anthropoid-like anterior dentition, fused symphysis, and similar size range. Because the issue of anthropoid origins requires a comparison of early anthropoid morphology with that of all potential anthropoid ancestors, we must defer consideration of the relationship between adapids and anthropoids until the next chapter.

Fossil Lorises and Galagos

In addition to the recently extinct Malagasy species (see Chapter 4), one group of fossil prosimians that can be linked clearly to living strepsirhines are the fossil lorisoids from the Oligocene(?), Miocene, Pliocene, and Pleistocene of Africa and Asia (Table 11.5). The earliest possible record of this group is a single upper molar from the early Oligocene of Egypt that Simons (Simons *et al.*, 1987) has identified as that of a loris. There are several genera and species of lorises and galagos from the early Miocene of Kenya and Uganda. One genus, *Mioeuoticus* (Fig. 11.12), seems to be related to the lorises, and two others, *Komba* and *Progalago*, seem to be closer to living galagos (Walker, 1978).

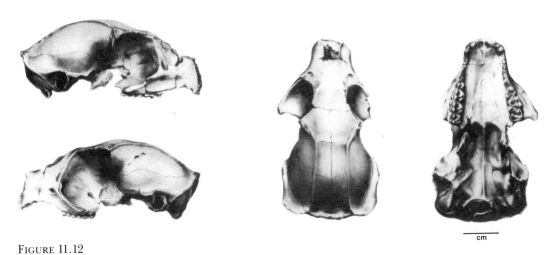

cm

FIGURE 11.12

The skull of a fossil loris, *Mioeuoticus*, from the Miocene of eastern Africa (from Le Gros Clark, 1956).

These Miocene prosimians are very similar to living African genera in their dental and cranial anatomy and probably had tooth combs, but none can be positively linked to any living genus or species. The galagos have elongated limbs, but their tarsals are not as elongated as those of living galagos. They are more similar to the tarsals of cheirogaleids.

Younger fossil galagos, from 2 to 3 million years ago in Ethiopia and Tanzania, are similar to the living *Galago* and *Otolemur*.

The earliest fossil record of the Asian lorises comes from the late Miocene of Pakistan, from which there is a fossil species, **Nycticeboides simpsoni**, that seems closely related to the living slow loris, *Nycticebus*.

Omomyids

Like adapids, omomyids first appeared in the earliest Eocene (Fig. 11.1) of North America, Europe, and possibly Asia (Dashzaveg and McKenna, 1977; Szalay and Li, 1986). Omomyids, like adapids, had a very different evolutionary history on the two continents where their evolution is well known. In North America they were very diverse taxonomically throughout most of the Eocene, with a few genera from the Oligocene. In Europe omomyids were less diverse, with only a single, poorly known genus, *Teilhardina*, from the early Eocene, and four genera from the middle and late Eocene. Omomyids are divided into three subfamilies: the Anaptomorphinae and the Omomyinae, both predominantly North American, and the European Microchoerinae.

Primitive omomyids are very similar to early adapids in their dental morphology, and it seems likely that they were derived from a primate with a dentition similar to that of *Cantius*. Most early omomyids have a dental formula of $\frac{2 \cdot 1 \cdot 3 \cdot 3}{2 \cdot 1 \cdot 3 \cdot 3}$, but many individuals of the early genus *Teilhardina* have four premolars (Fig. 11.13). The mandibular symphysis of omomyids is always unfused. The anterior dentition of omomyids is different from that of adapids. Most omomyids have a relatively large, procumbent lower central incisor and a smaller lateral one, and the canines are usually small—never large as in adapids or absent as in some plesiadapiforms (Figs. 11.2, 11.3, 11.14). The premolars vary considerably among subfamilies. In some, they are tall and pointed; in others, they are molariform. The upper molars are usually broad. Many early species have a prominent postprotocingulum (nannopithex fold) joining the protocone distally, and later species developed a hypocone from the lingual cingulum. The lower molars usually have relatively small, low, mesiodistally compressed trigonids and broadbasined talonids.

The skulls of most omomyids resemble extant tarsiers and galagos in their relatively short, narrow snout, posteriorly broadening palate, and large eyes (Figs. 11.15, 11.16).

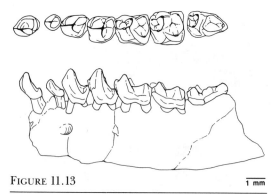

FIGURE 11.13   ⎯1 mm⎯

The lower jaw of *Teilhardina americana*, the oldest and most primitive omomyid in North America.

The auditory region of some species has an inflated auditory bulla and an tympanic that is fused to the bullar wall and extends laterally to form a bony tube. The internal carotid circulation is known in only two genera. In one, the stapedial and promontory canals are of similar caliber (Simons and Russell, 1960); in the other, the promontory is much larger (as in *Tarsius* and anthropoids).

FIGURE 11.14

Mandibles of representative omomyid primates from North America and Europe. The positions of the canine (C) and the first molar (M_1) are indicated.

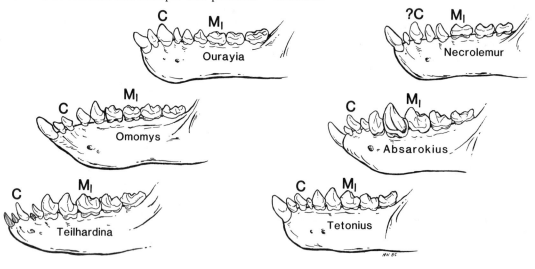

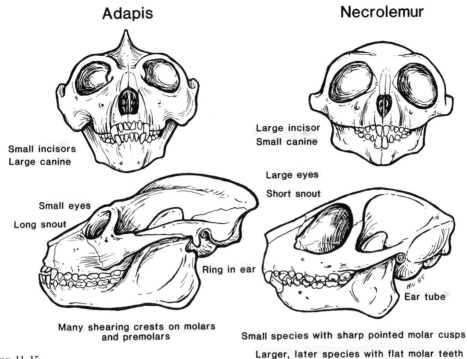

Adapis

Small incisors
Large canine

Small eyes

Long snout

Ring in ear

Many shearing crests on molars
and premolars

Necrolemur

Large incisor
Small canine

Large eyes

Short snout

Ear tube

Small species with sharp pointed molar cusps

Larger, later species with flat molar teeth

FIGURE 11.15

Comparative cranial morphology of an adapid and an omomyid.

FIGURE 11.16

Skulls of two small nocturnal living primates, *Microcebus murinus* and *Tarsius syrichta*, compared with reconstructed skulls of several omomyids, *Tetonius homunculus*, *Necrolemur antiquus*, and *Roo-* *neyia viejensis*. Note that *Tarsius* has relatively much larger eyes than the living strepsirhines and the fossil primates.

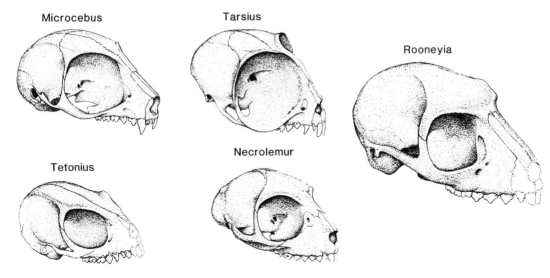

Microcebus

Tarsius

Rooneyia

Necrolemur

Tetonius

There are no complete skeletons and only a few skeletal elements known for omomyids (Szalay and Delson, 1979; Dagosto, 1985). In at least four genera, the calcaneus is moderately elongated, as in extant cheirogaleids, and in two European omomyids the distal tibia and fibula are appressed or fused, as in extant *Tarsius*. Most known skeletal elements indicate leaping, but not clinging, habits for these early prosimians.

Anaptomorphines

Anaptomorphines (Table 11.6) are the most primitive of the three subfamilies of omomyids. The earliest and most primitive genus, ***Teilhardina*** (Fig. 11.13), is from the early Eocene of both Europe and North America, and the remaining members of the subfamily (over a dozen genera) are from the early and middle Eocene of North America.

TABLE 11.6
Suborder Prosimii
Family Omomyidae
Subfamily ANAPTOMORPHINAE

Species	Body Weight (g)	Species	Body Weight (g)
Teilhardina (e. Eocene, North America, Europe)		*A. metoecus*	200
T. belgica	90	*A. gazini*	160
T. americana	120	*A. nocerai*	175
T. crassidens	90	*A. australis*	130
T. tenuicula	135	*Anaptomorphus* (m. Eocene, North America)	
Anemorhysis (e. Eocene, North America)		*A. aemulus*	275
A. sublettensis	70	*A. wortmani*	160
A. pearci	105	*A. westi*	465
A. wortmani	180	*Trogolemur* (m.–l. Eocene, North America)	
A. pattersoni	170	*T. myodes*	75
A. nettingi	100	*Aycrossia* (m. Eocene, North America)	
Chlororhysis (e. Eocene, North America)		*A. lovei*	325
C. knightensis	165	*Strigorhysis* (m. Eocene, North America)	
C. incomptus	—	*S. bridgeriensis*	500
Tetonius (e. Eocene, North America)		*S. rugosus*	—
T. homunculus	290	*S. huerfanensis*	?600
T. mckennai	100	*Gazinius* (m. Eocene, North America)	
T. matthewi	180	*G. amplus*	470
Pseudotetonius (e. Eocene, North America)		*Steinius* (e. Eocene, North America)	
P. ambiguus	170	*S. vespertinus*	310
Absarokius (e.–m. Eocene, North America)		*Loveina* (e. Eocene, North America)	
A. abbotti	200	*L. zephyri*	170
A. noctivagus	200	*L. minuta*	95
A. witteri	500		

Despite their systematic diversity, anaptomorphines are all relatively similar in many aspects of their morphology. All are very small, probably ranging from about 50 to 500 g. Later members of the subfamily are usually characterized by a tall pointed P_4 and a reduced M_3. Many species have only two premolars. Their lower molars have relatively low trigonids with bulbous cusps, and shallow talonids. The lower incisors are very large in some species but smaller in others.

The cranium is known from only one anaptomorphine, ***Tetonius homunculus***, from the early Eocene of Wyoming. It has a short snout, large eyes, and a relatively globular braincase (Fig. 11.16). Unfortunately, the auditory region is extremely damaged. The teeth of *Tetonius* suggest that it was probably largely insectivorous (Fig. 11.17). Its orbits are similar in size to those of a living cheirogaleid or a small galago, suggesting that it was nocturnal. Because the orbits are relatively smaller than those of *Tarsius*, it seems likely that it had a tapetum lucidum, like living strepsirhines.

Omomyines

The other subfamily of early North American omomyids, the omomyines (Table 11.7), was probably derived from an anaptomorphine-like ancestor. Indeed, the earliest members of the two subfamilies are virtually indistinguishable, and several early Eocene genera cannot be placed confidently in one subfamily rather than the other. The major adaptive radiation of omomyines was later than that of the more primitive anaptomorphines and followed the disappearance of the notharctines in the middle Eocene of North America (see Fig. 11.20). Omomyines were most abundant from the middle Eocene through the early Oligocene, with one late Oligocene genus. They ranged in size from about 100 g to over 2 kg.

FIGURE 11.17

Dentitions of several omomyids, showing different dietary adaptations.

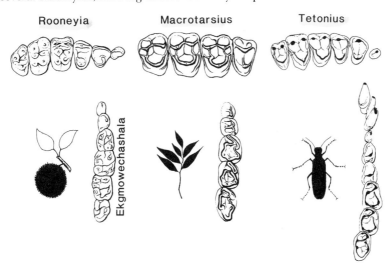

TABLE 11.7
Suborder Prosimii
Family Omomyidae
Subfamily OMOMYINAE

Species	Body Weight (g)	Species	Body Weight (g)
Arapahovius (e. Eocene, North America)		*Dyseolemur* (l. Eocene, North America)	
A. gazini	290	*D. pacificus*	165
Omomys (m. Eocene, North America)		*Stockia* (l. Eocene, North America)	
O. carteri	310	*S. powayensis*	475
O. lloydi	180	*Macrotarsius* (l. Eocene–e. Oligocene, North America)	
Chumashius (l. Eocene, North America)		*M. seigerti*	1,635
C. balchi	295	*M. montanus*	2,520
Ourayia (l. Eocene, North America)		*M. jepseni*	—
O. uintensis	2,170	*Uintanius* (?e.–m. Eocene, North America)	
O. hopsoni	1,150	*U. ameghini*	150
Shoshonius (e.–m. Eocene, North America)		*Jemezius* (e. Eocene, North America)	
S. cooperi	155	*J. szalayi*	155
Washakius (m.–l. Eocene, North America)		*Rooneyia* (e. Oligocene, North America)	
W. insignis	165	*R. viejaensis*	1,475
W. woodringi	130	*Ekgmowechashala* (l. Oligocene, North America)	
Utahia (l. Eocene, North America)		*E. philotau*	1,870
U. kayi	95		
Hemiacodon (m. Eocene, North America)			
H. gracilis	1,005		

Omomyines evolved a far greater range of dental adaptations than did the anaptomorphines. Their molars often have lower cusps and the trigonid cusps are less inflated; the last molar is usually elongated. Many later members of the family developed very flat molars with accessory cusps and crenulated enamel. Omomyines probably occupied a variety of dietary niches (Fig. 11.17). Whereas many of the earlier, smaller species with high trigonid crests and narrow talonid basins were probably insectivorous, later, larger species with broad, flat molars and rounded cusps (*Rooneyia* and *Ekgmowechashala*) were almost certainly frugivorous. One species, *Macrotarsius*, the largest known omomyid, has well-developed shearing crests and large stylar cusps, indicative of folivory.

The skull is known from only one omomyine, **Rooneyia viejaensis** (Fig. 11.16), a relatively late genus from the early Oligocene of Texas which may not be representative of the subfamily at all. *Rooneyia* has a relatively broad, short snout and moderately large orbits surrounded by a complete postorbital bar. On the basis of orbit size, it seems most likely that *Rooneyia* was diurnal. The braincase is relatively large, in the range of extant prosimians. The auditory region has an uninflated bulla with a tubular bony ectotympanic partly enclosed by the bulla. The details of its carotid circulation are unknown. The omomyid affinities of *Rooneyia* have been contested by several workers who suggest that it shows greater affinities with the European microchoerines.

The limb skeleton is well known for only one species of omomyine, **Hemiacodon gracilis**. All of the bones suggest an animal adapted for leaping. However, the individual limb elements seem to be more similar to the bones of living lemurs and galagos than to those of *Tarsius* (Simpson, 1940; Dagosto, 1985). The distal parts of the tibia and fibula are not fused, as in *Tarsius* or many European omomyids; they seem to have been firmly conjoined by connective tissues.

Microchoerines

The microchoerines (Table 11.8) were a small but diverse group of omomyids from the middle Eocene through the latest Eocene of western Europe. They were probably derived from an early anaptomorphine such as *Teilhardina*. The four genera vary in size from tiny **Pseudoloris** (50–120 g) to the medium-size **Microchoerus** (900–1800 g), and all are relatively abundant in the fossil record.

The dental formula for microchoerines has never been satisfactorily resolved, which complicates attempts to understand the relationships between this group of fossils and later primates (Fig. 11.18). The upper dentition has a formula of $\frac{2.1.3.3}{}$, with a large upper central incisor followed by a small

TABLE 11.8
Suborder Prosimii
Family Omomyidae
Subfamily MICROCHOERINAE

Species	Body Weight (g)
Nannopithex (m. Eocene, Europe)	
N. pollicaris	125
N. raabi	170
N. filholi	155
N. quaylei	—
Pseudoloris (l. Eocene, Europe)	
P. parvulus	45
P. isabenae	50
P. crusafonti	75
P. requanti	120
Necrolemur (l. Eocene, Europe)	
N. zitteli	290
N. antiquus	320
Microchoerus (l. Eocene–e. Oligocene, Europe)	
M. erinaceus	1,775
M. edwardsi	930
M. ornatus	915
M. wardi	—

lateral incisor, a moderate-size canine, three relatively simple premolars, and three molars. The most primitive genus, **Nannopithex**, has no hypocone on the upper molars but has a long postprotocingulum, or nannopithex fold. In the three other genera there is a hypocone derived from the lingual cingulum.

The lower dentition has one less tooth than the upper tooth row, but one of the teeth is so small that it does not occlude with anything, so the occlusal relationships cannot be used to interpret the homologies of the teeth (Fig. 11.18; Schmid, 1983). Thus, microchoerine lower dentitions contain a large procumbent tooth, probably I_1, followed by two small teeth (either I_2, C; C, P_2;

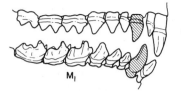

Necrolemur antiquus   *Tarsius syrichta*

FIGURE 11.18

A lateral view of the anterior dentition of *Necrolemur antiquus*, showing the dental proportions. Various authorities have identified each of the first three teeth as the canine. It seems most likely that the shaded tooth is the canine and that the teeth anterior to it are incisors. Note that regardless of how the dental formula of *Necrolemur* is interpreted, the dental proportions are very different from those of *Tarsius* (adapted from Schmid, 1983).

or P_1, P_2). The large tooth is roughly similar in shape to the canine in *Tarsius*, but it clearly functioned differently from that tooth in the living tarsier, which is used primarily to kill animal prey. The large tooth in both *Necrolemur* and *Microchoerus* developed heavy wear on the tips (from scraping and gouging) as well as fine parallel striations on its mesial surface, indicating that it also functioned as a grooming tooth (Schmid, 1983).

The cheek teeth of microchoerines vary considerably among the genera. The tiny *Nannopithex* has an enlarged, pointed premolar and anaptomorphine-like molars with a high trigonid and deep, narrow talonid, suggesting an insectivorous diet. *Pseudoloris* is similar. The larger genera, *Necrolemur* and *Microchoerus*, have molars with low, rounded cusps and elaborate crenulations of the enamel; these suggest a more frugivorous diet, or, considering their anterior dentition, perhaps a diet supplemented by gums.

There are many complete, usually crushed, skulls of *Necrolemur* (Figs. 11.15, 11.16) and cranial fragments of *Microchoerus*, *Pseudoloris,* and *Nannopithex*. All have a relatively short, narrow snout with a bell-shaped palate, a gap between the upper central incisors, large eyes, and a moderately large infraorbital foramen. The olfactory bulb apparently passes above the orbits as in all extant haplorhines, but the back of the orbit is not walled off from the temporal fossa as in *Tarsius* and anthropoids. In the ear region, the ectotympanic forms a ring within the bulla but extends laterally to form a bony tube. This unique condition resembles strepsirhines in the position of the ring and *Tarsius* in the tube (Cartmill, 1982). The canal for the stapedial artery and the groove for the promontory artery are similar in size. There is extensive inflation of the mastoid region behind the middle and inner ear. The large eyes of microchoerines suggest that they were all nocturnal animals. Like *Tetonius*, their large orbits are strepsirhine-like in proportions rather than *Tarsius*- or *Aotus*-like, suggesting that they probably had a tapetum lucidum.

Although there are no complete skeletons for microchoerines, numerous isolated hindlimb elements have been attributed to species of this subfamily; these include a nearly complete femur, a partly fused tibia-fibula, a talus (Godinot and Dagosto, 1983), and a calcaneus for *Necrolemur*, and isolated tarsal bones probably attributable to *Microchoerus*.

All of these postcranial elements indicate leaping abilities. In their elongation, however, the calcanei of microchoerines are more like those of cheirogaleids than those of *Tarsius* (Schmid, 1979).

Asian Omomyids

The evidence of omomyids in Asia (Table 11.9) is poor. **Altanius orlovi** is a tiny (30 g), insectivorous primate from the early Eocene of Mongolia known from a single jaw. **Kohatius** is a somewhat larger (200 g) species from the early to middle Eocene of Pakistan known from only a few teeth. Other fossil prosimians from the Eocene of China (e.g., *Hoanghonius*, Table 11.4) have been alternately identified as either adapids or omomyids by various authors. The omomyid affinities of all of the above, except *Kohatius*, have been questioned. Until we have more complete material, such issues will remain unresolved.

Tarsiids

One of the most recently discovered fossil prosimians is a small tarsiiform primate from the early Oligocene deposits of Fayum, Egypt (Simons and Bown, 1985). The new species, **Afrotarsius chatrathi** (Fig. 11.19, Table 11.10), is the first tarsier-like primate from Africa. Because the anterior dentition is not known, it is not possible to determine whether it is more closely related to the living *Tarsius* or to the European microchoerines, or even to the early anthropoids from Egypt (Ginsburg and Mein, 1987). In any case, it documents the presence of tarsier-like primates on the African continent.

One of the most exciting finds in recent years is **Tarsius thailandica**, a tarsier from the Miocene of Thailand (Ginsburg and

TABLE 11.9
Suborder Prosimii
Family Omomyidae
Subfamily *incertae sedis*

Species	Body Weight (g)
Altanius (e. Eocene, Asia)	
A. orlovi	30
Kohatius (e.–m. Eocene, Asia)	
K. coppensi	190

Mein, 1987). This fossil, known from only a lower molar, seems to place tarsiers in Asia by the early Miocene.

OMOMYIDS AND LATER PRIMATES

As small prosimians with large eyes, elongate calcanei, and in some species a fused tibia-fibula, omomyids have been traditionally linked with the extant *Tarsius*, just as their contemporaries, the adapids, have been allied with extant strepsirhines. Simons (1972) has even placed the European microchoerines into the family Tarsiidae. As with the adapid–lemur relationship discussed earlier, the omomyid–*Tarsius* connection has come under increased scrutiny in recent years, and it has become clear that omo-

TABLE 11.10
Suborder Prosimii
Family TARSIIDAE

Species	Body Weight (g)
Afrotarsius (e. Oligocene, Africa)	
A. chatrathi	100
Tarsius (e. Miocene–Recent, Asia)	
T. thailandica	—

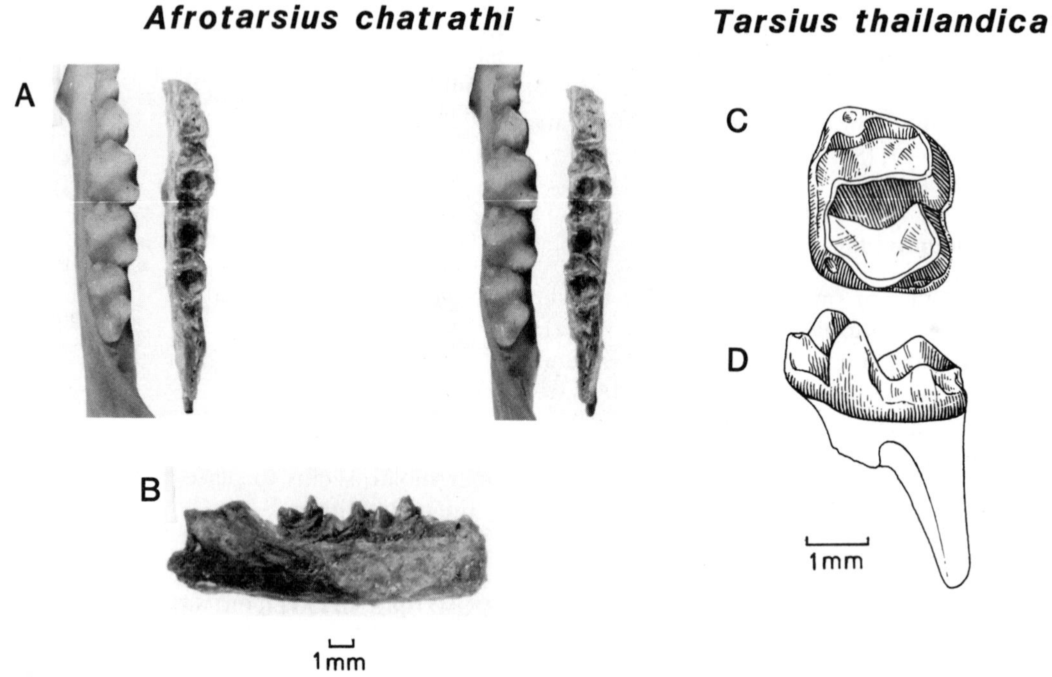

Afrotarsius chatrathi

Tarsius thailandica

FIGURE 11.19

Fossil tarsiid lower molars: A, occlusal stereoview of *Tarsius* (left) and *Afrotarsius chatrathi* (right); B, lateral view of *A. chatrathi*; C, occlusal view of *Tarsius thailandica*; D, lateral view of *T. thailandica*.

myids are not simply Eocene tarsiers. Many of their supposed tarsier-like resemblances are based on superficial comparisons, and omomyids undoubtedly lacked many of the distinguishing features that characterize the living *Tarsius*.

The most tarsier-like of the omomyids are the microchoerines. In particular, the genus *Pseudoloris* has been identified as an ancestor of *Tarsius*. However, although the molar teeth of *Pseudoloris* are strikingly like those of *Tarsius*, the teeth of other microchoerines are less obviously indicative of this relationship. Furthermore, the anterior dentition of all microchoerines is clearly different from that of *Tarsius* in both number of lower teeth

and relative sizes of teeth (Fig. 11.17). The large and procumbent incisors of microchoerines seem to preclude them from the ancestry of the Asian genus.

The construction of the auditory region in omomyids also seems to be only superficially like that of *Tarsius*, and, in some regards (the presence of the ring within the bulla), it is actually more like that of a strepsirhine. The carotid circulation of omomyids seems to be the primitive primate pattern rather than that seen among extant haplorhines.

Much of the tarsier-like appearance of omomyids derives from their large orbits. Again, however, the orbits are more similar to those of galagos in both construction (they lack any postorbital closure) and relative size. The relative size of omomyid orbits provides inferential evidence that these early prosimians were like strepsirhines in having an eye with a tapetum lucidum rather than lacking that structure as do *Tarsius* and anthropoids. Because all living haplorhines lack a tapetum, the light-catching efficiency of their eyes is less than that of strepsirhines, and both nocturnal haplorhines (*Tarsius* and *Aotus*) have eyes that are much larger than those of similar-size nocturnal strepsirhines. Although omomyids had large orbits and were almost certainly nocturnal, their orbits were more similar, in relative size (smaller), to those of a nocturnal strepsirhine such as a *Galago* than to *Tarsius* (Fig. 11.15). Thus they probably had a tapetum lucidum like living strepsirhines and lacked the derived haplorhine condition.

Aside from the similarities in molar teeth, the fused tibia-fibula in *Necrolemur*, a few features on the basicranium, and several primitive features of the ankle (Dagosto, 1988), the feature that most strongly suggests a link between omomyids and *Tarsius* (and anthropoids) is the apparent position of the olfactory bulb above the interorbital septum, as in all extant haplorhines. This organization of the anterior part of the cranium is unquestionably a derived condition and was probably present in *Necrolemur* and other omomyids as well (but this is difficult to confirm with crushed skulls). Assuming that this feature did not evolve in parallel in the various lineages, it suggests that omomyids are incipient haplorhines (see Fig. 11.20). Nevertheless, they appear to have retained many primitive primate features that are found today among strepsirhines, and they lacked many of the additional anatomical features that characterize the living tarsiers and some that unite

tarsiers with anthropoids. Behaviorally they were perhaps more like galagos than like tarsiers. At present, the arguments both for and against linking omomyids with later haplorhines are based on very little anatomical evidence. Future fossil finds and further studies should clarify the phyletic relationships of this group.

ADAPTIVE RADIATIONS OF EOCENE PROSIMIANS

The extinct prosimians from the Eocene and early Oligocene of North America and Europe were a diverse group of primates which occupied a wide range of ecological niches. There seem to be clear temporal trends in the adaptive radiations of these early prosimians. The adapids started out at a relatively large size compared to the omomyids. Throughout the Eocene and early Oligocene, adapids seem to have occupied adaptive niches—large size, diurnality, folivory—which characterize extant higher primates, while omomyids were perhaps more comparable to galagos. Only in the later part of the Eocene and the Oligocene do the omomyids appear to have expanded into the adaptive zones of large size and folivory. Even more striking, however, are the phyletic and adaptive differences between the Eocene prosimian faunas on the two continents from which they are well known.

In North America (Fig. 11.20), the omomyids of the early and middle Eocene were taxonomically diverse, but all were relatively small (less than 500 g). Their teeth suggest diets that were predominantly frugivorous and insectivorous. The single skull (*Tetonius*) is from a nocturnal species. In contrast, the North American adapids from the early and middle Eocene were much less taxonomi-

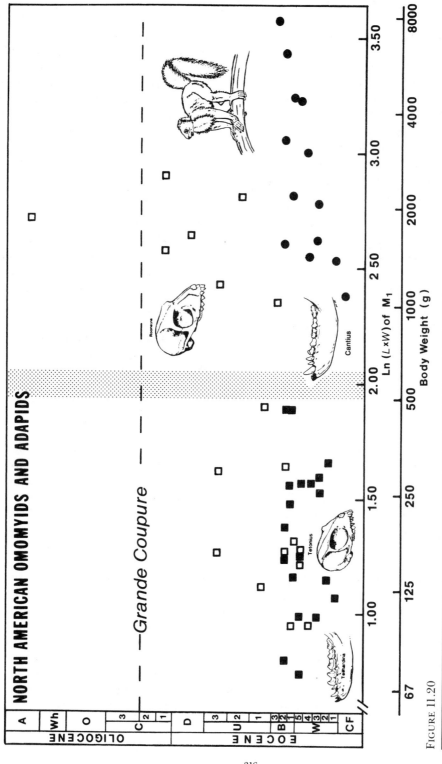

FIGURE 11.20

Size of North American adapids (●), anaptomorphine omomyids (■), and omomyine omomyids (□) through time. Note that adapids are all larger than contemporary anaptomorphines and that the radiation of larger omomyines takes place after the extinction of the adapids. Cross-hatching indicates Kay's threshold.

316

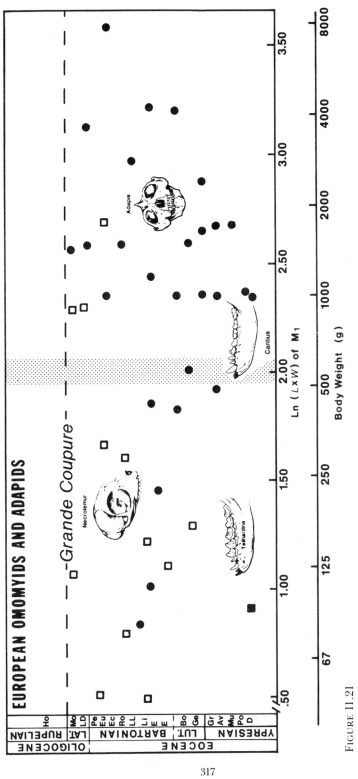

FIGURE 11.21

Size of European adapids (●) and microchoerine omomyids (□) through time. Note that there is considerable overlap in body size in the two radiations and that the adapids are more diverse. Compare to Figure 11.20.

cally diverse, with only five genera, and were all considerably larger (1.5 to 7 kg), frugivorous or folivorous, and probably diurnal. Only after the disappearance of notharctine adapids at the end of the middle Eocene do we find larger, probably frugivorous and folivorous omomyids in North America. The locomotor adaptations of Eocene prosimians are poorly known, but most remains indicate quadrupedal and leaping abilities for both omomyids and adapids.

In Europe (Fig. 11.21), the adapids were more diverse and the omomyids were limited to only four genera after the basal *Teilhardina*. Although the European adapids were generally larger than synchronic omomyids, the size range of the two groups overlapped somewhat in the late Eocene and early Oligocene with the evolution of very small adapids such as *Anchomomys gaillardi* and large microchoerines such as *Microchoerus*. Associated with their size diversity was considerable dietary diversity among the European adapids. There seem to have been insectivorous, frugivorous, and possibly carnivorous (*Pronycticebus*) species as well as many folivorous species. The microchoerines, although less diverse, included small insectivorous species and other species that probably specialized on fruits or gums. One ecological parameter that seems to have separated the two radiations was their activity cycle. Most microchoerines seem to have been nocturnal, and the adapines diurnal, judging from orbit size. Furthermore, the microchoerines seem to have been leapers (or cheirogaline-like arboreal runners), whereas the skeletal remains from adapines suggest slower quadrupedal climbing for some and leaping for others.

Our only information about the social organization of any Eocene prosimian is the sexual dimorphism in *Adapis* and *Leptadapis*, which suggests some sort of a polygynous social system (Gingerich, 1981).

PHYLETIC RELATIONSHIPS OF ADAPIDS AND OMOMYIDS

Although adapids and omomyids have traditionally been identified as Eocene lemurs and tarsiers, respectively, both Eocene families are decidedly more primitive in some respects than the recent prosimians. Moreover, it seems quite clear from the paleontological record that the earliest adapids and omomyids were extremely similar. Indeed, *Donrussellia* has been allocated by some authorities to the omomyids and by others to the adapines. As discussed earlier in this chapter, it is more appropriate to consider these Eocene taxa as basal "primates of modern aspect" from which the modern prosimians evolved. Both are "missing links" that have phyletic affinities with the modern taxa and preserve information about more primitive morphological stages in primate evolution (Fig. 11.22). Compared with later primate taxa, the adapids are clearly very primitive in virtually all aspects of their anatomy, but they may show a few derived features that link them with later strepsirhines. Although clearly distinct from the adapid radiation, omomyids are nevertheless very similar in retaining a more primitive morphology with respect to most later primate groups—and there are indications of a few features in the structure of the orbit, leg, and foot which link them with tarsiers and anthropoids and place them at the base of the haplorhine radiation. Thus it seems likely that the divergence between modern haplorhine and strepsirhine primates corresponds to the initial divergence of omomyids from adapids or a similarly primitive early prosimian group that subsequently gave rise to lemurs and lorises. Although the Eocene adapids and omomyids can be placed in this general phyletic position, more specific details concerning the divergence and radiation of modern prosimians are more difficult

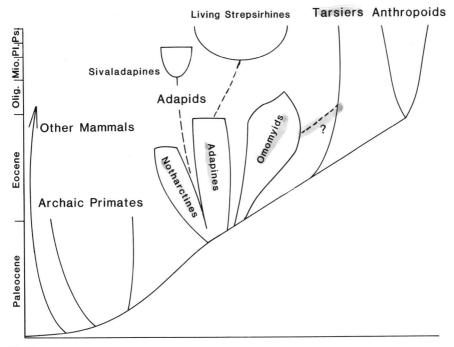

FIGURE 11.22

The phyletic relationships of adapids and omomyids.

to reconstruct. None of the North American or European prosimians from the Eocene seem very closely related to living strepsirhines, *Tarsius*, or anthropoids. The presence in Africa of early Miocene (and possibly Oligocene) lorises, as well as early Oligocene tarsier-like prosimians, suggests that many of the details of prosimian phylogeny lie on that continent.

BIBLIOGRAPHY

EOCENE EPOCH

Adams, C.G. (1981). An outline of Tertiary paleogeography. In *The Evolving Earth*, ed. L.R.M. Cocks, pp. 221–235. Cambridge: Cambridge University Press.

Rose, K.D. (1981) The Clarkforkian land-mammal age and mammalian faunal composition across the Paleocene-Eocene boundary. University of Michigan, Museum of Paleontology, Papers on Paleontology, no. 26.

———. (1984). Evolution and radiation of mammals in the Eocene, and the diversification of modern orders. In *Mammals: Notes for a Short Course*, ed. T.D. Broadhead, pp. 110–127. Knoxville: University of Tennessee, Dept. of Geological Sciences.

Wolfe, J.A. (1978). A paleobotanical interpretation of Tertiary climates in the Northern Hemisphere. *Am. Sci.* **66**:694–703.

THE FIRST MODERN PRIMATES

Covert, H.H. (1986). Biology of early Cenozoic primates. In *Comparative Primate Biology*, vol. 1: *Systematics, Evolution, and Anatomy*, ed. D.R. Swindler and J. Erwin, pp. 335–359. New York: Alan R. Liss.

Dagosto, M. (1988). Implications of postcranial evidence for the origin of euprimates. *J. Hum. Evol.* **17**:35–56.

Simons, E.L. (1972). *Primate Evolution: An Introduction to Man's Place in Nature*. New York: Macmillan.

Szalay, F.S. (1972). Paleobiology of the earliest primates. In *The Functional and Evolutionary Biology of Primates*, ed. R.H. Tuttle, pp. 3–35. Chicago: Aldine-Atherton.

ADAPIDS

Fleagle, J.G. (1978). Size distribution of living and fossil primate faunas. *Paleobiol.* **4**:67–76.

Gazin, C.L. (1958). A review of the middle and upper Eocene primates of North America. *Smithson. Misc. Coll.* **136**:1–112.

Gingerich, P.D. (1980). Eocene Adapidae, paleobiogeography and the origin of the South American Platyrrhini. In *Evolutionary Biology of the New World Monkeys and Continental Drift*, ed. R.L. Ciochon and A.B. Chiarelli, pp. 123–138. New York: Plenum Press.

———. (1984). Primate evolution. In *Mammals: Notes for a Short Course*, ed. T.D. Broadhead, pp. 167–181. Knoxville: University of Tennessee, Dept. of Geological Sciences.

Gingerich, P.D., and Martin, R.D. (1981). Cranial morphology and adaptations in Eocene Adapidae, II: The Cambridge skull of *Adapis parisiensis*. *Am. J. Phys. Anthropol.* **56**:235–257.

Gregory, W.K. (1920). On the structure and relation of *Notharctus*, an American Eocene primate. *Mem. Am. Mus. Nat. Hist. n.s.* **351**:243.

Moorman, S.J., and Fleagle, J.G. (1979). The nasal fossa in extinct strepsirhines. *Am. J. Phys. Anthropol.* **52**:260.

Simons, E.L. (1964). The Early Relatives of Man. *Sci. Am.* **211**:60.

Stehlin, H.G. (1916). Die Saugetiere des schweizerischen Eocaens. Siebenter Tel, zweite Halfte. *Caenopithecus—Necrolemur—Microchoerus—Nannopithex—Anchomomys—Periconodon—Heterochiromys—Nachtrade zu Adapis—Schlussbetrachtugen zu den Primaten. Abh. schweiz. palaeontol. Gessellsch.* **41**:1299–1552.

Szalay, F.S., and Delson, E. (1979). *Evolutionary History of Primates*. New York: Academic Press.

Notharctines

Ankel-Simons, F. (1974). Evolution of primate locomotor systems as seen in the fossil record. *Symp. Fifth Int. Con. Primatol. Soc.*, pp. 265–268.

Beard, K.C. (1988). New notharctine primate fossils from the early Eocene of New Mexico and southern Wyoming and the phylogeny of Notharctinae. *Am. J. Phys. Anthropol.* **75**:439–469.

Conroy, G.C., and Wible, J.R. (1978). Middle ear morphology of *Lemur variegatus*. *Folia Primatol.* **29**:81–85.

Gebo, D.L. (1985). The nature of the primate grasping foot. *Am. J. Phys. Anthropol.* **67**:269–277.

Gingerich, P.D. (1979). Phylogeny of middle Eocene Adapidae (Mammalia, Primates) in North America: *Smilodectes* and *Notharctus*. *J. Paleontol.* **53**(1):153–163.

———. (1984). Primate evolution. In *Mammals: Notes for a Short Course*, ed. T.D. Broadhead, pp. 167–181. Knoxville: University of Tennessee, Dept. of Geological Sciences.

———. (1986). Early Eocene *Cantius torresi*—oldest primate of modern aspect from North America. *Nature (London)* **319**:319–321.

Gingerich, P.D., and Haskin, R.A. (1981). Dentition of early Eocene *Pelycodus jarrovii* (Mammalia, Primates) and the generic attribution of species formerly referred to *Pelycodus*. *Contrib. Mus. Paleontol., Univ. Michigan* **25**(17):327–337.

Gingerich, P.D., and Simons, E.L. (1977). Systematics, phylogeny and evolution of early Eocene Adapidae (Mammalia, Primates) in North America. *Mus. Paleontol.* **24**(22):245–279.

Gregory, W.K. (1920). On the structure and relation of *Notharctus*, an American Eocene primate. *Mem. Am. Mus. Nat. Hist. n.s.* **351**:243.

Martin, R.D. (1972). Adaptive radiation and behavior of the Malagasy lemurs. *Phil. Trans. Royal Soc. London* **264**:295–352.

Napier, J., and Walker, A.C. (1967). Vertical clinging and leaping—a newly recognized category of locomotor behavior in primates. *Folia Primatol.* **6**:204–219.

Radinsky, L. (1975). Primate brain evolution. *Am. Sci.* **63**(6):656–663.

———. (1977). Early primate brains: Facts and fiction. *J. Hum. Evol.* **6**:79–86.

Rose, K.D., and Walker, A. (1985). The skeleton of early Eocene *Cantius*, oldest lemuriform primate. *Am. J. Phys. Anthropol.* **66**:73–89.

Rosenberger, A.L., Strasser, E., and Delson, E. (1985). Anterior dentition of *Notharctus* and the Adapid-Anthropoid hypothesis. *Folia Primatol.* **44**:15–39.

Simons, E.L. (1962). A new Eocene primate genus, *Cantius*, and a revision of some allied European lemuroids. *Bull. Brit. Mus. (Nat. Hist.) Geol.* **7**:1–30.

Adapines

Beard, K.C., Dagosto, M., Gebo, D.L., and Godinot, M. (1988). Interrelationships among primate higher taxa. *Nature* **331**:712–714.

Dagosto, M. (1983). Postcranium of *Adapis parisiensis* and *Leptadapis magnus* (Adapiformes): Adaptational and phylogenetic significance. *Folia Primatol.* **41**:49–101.

Decker, R.L., and Szalay, F.S. (1974). Origins and function of the pes in the Eocene Adapidae (Lemuriformes, Primates). In *Primate Locomotion*, ed. F.A. Jenkins, pp. 261–291. New York: Academic Press.

Filhol, H. (1883). Observations relatives au Memoire de M. Cope intitule: Relation des horizons renfermant des debris d'animaux vertebres fossiles en Europe et en Amerique. *Ann. Sci. Geol., Paris* **14**:1–51.

Gingerich, P.D. (1975). Dentition of *Adapis parisiensis* and the evolution of lemuriform primates. In *Lemur Biology*, ed. I. Tattersall and R.W. Susman, pp. 65–80. New York: Plenum Press.

———. (1977a). New species of Eocene primates and the phylogeny of European Adapidae. *Folia Primatol.* **28**:60–80.

———. (1977b). Radiation of Eocene Adapidae in Europe. *Geobios, Mem. Spec.* **1**:165–182.

———. (1980a). Dental and cranial adaptation in Eocene Adapidae. *Z. Morphol. Anthropol.* **71**(2):135–142.

———. (1980b). Eocene Adapidae, paleobiogeography and the origin of the South American Platyrrhini. In *Evolutionary Biology of the New World Monkeys and Continental Drift*, ed. R.L. Ciochon and A.B. Chiarelli, pp. 123–138. New York: Plenum Press.

———. (1981). Cranial morphology and adaptations in Eocene Adapidae, I: Sexual dimorphism in *Adapis magnus* and *Adapis parisiensis*. *Am. J. Phys. Anthropol.* **56**:217–234.

Gingerich, P.D., and Martin, R.D. (1981). Cranial morphology and adaptations in Eocene Adapidae, II: The Cambridge skull of *Adapis parisiensis*. *Am. J. Phys. Anthropol.* **56**:235–257.

Godinot, M. (1984). Un nouveau genre temoignant de la diversite des Adapines (Primates, Adapidae) a l'Eocene terminal. *C. R. Acad. Sci. (Paris), ser. II,* **299**(18):1291–1296.

Godinot, M., and Jouffroy, F.K. (1984). La main d'*Adapis* (Primates, Adapidae). In *Actes du Symposium Paleontologique G. Cuvier*, ed. E. Buffetaut, J.M. Mazin, and E. Salmion, pp. 221–242. Paris: Montbeliard.

Schwartz, J.H., and Tattersall, I. (1982a). A note on the status of *"Adapis" priscus* Stehlin, 1916. *Am. J. Primatol.* **3**:295–298.

———. (1982b). Relationships of *Microadapis sciureus* (Stehlin, 1916) and two new primate genera from the Eocene of Switzerland. *Folia Primatol.* **39**:178–186.

———. (1983). A review of the European primate genus *Anchomomys* and some allied forms. *Anthropol. Papers Am. Mus. Nat. Hist.* **57**(5):343–352.

Wilson, J.A., and Szalay, F.S. (1976). New adapid primate of European affinities from Texas. *Folia Primatol.* **25**:294–312.

———. (1977). *Mahgarita*, a new name for *Margarita* Wilson and Szalay, 1976 non Leach 1814. *J. Paleontol.* **51**:643.

von Koenigswald, W. (1979). Ein Lemurenrest aus dem eozanen Olschiefer der Grube Messel bei Darmstadt. *Palaeontol. Z.* **53**:63–76.

Asian Adapids

Chopra, S.R.K., and Vasishat, R.N. (1979). A new Mio-Pliocene *Indraloris* (Primate) material with comments on the taxonomic status of *Sivanasua* (Carnivora) from the Siwaliks of the Indian subcontinent. *J. Hum. Evol.* **9**:129–132.

———. (1980). Premiere indication de la presence dans le Mio-Pliocene des Siwaliks de l'Inde d'un Primate Adapidae, *Indoadapis shivaii*, nov. gen., nov. sp., *C.R. Acad. Sci. (Paris), ser. D,* **290**:511–513.

Gingerich, P.D. (1979). *Indraloris* and *Sivaladapis*: Miocene adapid primates from the Siwaliks of India and Pakistan. *Nature (London)* **279**(5712):415–416.

Gingerich, P.D., and Sahni. (1984). Dentition of *Sivaladapis nagrii* (Adapidae) from the late Miocene of India. *Int. J. Primatol.* **5**:63–69.

Pan, Y., and Wu, R. (1986). A new species of *Sinoadapis* from the Hominoid Site, Lufeng. *Acta Anthropol. Sinica* **5**:39–50.

Russell, D.E., and Gingerich, P.D. (1987). Nouveaux primates de l'Eocene du Pakistan. *C.R. Acad. Sci.* t. 304, ser. II (5):209–214.

Wu, R., Han, D., Xu, Q., Lu, Q., Pan, Y., Chen, W., Zhang, X., and Xiaa, M. (1982). More *Ramapithecus* skulls found from Lufeng, Yunnan—Report on the excavation of the site in 1981. *Acta Anthropol. Sinica* **1**(2):106–108.

Wu, R., and Pan, Y. (1985). A new adapid primate from the Lufeng Miocene, Yunnan. *Acta Anthropol. Sinica* **4**(1):1–6.

African Adapids

Gingerich, P.D. (1980). Eocene Adapidae, paleobiogeography and the origin of the South American Platyrrhini. In *Evolutionary Biology of the New World Monkeys and Continental Drift*, ed. R.L. Ciochon and A.B. Chiarelli, pp. 123–138. New York: Plenum Press.

Simons, E.L. (1972). *Primate Evolution: An Introduction to Man's Place in Nature*. New York: Macmillan.

Sudre, J. (1975). Un Prosimien du Paleogene ancien du Sahara nord-occidental: *Azibius trerki* n.g., n. sp. *C.R. Acad. Sci. (Paris)* **280**:1539–1542.

ADAPIDS AND LIVING PRIMATES

Beard, K.C., Dagosto, M., Gebo, D.L., and Godinot, M. (1988). Interrelationships among primate higher taxa. *Nature* **331**:712–714.

Dagosto, M. (1988). Implications of postcranial evidence for the origin of euprimates. *J. Hum. Evol.* **17**:35–56.

Gingerich, P.D. (1980). Eocene Adapidae, paleobiogeography and the origin of the South American Platyrrhini. In *Evolutionary Biology of the New World Monkeys and Continental Drift*, ed. R.L. Ciochon and A.B. Chiarelli, pp. 123–138. New York: Plenum Press.

Rasmussen, D.T. (1986). Anthropoid origins: A possible solution to the Adapidae-Omomyidae paradox. *J. Hum. Evol.* **15**:1–12.

Rosenberger, A.L., Strasser, E., and Delson, E. (1985). Anterior dentition of *Notharctus* and the adapid-anthropoid hypothesis. *Folia Primatol.* **44**:15–39.

Szalay, F.S., Rosenberger, A.L., and Dagosto, M. (1987). Diagnosis and differentiation of the order Primates *Yrbk. Phys. Anthropol.* **30**:75–105.

FOSSIL LORISES AND GALAGOS

Gebo, D.L. (1986). Miocene lorisids—the foot evidence. *Folia Primatol.* **47**:217–225.

Jacobs, L.L. (1981). Miocene lorisid from the Pakistan Siwaliks. *Nature (London)* **189**:585–587.

Le Gros Clark, W.E. (1956). A Miocene lemuroid skull from East Africa. Fossil mammals of Africa, no. 9. *Brit. Mus. (Nat. Hist.) London*, pp. 1–6.

Le Gros Clark, W.E., and Thomas D.P. (1952). The Miocene lemuroids of East Africa. Fossil mammals of Africa, no. 5. *Brit. Mus. (Nat. Hist.) London*, pp. 1–20.

MacPhee, R.D.E., and Jacobs, L.L. (1986). *Nycticeboides simpsoni* and the morphology, adaptations, and relationships of Miocene Siwalik Lorisidae. In *Vertebrates, Phylogeny, and Philosophy*, ed. K.M. Flanagen and J.A. Lillegraven. *Contrib. Geol. Univ. Wyoming, Special Papers* **3**:131–162.

Simons, E.L., Bown, T.M., and Rasmussen, D.T. (1987). Discovery of two additional prosimian primate families (Omomyidae, Lorisidae) in the African Oligocene. *J. Hum. Evol.* **15**:431–437.

Simpson, G.G. (1967). The Tertiary lorisiform primates of Africa. *Bull. Mus. Comp. Zool.* **136**:39–62.

Walker, A.C. (1970). Postcranial remains of the Miocene Lorisidae of East Africa. *Am. J. Phys. Anthropol.* **33**:249–262.

———. (1974). A review of the Miocene Lorisidae of East Africa. In *Prosimian Biology*, ed. R.D. Martin, G.A. Doyle, and A.C. Walker, pp. 435–447. London: Duckworth.

———. (1978). Prosimian primates. In *Evolution of African Mammals*, ed. V.J. Maglio and H.B.S. Cooke, pp. 90–99. Cambridge, Mass.: Harvard University Press.

———. (1987). Fossil galagines from Laetoli. In *Laetoli: A Pliocene Site in Northern Tanzania*, ed. M.D. Leakey and J.M. Harris, pp. 88–90. Oxford: Clarendon Press.

Wesselman, H.B. (1984). The Omo micromammals. *Contrib. Vert. Evol.* **7**:1–22.

OMOMYIDS

Bown, T.M., and Rose, K.D. (1984). Reassessment of some early Eocene Omomyidae with description of a new genus and three new species. *Folia Primatol.* **43**:97–112.

———. (1987). Patterns of dental evolution in early Eocene anaptomorphine primates (Omomyidae) from the Bighorn Basin, Wyoming. *J. Paleontol.* **61** (5.II, suppl.).

Dagosto, M. (1985). The distal tibia of primates with special reference to the Omomyidae. *Int. J. Primatol.* **6**:45–75.

Dashzaveg, D.T., and McKenna, M.C. (1977). Tarsioid primate from the early Tertiary of the Mongolian People's Republic. *Acta Palaeontol. Polonica* **22**(2):119–137.

Gingerich, P.D. (1981). Early Cenozoic Omomyidae and the evolutionary history of tarsiiform primates. *J. Hum. Evol.* **10**:345–374.

Godinot, M. (1982). Aspects nouveaux des echanges entre les faunes mammaliennes d'Europe et d'Amerique du Nord a la base de l'Eocene. *Geobios, Mem. Spec.* **6**:403–412.

Rose, K.D., and Krause, D.W. (1984). Affinities of the primate *Altanius* from the early Tertiary of Mongolia. *J. Mammal.* **65**(4):721–726.

Simons, E.L., and Russell, D.E. (1960). The cranial anatomy of *Necrolemur. Breviora* **127**:1–14.

Szalay, F.S. (1976). Systematics of the Omomyidae (Tarsiiformes, Primates): Taxonomy, phylogeny and adaptations. *Bull. Am. Mus. Nat. Hist.* **156**(3):157–450.

Szalay, F.S., and Li, C.-K. (1986). Middle Paleocene euprimate from southern China and the distribution of primates in the Paleogene. *J. Hum. Evol.* **15**:387–398.

Anaptomorphines

Bown, T.M. (1976). Affinities of *Teilhardina* (Primates, Omomyidae) with description of a new species from North America. *Folia Primatol.* **25**:62–72.

———. (1979). New omomyid primates (Haplorhini, Tarsiiformes) from middle Eocene rocks of west-central Hot Springs County, Wyoming. *Folia Primatol.* **31**:48–73.

Gingerich, P.D. (1981). Early Cenozoic Omomyidae and the evolutionary history of tarsiiform primates. *J. Hum. Evol.* **10**:345–374.

Rose, K.D., and Bown, T.M. (1984). Gradual phyletic evolution at the generic level in early Eocene omomyid primates. *Nature* **309**(5965):250–252.

Rose, K.D., and Fleagle, J.G. (1981). The fossil history of nonhuman primates in the Americas. In *Ecology and Behavior of Neotropical Primates*, vol. 1, ed. A.F. Coimbra-Filho and R.A. Mittermeier, pp. 111–167. Rio de Janeiro: Academia Brasiliera de Ciencias.

Rose, K.D., and Krause, D.W. (1984). Affinities of the primate *Altanius* from the early Tertiary of Mongolia. *J. Mammal.* **65**(4):721–726.

Szalay, F.S. (1976). Systematics of the Omomyidae (Tarsiiformes, Primates): Taxonomy, phylogeny and adaptations. *Bull. Am. Mus. Nat. Hist.* **156**(3):157–450.

———. (1982). A critique of some recently proposed Paleogene primate taxa and suggested relationships. *Folia Primatol.* **37**:153–162.

Omomyines

Beard, K.C. (1987). *Jemezius*: A new omomyid primate from the early Eocene of northwestern New Mexico. *J. Hum. Evol.* **16**:457–468.

Dagosto, M. (1985). The distal tibia of primates with special reference to the Omomyidae. *Int. J. Primatol.* **6**:45–75.

Rose, K.D., and Rensberger, J.M. (1983). Upper dentition of *Ekgmowechashala* (Omomyid, Primate) from the John Day Formation, Oligo-Miocene of Oregon. *Folia Primatol.* **41**:102–111.

Simpson, G.G. (1940). Studies on the earliest primates. *Bull. Am. Mus. Nat. Hist.* **77**:185–212.

Wilson, J.A. (1966). A new primate from the earliest Oligocene, west Texas, preliminary report. *Folia Primatol.* **4**:227–248.

Microchoerines

Cartmill, M. (1982). Basic primatology and prosimian evolution. In *Fifty Years of Physical Anthropology in North America*, ed. F. Spencer, pp. 147–186. New York: Academic Press.

Godinot, M., and Dagosto, M. (1983). The astragalus of *Necrolemur* (Primates, Microchoerinae). *J. Paleontol.* **57**:1321–1324.

Krishtalka, L., and Schwartz, J.H. (1978). Phylogenetic relationships of plesiadapiform-tarsiiform primates. *Ann. Carnegie Mus.* **47**:515–540.

Schmid, P. (1979). Evidence of microchoerine evolution from Dielsdorf (Zurich region, Switzerland)—a preliminary report. *Folia Primatol.* **31**:301–311.

———. (1982). Comparison of Eocene nonadapids and *Tarsius*. In *Primate Evolutionary Biology*, ed. A.B. Chiarelli and R.E. Corruccini, pp. 6–13. Berlin: Springer-Verlag.

———. (1983). Front dentition of the Omomyiformes (Primates). *Folia Primatol.* **40**:1–10.

Simons, E.L. (1961). Notes on Eocene tarsioids and a revision of some Necrolemurinae. *Bull. Brit. Mus. (Nat. Hist.) Geol.* **5**:43–49.

Simons, E.L., and Russell, D.E. (1960). The cranial anatomy of *Necrolemur. Breviora* **127**:1–14.

Szalay, F.S. (1975). Phylogeny, adaptations and dispersal of the tarsiiform primates. In *Phylogeny of the Primates*, ed. W.P. Luckett and F.S. Szalay, pp. 357–404. New York: Plenum Press.

———. (1976). Systematics of the Omomyidae (Tarsiiformes, Primates): Taxonomy, phylogeny and adaptations. *Bull. Am. Mus. Nat. Hist.* **156**(3):157–450.

Szalay, F.S., and Dagosto, M. (1980). Locomotor adaptations as reflected on the humerus of Paleogene primates. *Folia Primatol.* **34**:1–45.

American Microchoerines

Schwartz, J.H., Tattersall, I., and Eldredge, N. (1978). Phylogeny and classification of the primates revisited. *Yrbk. Phys. Anthropol.* **21**:95–133.

Asian Omomyids

Dashzaveg, D.T., and McKenna, M.C. (1977). Tarsioid primate from the early Tertiary of the Mongolian People's Republic. *Acta Palaeontol. Polonica* **22**(2):119–137.

Gingerich, P.D. (1981). Early Cenozoic Omomyidae and the evolutionary history of tarsiiform primates. *J. Hum. Evol.* **10**:345–374.

Rose, K.D., and Krause, D.W. (1984). Affinities of the primate *Altanius* from the early Tertiary of Mongolia. *J. Mammal.* **65**:721–726.

Russell, D.E., and Gingerich, P.D. (1980). Un noveau primate omomyidae dans l'Eocene du Pakistan. *C. R. Acad. Sci. (Paris)* **291**.

Szalay, F.S., and Li, C.K. (1986). Middle Paleocene euprimate from southern China and the distribution of primates in the Paleogene. *J. Hum. Evol.* **15**:387–398.

TARSIIDS

Ginsburg, L., and Mein, P. (1986). *Tarsius thailandica* nov. sp., Tarsiidae (Primates, Mammalia) fossile d'Asie. *C. R. Acad. Sci. (Paris),* t.304, ser. II, no. 19, pp. 1213–1215.

Simons, E.L., and Bown, T.M. (1985). *Afrotarsius chatrathi,* first tarsiiform primate (?Tarsiidae) from Africa. *Nature (London)* **313**:475–477.

OMOMYIDS AND LATER PRIMATES

Cartmill, M. (1980). Morphology, function and evolution of the anthropoid postorbital septum. In *Evolutionary Biology of New World Monkeys and Continental Drift,* ed. R.L. Ciochon and A.B. Chiarelli, pp. 243–274. New York: Plenum Press.

Cartmill, M., and Kay, R.F. (1978). Cranio-dental morphology, tarsier affinities, and primate suborders. In *Recent Advances in Primatology,* vol. 3, ed. D.J. Chivers and K.A. Joysey, pp. 205–213. London: Academic Press.

Dagosto, M. (1988). Implications of postcranial evidence for the origin of euprimates. *J. Hum. Evol.* **17**:35–56.

Fleagle, J.G., and Simons, E.L. (1983). The tibio-fibular articulation in *Apidium phiomense,* an Oligocene anthropoid. *Nature (London)* **301**(5897):238–239.

Gingerich, P.D. (1981). Early Cenozoic Omomyidae and the evolutionary history of tarsiiform primates. *J. Hum. Evol.* **10**:345–374.

Rosenberger, A.L., and Szalay, F.S. (1981). On the tarsiiform origins of Anthropoidea. In *Evolutionary Biology of New World Monkeys and Continental Drift,* ed. R.L. Ciochon and A.B. Chiarelli, pp. 139–157. New York: Plenum Press.

Schmid, P. (1982). Comparison of Eocene nonadapids and *Tarsius.* In *Primate Evolutionary Biology,* ed. A.B. Chiarelli and R.L. Corruccini, pp. 6–13. Berlin: Springer-Verlag.

———. (1983). Front dentition of the Omomyiformes (Primates). *Folia Primatol.* **40**:1–10.

Simons, E.L. (1961). The dentition of *Ourayia:* Its bearing on relationships of omomyid prosimians. *Postilla* **54**:1–20.

———. (1972). *Primate Evolution.* New York: Macmillan.

Szalay, F.S., Rosenberger, A.L., and Dagosto, M. (1987). Diagnosis and differentiation of the order Primates. *Yrbk. Phys. Anthropol.* **30**:75–105.

Wortman, J.L. (1903). Classification of the primates. *Am. J. Sci.* **15**:399–414.

Early Anthropoids and Fossil Platyrrhines

OLIGOCENE EPOCH

In the Oligocene epoch, approximately 37 to 23 million years ago, the continents were beginning to look as they do today except for the lack of a connection between North America and South America. India was colliding with the Asian mainland to close off the Tethys Seaway on the east, and both South America and Australia were separating from Antarctica. These last events made possible the first deep water currents around Antarctica. As a result, the beginning of the Oligocene was marked by a major drop in global temperatures from the more tropical climates of the preceding Eocene epoch, and the middle of the Oligocene saw a dramatic lowering of sea level (see Fig. 9.3), probably as a result of glaciations at the poles. The primate fossil record of the Oligocene is strikingly different from that of previous epochs. In the Northern Hemisphere, the prosimians that had been abundant in the Eocene disappeared at the beginning of the Oligocene in Europe, and they became increasingly rare in North America, so that primates are virtually unknown from northern continents during that epoch. The Oligocene does, however, provide us with the earliest record of fossil primates in Africa and South America and the first substantial record of fossil anthropoids (Fig. 12.1).

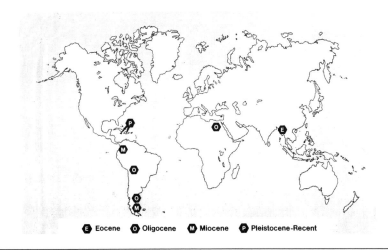

FIGURE 12.1

Geographic distribution of early fossil anthropoids and fossil platyrrhines.

Possible Early Higher Primates

The earliest indication of higher primates in the fossil record are two poorly known species from the late Eocene of Burma, *Amphipithecus mogaungensis* and *Pondaungia cotteri* (Fig. 12.2, Table 12.1). A few tantalizing fossils of each were recovered earlier this century and additional material has come to light in recent years. In both species, the broad, low-crowned molars and deep mandibles suggest higher primate rather than either adapid or omomyid affinities, but the material presently available is insufficient to confirm this suggestion. Both were moderate-size primates (6–10 kg) with molars that suggest a frugivorous diet.

Fossil Primates from Fayum, Egypt

The Burmese fossils offer hints that higher primates may have evolved by late Eocene times, but most of our knowledge of early higher primate evolution in the Old World

TABLE 12.1
Suborder Anthropoidea
Infraorder *incertae sedis*

Species	Body Weight (g)
Amphipithecus mogaungensis	8,600
Pondaungia cotteri	7,000

FIGURE 12.2

Amphipithecus and *Pondaungia*, possible early anthropoids from the Eocene of Burma (courtesy of R.L. Ciochon).

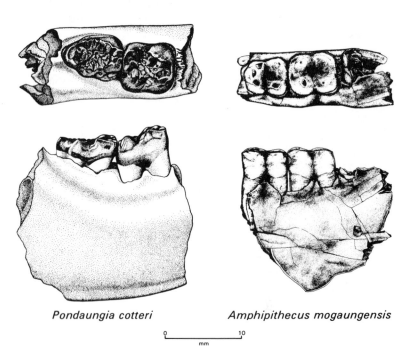

Pondaungia cotteri   *Amphipithecus mogaungensis*

0 ——————— 10
mm

comes from an area in Egypt known as the Fayum Depression. Here, in an expanse of eroded badlands on the western edge of the Sahara Desert (Fig. 12.3), is a sequence of very fossiliferous sedimentary deposits, the Jebel Qatrani Formation, from the early part of the Oligocene, sometime between approximately 37 and 31 million years ago (Fig. 12.4). This sequence of freshwater deposits is covered by volcanic lava flows approximately 31 million years old, but it is not possible to determine whether the sediments were laid down in the early part of this period, in the last part, or evenly over the entire span.

We do know much about the environment under which the sediments and the fossil primates within them were deposited. From the sediments, we know that the climate was warm, wet, and somewhat seasonal. The fossil plants were most similar to species currently found in the tropical forests of Southeast Asia. The quarries yielding most of the primate fossils were laid down as sandbars in river channels and show repeated sequences of standing water (probably oxbow lakes) as well as roots of mangrovelike plants. This evidence, together with abundant fossil remains of water birds, indicates a swampy environment at the time of deposition (Fig. 12.5).

Most of the mammals that are found with the primates are different from Oligocene mammals in other parts of the world, and many are unique to this locality. The rodents are the earliest members of the porcupine suborder (hystricomorphs), which includes the guinea-pig-like rodents of South America as well as several species from Africa. These Oligocene rodents were more arbo-

FIGURE 12.3

The desert landscape of the Fayum Depression in Egypt, a site that has yielded many early Oligocene fossil primates.

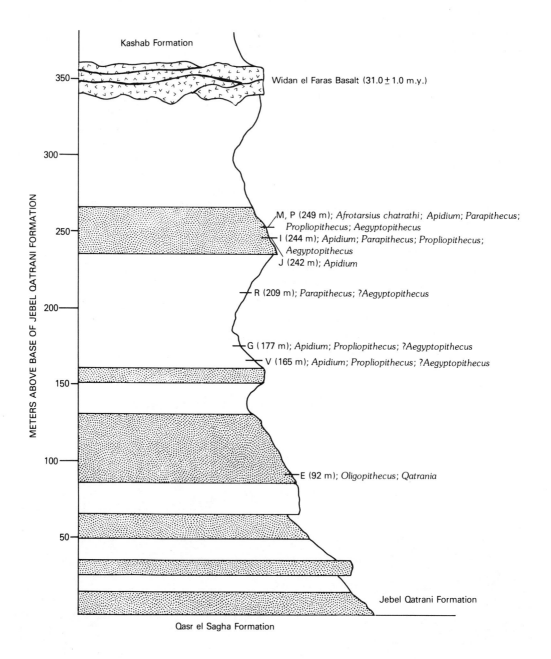

FIGURE 12.4

A geological section of the Jebel Qatrani Formation in Fayum, Egypt, showing stratigraphic levels at which the fossil primates have been found.

FIGURE 12.5

A reconstruction of the Fayum environment in the early Oligocene.

real than their living African relatives. There were opossums, as well as insectivores, bats, carnivores, and an archaic group of artiodactyls (the anthracotheres) related to the hippopotamus. In addition, the Fayum provides the first substantial record of several African groups of mammals such as hyraces, elephants, and elephant shrews, as well as numerous fossil primates.

The fossil primates have been recovered primarily from three levels within the overall sequence of sediments (Fig. 12.4). In the uppermost level, primates are the most common mammals, and one species, *Apidium phiomense*, is known from hundreds of fossils. There are at least five different groups of primates known from the Fayum. There are the two prosimians discussed in Chapter 11—*Afrotarsius* and a lorislike species known from a single tooth—and three types of higher primates, the parapithecids, the "apelike" propliopithecids, and *Oligopithecus*, a poorly known genus whose affinities are rather uncertain (Fig. 12.6). This diverse array of prosimians and primitive anthropoids provides many insights into the initial diversification of higher primates.

FIGURE 12.6

Three Fayum anthropoids: above, the propliopithecids *Aegyptopithecus zeuxis* (left) and *Propliopithecus chirobates* (right); below, the parapithecid *Apidium phiomense*.

Parapithecids

Although the first parapithecid was discovered near the turn of the century, an appreciation of their diversity has come only in recent years through the discovery of abundant new fossils by Elwyn Simons. The six parapithecid species ranged in size from the marmoset-size *Qatrania wingi*, the smallest Old World higher primate, to the guenon-size *Parapithecus grangeri* (Table 12.2). These early Oligocene anthropoids are the most primitive of all known higher primates and have a number of anatomical features that distinguish them from all other Old World primates.

Parapithecids have a primitive dental formula of $\frac{2.1.3.3.}{2.1.3.3.}$, as in New World monkeys. This is probably the primitive dental formula for all higher primates, but it is unique in Africa. In the best-known species, the lower incisors are small and spatulate, but one species, *Parapithecus grangeri*, lost its permanent lower incisors altogether (Fig. 12.7). Upper incisors are poorly known. The canines in *Apidium* are similar to those of most platyrrhines, but in *P. grangeri* they are large and tusklike (Simons, 1986). The three lower premolars increase in size and complexity from front to back, but in all species the last premolar resembles the premolars of

earlier prosimians rather than later anthropoids in having a metaconid that is smaller and distally positioned relative to the protoconid. The upper premolars of parapithecids are broad, with three cusps rather than two as in other higher primates.

Parapithecid molars are characterized by low rounded cusps. The upper molars are quadrate with well-developed conules and a large hypocone. The lower molars have a small trigonid (often with paraconid) and a broad talonid basin. In some species accessory cusps are common and often there is a buccolingual alignment of the molar cusps and a narrowing in the center of the tooth, giving parapithecid molars a "waisted" shape superficially similar to that seen in cercopithecoid monkeys. The mandible is fused at the symphysis.

FIGURE 12.7

Lower dentitions of parapithecids (courtesy of Richard Kay).

Qatrania wingi

Parapithecus grangeri

Parapithecus fraasi

Apidium phiomense

TABLE 12.2
Suborder Anthropoidea
Superfamily Parapithecoidea
Family PARAPITHECIDAE

Species	Body Weight (g)
Qatrania wingi	300
Apidium phiomense	1,600
A. moustafai	850
Parapithecus fraasi	1,700
P. grangeri	3,000

The skull of parapithecids is known only from fragments (Fig. 12.8), but these show higher primate features such as fused frontal sutures and postorbital closure. The arrangement of the cranial sutures in the pterion region of the skull seems to be similar to that of platyrrhines in having a zygomatic-parietal contact exposed on the skull wall. Several frontal fragments that preserve an endocast of the anterior part of the brain show a relatively large olfactory bulb. The auditory region in parapithecids is poorly known but seems to be characterized by a large promontory artery as in anthropoids (and *Tarsius*) and the lack of a tubular tympanic.

Dozens of parts of the limb skeleton have been recovered for one species, *Apidium phiomense* (Fig. 12.9). In many features of their limbs, parapithecids are more primitive than any later Old World higher primates and resemble platyrrhines or omomyid prosimians. In *Apidium*, the tibia and fibula are joined for approximately 40 percent of their length, a similarity to some microchoerines, some platyrrhines, and *Tarsius*.

There are five species and four genera of parapithecids. **Qatrania wingi**, from the lower part of the Jebel Qatrani Formation, is the earliest and most primitive parapithecid. This tiny primate (less than 300 g), known from only two lower jaws and a few isolated teeth, is the smallest known catarrhine. The absence of shearing crests on the teeth indicates that its diet was probably fruits or gums rather than insects. A second species of *Qatrania* is from the upper part of the formation.

There are two species of *Apidium*, the best-known parapithecid. A smaller one, **A. moustafai**, is more common in the intermediate zone of the formation, and a larger one, **A. phiomense**, is more abundant in the upper zone. The former is known only from jaws and teeth; the latter is known from hundreds of specimens. *Apidium* has tiny incisors, moderate-size, sexually dimorphic canines, and molars with numerous low, rounded cusps and very few shearing crests (Fig. 12.7). Both species of *Apidium* have a fused mandibular symphysis. Functionally, the teeth indicate a diet of predominantly fruit, but the very thick enamel on the molars suggest that seeds also may have been an important component. The canine dimorphism, unusual in a primate this small, suggests that *Apidium* lived in polygynous social groups.

The few cranial remains of *A. phiomense* indicate a short snout, a small infraorbital foramen, and relatively small eyes (Fig. 12.8). It was a diurnal monkey.

The many postcranial bones attributed to *Apidium* show it was an excellent leaper (Fig. 12.9). The hindlimb is relatively long compared with the forelimb (intermembral index = 70), the ischium is extremely long,

FIGURE 12.8

A reconstructed facial skeleton of *Apidium phiomense* (after Simons, 1971).

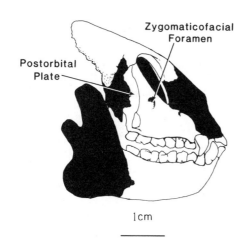

Zygomaticofacial Foramen

Postorbital Plate

1cm

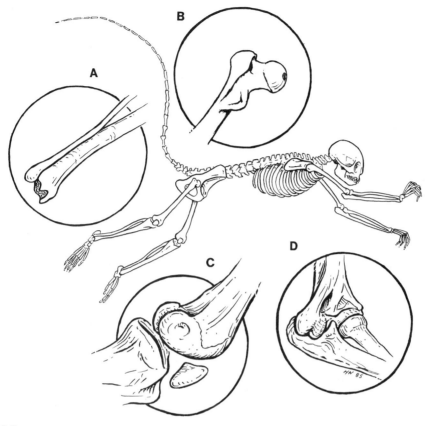

FIGURE 12.9

A restored skeleton of *Apidium phiomense* showing many of the distinctive features of this species: A, the tibia and fibula nearly fused for the distalmost 40 percent of their length; B, the large lesser trochanter of the femur; C, the deep distal condyles on the femur; D, the entepicondylar foramen and elongate capitulum of the humerus. In these features, *Apidium* is more like omomyid prosimians or small platyrrhines than modern catarrhines.

the femoral neck is oriented at a right angle to the shaft, and the distal femoral condyles are very deep, more so than in any other higher primate. The tibia is extremely long and laterally compressed, and the fibula is attached to it for nearly 40 percent of its length. The ankle joint is hinged for rapid flexion and extension. *Apidium* probably had a divergent hallux. The scapula is similar to that seen in many living anthropoid quadru- pedal leapers such as *Saimiri*, and the short forelimb bones indicate quadrupedal rather than clinging habits. In many details of limb structure, *Apidium* shows greatest similarities to platyrrhines and to Eocene prosimians rather than to later Old World anthropoids.

The most unusual primate from the Fa- yum is **Parapithecus grangeri**, the largest parapithecid, often placed in a separate genus, *Simonsius*. Like *Apidium*, this species has three premolars and three molars. The cusps on the lower molars are arranged in two lophs, superficially similar to the condition found in cercopithecoid monkeys. The lower premolars are short with bulbous cusps, and the upper premolars have three major prominent cusps. A most unusual feature of *P. grangeri* is the anterior dentition—large, tusklike canines and no permanent incisors (Kay and Simons, 1983; Simons, 1986). The function of this tusklike arrangement is unclear. The molars suggest that *P. grangeri* may have been partly folivorous. The few facial parts known indicate a short, pointed snout.

The most enigmatic parapithecid is ***Parapithecus fraasi***. The type specimen was described earlier this century from an unknown site in the Fayum, and this medium-size species is known from only a few jaws, none of which preserve the incisors intact. The dental formula of *P. fraasi* has been debated since its initial discovery. Because *Apidium* has a dental formula of $\frac{}{2.1.3.3.}$, it was assumed that *Parapithecus* is similar and that the lateral incisors of the type specimen were lost during collecting. But it is now known that *P. grangeri* lacks permanent incisors altogether, so it is quite possible that *P. fraasi* also lacks permanent incisors and that the tiny anterior teeth preserved in the type specimen are deciduous incisors. More complete fossils are needed to resolve this question. *Parapithecus fraasi* has distinct rounded cusps on its molars, suggesting a frugivorous diet, relatively simple premolars, and a reduced third molar.

PHYLETIC RELATIONS   The phyletic position of parapithecids in anthropoid evolution has long been debated, but new fossils and comparative analyses have greatly expanded our understanding of this group. Parapithecids are among the earliest and most primi-

tive fossil higher primates. They have many primitive features in their dentition, including three simple premolars and occasional paraconids on their molars. Many skeletal features of *Apidium*, such as lack of expanded ischial tuberosities, a large greater trochanter, deep condyles on the femur, and retention of an entepicondylar foramen on the humerus, are also primitive features not found in most later Old World higher primates. The arrangement of the cranial bones on the skull wall and the morphology of the ear region seem to be similar to that in platyrrhines. Although some authorities (see, e.g., Hoffstetter, 1977) have advocated linking parapithecids with platyrrhines, most of the similarities are likely to be primitive anthropoid features retained in the two groups.

Many authors have considered the parapithecids, and especially *P. grangeri*, to be directly ancestral to Old World monkeys (Simons, 1970, 1972; Kay, 1977; Gingerich, 1978). Although some parapithecids, particularly *P. grangeri*, have lower molars and canines that are superficially similar to those of cercopithecoid monkeys, parapithecids lack many anatomical features characteristic of catarrhines, such as presence of two rather than three premolars, broad ischial tuberosities, and a tubular tympanic. If parapithecids are uniquely ancestral to cercopithecoids, then many of the bony features that living apes and monkeys have in common must have evolved independently. In addition, the species that shows the greatest similarity to cercopithecoids in its molar morphology, *P. grangeri*, is the species with the most aberrant anterior dentition. It seems more likely that the bilophodont appearance of the parapithecid molars is an evolutionary convergence with later monkeys rather than an indication of a phyletic relationship (Delson, 1975).

The more difficult question is whether parapithecids preceded or followed the divergence of platyrrhines and catarrhines (Fig. 12.10). In contrast with the large number of primitive prosimian and platyrrhine features in parapithecids, there are few, if any, derived features shared by parapithecids and catarrhines. The presence of a hypoconulid on the lower molars is the main feature linking parapithecids with extant catarrhines to the exclusion of other anthropoids. There is, however, reason to suspect that this feature may well be a primitive anthropoid feature lost in platyrrhines. Moreover, platyrrhines and undoubtedly catarrhines share a number of features lacking in parapithecids, including shallow femoral condyles and broad lower fourth premolars with a crest joining the protoconid and metaconid. Thus it seems more likely that parapithecids preceded the divergence of platyrrhines and lie near the origin of anthropoids (Fleagle and Kay, 1987; Harrison, 1987).

FIGURE 12.10

The phyletic position of parapithecids in anthropoid evolution.

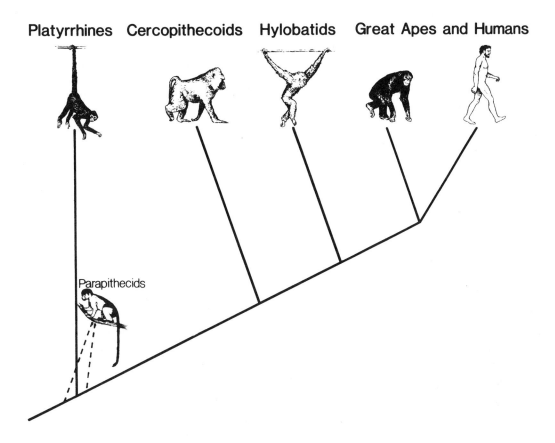

Platyrrhines Cercopithecoids Hylobatids Great Apes and Humans

Parapithecids

Propliopithecids

The other group of early anthropoids from the Fayum, the "apelike" propliopithecids, were as large as or larger than the largest parapithecid. They have a dental formula of $\frac{2.1.2.3.}{2.1.2.3.}$ and a dental morphology more like that of later apes than of cercopithecoid monkeys in that they lack bilophodont molars, but in details of their dental, cranial, and postcranial anatomy they are much more primitive than any living catarrhines. There are two genera (Table 12.3).

The first fossil "ape" described from the Fayum is **Propliopithecus**. There are four species, the best known being *P. chirobates*, a medium-size (4 kg) anthropoid. *Propliopithecus* has relatively broad, spatulate lower incisors and large, sexually dimorphic canines (Figs. 12.6, 12.11). As in most living anthropoids, the anterior lower premolar shears

Figure 12.11

A mandible of *Propliopithecus chirobates*. Note the dental formula of $\overline{2.1.2.3.}$, the apelike arrangement of cusps on the lower molars, the elongate anterior premolar that shears against the upper canine, and the fused mandibular symphysis (courtesy of Richard Kay).

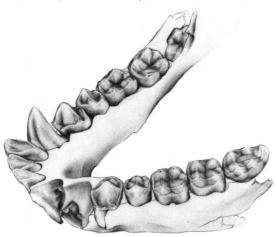

TABLE 12.3
Infraorder Catarrhini
Superfamily HOMINOIDEA

Species	Body Weight (g)
Family PROPLIOPITHECIDEA (e. Oligocene, Africa)	
Propliopithecus haeckeli	4,000
P. chirobates	4,200
P. markgrafi	4,000
P. ankeli	5,700
Aegyptopithecus zeuxis	6,700
Family *incertae sedis*	
Oligopithecus savagei (e. Oligocene, Africa)	1,500

against the posterior surface of the upper canine to sharpen it; the posterior premolar is semimolariform with protoconids and metaconids of equal size. The lower molars resemble those of later apes in that they are formed by a broad talonid basin surrounded by five rounded cusps. There is no paraconid, and the trigonid is small. The three lower molars are similar in size. The upper premolars are bicuspid and the upper molars are broad and quadrate, with a small hypocone connected to a pronounced lingual cingulum. There are no conules or stylar cusps on the upper molars. The simple molars with low rounded cusps and the broad incisors suggest that *Propliopithecus* was frugivorous.

There are no described cranial remains of *Propliopithecus*. Several isolated limb elements indicate that *Propliopithecus* was an arboreal quadruped with a strong grasping foot and was probably capable of hindlimb suspension.

Aegyptopithecus zeuxis, from the same quarries in the Fayum as *P. chirobates*, was a much larger animal (6–8 kg) and is one of the best known of all fossil anthropoids (Figs. 12.6, 12.12, 12.13). Dentally, *Aegyptopithecus* differs from *Propliopithecus* in having narrower incisors, lower molars with larger cusps and a more restricted talonid basin, and upper molars with better developed conules and stylar cusps. In contrast to *Propliopithecus*, in which the three lower molars are similar in length, the molars of *Aegyptopithecus* increase in size posteriorly. Overall, the dental differences suggest that *A. zeuxis* was largely frugivorous but probably more folivorous than *Propliopithecus*. Like

FIGURE 12.12

Cranial remains of *Aegyptopithecus zeuxis*. Note the long snout, small orbits, sagittal and nuchal crests, and converging temporal lines in older individuals (courtesy of E.L. Simons).

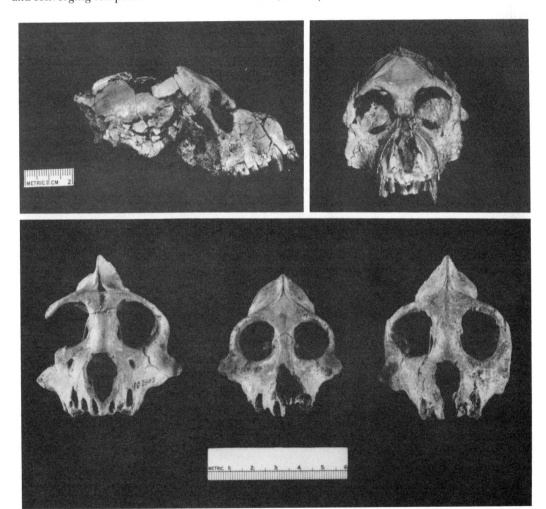

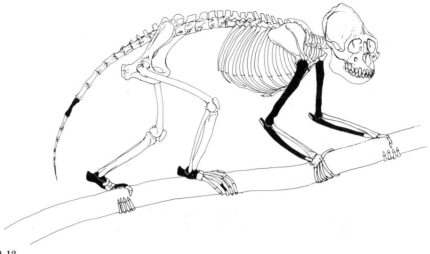

FIGURE 12.13

A reconstructed skeleton of *Aegyptopithecus zeuxis*, showing (in black) the bones that have been recovered for this species.

Propliopithecus, Aegyptopithecus has sexually dimorphic canines; it probably lived in polygynous social groups.

The cranial anatomy of *Aegyptopithecus* is more primitive than that of any living Old World anthropoid but more advanced than that of any Eocene prosimian (Fig. 12.12). The skull resembles other anthropoids in that the lacrimal bone lies within the orbit, and the relatively small orbits (indicating diurnal habits) are completely walled off posteriorly with a bony configuration similar to that found in extant catarrhines rather than in platyrrhines (Fleagle and Rosenberger, 1983). *Aegyptopithecus* has a premaxillary bone that is very large for an anthropoid, and the superficial cranial morphology changes dramatically with age (Simons, 1987). Older individuals develop a pronounced sagittal crest that divides anteriorly and extends over the brow ridges. There is also a large nuchal crest along the posterior border of the occiput. The auditory region is most similar to that in platyrrhines; the

tympanic is a bony ring fused to the lateral surface of the bulla, with no bony tube.

The brain of *Aegyptopithecus* was relatively small compared with the brains of living anthropoids and more like a prosimian brain, but compared with contemporaneous Oligocene mammals or Eocene prosimians it was relatively large, with an expanded parietal region (Radinsky, 1974).

The forelimb of *Aegyptopithecus* is known from the humerus and ulna, and the hindlimb is known only from the talus, calcaneus, and first metatarsal (Fig. 12.13). All of these elements indicate that *Aegyptopithecus* was a robust arboreal quadruped (see Fig. 12.6). The foot bones indicate that it had a grasping hallux and was capable of considerable inversion of the foot. In many anatomical details, the limb elements of *Aegyptopithecus* are more similar to those of platyrrhines and prosimians than to either living apes or cercopithecoid monkeys. This early ape retained many primitive features lost in later catarrhines.

PHYLETIC RELATIONS   Since their initial discovery, *Propliopithecus* and *Aegyptopithecus* have been identified as early apes on the basis of their dental similarities to living hominoids and to later fossil apes from Europe and Africa (Schlosser, 1911; Simons, 1967a, 1972, 1985; Szalay and Delson, 1979). The similarities to living apes are, however, primitive anthropoid features rather than specializations, and increasing knowledge of their cranial and postcranial anatomy has shown that these early anthropoids were more like platyrrhines than catarrhines in many aspects of their anatomy. They have all of the characteristic features of anthropoids (fused mandibular symphysis, postorbital closure, lacrimal bone within the orbit) but are linked with living catarrhines only by their dental formula of $\frac{2.1.2.3}{2.1.2.3}$. In the anatomy of their auditory region and limbs, they lack common specializations found both in living apes and in living Old World monkeys, and they have the more primitive platyrrhine morphology. Thus the Fayum "apes" are neither Old World monkeys nor apes but a primitive group of catarrhines that preceded the evolutionary divergence and subsequent radiations of both living groups (Fig. 12.14). They are usually placed in a primitive family of catarrhines, Propliopithecidae. However, because these early catarrhines share more primitive features with later apes than with the specialized cercopithecoids, this family is most conveniently placed in the ape superfamily, Hominoidea, despite preceding the monkey–ape divergence.

Oligopithecus

One of the earliest and the most enigmatic of the Fayum primates is **Oligopithecus savagei** (see Table 12.3). This species is known from only one jaw (Fig. 12.15). *Oligopithecus*

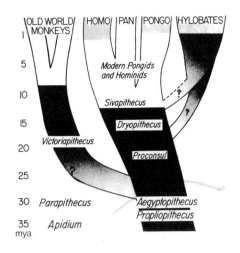

FIGURE 12.14

A phyletic tree showing the relation of *Aegyptopithecus* and *Propliopithecus* to later catarrhines.

is about the size of a titi monkey (*Callicebus*) and has a deep mandible. Its dental formula is $\overline{?.1.2.3.}$, as in *Propliopithecus*, but its teeth show an odd mixture of features quite different from those of other Fayum primates. The canine is small and mesiodistally compressed, and the simple P_3 is narrow and more similar to that tooth in callitrichines than in catarrhines. The last premolar, however, is strikingly similar to the same tooth in propliopithecids. The molars are very primitive compared with those of other anthropoids in having a relatively high trigonid and a small paraconid on the first molar. On the second molar, the trigonid is compressed anteroposteriorly with no paraconid, and the talonid basin is relatively broad. On both molars there is a large hypoconulid near the entoconid and no posterior fovea.

PHYLETIC RELATIONS   The phyletic affinities of *Oligopithecus* have been much debated since its discovery in the early 1960s. Some authorities (Simons and Pilbeam, 1972) have

FIGURE 12.15

Jaw of *Oligopithecus savagei*, an enigmatic fossil primate from the Fayum (courtesy of E.L. Simons).

considered it a primitive propliopithecid related to *Propliopithecus* and *Aegyptopithecus* because of the similar dental formula. Others have argued that the dentition is more suggestive of adapid affinities. The premolar similarities to *Aegyptopithecus* and *Propliopithecus* together with the catarrhine dental formula suggest that *Oligopithecus* is more probably an early catarrhine. Only more fossils of this species will resolve its phyletic position.

THE FAYUM PRIMATES IN ANTHROPOID EVOLUTION

The fossil primates from the early Oligocene of Egypt provide our only record of Old World higher primate evolution from that entire epoch. The Fayum is a very rich site and has yielded an impressive array of different primates with diverse adaptive and phyletic affinities. In its adaptations, the Fayum primate community is distinctly different from later Old World primate communities composed of more modern catarrhines (Fig. 12.16). These Oligocene anthropoids were all small to medium in size, comparable to extant platyrrhines. Their dentitions indicate that they ate fruits, seeds, and perhaps gums, but there is no evidence of predominantly folivorous species. From the available limb bones, they seem to have been arboreal quadrupeds and leapers; there is no evidence of either terrestrial quadrupeds or suspensory species. Overall, the adaptive breadth of these early

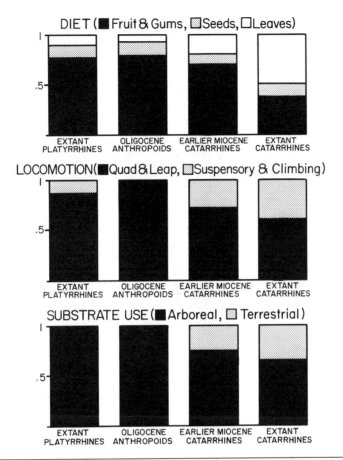

FIGURE 12.16

A comparison of the adaptive characteristics of the early Oligocene anthropoids from Egypt with those of extant platyrrhines, extant catarrhines, and early Miocene fossil catarrhines. Note that the early Oligocene higher primates are most like platyrrhines in their adaptive diversity.

Oligocene primates is more like that of extant platyrrhines than that found among later catarrhine primates of the Old World. It seems likely that these are the primitive anthropoid adaptations.

The Fayum primates are all more primitive than later Old World higher primates, but they seem to preserve several different "stages" of anthropoid evolution (Fig. 12.17). The parapithecids are the most primitive and closest to the origin of anthropoids.

They share features with omomyid prosimians as well as with platyrrhines, suggesting that higher primates as a group probably originated in Africa. *Aegyptopithecus* and *Propliopithecus* are more advanced than platyrrhines but more primitive than later catarrhines, and *Oligopithecus* seems to fall between the two groups. The various Fayum primates are intermediate forms that fill in many of the morphological gaps between the major radiations of extant anthropoids.

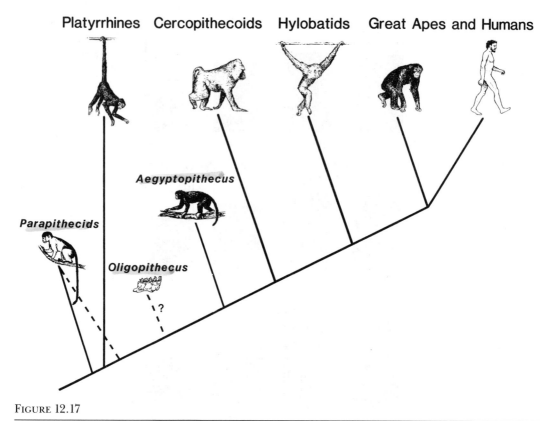

Platyrrhines   Cercopithecoids   Hylobatids   Great Apes and Humans

Aegyptopithecus

Parapithecids

Oligopithecus

?

FIGURE 12.17

The phyletic position of the early anthropoids from the Fayum relative to extant anthropoid groups.

FOSSIL PRIMATES OF SOUTH AMERICA

The first record of platyrrhines in the fossil record of South America comes from the late Oligocene, 5 to 10 million years later than the Fayum primates (Fig. 12.18). Whereas the Fayum is the earliest significant fossil mammal locality for the early Cenozoic of Africa, South America has an extensive record of Paleocene and Eocene deposits, mostly in southern Argentina. For most of the Cenozoic, South America was an island with no connections to other continents except possibly Antarctica (Tarling, 1980). The early mammalian fossil record of South America reflects this isolation. It contains many unusual mammals unique to that continent, such as armadillos, many types of marsupials, and a large radiation of endemic ungulates, rather than the mammals common to the Paleocene and Eocene of North America and Europe (Patterson and Pascual, 1972; Simpson, 1980). There are, however, no remains of primates in South America from either the Paleocene or Eocene. Monkeys first appear, along with another Old World group of mammals, the porcupine-like rodents, in the late Oligocene.

The first appearance of primates and rodents during the Oligocene marks novel additions to the South American fauna. Where they came from and how they got there are two of the most fascinating and difficult questions in primate evolution (see, e.g., Ciochon and Chiarelli, 1980; Rose and Fleagle, 1981). Before we tackle these questions, we examine the fossil record.

FIGURE 12.18

A map of the neotropics showing primate fossil localities.

Fossil Platyrrhines

Considering the extensive radiation of living primates found in the neotropics today and the relatively good fossil record for other South American mammals, the fossil record of New World monkeys is exceptionally poor. There is only a shoebox full of primate fossils from the entire continent of South America from the last 30 million years. The scarcity of primates among the well-documented mammalian faunas of South America presumably indicates that much of the evolution of this group took place in areas from which there are very few fossil mammals at all, such as the vast Amazonian Basin. Although it is not extensive, the fossil record nevertheless provides us with tantalizing hints about the evolutionary history of the group.

Fossil platyrrhines can be conveniently divided into four groups on the basis of age and origin: (a) *Branisella boliviana*, the earliest species, from the late Oligocene of Bolivia; (b) several genera from the latest Oligocene or early Miocene of southern Argentina; (c) several genera from the middle Miocene of Colombia; and (d) two genera from Pleistocene or Recent caves in the Caribbean (Table 12.4).

The earliest platyrrhine from South America is ***Branisella boliviana*** (Fig. 12.19), from late Oligocene (Deseadan) deposits in Bolivia. The four specimens of this species are probably from a single individual. *Branisella* has three premolars and three molars. Its upper molar morphology is similar to that seen in the living squirrel monkey (*Saimiri*) or the night monkey (*Aotus*), with a small hypocone and a well-developed lingual cingulum, but the lower, more rounded cusps suggest a more frugivorous diet. The small P^2 and the shape of the mandible suggest that it was a short-faced monkey. The rela-

TABLE 12.4
Suborder Anthropoidea
Infraorder PLATYRRHINI

Species	Body Weight (g)
Subfamily PITHECIINAE	
Cebupithecia sarmientoi (m. Miocene, Colombia)	2,200
Mohanamico hershkovitzi (m. Miocene, Colombia)	1,000
Subfamily AOTINAE	
Tremacebus harringtoni (e. Miocene, Argentina)	1,800
Aotus dindensis (m. Miocene, Colombia)	1,000
Homunculus patagonicus (e. Miocene, Argentina)	2,700
Subfamily CEBINAE	
Neosaimiri fieldsi (m. Miocene, Colombia)	840
"Saimiri" bernensis (Recent, Dominican Republic)	—
Subfamily ATELINAE	
Stirtonia tatacoensis (m. Miocene, Colombia)	5,800
S. victoriae (m. Miocene, Colombia)	10,000
Subfamily CALLITRICHINAE	
Micodon kiotensis (m. Miocene, Colombia)	—
Subfamily *incertae sedis*	
Branisella boliviana (l. Oligocene, Bolivia)	1,000
Dolichocebus gaimanensis (e. Miocene, Argentina)	2,700
Soriacebus ameghinorum (e. Miocene, Argentina)	2,000
Xenothrix mcgregori (Recent, Jamaica)	—

FIGURE 12.19

Dental remains of *Branisella boliviana*, the earliest fossil platyrrhine (courtesy of Ronald Wolff).

tionship of *Branisella* to later platyrrhines is obscure.

Tremacebus (Fig. 12.20), from the late Oligocene or early Miocene locality of Sacanana in south central Argentina, was of medium size (1–2 kg). There is one nearly complete but broken skull and a lower jaw with two teeth. *Tremacebus* has relatively small canines, three premolars, and three molars. The broken upper molars on the

skull and the single known lower molar are most similar to the teeth of *Callicebus* or *Aotus*. The upper molars are quadrate with a large hypocone and a broad lingual cingulum. The lower molar has a narrow trigonid lacking distinct cusps and a broader talonid with a strong cristid obliqua. In addition to a relatively short, broad snout, *Tremacebus* has larger orbits than most diurnal platyrrhines but smaller ones than the

FIGURE 12.20

Reconstructed skulls of *Tremacebus harringtoni* and *Dolichocebus gaimanensis*, two fossil platyrrhines from the latest Oligocene or early Miocene of Argentina.

Tremacebus harringtoni

Dolichocebus gaimanensis

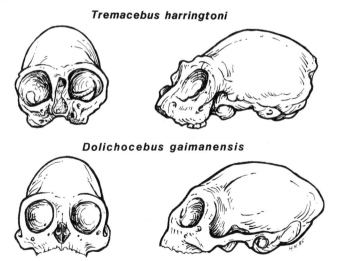

nocturnal *Aotus*, suggesting to Hershkovitz (1974) that the species was possibly crepuscular. The posterior wall of the orbit is not completely walled off in the type specimen. Hershkovitz (1974) has argued from this evidence that *Tremacebus* was more primitive than any known anthropoid, but there are other indications that the large opening in the back of the orbit is due to breakage of the fossil, and that *Tremacebus* is similar to living platyrrhines in its postorbital wall.

Tremacebus shows greatest dental and cranial similarities to the extant platyrrhines *Callicebus* and *Aotus*. Rosenberger (1984) has suggested that it is an ancestor of the living owl monkey and other authorities have noted similarities to *Callicebus*. The mandibular dentition is similar to that of fossil monkeys from the slightly younger Miocene deposits of Argentina, suggesting that they may be part of a single radiation.

Dolichocebus gaimanensis (Fig. 12.20) is also from the latest Oligocene or earliest Miocene of southern Argentina. It is known from a nearly complete but damaged skull, several isolated teeth, and a talus. *Dolichocebus* was twice as large as *Tremacebus*, probably weighing nearly 3 kg. It has small canines, three premolars, and three broad upper molars with a moderate-size hypocone and a broad lingual cingulum. The molar morphology is much like that of *Saimiri* or *Aotus* but more primitive than either in some respects. In retaining many primitive anthropoid features, such as broad molars and a paraconule, *Dolichocebus* resembles the Oligocene anthropoids from Egypt. The molar morphology and size of *Dolichocebus* suggest a frugivorous diet.

The skull of *Dolichocebus* has a narrower snout than that of *Tremacebus*, complete postorbital closure, moderate-size eyes, and a very narrow interorbital dimension. The brain size is similar to that of extant platyrrhines. The distortion of the cranium suggests that in *Dolichocebus*, as in *Tremacebus* and many living platyrrhines, the cranial sutures fused late in adulthood (Chopra, 1957). The cranial morphology of *Dolichocebus*, like that of *Tremacebus*, has been the subject of considerable debate. Rosenberger (1979, 1982) has argued that *Dolichocebus* had an interorbital foramen linking the right and left orbits—an unusual cranial feature found only in *Saimiri* among living primates. Hershkovitz (1970, 1982) has argued that the supposed interorbital fenestra is an artifact of breakage. The talus of *Dolichocebus* is most similar to that of *Cebus* or *Saimiri*, suggesting either a rapid arboreal quadruped or a leaper.

On the basis of the supposed interorbital foramen and several other aspects of the cranial morphology of *Dolichocebus*, Rosenberger (1979) has argued that this genus is uniquely related to the living squirrel monkey. Hershkovitz has argued that the Oligocene monkey is too distinctive to bear any relationship to living platyrrhines. The present material of *Dolichocebus* is too fragmentary to resolve this issue.

Homunculus patagonicus, from the earliest Miocene of southern Argentina, was the first fossil platyrrhine discovered (Ameghino, 1891). It was a medium-size monkey, with the largest individuals probably weighing nearly 3 kg. The dental formula is $\frac{2.1.3.3.}{2.1.3.3.}$. Only the lower dentition is well known. The lower incisors are narrow and spatulate; the canines are probably sexually dimorphic. The lower premolars increase in complexity from front to back—P_2 is a small pyramid-shaped tooth; P_3 and P_4 are semimolariform. The molars are characterized by relatively small cusps connected by long shearing crests; they have a small, square trigonid and a broader talonid with a prominent cristid obliquid. *Homunculus* was probably frugivorous and folivorous.

The facial fragment attributed to *Homunculus* has a relatively short snout and moderate-size orbits (indicating diurnal habits) with complete postorbital closure. The lacrimal bone is well within the orbit margin. The cranium appears relatively gracile with no sagittal crest.

The limb elements resemble those of a callitrichid (Ciochon and Corruccini, 1975) and suggest that *Homunculus* was possibly saltatory in its locomotion. In some details of its limbs, such as the size of the lesser trochanter, *Homunculus* resembles the parapithecids from Egypt.

As the name indicates, Ameghino (1891) originally thought *Homunculus* was in the ancestry of humans; it is not. Most later studies have noted either the unique features of the genus (Hershkovitz, 1970) or dental similarities to *Aotus*, *Callicebus*, or *Alouatta*. There are numerous broken remains, mostly jaws and teeth that have been attributed to the species *Homunculus patagonicus*, but there are no unworn dentitions. As a result, it is presently impossible to sort out either the adaptations or phyletic relationships of *Homunculus*.

Recent paleontological expeditions in the southernmost part of Argentina have uncovered an abundant fauna of primates from early Miocene (Santacrucian) deposits. The new finds—many of which have not yet been named—vary from tiny tamarin-size species to larger *Cebus*-size species. The most unusual is **Soriacebus ameghinorum** (Fig. 12.21). This saki-size monkey has a dental formula of $\frac{2.1.3.3.}{2.1.3.3.}$, with three narrow, marmoset-like molars, large procumbent incisors, large premolars, and a deep jaw. It was probably frugivorous and insectivorous and

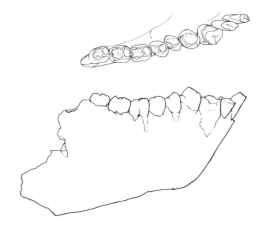

FIGURE 12.21

The mandible and lower dentition of *Soriacebus ameghinorum*, an early Miocene platyrrhine from southern Argentina.

used its large front teeth for some type of gnawing.

Other fossil platyrrhines from the early Miocene of Argentina include a tiny species similar to *Soriacebus* and a very large species (4 kg) that shows dental and skeletal similarities to *Alouatta* or *Callicebus*. Work on this new material is still in progress.

Compared with the Argentine fossil platyrrhines, which are difficult to place in extant platyrrhine subfamilies, the fossil monkeys from the middle Miocene of La Venta, Colombia, are strikingly similar to modern platyrrhines and clearly belong in living subfamilies or even genera (Fig. 12.22).

In size and all known details of dental anatomy, the single mandible of **Neosaimiri fieldsi** is virtually identical with the living squirrel monkey. Like *Saimiri*, it was insectivorous and frugivorous.

Cebupithecia sarmientoi was similar in size (2–3 kg) and many aspects of skeletal morphology to the living saki, *Pithecia pithecia*. This species is known from a mandible, several cranial fragments, and parts of a skeleton. In all aspects of its dental morphology, such as the stout canines, procumbent incisors, and flat cheek teeth with little cusp relief, the Miocene genus is very similar to the living pithecines. Like living pithecines, *Cebupithecia* probably ate mainly fruit and used its large anterior dentition for opening

seeds. The *Cebupithecia* skeleton shows more similarities to the saltatory *Pithecia* than to the more quadrupedal sakis such as *Chiropotes*. There are also indications of vertical clinging habits in the morphology of the elbow (Meldrum and Fleagle, 1988).

Mohanamico hershkovitzi is a small (1 kg), recently named fossil monkey from La

FIGURE 12.22

Several fossil primates from the middle Miocene locality of La Venta, Colombia.

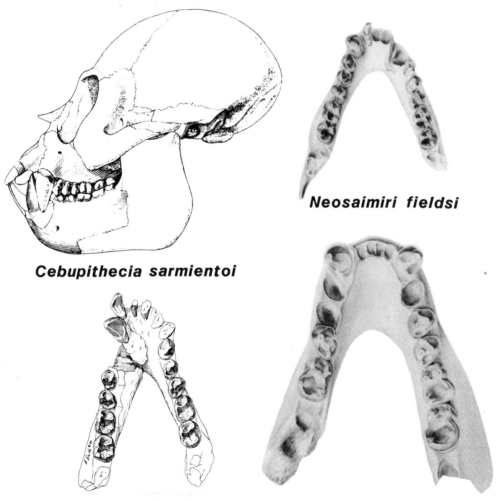

Cebupithecia sarmientoi

Neosaimiri fieldsi

Mohanamico hershkovitzi

Stirtonia tatacoensis

Venta, known from a single mandible (Luchterhand *et al.*, 1986). It has been placed near the base of the evolutionary radiation of the pithecines on the basis of its large lateral incisor and the structure of the canine and anterior premolar. It was probably frugivorous.

Setoguchi and Rosenberger (1987) have recently described a new species of owl monkey from the middle Miocene deposits of La Venta. *Aotus dindensis* is virtually identical to the extant *Aotus* in molar and premolar morphology, but it has narrower lower incisors. A small facial fragment demonstrates that the Miocene species also has large orbits similar to those of the nocturnal owl monkey.

Stirtonia tatacoensis is the largest La Venta primate (6 kg). It has many dental similarities in its upper and lower dentition to the living howling monkey (*Alouatta*). Like *Alouatta*, *Stirtonia* has long molars with a relatively small trigonid and large talonid and very large upper molars with well-developed shearing crests and styles. It was a folivore. A larger species, *Stirtonia victoriae*, is known from slightly older deposits within the same area (Kay *et al.*, 1987).

Micodon kiotensis is a new species from La Venta that is based on three small, isolated teeth (Setoguchi and Rosenberger, 1985). It has been described as a fossil marmoset, primarily on the basis of size. The type specimen, an upper molar, lacks any marmoset features and resembles that of a small pithecine in occlusal morphology. Any determination regarding either the validity of the species or its affinities must await more fossil remains.

The primate fauna from the middle Miocene of Colombia is clearly modern compared with other fossil platyrrhines. This modernity may reflect a relatively late age, or it may reflect the fact that La Venta is closer than other fossil localities to the Amazon Basin, where living New World monkeys are most abundant. In any case, the La Venta fauna provides excellent documentation for the presence of most of the modern subfamilies of platyrrhines by the middle Miocene. Oddly, the most unusual fossil platyrrhines are the youngest, those from Pleistocene and Recent caves of the Caribbean.

Xenothrix mcgregori is a latest Pleistocene or Recent primate from the island of Jamaica, where there are no extant nonhuman primates. It is known only from a mandible with two molars. It was a relatively large monkey (2 kg) with a dental formula of $\frac{2.1.3.2}{2.1.3.2}$, the same as in marmosets and tamarins. The molars, however, are very different from those of extant callitrichines in both cusp morphology and proportions. They have large, bulbous cusps, and M_2 is longer than M_1. *Xenothrix* was probably a frugivorous species, or it may have specialized on insect larvae, like the aye-aye of Madagascar.

There is also a fossil primate femur from another, older cave site in Jamaica. This bone is of the right size to belong to *Xenothrix* and shows greatest morphological similarities to the femora of callitrichines.

Two dental specimens and a tibia from Recent cave deposits in Haiti and the Dominican Republic have been assigned to the species *"Saimiri" bernensis*. The dental remains suggest a large primate (2–3 kg) with a diet of hard fruit or seeds. The species apparently shows greater phyletic similarities to *Cebus* than to *Saimiri*. The tibia is from a similar-size platyrrhine, but it shows greatest similarity to that bone in callitrichines.

SUMMARY OF FOSSIL PLATYRRHINES

Despite the scarcity of fossils from Central and South America, the available remains of fossil platyrrhines provide a number of insights into the history of the group and timing of appearance of many modern groups of New World monkeys. Perhaps the most striking feature of the record of fossil platyrrhines is the overall similarity of the extinct species to modern lineages, especially those from the Miocene. Although it is important to remember that our knowledge of fossil New World monkeys is based largely on fragmentary dental remains, much of the fossil record seems to accord with the view that many lineages of extant platyrrhines have been distinct since at least the Miocene (Fig. 12.23). Fossil species related to the extant owl monkey (*Aotus*), the squirrel monkey (*Saimiri*), the pitheciines, and the howling monkey (*Alouatta*) were definitely present in the middle Miocene of Colombia, and there is evidence suggesting that some of these lineages can be traced back to late Oligocene (Colhuehuapian) or early Miocene (Santacrucuian) times. New fossil discoveries from these earlier periods should help resolve both the age of these Miocene lineages and the relationships of the modern subfamilies.

The fossil record also provides indications of some taxa, such as the Jamaican *Xenothrix*, that are very different from any extant taxa. The early Miocene *Soriacebus*, with its narrow jaw and procumbent incisors, shows dental adaptations very different from those

FIGURE 12.23

A speculative phylogeny of platyrrhines, showing the position of the best-known fossils relative to extant genera and subfamilies (modified from Rosenberger, 1984).

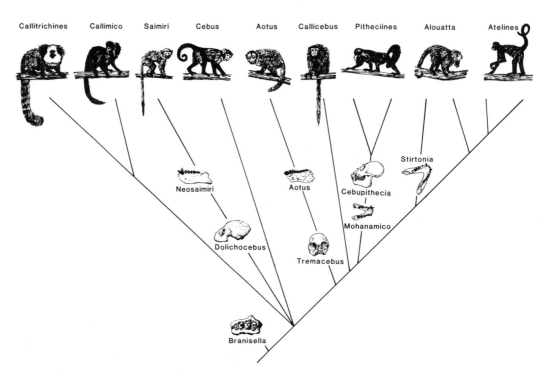

found in any extant platyrrhines. Perhaps someday we will find evidence of a terrestrial lineage of platyrrhines in the fossil record.

PLATYRRHINE ORIGINS

The most unsettled question surrounding platyrrhine origins is the geographic one: How did platyrrhines get to South America? The issue is a particularly complex one involving not only paleontological information about fossil platyrrhines but also information about paleogeography and the faunas of other continental areas. South America was an island continent throughout most of the early Cenozoic, separated from Africa by the South Atlantic and from North America by the Caribbean Sea. Debate over the origin of neotropical primates has focused on whether North America or Africa is the most likely source of the immigrating primates (Fig. 12.24).

Most geophysical studies indicate that the positions of North and South America and Africa relative to one another were much the same in the Eocene and Oligocene as they are now (Fig. 12.25); the rifting of the South Atlantic had taken place much earlier, during the Mesozoic era. There was, then, a considerable body of water for migrating primates to cross, from either North America or Africa. During the early Cenozoic, though, there were probably large areas of relatively shallow water in the South Atlantic, and possibly a series of islands in the areas of the Walvis Ridge and the Sierra Leon Rise. In periods of low sea level, such as the middle Oligocene, these areas and the continental shelves of Africa were probably dry land, which would appreciably shorten the open-water distances between the conti-

FIGURE 12.24

How did the ancestral platyrrhines reach South America?

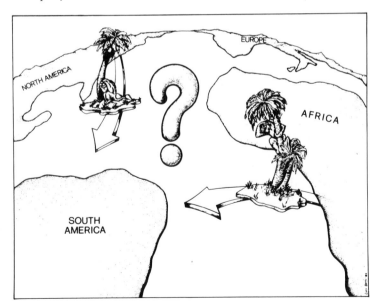

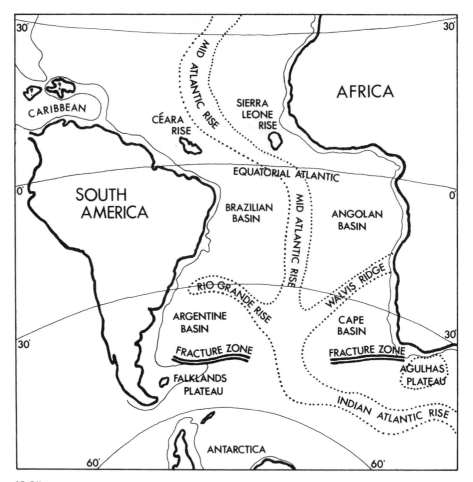

FIGURE 12.25

The South Atlantic in the Oligocene (after Tarling, 1982).

nents. The reconstructed currents seem to favor a crossing from Africa to South America (Tarling, 1982). On the other hand, most reconstructions show the distance between Oligocene North America and South America to be longer than that between Africa and South America, and they also indicate that there were no favorable currents to facilitate a north–south crossing. The geological history of the Caribbean region is, however, very poorly known (Stehli and Webb, 1985). Overall, the geophysical evidence does not seem to favor one continental source over the other.

Because all available evidence indicates that the immigrant primates that rafted to South America and gave rise to living platyrrhines were anthropoids rather than prosimians, we must also consider the nature of the fossil primates known from the potential source continents. North America or Africa could only be source areas for the earliest platyrrhines if there were suitable primates on those continents to be the ancestral platyrrhines. In this respect, Africa is unquestionably the more likely source of early platyrrhines. The only undoubted Oligocene anthropoids are those of Africa. There are many similarities between the Fayum anthropoids and platyrrhines, and the parapithecids seem to be basal anthropoids that preceded the evolutionary divergence of platyrrhines and catarrhines. The closest relatives of South American rodents are the African porcupines (Hoffstetter and Lavocat, 1970), which supports the suggestion of a faunal connection between South America and Africa.

In contrast, there is no evidence of either Eocene or Oligocene anthropoids in North America. Arguments for a North American origin for platyrrhines must postulate either a separate prosimian ancestry for platyrrhines and catarrhines, which seems unlikely in view of the morphological similarities shared by all higher primates, or colonization of South America by a group of unknown North American or Central American anthropoids. Certainly the discovery of prosimians with European affinities (the adapine *Mahgarita*) in the late Eocene and Oligocene of Texas demonstrates the likelihood that there were other primates in North America that remain to be discovered. Nevertheless, on the basis of the known North American Eocene primate fauna, that continent is a very unlikely source of platyrrhines.

Finally, it must be noted that several authors (Wood and Patterson, 1970; Simons, 1976; Cartmill *et al.*, 1981) have questioned whether long-distance rafting between any continents is a likely method for biogeographic dispersal of animals with the dietary and climatic requirements of primates. Anthropoids seem less suited to dispersal on floating masses of vegetation than animals that can hibernate or have long periods of inactivity (such as cheirogaleids or rodents).

But regardless of how unlikely rafting may seem, it is presently the only suggested mechanism for transporting terrestrial animals between continents separated by open ocean. If South America was indeed an island continent during the period in question, we must assume that primates rafted from some other continental area. Only a revision of the paleocontinental maps could eliminate the need for rafting in the origin of platyrrhines. However, rafting could be feasible; the largest drop in sea level during the entire Cenozoic occurred in the middle Oligocene, before the first appearance of *Branisella* but after the Fayum deposits were accumulated (see Fig. 9.3). Such a drop would have facilitated intercontinental dispersal to South America from any continent.

Although most discussions of platyrrhine origins are restricted to the question of whether a North American or African origin is more probable, there are other scenarios that cannot be eliminated from consideration. One possible source that has not been seriously discussed is Antarctica. Although South America was apparently connected to Antarctica throughout much of the early Cenozoic, virtually nothing is known about the fauna (if any) of Antarctica during this time. The presence of monkeys near the southern tip of South America during the early Miocene clearly indicates that present climates are no indicator of the past in this region, but until we know something about mammal evolution on Antarctica this possibility cannot be evaluated. The same is largely true for speculations about Central America or Southeast Asia as source areas. Finally, Szalay (1975) has suggested that perhaps anthropoids originated in the neotropics and then dispersed to the Old World. The more primitive nature and greater age of anthropoids in Egypt than in South America argues (weakly) against this view. In any case, this theory generates the same problem of dispersion between Africa and South America (in reverse) and does not seem to be supported by evidence from the biogeography of other mammalian groups (in the way that the distribution of rodents suggests migration of primates from Africa to South America).

At present there is no convincing explanation of the origin of South American monkeys, but dispersal across the South Atlantic from Africa seems to be the least unlikely method. Future discoveries of early anthropoids should help clarify this question.

PROSIMIAN ORIGINS OF ANTHROPOIDS

One question that has not been discussed in either the preceding chapter or this one is which group of prosimians gave rise to higher primates. Among living primates, anthropoids seem to be more closely related to the living *Tarsius* than to living lemurs and lorises. When we consider fossil prosimians, however, the possible phyletic relationships between anthropoids and various groups of living and fossil prosimians are much more complicated. Many of these complications arise from the conflicting views regarding the phyletic relationships between Eocene prosimians and living lemurs, lorises, and tarsiers.

The debate over the prosimian origins of anthropoids has traditionally centered on whether the "tarsier-like" omomyids or the lemurlike adapids are closer to the origin of anthropoids. Many authors (Rosenberger and Szalay, 1981) have argued that the Eocene omomyids are the prosimian group ancestral to anthropoids, largely on the basis of their reputed affinities with the living *Tarsius*. In contrast, others (e.g., Gingerich,

1980; Rasmussen, 1986) have argued that the Eocene adapids are more suitable ancestors for anthropoids because of similarities in their body size, canine and incisor proportions, fused mandibular symphysis, and more primitive ear structure and limb skeleton compared with omomyids (and *Tarsius*).

Despite all attempts to find "protoanthropoid" features in both adapids and omomyids, neither group is especially convincing as an ancestral anthropoid. Identification of either group as the one that gave rise to anthropoids has often been heavily influenced by associating the fossil group with a presumed living descendent (in the case of omomyids) or by choosing the least specialized alternative (adapids). It is not surprising that a large number of workers have found, on closer inspection, that there is little to recommend one group over the other (Kay, 1980). If we add to this the many questions regarding the phyletic relationship between omomyids and tarsiers and the biogeographic discrepancy between the known distribution of Eocene prosimians and the appearance of the first anthropoids on continents with a poorly known fossil prosimian fauna, it becomes a very murky area indeed. The most intriguing evidence on this topic comes from the recent recovery of tarsier- and omomyid-like remains from the Fayum, which, together with the increasing evidence of omomyid-like features in the parapithecids, supports a tarsiiform origin of anthropoids, probably in Africa, where the Eocene fossil record is presently devoid of primates.

EARLY ANTHROPOID EVOLUTION

The Oligocene was a critical period in primate evolution, since it marks the earliest appearance of higher primates in both Africa and South America. We have a number of unusual early anthropoids from both continents—enough to raise fascinating questions but too few to provide totally satisfying answers. The fossil anthropoids from the Oligocene of Egypt and South America cannot be unequivocally linked with either of the common prosimian groups from the previous Eocene epoch, and the alternatives of platyrrhine immigration from either Africa or North America to South America are not easy to demonstrate or defend.

The relationships between the Oligocene anthropoids and living higher primates are more evident, at least on a general level. The earliest New World anthropoids are similar overall to the living platyrrhines in most anatomical features. Several extant subfamilies can be traced to the middle Miocene and perhaps earlier. In contrast, the Fayum Oligocene anthropoids are far more primitive than living catarrhines. The parapithecids are early Old World anthropoids that retained many primitive anthropoid or platyrrhine features, including three premolars, a platyrrhine-like pterion region, and an omomyid- or platyrrhine-like skeleton. They are probably similar to the early anthropoids that gave rise to New World monkeys. The propliopithecids have a catarrhine-like dental formula but a primitive anthropoid- or platyrrhine-like cranial and dental anatomy. They were incipient catarrhines, but they preceded the divergence of the lineages leading to Old World monkeys and to living hominoids. The Oligocene record of Old World higher primates shows a remarkable array of intermediate forms and missing links that provides us with many insights into the step-by-step process by which later Old World anthropoids came to acquire their present features.

BIBLIOGRAPHY

OLIGOCENE EPOCH

Gingerich, P.D. (1984). Primate evolution. In *Mammals: Notes for a Short Course*, ed. T.D. Broadhead, pp. 167–181. Knoxville: University of Tennessee, Dept. of Geological Sciences.

Robert, C. (1980). Santonian to Eocene palaeogeographic evolution of the Rio Grande Rise (South Atlantic) deduced from clay-mineralogical data (DSDP LEGS 3 and 39). *Palaeogeogr., Palaeoclimatol, Palaeoecol.* **33**:311–325.

Wolfe, J.A. (1978). A paleobotanical interpretation of Tertiary climates in the Northern Hemisphere. *Am. Sci.* **66**:694–703.

EARLY HIGHER PRIMATES

Ciochon, R.L. (1985). Fossil ancestors of Burma. *Nat. Hist.* **10**:26–36.

Ciochon, R.L., Savage, D.E., Thaw, T., and Maw, B. (1985). Anthropoid origins in Asia? New discovery of *Amphipithecus* from the Eocene of Burma. *Science* **229**(4715):756–759.

Maw, B., Ciochon, R.L., and Savage, D.E. (1979). Late Eocene of Burma yields earliest anthropoid primate *Pondaungia cotteri*. *Nature (London)* **282**:65–67.

FAYUM FOSSIL ANTHROPOIDS

Bown, T.M., Kraus, M.J., Wing, S.L., Fleagle, J.G., Tiffany, B., Simons, E.L., and Vondra, C.F. (1982). The Fayum forest revisited. *J. Hum. Evol.* **11**(7):603–632.

Fleagle, J.G., Bown, T.M., Obradovich, J.O., and Simons, E.L. (1986a). How old are the Fayum primates? In *Primate Evolution*, ed. J.G. Else and P.C. Lee, pp. 133–142. Cambridge: Cambridge University Press.

———. (1986b). Age of the earliest African anthropoids. *Science* **234**:1247–1249.

Kay, R.F., and Simons, E.L. (1980). The ecology of Oligocene African Anthropoidea. *Int. J. Primatol.* **1**:21–37.

Parapithecids

Anapol, F. (1983). Scapula of *Apidium phiomense*: A small anthropoid from the Oligocene of Egypt. *Folia Primatol.* **40**:11–31.

Cartmill, M., MacPhee, R.D.E., and Simons, E.L. (1981). Anatomy of the temporal bone in early anthropoids, with remarks on the problem of anthropoid origins. *Am. J. Phys. Anthropol.* **56**:3–21.

Conroy, G.C. (1976). Primate postcranial remains from the Oligocene of Egypt. *Contrib. to Primatol.* **8**:1–134.

Fleagle, J.G., and Kay, R.F. (1987). The phyletic position of the Parapithecidae. *J. Hum. Evol.* **16**:483–531.

Fleagle, J.G., Kay, R.F., and Simons, E.L. (1980). Sexual dimorphism in early anthropoids. *Nature (London)* **287**:328–330.

Fleagle, J.G., and Simons, E.L. (1979). Anatomy of the bony pelvis in parapithecid primates. *Folia Primatol.* **31**:176–186.

———. (1983). The tibio-fibular articulation in *Apidium phiomense*, an Oligocene anthropoid. *Nature (London)* **301**(5897):238–239.

Gingerich, P.D. (1973). Anatomy of the temporal bone in the Oligocene anthropoid *Apidium* and the origin of the Anthropoidea. *Folia Primatol.* **19**:329–337.

———. (1978). The Stuttgart collection of Oligocene primates from the Fayum Province of Egypt. *Paleontol. Z.* **52**:82–92.

Hoffstetter, R. (1982). Les primates simiiformes (= Anthropoidea) (Comprehension, Phylogenie, Histoire Biogeographique). *Ann. Paleontol. (Vert.–Invert.)* **68**(3):241–290.

Kay, R.F., and Simons, E.L. (1980). The ecology of Oligocene African Anthropoidea. *Int. J. Primatol.* **1**:21–37.

———. (1983). Dental formulae and dental eruption patterns in Parapithecidae (Primates, Anthropoidea). *Am. J. Phys. Anthropol.* **62**:363–375.

Osborn, H.F. (1908). New fossil mammals from the Fayum Oligocene, Egypt. *Bull. Am. Mus. Nat. Hist.* **24**:265–272.

Schlosser, M. (1910). Uber einige fossile Saugertiere aus dem Oligocan von Agypten. *Zool. Anz.* **34**:500–508.

———. (1911). Beitrage zur Kenntnis der Oligozanen Lansaugetiere aus dem Fayum, Aegypten. *Beitr. Palaeontol. Oesterreich-Ungarns Orients* **6**:1–227.

Simons, E.L. (1962). Two new primate species from the African Oligocene. *Postilla* **64**:1–12.

———. (1974). *Parapithecus grangeri* (Parapithecidae, Old World Higher Primates): New species from the Oligocene of Egypt and the initial differentiation of Cercopithecoidea. *Postilla* **166**:1–12.

———. (1986). *Parapithecus grangeri* of the African Oligocene: An archaic catarrhine without lower incisors. *J. Hum. Evol.* **15**:205–213.

Simons, E.L., and Kay, R.F. (1983). *Qatrania*, new basal anthropoid primate from the Fayum, Oligocene of Egypt. *Nature (London)* **304**:624–626.

Parapithecids and Later Anthropoids

Delson, E. (1975). Toward the origin of the Old World monkeys. *Evolution des vertebres: Problemes actuels de palaeontologie. Actes CNRS Coll. Int.* **218**:839–850.

Delson, E., and Andrews, P. (1975). Evolution and interrelationships of the catarrhine primates. In *Phylogeny of the Primates: A Multidisciplinary Approach*, ed. W.C. Luckett and F.S. Szalay, pp. 405–446. New York: Plenum Press.

Fleagle, J.G. (1986). The fossil record of early catarrhine evolution. In *Major Topics in Primate and Human Evolution*, ed. B. Wood, L. Martin, and P. Andrews, pp. 130–149. Cambridge: Cambridge University Press.

Fleagle, J.G., and Kay, R.F. (1987). The phyletic position of the Parapithecidae. *J. Hum. Evol.* **16**:483–531.

Fleagle, J.G., and Rosenberger, A.L. (1983). Cranial morphology of the earliest anthropoids. In *Morphologie Evolutive, Morphogenese du Crane, et Origine de l'Homme*, ed. M. Sakka, pp. 141–153. Paris: CNRS.

Gingerich, P.D. (1978). The Stuttgart collection of Oligocene primates from the Fayum Province of Egypt. *Paleontol. Z.* **52**:82–92.

Harrison, T. (1987). The phyletic relationships of the early catarrhine primates: A review of the current evidence. *J. Hum. Evol.* **16**:41–80.

Hoffstetter, R. (1977). Primates: Filogenia e historia biogeographica. *Studia Geol.* **13**:211–253.

Kay, R.F. (1977). The evolution of molar occlusion in the Cercopithecoidea and early catarrhines. *Am. J. Phys. Anthropol.* **46**:327–352.

Simons, E.L. (1970). The deployment and history of Old World monkeys (Cercopithecoidea, Primates). In *Old World Monkeys*, ed. J.R. Napier and P.H. Napier, pp. 92–147. New York: Academic Press.

———. (1972). *Primate Evolution: An Introduction to Man's Place in Nature*. New York: Macmillan.

Propliopithecids

Conroy, G.C. (1976). Hallucial tarsometatarsal joint in an Oligocene anthropoid, *Aegyptopithecus zeuxis*. *Nature* **262**:684–686.

Feldesman, M.R. (1982). Morphometric analysis of the distal humerus of some Cenozoic catarrhines: The late divergence hypothesis revisited. *Am. J. Phys. Anthropol.* **59**:173–195.

Fleagle, J.G. (1983). Locomotor adaptations of Oligocene and Miocene hominoids and their phyletic implications. In *New Interpretations of Ape and Human Ancestry*, ed. R.L. Ciochon and R. Corruccini, pp. 301–324. New York: Plenum Press.

Fleagle, J.G., and Kay, R.F. (1983). New interpretations of the phyletic position of Oligocene hominoids. In *New Interpretations of Ape and Human Ancestry*, ed. R.L. Ciochon and R. Corruccini, pp. 181–210. New York: Plenum Press.

———. (1985). The paleobiology of catarrhines. In *Ancestors: The Hard Evidence*, ed. E. Delson, pp. 23–36. New York: Alan R. Liss.

Fleagle, J.G., and Rosenberger, A.L. (1983). Cranial morphology of the earliest anthropoids. In *Morphologie, Evolution, Morphogenese du Crane et Anthropogenese*, ed. M. Sakka, pp. 141–153. Paris: CNRS.

Fleagle, J.G., and Simons, E.L. (1978). Humeral morphology of the earliest apes. *Nature (London)* **276**:705–707.

———. (1982a). Skeletal remains of *Propliopithecus chirobates* from the Egyptian Oligocene. *Folia Primatol.* **39**:161–177.

———. (1982b). The humerus of *Aegyptopithecus zeuxis*, a primitive anthropoid. *Am. J. Phys. Anthropol.* **59**:175–193.

Fleagle, J.G., Kay, R.F., and Simons, E.L. (1980). Sexual dimorphism in early anthropoids. *Nature (London)* **287**:328–330.

Fleagle, J.G., Simons, E.L., and Conroy, G.C. (1975). Ape limb bones from the Oligocene of Egypt. *Science* **189**:135–137.

Kay, R.F., Fleagle, J.G., and Simons, E.L. (1981). A revision of the Oligocene apes from the Fayum Province, Egypt. *Am. J. Phys. Anthropol.* **55**:293–322.

Radinsky, L. (1974). The fossil evidence of anthropoid brain evolution. *Am. J. Phys. Anthropol. n.s.* **41**:15–27.

Schlosser, M. (1910). Uber einige fossile Saugertiere aus dem Oligocan von Agypten. *Zool. Anz.* **34**:500–508.

———. (1911). Beitrage zur Kenntnis der Oligozanen Lansaugetiere aus dem Fayum, Aegypten. *Beitr. Palaeontol. Oesterreich-Ungarns Orients* **6**:1–227.

Simons, E.L. (1965). New fossil apes from Egypt and the initial differentiation of Hominoidea. *Nature (London)* **205**:135–139.

———. (1967). The earliest apes. *Sci. Am.* **217**:28–35.

———. (1984). Ancestors—dawn ape of the Fayum. *Nat. Hist.* **93**(5):18–20.

———. (1987). New faces of *Aegyptopithecus* from the Oligocene of Egypt. *J. Hum. Evol.* **16**:273–289.

Simons, E.L., and Pilbeam, D.R. (1972). Hominoid paleoprimatology. In *The Functional and Evolutionary Biology of Primates*, ed. R.H. Tuttle, pp. 36–62. Chicago: Aldine-Atherton.

Propliopithecids and Later Primates

Andrews, P. (1985). Family group systematics and evolution among catarrhine primates. In *Ancestors: The Hard Evidence*, ed. E. Delson, pp. 14–22. New York: Alan R. Liss.

Delson, E., and Andrews, P. (1975). Evolution and interrelationships of the catarrhine primates. In *Phylogeny of the Primates: A Multidisciplinary Approach*, ed. W.C. Luckett and F.S. Szalay, pp. 405–446. New York: Plenum Press.

Fleagle, J.G. (1986). The fossil record of early catarrhine evolution. In *Major Topics in Primate and Human Evolution*, ed. B. Wood, L. Martin, and P. Andrews, pp. 130–149. Cambridge: Cambridge University Press.

Fleagle, J.G., and Kay, R.F. (1983). New interpretations of the phyletic position of Oligocene hominoids. In *New Interpretations of Ape and Human Ancestry*, ed. R.L. Ciochon and A.B. Corruccini, pp. 181–210. New York: Plenum Press.

———. (1985). The paleobiology of catarrhines. In *Ancestors: The Hard Evidence*, ed. E. Delson, pp. 23–36. New York: Alan R. Liss.

Harrison, T. (1987). The phyletic relationships of the early catarrhine primates: A review of the current evidence. *J. Hum. Evol.* **16**:41–80.

Kay, R.F., Fleagle, J.G., and Simons, E.L. (1981). A revision of the Oligocene apes from the Fayum Province, Egypt. *Am. J. Phys. Anthropol.* **55**:293–322.

Schlosser, M. (1911). Beitrage zur Kenntnis der Oligozanen Lansaugetiere aus dem Fayum, Aegypten. *Beitr. Palaeontol. Oesterreich-Ungarns Orients* **6**:1–227.

Simons, E.L. (1967a). Review of the phyletic interrelationships of Oligocene and Miocene Old World Anthropoidea. In *Evolution des vertebres: Problemes actuels de palaeontologie. Actes CNRS Coll. Int.* **163**:597–602.

———. (1967b). The earliest apes. *Sci. Am.* **217**:28–35.

———. (1972). *Primate Evolution: An Introduction to Man's Place in Nature*. New York: Macmillan.

———. (1985). Origins and characteristics of the first hominoids. In *Ancestors: The Hard Evidence*, ed. E. Delson, pp. 37–41. New York: Alan R. Liss.

Simons, E.L., and Pilbeam, D.R. (1972). Hominoid paleoprimatology. In *The Functional and Evolutionary Biology of Primates*, ed. R.H. Tuttle, pp. 36–62. Chicago: Aldine-Atherton.

FOSSIL PLATYRRHINES

Ameghino, F. (1891). Nuevos restos de mamiferos fosiles descubiertos por Carlos Ameghino en al Eoceno inferior de la Patagonia Austral. *Rev. Argentina Hist. Nat.* **1**:289–328.

———. (1893). New discoveries of fossil mammalia of southern Patagonia. *Am. Nat.* (May) pp. 439–449.

———. (1906). Les formations sedimentaires du Cretace superieur et du Tertiaire de Patagonie, avec un parallele entre leurs faunes mammalogiques et celles de l'ancien continent. *An. Mus. Nac. Hist. Buenos Aires, ser. 3*, **8**:1–568.

Ciochon, R.L., and Chiarelli, A.B. (1980). Paleobiogeographic perspectives on the origin of the Platyrrhini. In *Evolutionary Biology of the New World Monkeys and Continental Drift*, ed. R.L. Ciochon and A.B. Chiarelli, pp. 459–493. New York: Plenum Press.

Delson, E., and Rosenberger, A.L. (1980). Phyletic perspectives on platyrrhine origins and anthropoid relations. In *Evolutionary Biology of the New World Monkeys and Continental Drift*, ed. R.L. Ciochon and A.B. Chiarelli, pp. 445–458. New York: Plenum Press.

Hoffstetter, R. (1982). Les primates simiiformes (=Anthropoidea) (Comprehension, Phylogenie, Histoire Biogeographique). *Ann. Paleontol. (Vert.–Invert.)* **68**(3):241–290.

MacFadden, B.J., Campbell, K.E., Cifelli, R.L., Siles, O., Johnson, N., Naeser, C.W., and Zeitler, P.K. (1985). Magnetic polarity and mammalian biostratigraphy of the Deseadan (late Oligocene–early Miocene) Salla Beds of northern Bolivia. *J. Geol.* **93**:223–250.

Marshall, L.G. (1985). Geochronology and land-mammal biochronology of the transamerican faunal interchange. In *The Great American Biotic Interchange*, ed. F.E. Stehli and S.D. Webb, pp. 49–85. New York: Plenum Press.

Marshall, L.G., Drake, R.E., Curtis, G.H., Butler, R.F., Flanagan, K.M., and Naeser, C.W. (1986). Geochronology of type Santacrucian (Middle Tertiary) land mammal age, Patagonia, Argentina. *J. Geol.* **94**:449–457.

Patterson, B., and Pascual, R. (1972). The fossil mammal fauna of South America. In *Evolution, Mammals and Southern Continents*, ed. A. Keast, F.C. Erk, and B. Glass, pp. 247–309. Albany, N.Y.: SUNY Press.

Robert, C. (1980). Santonian to Eocene paleogeographic evolution of the Rio Grande Rise (South Atlantic) deduced from clay-mineralogical data (DSDP LEGS 3 and 39). *Palaeogeogr., Palaeoclimatol., Palaeoecol.* **33**:311–325.

Rose, K.D., and Fleagle, J.G. (1981). The fossil history of nonhuman primates in the Americas. In *Ecology and Behavior of Neotropical Primates*, ed. A.F. Coimbra-Filho and R.A. Mittermeier, pp. 111–167. Rio de Janeiro: Academia Brasiliera de Ciencias.

Simpson, G.G. (1980). *Splendid Isolation*. New Haven, Conn.: Yale University Press.

Tarling, D.H. (1980). The geologic evolution of South America during the last 200 million years. In *Evolutionary Biology of the New World Monkeys and Continental Drift*, ed. R.L. Ciochon and A.B. Chiarelli, pp. 1–41. New York: Plenum Press.

Branisella

Hoffstetter, R. (1969). Un primate de l'Oligocene inferieur sud-Americain: *Branisella boliviana* gen. et sp. nov. *C. R. Acad. Sci. (Paris) ser. D.* **269**:434–437.

Orlosky, F. (1973). Comparative dental morphology of extant and extinct Cebidae. University Microfilms, Ann Arbor, Michigan.

Rosenberger, A.L. (1981). A mandible of *Branisella boliviana* (Platyrrhini, Primates) from the Oligocene of South America. *Int. J. Primatol.* **2**:1–7.

Wolff, R.G. (1984). New specimens of the primate *Branisella boliviana* from the early Oligocene of Salla, Bolivia. *J. Vert. Paleontol.* **4**(4):570–574.

Tremacebus

Fleagle, J.G., and Bown, T.M. (1983). New primate fossils from late Oligocene (Colhuehuapian) localities of Chubut Province, Argentina. *Folia Primatol.* **41**:240–266.

Hershkovitz, P. (1974). A new genus of late Oligocene monkey (Cebidae, Platyrrhine) with notes on post orbital closure and platyrrhine evolution. *Folia Primatol.* **21**:1–35.

Rosenberger, A.L. (1984). Fossil New World monkeys dispute the molecular clock. *J. Hum. Evol.* **13**:737–742.

Rusconi, C. (1933). Nuevos restos de monos del terciario antiguo de la Patagonia. *Anal. Soc. Cient. Argentina* **116**:286–289.

Dolichocebus

Bordas, A.F. (1942). Anotaciones sobre un "Cebidae" fosil de Patagonia. *Physis.* **19**:265–269.

Chopra, S.R.K. (1957). The cranial sutures in monkeys. *Proc. Zool. Soc. London* **128**:67–112.

Fleagle, J.G., and Bown, T.M. (1983). New primate fossils from late Oligocene (Colhuehuapian) localities of Chubut Province, Argentina. *Folia Primatol.* **41**:240–266.

Hershkovitz, P. (1970). Notes of Tertiary platyrrhine monkeys and description of a new genus from the late Miocene of Colombia. *Folia Primatol.* **12**:1–37.

———. (1982). Supposed squirrel monkey affinities of the late Oligocene *Dolichocebus gaimanensis. Nature (London)* **298**:201–202.

Kraglievich, J.L. (1951). Contribuciones al concimiento de los primates fosilles de la Patagonia. I. Diagnosis previa de un nuevo primate fosil de Oligoceno superior (Colhuehuapiano) de Gaiman, Chubut. *Comm. Inst. Nac. Cient. Nat.* **2**:57–82.

Rosenberger, A.L. (1979). Cranial anatomy and implications of *Dolichocebus,* a late Oligocene ceboid primate. *Nature (London)* **279**:416–418.

———. (1982). Supposed squirrel monkey affinities of the late Oligocene *Dolichocebus gaimanensis. Nature (London)* **298**:202.

Homunculus

Ameghino, F. (1891). Nuevos restos de mamiferos fosiles descubiertos por Carlos Ameghino en al Eoceno inferior de la Patagonia Austral. *Rev. Argentina Hist. Nat.* **1**:289–328.

Bluntschili, H. (1931). *Homunculus patagonicus* und die ihm zugereihten Fossil funde aus den Santa-Cruz-Schichten Patagoniens. *Morphol. Jahr.* **67**:811–892.

Ciochon, R.L., and Corrucini, R. (1975). Morphometric analysis of platyrrhine femora with taxonomic implications and notes on two fossil forms. *J. Hum. Evol.* **4**:193–217.

Hershkovitz, P. (1970). Notes of Tertiary platyrrhine monkeys and description of a new genus from the late Miocene of Colombia. *Folia Primatol.* **12**:1–37.

———. (1981). Comparative anatomy of platyrrhine mandibular cheek teeth dPM_4, PM_4, M_1 with particular reference to those of *Homunculus* (Cebidae) and comments on platyrrhine origins. *Folia Primatol.* **35**:179–217.

———. (1984). More on the *Homunculus* dPM_4 and M_1 and comparisons with *Alouatta* and *Stirtonia* (Primates, Platyrrhini, Cebidae). *Am. J. Primatol.* **7**:261–283.

Soriacebus

Fleagle, J.G., Powers, D.W., Conroy, G.C., and Watters, J.P. (1987). New fossil platyrrhines from Santa Cruz Province, Argentina. *Folia Primatol.* **48**:65–77.

Cebupithecia

Meldrum, J., and Fleagle, J. (1988). Morphological affinities of the postcranial skeleton of *Cebupithecia sarmientoi. Am. J. Phys. Anthropol.* **75**:249–250.

Stirton, R.A. (1951). Ceboid monkeys from the Miocene of Colombia. *Bull. Univ. Calif. Pub. Geol. Sci.* **28**(11):315–356.

Stirton, R.A., and Savage, D.E. (1951). A new monkey from the La Venta Miocene of Colombia. *Compilacion de los Estudios Geol. Oficiales en Columbia, Serv. Geol. Nac. Bogota* **7**:345–356.

Mohanamico

Luchterhand, K., Kay, R.F., and Madden, R.H. (1986). *Mohanamico hershkovitzi,* gen. et ap. nov., un primate du Miocene moyen d'Amerique du Sud. *C.R. Acad. Sci. (Paris) ser. 3,* **303**(19):1753–1758.

Aotus

Setoguchi, T., and Rosenberger, A.L. (1987). A fossil owl monkey from La Venta, Colombia. *Nature (London)* **326**:692–694.

Stirtonia

Hershkovitz, P. (1970). Notes on Tertiary platyrrhine monkeys and description of a new genus from the late Miocene of Colombia. *Folia Primatol.* **12**:1–37.

———. (1984). More on the *Homunculus* dPM_4 and M_1 and comparisons with *Alouatta* and *Stirtonia* (Primates, Platyrrhini, Cebidae). *Am. J. Primatol.* **7**:261–283.

Kay, R.F., Madden, R., Plavcan, J.M., Cifelli, R.L., and Díaz, J.G. (1987). *Stirtonia victoriae,* a new species of Miocene Colombian primate. *J. Hum. Evol.* **16**:173–196.

Setoguchi, T. (1985). *Kondous laventicus,* a new ceboid primate from the Miocene of La Venta, Colombia, South America. *Folia Primatol.* **44**:96–101.

Setagouchi, T., Watanabe, T., and Mouri, T. (1981). The upper dentition of *Stirtonia* (Ceboidea, Primates) from the Miocene of Colombia, South America, and the origin of the postero-internal cusps of upper molars of howler monkeys (*Alouatta*). *Kyoto Univ. Reports of New World Monkeys,* (1981) pp. 51–60.

Stirton, R.A. (1951). Ceboid monkeys from the Miocene of Colombia. *Bull. Univ. Calif. Pub. Geol. Sci.* **28**(11):315–356.

Stirton, R.A., and Savage, D.E. (1951). A new monkey from the La Venta Miocene of Colombia. *Compilacion de los Estudios Geol. Oficiales en Columbia, Serv. Geol. Nac. Bogota* **7**:345–356.

Micodon

Setoguchi, T., and Rosenberger, A.L. (1985). Miocene marmosets: First fossil evidence. *Int. J. Primatol.* **6**:615–625.

Xenothrix

Ford, S. (1986). Subfossil platyrrhine tibia (Primates: Callitrichidae) from Hispanola: A possible further example of island gigantism. *Am. J. Phys. Anthropol.* **70**:47–62.

Ford, S., and Morgan, G.S. (1986). A new ceboid femur from the late Pleistocene of Jamaica. *J. Vert. Paleontol.* **6**:281–289.

Rosenberger, A.L. (1977). *Xenothrix* and ceboid phylogeny. *J. Hum. Evol.* **6**:461–481.

Williams, E.E., and Koopman, K.E. (1952). West Indian fossil monkeys. *Am. Mus. Nov.* **1546**:1–16.

Saimiri bernensis

MacPhee, R.D.E., and Woods, C.A. (1982). A new fossil cebine from Hispaniola. *Am. J. Phys. Anthropol.* **58**:419–436.

Rimoli, R. (1977). Una nueva especie de Monos (Cebidae: Saimirinae: *Saimiri*) de la Hispaniola. *Cuadernos del Cendia, Univ. Autonoma de Santo Domingo* **242**:1–14.

Rosenberger, A.L. (1978). New species of Hispaniolan monkey. *An. Cient. Univ. Cent. Este D.R.* **3**:249–251.

SUMMARY OF FOSSIL PLATYRRHINES

Conroy, G.C. (1981). Review of *Evolutionary Biology of the New World Monkeys and Continental Drift*, edited by R.L. Ciochon and A.B. Chiarelli. *Folia Primatol.* **36**:155–156.

Delson, E., and Rosenberger, A.L. (1980). Phyletic perspectives on platyrrhine origins and anthropoid relations. In *Evolutionary Biology of the New World Monkeys and Continental Drift*, ed. R.L. Ciochon and A.B. Chiarelli, pp. 445–458. New York: Plenum Press.

———. (1984). Are there any anthropoid primate living fossils? In *Living Fossils*, ed. N. Eldridge and S.M. Stanley, pp. 50–61. New York: Springer-Verlag.

Ford, S.M. (1980). Callithricids as phyletic dwarfs and the place of the Callithricidae in Platyrrhini. *Primates* **21**:31–34.

Gregory, W.K. (1922). *The Origin and Evolution of the Human Dentition*. Baltimore: Williams and Wilkins.

Hershkovitz, P. (1977). *Living New World Monkeys (Platyrrhini)*. Chicago: University of Chicago Press.

Kay, R.F. (1980). Platyrrhine origins: A reappraisal of the dental evidence. In *Evolutionary Biology of the New World Monkeys and Continental Drift*, ed. R.L. Ciochon and A.B. Chiarelli, pp. 154–188. New York: Plenum Press.

Orlosky, F. (1980). Dental evolutionary trends of relevance to the origin and dispersal of the platyrrhine monkeys. In *Evolutionary Biology of the New World Monkeys and Continental Drift*, ed. R.L. Ciochon and A.B. Chiarelli, pp. 189–200. New York: Plenum Press.

Rosenberger, A.L. (1984). Platyrrhines contradict the molecular clock. *J. Hum. Evol.* **13**:737–742.

PLATYRRHINE ORIGINS

Cartmill, M., MacPhee, R., and Simons, E.L. (1981). Anatomy of the temporal bone in early anthropoids, with remarks on the problem of anthropoid origins. *Am. J. Phys. Anthropol.* **56**:3–22.

Fleagle, J.G. (1986). Early anthropoid evolution in Africa and South America. In *Primate Evolution*, ed. J.G. Else and P.C. Lee, pp. 133–141. Cambridge: Cambridge University Press.

Gingerich, P.D. (1980). Eocene Adapidae: Paleobiogeography and the origin of South American Platyrrhini. In *Evolutionary Biology of the New World Monkeys and Continental Drift*, ed. R.L. Ciochon and A.B. Chiarelli, pp. 123–138. New York: Plenum Press.

Hoffstetter, R. (1972). Relationships, origins and history of the ceboid monkeys and caviomorph rodents: A modern reinterpretation. In *Evolutionary Biology*, ed. T. Dobzhansky, M.K. Hecht, and W.C. Steere, pp. 323–347. New York: Appleton-Century-Crofts.

———. (1980). Origin and deployment of New World monkeys emphasizing the southern continents route. In *Evolutionary Biology of the New World Monkeys and Continental Drift*, ed. R.L. Ciochon and A.B. Chiarelli, pp. 103–138. New York: Plenum Press.

Hoffstetter, R., and Lavocat, R. (1970). Decouverte dans le Deseadien de Bolivie de genres pentalophodentes appuyant les affinites africaines des Rongeurs Caviomorphes. *C. R. Acad. Sci. (Paris), ser. D.,* **271**:172–175.

Lavocat, R. (1980). The implication of rodent palaeontology and biogeography to the geographical sources and origins of platyrrhine primates. In *Evolutionary Biology of the New World Monkeys and Continental Drift,* ed. R.L. Ciochon and A.B. Chiarelli, pp. 93–102. New York: Plenum Press.

Rand, H.M., and Mabesoone, J.M. (1982). Northeast Brasil and the final separation of South America and Africa. *Palaeogeogr., Palaeoclimatol., Palaeoecol.* **38**:163–183.

Rosenberger, A.L., and Szalay, F.S. (1981). On the tarsiiform origins of Anthropoidea. In *Evolutionary Biology of the New World Monkeys and Continental Drift,* ed. R.L. Ciochon and A.B. Chiarelli, pp. 139–157. New York: Plenum Press.

Simons, E.L. (1976). The fossil record of primate phylogeny. In *Molecular Anthropology,* ed. M. Goodman, R.E. Tashian, and J.H. Tashian, pp. 35–62. New York: Plenum Press.

Stehli, F.G., and Webb, S.D., eds. (1985). *The Great American Biotic Exchange. Topics in Geobiology,* vol. 4. New York: Plenum Press.

Szalay, F.S. (1975). Phylogeny, adaptations and dispersal of the tarsiiform primates. In *Phylogeny of the Primates: A Multidisciplinary Approach,* ed. W.C. Luckett and F.S. Szalay, pp. 357–404. New York: Plenum Press.

Tarling, D.H. (1980). The geologic evolution of South America during the last 200 million years. In *Evolutionary Biology of the New World Monkeys and Continental Drift,* ed. R.L. Ciochon and A.B. Chiarelli, pp. 1–41. New York: Plenum Press.

———. (1982). Land bridges and plate tectonics. In *Phylogenie et paleobiogeographie,* ed. E. Buffetaut, P. Janvier, J.C. Rage, and P. Tassy. *Geobios, Mem. Spec.* **6**:361–374.

Wood, A.E., and Patterson, B. (1970). Relationships among hystricognathous and hystricomorphous rodents. *Mammalia* **34**:628–639.

PROSIMIAN ORIGINS OF ANTHROPOIDS

Cartmill, M. (1980). Morphology, function and evolution of the anthropoid postorbital septum. In *Evolutionary Biology of the New World Monkeys and Continental Drift,* ed. R.L. Ciochon and A.B. Chiarelli, pp. 243–274. New York: Plenum Press.

Cartmill, M., and Kay, R.F. (1978). Cranio-dental morphology, tarsier affinities, and primate suborders. In *Recent Advances in Primatology,* vol. 3, ed. D.J. Chivers and K.A. Joysey, pp. 205–213. London: Academic Press.

Cartmill, M., MacPhee, R., and Simons, E.L. (1981). Anatomy of the temporal bone in early anthropoids, with remarks on the problem of anthropoid origins. *Am. J. Phys. Anthropol.* **56**:3–22.

Conroy, G.C. (1978). Candidates for anthropoid ancestry: Some morphological and paleozoogeographical considerations. In *Recent Advances in Primatology,* vol. 3, ed. D.J. Chivers and K.A. Joysey, pp. 27–41. London: Academic Press.

———. (1981). Review of *Evolutionary Biology of the New World Monkeys and Continental Drift,* edited by R.L. Ciochon and A.B. Chiarelli. *Folia Primatol.* **36**:155–156.

Gingerich, P.D. (1980). Eocene Adapidae: Paleobiogeography and the origin of South American Platyrrhini. In *Evolutionary Biology of the New World Monkeys and Continental Drift,* ed. R.L. Ciochon and A.B. Chiarelli, pp. 123–138. New York: Plenum Press.

Kay, R.F. (1980). Platyrrhine origins: A reappraisal of the dental evidence. In *Evolutionary Biology of the New World Monkeys and Continental Drift,* ed. R.L. Ciochon and A.B. Chiarelli, pp. 159–188. New York: Plenum Press.

MacPhee, R.D.E., and Cartmill, M. (1986). Basicranial structures and primate systematics. In *Comparative Primate Biology,* Vol. 1: *Systematics, Evolution and Anatomy,* ed. D.R. Swindler and J. Erwin, pp. 219–275. New York: Alan R. Liss.

Rasmussen, D.T. (1986). Anthropoid origins: A possible solution to the Adapidae-Omomyidae paradox. *J. Hum. Evol.* **15**:1–12.

Rosenberger, A.L. (1986). Platyrrhines, catarrhines, and the anthropoid transition. In *Major Topics in Primate and Human Evolution,* ed. B. Wood, L. Martin, and P. Andrews, pp. 66–88. Cambridge: Cambridge University Press.

Rosenberger, A.L., and Szalay, F.S. (1981). The tarsiiform origins of Anthropoidea. In *Evolutionary Biology of the New World Monkeys and Continental Drift,* ed. R.L. Ciochon and A.B. Chiarelli, pp. 139–157. New York: Plenum Press.

Szalay, F.S., Rosenberger, A.L., and Dagosto, M. (1987). Diagnosis and differentiation of the order Primates. *Yrbk. Phys. Anthropol.* **30**:75–105.

Fossil Apes

The Miocene is a relatively long epoch that began approximately 23 million years ago and ended about 5 million years ago. In the early Miocene, world temperatures seem to have warmed appreciably from the cooler Oligocene, and there were minor fluctuations of warming and cooling periods and increasing aridity throughout much of the epoch (see Fig. 9.3). Several major geophysical events took place during this epoch that affected both the climate and the biogeography of mammals throughout the Old World. The Tethys Sea contracted and was cut off from the Indian Ocean by the emergence of the Arabian peninsula (Whybrow, 1984). On at least one occasion in the late Miocene, the Mediterranean remnant of the Tethys dried up completely. Farther east, India continued to crash into Asia, leading to the rise of the Himalayas.

In East Africa, the Miocene was characterized by considerable volcanic activity in conjunction with the developing rift system. It is here, in the early Miocene sediments of Kenya and Uganda, that we find the earliest fossil Old World monkeys and an impressive array of primitive apes. The monkeys of the early Miocene are not very diverse, and fossil evidence for their major radiation appears only in the latest part of this epoch and in the succeeding Pliocene (see Chapter 14). In contrast, the Miocene deposits of Africa and Eurasia hold an extraordinary abundance and diversity of fossil apes (Hominoids), or non-cercopithecoid-like catarrhines (Fig. 13.1).

Early and Middle Miocene Apes from Africa

In the early and middle Miocene sediments of Kenya and Uganda (23 to 15 million years ago) we find evidence of an extensive radiation of primitive apes, the proconsulids (Fig. 13.2, Table 13.1). Although cranial and postcranial remains are available for only a few of the genera and species, these indicate that proconsulids were more advanced than *Aegyptopithecus* and *Propliopithecus* from the early Oligocene. They seem to have all of the anatomical features that characterize living catarrhines, not just a few as found in the early Oligocene taxa. These Miocene apes ranged in size from the small, capuchin-size (3.5 kg) *Micropithecus clarki* to the female-gorilla-size (50 kg) *Afropithecus* and *Proconsul major*. These fossil apes have been found in association with a variety of paleoenviron-

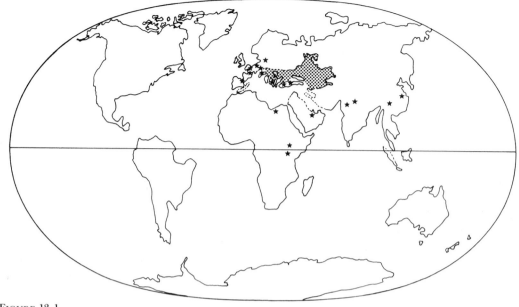

FIGURE 13.1

Map of the early Miocene world showing fossil ape localities.

FIGURE 13.2

East African early Miocene fossil localities.

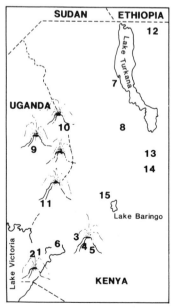

PRIMATE LOCALITIES

1. Rusinga
2. Mfwangano
3. Songhor
4. Koru, Legetet, Meswa, Chamtwara
5. Fort Ternan
6. Maboko, Majiwa, Kaloma
7. Losidok, Morourot, Kalodirr
8. Loperot
9. Napak
10. Moroto
11. Mt. Elgon
12. Buluk
13. Nachola, Namurungule
14. Nakali
15. Tugen Hills, Lukeino, Tabarin

AGES OF PRIMATE LOCALITIES
(from Pickford, 1986)

	FAUNAL SETS	LOCALITIES
PLIO PS	1	
	2	
	3	
	4	
	5	
MIOCENE	6 VII	Lukeino
	7	
	8	
	9 VI	Namurungule
	10	
	11 V	Nachola
	12	Ngorora
	13 IV	Fort Ternan, Serek, Kapsibor
	14	Moroto
	15 III B	Maboko, Majiwa, Kaloma,
	16 III A	Moruorot, Loperot, Losidok, Buluk
	17 II	
	18	
	19 I	Songhor, Chamtwara
	20	Legetet, Koru, Napak
	21	
	22	
	23 Pre I	Meswa
OLIGO	m.y.	

TABLE 13.1
Infraorder Catarrhini
EARLY AND MIDDLE MIOCENE APES

Species	Body Weight (g)
Family PROCONSULIDAE	
Proconsul (e. Miocene, Africa)	
P. africanus	18,000
P. nyanzae	28,000
P. major	50,000
Rangwapithecus (e. Miocene, Africa)	
R. gordoni	15,000
Limnopithecus (e. Miocene, Africa)	
L. legetet	5,000
Dendropithecus (e. Miocene, Africa)	
D. macinnesi	9,000
Simiolus (e. Miocene, Africa)	
S. enjiessi	7,000
Micropithecus (e. Miocene, Africa)	
M. clarki	3,500
M. songhorensis	5,000
Dionysopithecus (?e. Miocene, Asia)	
D. shuangouensis	3,300
Platydontopithecus (?e. Miocene, Asia)	
P. jianghuaiensis	15,000
Family OREOPITHECIDAE[a]	
Nyanzapithecus (e.–m. Miocene, Africa)	
N. vancouveringi	9,000
N. pickfordi	10,000
Family incertae sedis	
Afropithecus (e.–?m. Miocene, Africa, Saudi Arabia)	
A. turkanensis	50,000
A. leakeyi	—
Turkanapithecus (e. Miocene, Africa)	
T. kalakolensis	10,000
Kenyapithecus (m.–l. Miocene, Africa)	
K. africanus	30,000
K. wickeri	—

[a]See also Table 13.3.

ments, ranging from tropical rain forests to open woodlands (Pickford, 1983), and they seem to have spanned a range of ecological niches comparable to those occupied today by both Old World monkeys and apes.

Despite their adaptive diversity, associated with dietary and locomotor differences, the proconsulids from East Africa have many dental features in common, which suggests that they were the result of a single evolutionary radiation (Figs. 13.3, 13.4). All proconsulids share a number of primitive catarrhine features with the earlier propliopithecids. They have a dental formula of $\frac{2.1.2.3}{2.1.2.3}$, with broad upper central incisors and smaller upper laterals. The lower incisors of most species are taller and narrower than those of most living apes (Pilbeam, 1972). All species have relatively large, sexually dimorphic canines that shear against the lower anterior premolar. The upper premolars are relatively broad and bicuspid; the posterior lower premolar is a broad semimolariform tooth.

The upper molars of the group are most diagnostic and are characterized by their quadrate shape with a relatively large hypocone, a pronounced, often beaded, lingual cingulum, and some details of the conules (Kay, 1977). The lower molars have a broad talonid basin surrounded by five prismlike cusps including a large hypoconulid. The major dental differences between the many genera and species are in overall size, in the relative proportions of the anterior dentition, and in the development of shearing crests on the molars, all of which seem related to dietary differences.

There are no complete crania for any of the Miocene apes of East Africa, but parts of the facial skeleton and other cranial bones are known for several species (Figs. 13.5, 13.6). The shape of face varies considerably, being short in some (*Micropithecus*), moder-

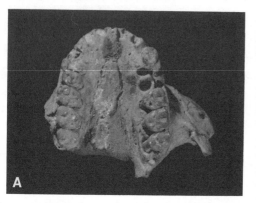

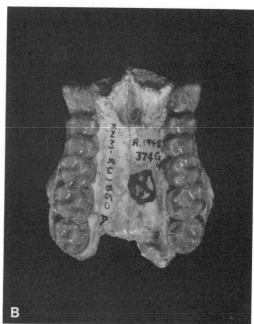

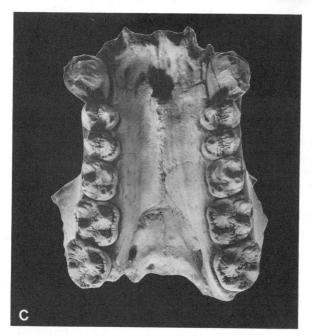

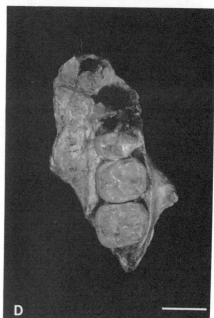

FIGURE 13.3

Upper dentitions of fossil apes from the early Miocene of East Africa: A, *Micropithecus clarki*; B, *Dendropithecus macinnesi*; C, *Rangwapithecus gordoni*; D, *Proconsul africanus*. Note the large lingual cingulum on the upper molars of all except *Micropithecus* and the very long *Rangwapithecus* molars (photographs courtesy of R.L. Ciochon and P. Andrews).

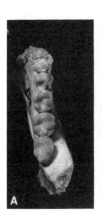
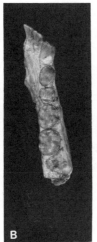

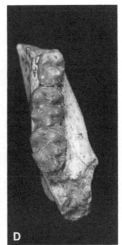
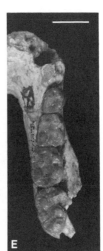

FIGURE 13.4

Lower dentitions of fossil apes from the early Miocene of East Africa: A, *Micropithecus songhorensis*; B, *Limnopithecus legetet*; C, *Dendropithecus macinnesi*; D, *Rangwapithecus gordoni*; E, *Proconsul africanus*. Note the high cusps and well-developed crests on the *Rangwapithecus* molars (photographs courtesy of R.L. Ciochon and P. Andrews).

ately long and broad in others (*Turkanapithecus*), and long and narrow in still others (*Afropithecus*). In most species, the nasal opening has been described as tall and relatively narrow, as in cercopithecoid monkeys, rather than broad and rounded, as in living apes. Orbit size suggests diurnal habits for all species, but the orbits are relatively larger in the smaller species. The only known auditory region, that for *Proconsul africanus*, is identical to that in living catarrhines (but unlike *Aegyptopithecus*) in having a tubular tympanic extending laterally from the side of the bulla. Relative brain size in *P. africanus* and also in *Turkanapithecus* seems to have been similar to that of living Old

FIGURE 13.5

Reconstructed faces of four early Miocene fossil apes from East Africa.

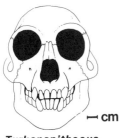
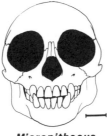

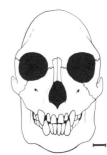

Turkanapithecus   **Micropithecus**   **Afropithecus**   **Proconsul**

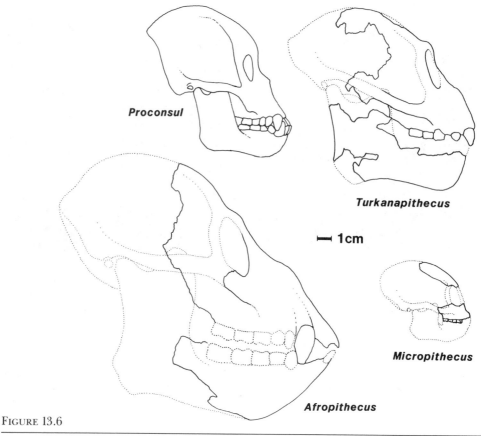

Proconsul

Turkanapithecus

⊢ **1cm**

Micropithecus

Afropithecus

FIGURE 13.6

Reconstructed skulls of four early Miocene fossil apes from East Africa.

World monkeys or perhaps a little larger. The external surface of the cerebrum retains a number of primitive features lacking in extant apes, although in *Afropithecus* the brain seems to have been relatively small.

Hundreds of isolated skeletal elements are known for these primitive apes, and relatively complete skeletons are available for several individuals of *P. africanus* (Fig. 13.7). In limb proportions and many skeletal details these apes resemble living platyrrhines, and they lack many specialized skeletal features of the elbow or wrist, for example, that

characterize either Old World monkeys or living hominoids. They have a more primitive, in some ways more behaviorally versatile, locomotor skeleton. The interspecific skeletal differences indicate considerable locomotor diversity.

The systematics of the early and middle Miocene apes from East Africa has been in a state of flux for many years, partly because of repeated attempts to break the radiation into unnatural groups on the basis of size, but largely because of the continued discovery and recognition of new species. The

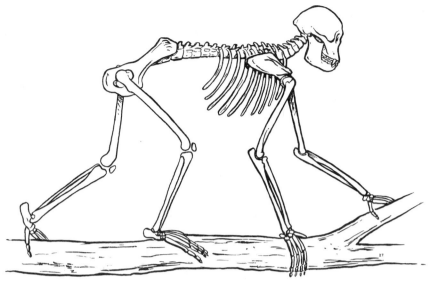

FIGURE 13.7

A reconstructed skeleton of *Proconsul africanus*. Note the monkeylike limb proportions.

many genera and species discussed here almost certainly underestimate the diversity of fossil apes from the earlier part of the Miocene of Kenya and Uganda (Fig. 13.8).

Proconsul is the best-known genus of Miocene ape from East Africa. There are at least three species generally recognized: *Proconsul major* (50 kg), *P. nyanzae* (20–30 kg), and *P. africanus* (15–20 kg). In their dentition, these three species differ mainly in size, and there is considerable debate regarding how many species of *Proconsul* are represented at several sites (Andrews, 1978; Kelley, 1986; Pickford, 1986). All have sexually dimorphic canines and a molar morphology indicating a predominantly frugivorous diet (Kay, 1977).

Many cranial parts are known for the smallest species, *P. africanus* (Fig. 13.6). It has a pronounced snout with prominent canine jugae and a relatively robust zygo-

matic bone. The brain is similar in size to that of a large monkey. As noted above, the auditory region in *P. africanus* is identical to that of extant apes and cercopithecoid monkeys. The external surface of the brain has a primitive sulcal morphology similar to that seen in gibbons and cercopithecoids, but it lacks many features seen in the brain of living great apes (Falk, 1983).

There is a nearly complete juvenile skeleton known for *P. africanus* (Fig. 13.7). The limb proportions are monkeylike, with an intermembral index of 89. Compared with living catarrhines, *P. africanus* has short limbs for its estimated body size. It has a mixture of apelike and more primitive monkeylike features throughout the skeleton (Walker and Pickford, 1983; Beard *et al.*, 1986), resembling living apes in such features as the shape of the distal part of the humerus (Rose, 1988), the robustness of the

FIGURE 13.8

A fossil ape community from Rusinga Island, Kenya, approximately 18 million years ago: upper left, *Proconsul africanus*; upper right, *Dendropithe-* *cus macinnesi*; center, *Limnopithecus legetet*; lower, *Proconsul nyanzae*.

fibula, the conformation of the tarsal bones, and the absence of a tail. It also lacks many characteristic features of living apes, such as a reduced ulnar styloid process, a short ulnar olecranon, and long curved digits. At the same time, *Proconsul* has none of the detailed skeletal features, such as a narrow elbow region, that characterize cercopithecoid monkeys. The skeleton indicates that *P. africanus* was quadrupedal and probably arboreal but lacked both the suspensory abilities of many living apes and the rapid running or leaping habits of Old World monkeys.

Proconsul nyanzae resembles *P. africanus* in many general aspects of its skeleton but shows adaptations for a more terrestrial locomotion. For example, the olecranon process on the ulna extends posteriorly rather than proximally. *Proconsul major* is a very poorly known species, one based primarily on dental remains.

The canine sexual dimorphism of all *Proconsul* species suggests that they did not live in monogamous social groups, but, in view of the diverse types of social groups found among extant apes, we have little evidence about what type of social groups characterized these Miocene species.

Afropithecus turkanensis is a very large fossil ape from the early to middle Miocene of Kenya and Uganda (Leakey and Leakey, 1986a). Compared to *Proconsul, Afropithecus* has a long narrow snout, small orbits, and a broad interorbital area. The dentition is characterized by robust, procumbent incisors, short, round, tusklike canines, and extremely broad upper premolars. A palate from Moroto, Uganda (Pilbeam, 1969), which was originally assigned to *Proconsul major* almost certainly belongs to this species, and a maxilla from Saudi Arabia classified as "*Heliopithecus*" (Andrews and Martin, 1987b) seems to be of the same genus.

Only a preliminary description of this giant ape has been published thus far, and its affinities are far from resolved. The long snout, straight facial profile, and small frontal bone have suggested to some (Simons, 1987) that *Afropithecus* is more closely related to primitive catarrhines such as *Aegyptopithecus*. However, the premolar morphology seems to indicate that this genus is more advanced than other proconsulids and closer to the origins of great apes and humans (Andrews and Martin, 1987a,b).

Turkanapithecus kalakolensis (Figs. 13.5, 13.6) is a medium-size fossil ape from Kalodirr, the same site in northern Kenya that yielded *Afropithecus* (Leakey and Leakey, 1986b). *Turkanapithecus* has relatively long upper molars with many extra cusps and relatively large anterior upper premolars. The mandible is relatively shallow with a broad ascending ramus. The cranium shows a broad, square snout, a broad interorbital region, large rimmed orbits, and flaring zygomatic arches. The phyletic affinities of this ape are very uncertain.

One of the most distinctive of the early Miocene apes is *Rangwapithecus gordoni*. This medium-size (15 kg) species has relatively long and narrow molar teeth with numerous shearing crests that indicate a more folivorous diet than that of other early Miocene apes (Kay, 1977). It also has a very deep mandible (Hill and Odhiambo, 1987). *Rangwapithecus gordoni* seems to be found primarily in rain forest environments.

Nyanzapithecus is a small fossil ape known almost exclusively from dental remains (Harrison, 1987). There are two species, *N. vancouveringi*, from the early Miocene of Rusinga Island, and *N. pickfordi*, from the middle Miocene of Maboko Island. *Nyanzapithecus* is characterized by long upper premolars with similar-size buccal and lingual cusps, long upper molars, and lower molars with deep notches. It was a folivorous primate. Compared with the other fossil apes from the Miocene of East Africa, *Nyanzapithecus* shows greatest similarities to *Rangwapithecus* and is almost certainly derived from that genus. More interesting, however, are the distinctive dental features that *Nyanzapithecus* shares with the European *Oreopithecus* in molar and premolar anatomy; these seem to indicate an African origin for *Oreopithecus*, which is usually placed in the family Oreopithecidae (discussed later in this chapter).

The two species of **Limnopithecus,** *L. legetet* and *L. evansi*, were each about the size of a living gibbon (4–5 kg). They had a frugivorous diet. The few skeletal elements of these species indicate that they were arboreal quadrupeds.

Dendropithecus macinnesi is known from numerous jaws and teeth and much of a skeleton. It has tall, narrow incisors and broad molars with numerous crests, suggesting a frugivorous-folivorous diet. It was a medium-size (9 kg) animal with long, slender limbs similar to those of the neotropical spider monkey (*Ateles*). It was probably mainly quadrupedal but was the most suspensory of the earlier Miocene apes. Although there is striking canine dimorphism in this species, both sexes nevertheless have relatively long, sharp canines, suggesting that the species was possibly monogamous.

Simiolus enjiessi is a newly described small ape (7,000 g) from the locality of Kalodirr in northern Kenya which differs from other small early Miocene apes in its very narrow upper canines, triangular P^3, and long upper molars. It has a mosaic of characteristics found singly in various other genera but shows greatest similarities to *Dendropithecus* and *Rangwapithecus*. The humerus of *Simiolus* is very similar to that of *Dendropithecus*.

Micropithecus clarki is the smallest known

ape, with an estimated body weight of 3 to 4 kg. A second species, **M. songhorensis**, was slightly larger. *Micropithecus* (and especially *M. clarki*) has distinctive dental proportions compared with the other early Miocene apes. The dentition is characterized by relatively large incisors and canines and relatively small cheek teeth. There is a reduced cingulum on the upper molars. The face of *M. clarki* (Figs. 13.5, 13.6) has a very short snout, broad nasal opening, and large orbits, giving it a very gibbonlike appearance (Fleagle, 1975). A frontal bone attributed to this species has a smooth cranial surface and lacks any brow ridges. The endocast of the brain indicates a gibbonlike sulcal pattern (Radinsky, 1975). *Micropithecus songhorensis* was a larger species with much broader lower premolars.

Micropithecus is virtually identical to another small ape, **Dionysopithecus**, from the Miocene of China and other parts of Asia. Both have been frequently identified as possible gibbon ancestors because of their reduced upper molar cingulum and geographic distribution. Although there is nothing in the known dental and cranial anatomy of *Micropithecus* to preclude such a phyletic relationship, there are also few derived anatomical features of either *Micropithecus* or *Dionysopithecus* which would strongly support a unique link with living gibbons.

Most of the African fossil apes we have discussed so far are known primarily from the early Miocene. In contrast, **Kenyapithecus** is from the middle Miocene and seems to come from deposits representing drier, more open, woodland environments. There are two described species: *K. africanus*, primarily from Maboko Island, and *K. wickeri*, from Fort Ternan. Other specimens are known from later Miocene deposits. *Kenyapithecus* differs from the earlier Miocene apes in having thicker molar enamel, a more robust mandible, and large upper premolars, and it seems to be closer to the radiation of great apes and humans than to any other African taxon, including *Afropithecus.*

ADAPTIVE RADIATION OF EAST AFRICAN FOSSIL APES

The fossil apes from the earlier Miocene of East Africa exhibit a diversity in size comparable to that of the living Old World monkeys and of the living apes (Fig. 13.9). Although their dental morphology indicates that most species were predominantly frugivorous (Kay, 1977), there were also more folivorous genera such as *Rangwapithecus, Nyanzapithecus,* and probably *Dendropithecus.* The skeletal anatomy of the early Miocene apes is well known for only a few species, but the many isolated skeletal elements indicate that this radiation included arboreal quadrupeds, suspensory species, and more terrestrial species. There is no evidence of either the extreme fast running or leaping abilities of living cercopithecoids or the brachiating habits of the lesser apes; rather, the skeletal anatomy suggests less specialized but more versatile locomotor abilities such as those of the living spider monkeys or chimpanzees.

There is also evidence of diversity in the habitat preferences of different East African species (Bishop, 1967; Pickford, 1983). On the basis of the associated mammals, gastropods, and sediments, it seems that *Limnopithecus, Micropithecus,* and *Proconsul major* were more common in rain forest environments, whereas *Dendropithecus, Proconsul africanus,* and *P. nyanzae* were more common in dry forest localities. In the overall breadth of their ecological adaptations, the early Miocene apes seem to have filled most of the locomotor and dietary niches found among extant catarrhines, more so than did

FIGURE 13.9

Adaptive diversity of early Miocene apes.

the early Oligocene propliopithecids from Egypt (see Fig. 12.16). Compared with the Egyptian fauna, the early Miocene catarrhines were larger and had more terrestrial, suspensory, and folivorous species. Differences in size dimorphism and also differences between species in the type of sexual canine dimorphism suggest a diversity of social structures.

PHYLETIC RELATIONSHIPS OF EARLY MIOCENE APES

The early Miocene apes were much more similar to extant catarrhines in their dentition, cranium, and skeleton than were the early Oligocene propliopithecids. The one species that is well known, *Proconsul africanus*, has a tubular tympanic, a larger braincase, and more modern skeletal anatomy than *Aegyptopithecus*. Although cranial and skeletal remains are rare for most other species, there is no evidence from the available remains to indicate that any of the other genera were less advanced. All seem to be full-fledged catarrhines compared with the Oligocene taxa but more primitive than any living apes (Beard *et al.*, 1986).

When these fossil apes were known primarily from dental remains, it seemed to many authorities that they were the direct lineal ancestors of living hominoids of similar size (see, e.g., Simons, 1967, 1972; Pilbeam, 1969): the smaller apes *Limnopithecus* and *Dendropithecus* were widely regarded as early gibbons, the medium-size *Proconsul africanus* was linked with the chimpanzee, and the large *Proconsul major* was identified as the ancestor of the gorilla. But a careful consideration of the primitive skeletons, as well as numerous dental and cranial features, has cast doubt on these earlier views. In most skeletal features these early Miocene "apes" are primitive compared with all living hominoids. In the same way that *Aegyptopithecus* seems to be an incipient catarrhine, the proconsulids seem to be incipient apes. *Proconsul africanus*, for example, resembles living hominoids in having a spool-shaped articulation on the distal end of the humerus and possibly in lacking a tail, but at the same time it retains an articulation between the ulna and carpal bones. Thus *Proconsul* possesses some of the derived features that characterize the ape lineage, but it lacks other unique specializations found in all living apes. The common anatomical specializations of extant apes presumably char-

acterize their last common ancestor, indicating that the radiation of living apes came from a type of hominoid more advanced than *Proconsul* and probably most of the early Miocene genera. However, *Afropithecus* among the early Miocene apes and *Kenyapithecus* from the later Miocene show greater similarities to extant hominoids.

Although these Miocene catarrhines are more primitive than living apes, they are also quite different from any cercopithecoid monkeys. What they and the earlier Oligocene hominoids such as *Aegyptopithecus* demonstrate is that Old World monkeys are a very specialized group of higher primates. In their dentition, skull, and some aspects of their skeleton, the living apes have retained many more features from the early catarrhines than have cercopithecoids. For this reason, primitive catarrhines such as *Aegyptopithecus* and *Proconsul* have been traditionally identified as fossil apes, when they are actually much closer to the common ancestry of both Old World monkeys and apes than to the radiation of modern hominoids. Like many fossil primates, they are more appropriately seen, not as extinct members of extant lineages, but rather as missing links that fill in the morphological gaps between more distinct living primate lineages (Fig. 13.10; see also Fig. 13.22).

Although the fossil apes from the early and middle Miocene of East Africa are treated as a single radiation and a single taxonomic group, the proconsulids, this is undoubtedly a gradistic classification. It is in fact quite likely that several more derived lineages of higher primates can be traced back to one of these Miocene apes. The lineage leading to *Oreopithecus*, for example, seems to have originated via *Nyanzapithecus* from a genus such as *Rangwapithecus*. It has also been suggested that the lineage

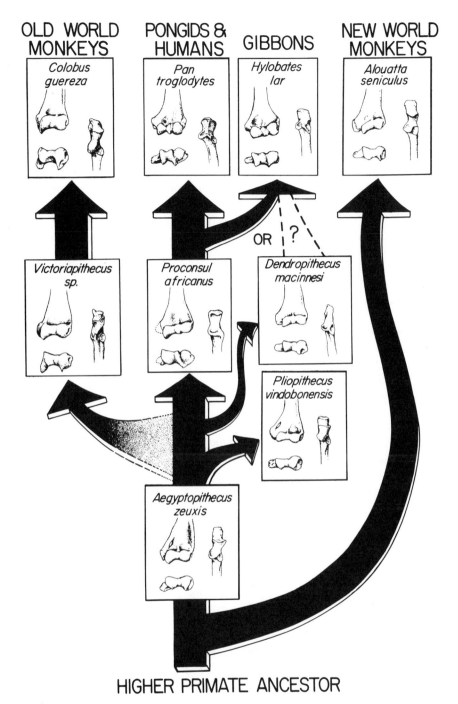

FIGURE 13.10

Phyletic relationships of Oligocene and Miocene fossil catarrhines based on the elbow region.

leading to great apes and humans can be traced back to *Afropithecus*. The position of the proconsulid radiation relative to the origin of the gibbon lineage is more difficult to determine precisely from present evidence. If the great ape and human lineage can be traced to a genus from the Miocene of East Africa, it is quite likely that the more primitive gibbon lineage originated somewhere within this radiation as well.

It has become clear in recent years that the radiation of proconsulids extended far beyond East Africa (see Fig. 13.1). Fossil apes from the middle Miocene of China are remarkably similar to several African genera (Gu and Lin, 1983). One genus in particular, *Dionysopithecus* (Li, 1978), is virtually identical in its molar morphology to *Micropithecus*, despite the 10,000 km separating them, and a similar small ape has been found in the Miocene of Pakistan (Barry *et al.*, 1986). Unfortunately, these Asian proconsulids are known only from a few dental specimens, so the similarities to the East African apes, although striking, are based on very few features. Further discoveries in the Miocene of Asia will surely increase our knowledge of early ape evolution and undoubtedly bring many surprises.

Eurasian Fossil Apes

In contrast with the abundance of fossil apes from the early Miocene of East Africa, fossil primates seem to be generally absent from this time span in Europe and Asia. Only in the middle and late Miocene do we find abundant fossil apes from Europe and Asia, and even then most remains are fragmentary or badly crushed and many come from sites for which there are no absolute dates. Thus both the morphology and the relative ages of European fossil apes are more poorly known than those of the African fossil apes. Never-

theless, it has become increasingly apparent that the less diverse Miocene apes from the more northern latitudes are far more distinct in their phyletic relationships than are the proconsulids from Africa. Rather than documenting different dimensions of a single evolutionary radiation like the proconsulids, the European and Asian apes seem to include representatives of several distinct lineages of relatively primitive and relatively advanced catarrhines.

There are four major groups of fossil apes from the middle and late Miocene of Europe and Asia, many of which are widespread genera with many different species: (a) pliopithecids (*Pliopithecus*, *Crouzelia*, and *Laccopithecus*); (b) oreopithecids (*Oreopithecus*); (c) dryopithecines (*Dryopithecus*, *Lufengpithecus*); and (d) pongines (*Sivapithecus*, *Graecopithecus*, and *Gigantopithecus*).

Pliopithecids

Pliopithecus (Table 13.2) is the oldest and most primitive of the so-called fossil apes from Europe. This gibbon-size primate has

TABLE 13.2
Infraorder Catarrhini
Family PLIOPITHECIDAE

Species	Body Weight (g)
Pliopithecus (m.–l. Miocene, Europe)	
P. antiquus	6,000
P. vindobonensis	7,000
P. lockeri	5,000
P. piveteaui	—
Crouzelia (m. Miocene, Europe)	
C. auscitanensis	5,000
C. rhodanica	—
C. hernyaki	12,000
Laccopithecus (l. Miocene, Asia)	
L. robustus	12,000

been found in fossil sites spanning much of the middle and late Miocene of Europe (Ginsburg, 1986). A closely related genus, **Crouzelia**, is from roughly the same period. The numerous species of *Pliopithecus* and *Crouzelia* ranged in size from about 6 to 10 kg. The earliest and best-known species, *Pliopithecus vindobonensis*, was the size of a siamang (10 kg).

The teeth of *Pliopithecus* (Fig. 13.11) are quite primitive compared with those of other Eurasian apes. The anterior dentition has broad upper central incisors, smaller upper laterals, and tall narrow lower incisors. The canines are very sexually dimorphic, long and daggerlike in some individuals and short in others. The lower anterior premolar is similarly variable—narrow and sectorial in some individuals (presumably males) and broad in others (presumably females). The upper molars are broad and have a large lingual cingulum. The lower cheek teeth, including the posterior premolar, usually have a long and narrow occlusal surface and a prominent buccal cingulum. All of the teeth have extraordinary development of shearing crests, indicating that *Pliopithecus* was folivorous.

Parts of the skull and much of the skeleton from several individuals of *Pliopithecus vindobonensis* are known from a fissure fill at Neudorf an der Marche, Czechoslovakia. The lower jaw is shallow with a broad ascending ramus, similar to that in extant gibbons. The skull is similar to that of a living gibbon in overall appearance but is more primitive in many details. The face has a short, narrow snout. The interorbital region is very broad and the orbits are large and circular. The zygomatic region is relatively gracile. The frontal bone is high and rounded, suggesting a relatively large brain. Posteriorly the temporal lines converge to form a sagittal crest in some individuals. The

FIGURE 13.11

Cranial and dental remains of *Pliopithecus vindobonensis*, from the middle Miocene of Czechoslovakia. Note the gibbonlike face (from Zapfe, 1961).

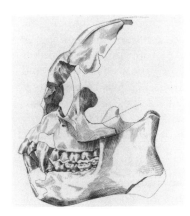

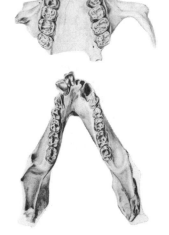

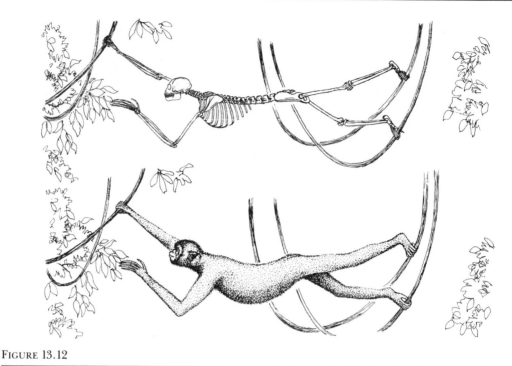

FIGURE 13.12

The skeleton of *Pliopithecus* and a reconstruction of its locomotor habits.

structure of the ear region is intermediate between that of *Aegyptopithecus* and that found in modern catarrhines (and *Proconsul*). The tympanic bone forms a ring at the lateral surface of the bulla, but it does not form a complete tube as in living catarrhines. The inferior half of the tube is not ossified, suggesting a more primitive morphological condition than that found in either Old World monkeys or apes.

The skeleton of *Pliopithecus* (Fig. 13.12) is much like that of a large living platyrrhine such as *Ateles* or *Lagothrix*. The intermembral index is 94. Both the forelimb and the hindlimb show adaptations in the joint surfaces that are characteristic of suspensory primates. Like the Fayum propliopithecids, *Pliopithecus* lacks the distinguishing skeletal features of either living apes or living cercopithecoids, and it has many primitive

skeletal features, such as an entepicondylar foramen on the humerus, a long ulnar styloid process, and a prehallux bone in the foot. It is not clear from the available skeletal material whether *Pliopithecus* had a tail. Ankel (1965) has demonstrated that the sacral canal has the proportions of a monkey with a small tail, but the overall shape and development of the sacrum is most comparable to that of tailless apes. *Pliopithecus* was an arboreal quadruped with suspensory abilities like those of the larger platyrrhines.

PHYLETIC RELATIONS Because of its size and the gibbonlike features in its face, *Pliopithecus* has traditionally been considered an ancestral gibbon. However, in many details of its dentition, skull (particularly the auditory region), and skeleton, *Pliopithecus*, like *Aegyptopithecus*, was more primitive than any

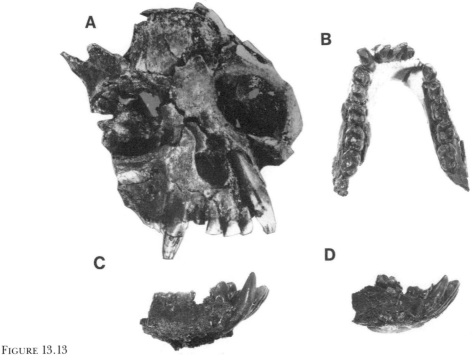

FIGURE 13.13

Cranial and dental remains of *Laccopithecus robustus*, from the latest Miocene of Lufeng, China: A, cranium; B, mandible; C and D, lateral views of male and female mandibles, showing sexual dimorphism (courtesy of Pan Yuerong).

living catarrhine (or *Proconsul*) and lacked the specializations that would be expected in the last common ancestor of Old World monkeys and apes. If gibbons evolved from *Pliopithecus*, then the anatomical features that are shared by living hominoids and Old World monkeys, such as a tubular tympanic and absence of the entepicondylar foramen on the humerus, must be the result of parallel evolution rather than inheritance from a common ancestor. At the same time, *Pliopithecus* shares no unique features with the extant lesser apes that are not also found in several other Miocene or Oligocene fossil anthropoids. Like the early Oligocene *Aegyptopithecus*, the middle Miocene *Pliopithecus* does not appear to be a full-fledged catarrhine and is more primitive than *Proconsul* from the early Miocene. *Pliopithecus* seems to be a late member of the early catarrhine radiation that includes the Fayum "apes" and is usually placed in a separate family of primitive catarrhines, the Pliopithecidae.

Laccopithecus robustus (Fig. 13.13), from the latest Miocene site of Lufeng, China, is a large fossil ape (12 kg) that is virtually identical to *Pliopithecus* in dental morphology. It has sexually dimorphic canines and anterior premolars. Like *Pliopithecus*, *Laccopithecus* has large orbits and a short snout, but the zygomatic region is more robust. The striking dental similarities to *Pliopithecus* suggest that *Laccopithecus* is a late-surviving member of the pliopithecids. Like *Pliopithecus*, *Laccopithecus* has been considered a fossil gibbon (Wu and Pan, 1984, 1985), and its Asian location is compatible with such a relationship. Moreover, the parts of the anatomy in which *Pliopithecus* is more primitive than modern catarrhines (the auditory region and the skeleton) are not known for *Laccopithecus*. Should *Laccopithecus* turn out to be more like modern gibbons in further aspects of cranial and skeletal anatomy, this would necessitate revision of the position of the pliopithecids and reconsideration of the amount of parallel evolution in hominoid evolution.

TABLE 13.3
Infraorder Catarrhini
Family OREOPITHECIDAE

Species	Body Weight (g)
Oreopithecus (l. Miocene, Europe)	
O. bambolii	30,000

Oreopithecus

This fossil ape from the late Miocene of Europe has been an enigma to paleontologists since its initial discovery in the latter part of the last century. The single species, ***Oreopithecus bambolii*** (Table 13.3), is known only from sites in northern Italy, particularly from coal mines. Numerous remains, including cranial and skeletal elements, have been recovered (Figs. 13.14, 13.15), but the most complete remains are crushed, making interpretation of their morphology rather difficult.

Oreopithecus has a dental formula of $\frac{2.1.2.3}{2.1.2.3}$, like all catarrhines, but many aspects of its dentition are quite unique—hence the long-standing difficulties in determining its phyletic position among catarrhines. The upper central incisor is relatively large and round, and the lateral is a smaller peglike tooth; the lower incisors are narrow spatulate teeth. The canines are quite dimorphic, with presumed males having tall upper and lower canines and the females very small canines. In the males, the upper canine shears against the anterior surface of the anterior lower premolar; in females, the lower premolars are more semimolariform. Upper premolars are characterized by two relatively tall cusps of similar size. The upper molars are long and narrow, with a well-formed trigon, a large hypocone, and a

lingual cingulum as in other Miocene catarrhines. The paraconule is particularly well developed. The lower molars have the characteristic basic cusps found in all non-cercopithecoid catarrhines but also have an additional sixth cusp, the centroconid. The well-developed shearing crests clearly indicate a folivorous diet. Overall, the dentition is a more specialized version of that found in the African early and middle Miocene hominoids *Nyanzapithecus* and *Rangwapithecus*.

The skull has a relatively short snout, a small brain, and a pronounced sagittal crest in some individuals. The auditory region indicates the presence of a tubular ectotympanic as in extant catarrhines.

The skeleton of *Oreopithecus* has several indications of suspensory locomotor habits, including a relatively short trunk, a broad thorax, relatively long forelimbs, short hindlimbs, long, slender manual digits, and evidence of extensive mobility in virtually all joints. The elbow region is identical to that of extant great apes.

PHYLETIC RELATIONS Since its initial discovery, *Oreopithecus* has been identified by various authorities as being closely related to parapithecids, cercopithecoids, pongids, hominids, or an ancient higher primate lineage not closely related to any modern group of anthropoids. Many of these diverse interpretations are still championed by one

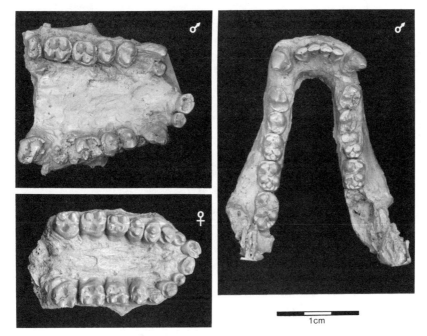

FIGURE 13.14

Upper (left) and lower (right) dentition of *Oreopithecus bambolii*, from the Pliocene of Europe (courtesy of Eric Delson).

FIGURE 13.15

Skeleton of *Oreopithecus* and a reconstruction of its locomotor habits.

or more authorities. Nevertheless, recent analyses of *Oreopithecus* demonstrate that any similarities to cercopithecoid monkeys are almost certainly primitive hominoid retentions, and that *Oreopithecus* is more closely related to extant apes (Harrison, 1986; Sarmiento, 1987). The lineage leading to *Oreopithecus* seems to have arisen among the proconsulids in East Africa and probably diverged from the hominoid lineage near the origin of gibbons.

Dryopithecines and Pongines

The systematics and evolutionary relationships of the other, more widespread, large fossil apes from the middle and late Miocene of Eurasia are very unsettled (see, e.g., Kelley and Pilbeam, 1986). Most current authorities divide the various species of middle and late Miocene apes into two genera (or species groups): *Dryopithecus*, primarily from Europe and possibly China; and *Sivapithecus*, primarily from western and southern Asia (Table 13.4). The exact allocation of species to one or the other of these groups varies somewhat from authority to authority. Because teeth and jaws are all that is known for many species, the division is usually made on the basis of the relative thickness of dental enamel, development of the lingual cingulum on the upper molars, premolar proportions, mandible shape, and subnasal morphology (Martin, 1986). The actual evolutionary diversity was certainly much greater than two genera, and the evolutionary relationships among the species were probably much more complicated than can be determined from present evidence. It is quite likely that both the *Dryopithecus* and *Sivapithecus* (pongine) groups contain more than one clade (Kelley and Pilbeam, 1986), but the two-group scheme is appropriate to the dental material that currently provides most of our knowledge of most taxa.

TABLE 13.4
Infraorder Catarrhini
Family PONGIDAE

Species	Body Weight (g)
Dryopithecus (m.–l. Miocene, Europe)	
D. fontani	35,000
D. laietanus	20,000
Lufengpithecus (l. Miocene, Asia)	
L. lufengensis	40,000
Sivapithecus (l. Miocene, Europe, Asia)	
S. sivalensis (= indicus)	58,000
S. punjabicus	40,000
S. meteai	82,000
?S. alpani	—
?S. darwini	—
Gigantopithecus (l. Miocene–Pleistocene, Asia)	
G. giganteus (= bilaspurensis)	166,000
G. blacki	300,000
Graecopithecus (l. Miocene, Europe)	
G. freybergi	—
G. macedoniensis	110,000

Dryopithecus is the more primitive genus and seems to have lived slightly earlier. It is known only from Europe and possibly China. It was intermediate in size between a siamang and a chimpanzee. *Dryopithecus* is known almost totally from dental remains in western Europe, but various cranial and skeletal remains are known for fossil ape species from Hungary and China that are similar to *Dryopithecus* (Wu, 1985). The exact number of species is presently under study.

The lower premolars of *Dryopithecus* are broader than those of either primitive catarrhines or extant gibbons and the upper premolars are longer. The molar morphology is roughly intermediate between that of the early Miocene *Proconsul* from Africa and the later *Sivapithecus* from Asia. The upper molars of *Dryopithecus* are not as broad as

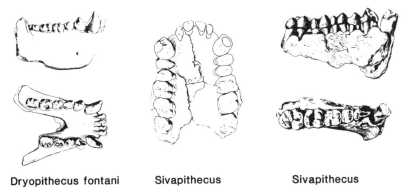

Dryopithecus fontani   **Sivapithecus**   **Sivapithecus**

FIGURE 13.16

Dental remains of two large, middle and late Miocene fossil apes from Eurasia, *Dryopithecus* and *Sivapithecus*.

those of the Early Miocene apes or *Pliopithecus*, and they often have only a partly formed lingual cingulum. *Dryopithecus* differs from *Sivapithecus* in having thin rather than thick enamel on the cheek teeth, gracile canines, a relatively short premaxilla, and a relatively gracile mandible (Fig. 13.16). The broad, rounded cusps on the cheek teeth indicate a predominantly frugivorous diet.

The cranial remains attributed to *Dryopithecus* have not yet been described. The few skeletal elements indicate a postcranial anatomy that is more similar to that of living hominoids than that of any of the proconsulids or *Pliopithecus* on the basis of their reduced olecranon process, deep humeral trochlea, and loss of the entepicondylar foramen (Morbeck, 1983). These limbs also suggest that some species were suspensory.

In contrast with *Dryopithecus*, which was most common in the middle and late Miocene of Europe, **Sivapithecus** (including *Ramapithecus*) is best known from the later half of the Miocene in eastern Europe and especially Asia. The genus *Sivapithecus* (Figs. 13.16, 13.17) contains a very diverse group of species, with some as large as a male orangutan or a female gorilla (probably greater

FIGURE 13.17

Cranial remains of *Sivapithecus* and crania of *Pan* (left) and *Pongo* (right) (photograph courtesy of William and David Pilbeam).

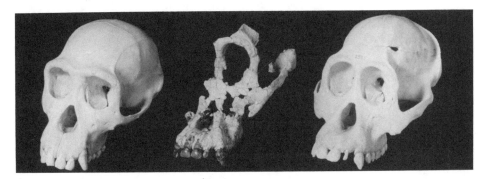

than 75 kg). Most species are larger than earlier apes. The boundaries of the genus *Sivapithecus* are subject to considerable disagreement. Many authorities recognize a species group with several genera within Eurasia, in addition to related taxa in Africa (Kelley and Pilbeam, 1986); others recognize a single genus with various numbers of species (Kay, 1982b; Martin, 1986). Table 13.4 represents an intermediate arrangement. Most species are attributed to *Sivapithecus*, but the large fossil ape from Greece, ***Graecopithecus (= Ouranopithecus) macedoniensis*** (Fig. 13.18), is separated as a distinct genus because it seems to lack the distinctive nasal morphology found in the fossils from Turkey and Pakistan.

Sivapithecus (and *Graecopithecus*) are characterized by thick enamel on the cheek teeth and the common absence of any cingulum on the molars, very broad lower premolars, robust canines, and thick mandible. In most species, the upper central incisors are broad and the laterals are very small (Andrews,

1983). On the basis of the available material, *Sivapithecus* shows relatively little canine dimorphism compared with that found among living apes and monkeys (Kay and Simons, 1983; cf. Kelley and Pilbeam, 1986). In some individuals, the lower anterior premolar is elongated and sectorial (apelike) and the canine is aligned anteroposteriorly. In others, the canines and premolars are broad and oriented buccolingually as in living hominids (Kay, 1982b). In addition to the thick enamel and absence of a cingulum, the molars of *Sivapithecus* have relatively low cusp relief, so the teeth wore flat (Fig. 13.16). This combination of thick-enameled molars and low cusp relief is characteristic of living primates that eat seeds and nuts. It has been suggested that *Sivapithecus* had a diet of hard nuts, bark, or fruits with hard pits.

Several partial skulls are known for *Sivapithecus*. Those from Pakistan and Turkey show a striking resemblance to the living orangutan in such features as a narrow snout with a very large procumbent premax-

FIGURE 13.18

Male and female lower jaws of a late Miocene fossil ape from Greece, *Graecopithecus macedoniensis*.

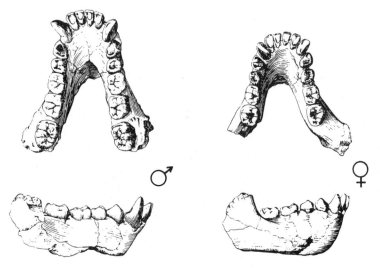

illa, a small incisive foramen, broad zygo-matic arches, a tall, narrow nasal aperture, and high orbits (Fig. 13.17). The dentally similar fossil ape from Greece, *Graecopithecus macedoniensis* (Fig. 13.18), seems to lack the distinctive orangutan features in the nasal region (DeBonis and Melentis, 1985, 1987). Fossils from Hungary (Kretzoi, 1975) often included in *Sivapithecus* are probably best placed in *Dryopithecus*. Chinese fossils fre-quently placed in *Sivapithecus* have been recently given a new generic name, **Lufeng-pithecus** (Wu, 1987). These fossils also show greater similarity to *Dryopithecus* than to *Sivapithecus*.

There are only a few skeletal remains of *Sivapithecus*. At least one species has an opposable hallux and an elbow that resem-bles the living gorilla, suggesting terrestrial habits. There was probably considerable lo-comotor diversity in these late Miocene apes.

A close relative of *Sivapithecus* is **Giganto-pithecus**, the largest primate that ever lived (Figs. 13.19, 13.20). The two species of *Gigantopithecus* were almost certainly derived from a large Asian species of *Sivapithecus*. The earlier, smaller *G. giganteus* (= *G. bilaspurensis*) is from the latest Miocene of India and Pakistan; the larger *G. blacki* is from Pleistocene caves in China and Viet-nam. The smaller species probably weighed as much as a living gorilla (125 kg), and the Pleistocene species has an estimated weight several times that (perhaps as much as 300 kg—based primarily on the large mandible).

These extraordinary primates are known only from lower jaws and isolated teeth. They were initially discovered in Chinese drugstores where the teeth were being sold as medicine (von Koenigswald, 1983). The lower incisors are very small and vertical. The canines are thick but relatively short.

FIGURE 13.19

Lower jaws of *Gigantopithecus* and *Sivapithecus* compared with that of a male mountain gorilla, the largest living primate.

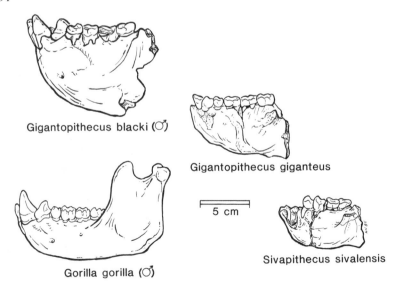

Gigantopithecus blacki (♂)

Gigantopithecus giganteus

Gorilla gorilla (♂)

5 cm

Sivapithecus sivalensis

FIGURE 13.20

A reconstruction of *Gigantopithecus blacki* from the Pleistocene of China.

The lower anterior premolar is relatively broad, as in *Homo sapiens*, rather than elongated. Like those of *Sivapithecus*, the teeth of *Gigantopithecus* have thick enamel and low, flat cusps. In *G. blacki* there are often accessory cusps. In both species, the mandible is very thick and extremely deep compared with the jaws of living apes (Fig. 13.19). The dental proportions, cheek tooth morphology, and robust mandibles indicate that *Gigantopithecus* ate some type of hard fibrous material. One worker has suggested that they ate bamboo, like the living panda. Their enormous size would seem to have precluded anything except a folivorous diet and terrestrial locomotion.

PHYLETIC RELATIONS Although *Dryopithecus* and *Sivapithecus* have played central roles in interpretations of ape and human evolution for over a century, our understanding of the cranial and postcranial anatomy of these large Eurasian apes is remarkably scanty compared with our knowledge of other fossil primates. Accordingly, interpretations of their systematics and phyletic relationships have changed considerably from decade to decade, depending on which of the few available anatomical features have been used to evaluate relationships (see, e.g., Pilbeam, 1966, 1972, 1978; Kelley and Pilbeam, 1986; Martin, 1986). Interpretations based on canine size differ considerably from those based on enamel thickness, subnasal morphology, or relative premolar size. With the recovery of more complete remains of more taxa, these relationships will continue to be modified for many years.

Dryopithecus is the more primitive of the two genera. This European ape is more derived than proconsulids or gibbons in some dental features, but it still lacks many features characteristic of the living great apes and humans such as long upper premolars, robust canines, and thick molar enamel (Andrews and Martin, 1987a). Most of the skeletal remains attributed to *Dryopithecus* accord with this assessment (Morbeck, 1983), but a few isolated bones from Europe that are normally attributed to *Dryopithecus* are more primitive and similar to remains of *Proconsul* (Rose, 1983; Kelley and Pilbeam, 1986).

Among the *Dryopithecus* species, those from Hungary and China are the most completely known and the most difficult to assess. Both have been considered to be *Sivapithecus* (or *Ramapithecus*) by some authorities and *Dryopithecus* by others. Further description and analysis of these relatively complete but crushed remains should help clarify and undoubtedly complicate our understanding of the relationship between *Dryopithecus* and *Sivapithecus*.

The phyletic position of *Sivapithecus* has been the subject of more controversy. Many of the fossils now considered to belong to *Sivapithecus* were for many years placed in a separate genus, *Ramapithecus*, which was widely regarded as an early hominid (Simons, 1961, 1975; Pilbeam, 1968). This view has been generally abandoned by almost all authorities (cf. Kay and Simons, 1983; Schwartz, 1986), and the most hominid-like fossils have been identified as female apes with small canines or old individuals with very worn teeth.

More complete cranial remains of several species of *Sivapithecus* have been shown to have striking similarities to the orangutan, *Pongo* (Andrews and Tekkaya, 1980; Lipson and Pilbeam, 1982; Preuss, 1982), in such features as the shape of the orbits, proportions of the upper incisors, flaring of the zygomatic bone, and particularly the size and shape of the premaxilla and incisive foramen (Ward and Pilbeam, 1983). Thus it seems most likely that some species of *Sivapithecus* are uniquely related to the orangutan (Ward and Brown, 1986). Still, the evolutionary relationships of *Sivapithecus* and other taxa often included in a "*Sivapithecus* group" are almost certainly not resolved, for several reasons. For one, many of the distinctive cranial features shared by *Sivapithecus* and orangutans, such as tall orbits, flaring zygomatic bones, and absence of continuous brow ridges, are quite common among Miocene or even Oligocene fossil apes and may well be primitive hominoid features retained by orangutans rather than unique attributes of that lineage. Likewise, many of the features that suggest a phyletic link between *Sivapithecus* (or *Ramapithecus*) and early hominids, such as thick molar enamel or robust mandibles, seem to be primitive features retained by both groups (Fig. 13.21; Martin, 1985, 1986). Finally, there is considerable morphological variability among the fossil apes normally grouped with *Sivapithecus* on the basis of presumed thick enamel. The morphology of *Sivapithecus* varies both through time and from region to region (Kelley and Pilbeam, 1986). *Graecopithecus macedoniensis* has a dental morphology similar to that of *Sivapithecus*, but in the shape of its premaxilla and nasal region it is more like African apes and hominids than orangutans and *Sivapithecus*. Many authorities still consider *Kenyapithecus* to be a part of the *Sivapithecus* group. The thick-enameled

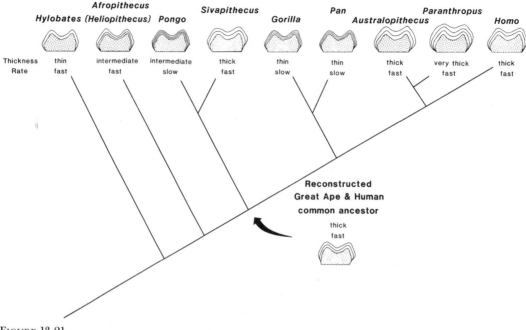

FIGURE 13.21

Changes in molar enamel thickness and rates of enamel deposition in fossil and extant apes. Thick enamel is best interpreted as the ancestral condi- tion for great apes and humans, with chimpan- zees and gorillas showing a secondary reduction in thickness (after Martin, 1985).

Sivapithecus group will probably turn out to be a broad radiation (Wolpoff, 1983) rather than a single widespread taxon uniquely related to the orangutan. At present, how- ever, it is impossible to sort out the number of lineages present within this group, or to evaluate their likely adaptive diversity.

THE EVOLUTION OF LIVING HOMINOIDS

In the preceding pages we have reviewed the fossil apes from the Miocene epoch. Like the Oligocene anthropoids from Egypt, the Mio- cene genera and species can be ordered on the basis of a suite of mostly dental features into more primitive and more advanced species (Fig. 13.22). It is quite evident that the radiation of hominoids during the Mio- cene was much more extensive and pro-

duced many more lineages than many earlier workers imagined (e.g., Simons and Pilbeam, 1965). A corollary of this increas- ingly complex picture of ape evolution dur- ing the Miocene is that the identification of unique lineages leading to particular extant genera is far more difficult than was previ- ously thought. Attempts to find the ancestry of unique lineages leading to gibbons, to the great apes, or to hominids have been compli- cated repeatedly by the discovery of more complete fossils with unsuspected primitive features, by more careful consideration of comparative anatomy, and by more refined understanding of stratigraphic relationships. Until very recently, the great temporal ex- panse of the Miocene epoch, the diversity of Miocene environments and faunas, and the morphological diversity of the fossil apes from that epoch were all unknown and

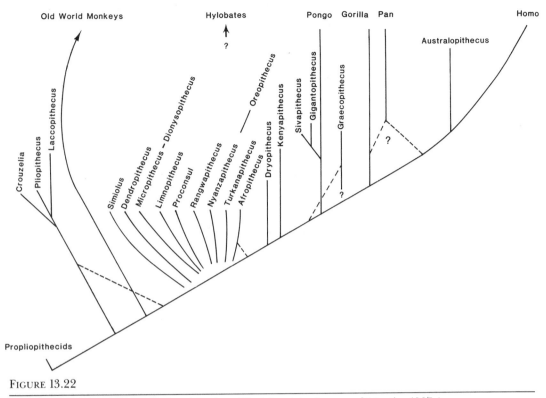

Figure 13.22

Summary of Miocene ape relationships (modified from Andrews and Martin, 1987a).

largely unsuspected. Furthermore, the extraordinary diversity of Miocene apes that has been discovered or identified in recent years (e.g., Leakey and Leakey, 1986a,b) demonstrates how little we really know about ape evolution during this epoch and how many early apes are yet to be uncovered. To put our current understanding of ape evolution into perspective and to contrast it with earlier views, we now examine the fossil evidence specifically for what it tells about the evolution of extant hominoids, consider the alternate ways in which they could have evolved from the diverse radiations of Miocene apes, and compare these results with predictions about ape and human evolution derived from biomolecular studies.

EVOLUTION OF GIBBONS   In each of the successive radiations of Oligocene and Miocene hominoids there were small apes that at one time or another have been identified as fossil gibbons. *Propliopithecus (=Aeolopithecus)*, *Pliopithecus, Dendropithecus, Micropithecus*, and *Dionysopithecus* all show various features (such as small size, short snouts, or large orbits) that cause them to resemble living lesser apes. As discussed above, however, most of these supposed fossil gibbons were extremely primitive in many detailed aspects of their cranial and skeletal anatomy—more so than we would expect in an ancestral gibbon based on the comparative anatomy of extant higher primates. For example, although *Propliopithecus* and *Pliopithecus* were similar to living gibbons in their size and (in some species) had simple, gibbonlike lower molars, they lacked such features as the tubular ectotympanic bone found in all living apes and Old World monkeys and they retained primitive features in their limb bones that are lacking in the limbs of all living catarrhines. For other genera, such as *Micropithecus*, *Dendropithecus*, and *Dionysopithecus*, the critical cranial and skeletal material is not available. As a result, we have little unassailable evidence for fossil gibbons from the Miocene, only a series of possibilities.

All of the small apes were probably to some extent ecological vicars of the living lesser apes, but none can be clearly shown to be uniquely related to the living Asian gibbon genus, which has a fossil record extending back only to the middle Pleistocene of China and Indonesia. Molecular estimates of the dating of the divergence of gibbons from the hominoid lineage are quite variable. Estimates from DNA studies are between 17 and 20 million years ago. This time range includes virtually all of the gibbonlike primitive catarrhines from the early Miocene, as well as the fossil apes that seem to mark the appearance of the great ape and human clade, coincident with the origin of gibbons (Fig. 13.22). Immunological studies suggest a more recent divergence of gibbons, closer to 12 million years ago. In the absence of more definitely gibbonlike fossils from the Miocene of Asia, we cannot resolve the question of their ancestry.

EVOLUTION OF THE ORANGUTAN The one living ape whose evolutionary history is now generally considered to be well established is the orangutan. The Asian *Sivapithecus* gave rise to at least two lineages, one leading to *Gigantopithecus*, the other to the orangutan. The late Miocene specimens of *Sivapithecus* and the living *Pongo* are so similar in many details of dental and facial morphology that

the latter is almost certainly derived from the former. The geographic and temporal gap between the late Miocene fossils and the living great ape of Borneo and Sumatra is partly bridged by fossil teeth from the Pleistocene of China and Java, but the extent to which fossil orangutans resembled the living species in such things as locomotor behavior or social structure cannot be determined from their teeth alone.

Precise dating of the divergence of the orangutan lineage from that leading to African apes and humans is complicated by doubts as to whether the similarities between the living orangutan and *Sivapithecus* are specializations unique to only the latest species of that genus, characteristic of all species of *Sivapithecus*, or remnants of the primitive hominoid morphology that also characterizes the ancestors of all living apes and hominids. The *Sivapithecus* fossils that show the greatest similarity to orangutans are from the late Miocene, 9 to 12 million years ago (Andrews, 1986; Kelley and Pilbeam, 1986). Earlier fossils allied with *Sivapithecus* are known mainly from dental remains. Molecular studies have indicated dates of 10 to 16 million years ago for the orangutan divergence, all more or less concordant with the fossil data (Andrews and Cronin, 1982; Andrews, 1986).

EVOLUTION OF AFRICAN APES The evolutionary history of gorillas and chimpanzees is one of the most notable gaps in our current understanding of ape and human evolution. For many years, the early Miocene species of *Proconsul* were generally recognized as direct ancestors of the living African apes; one species (*P. africanus*) was identified as the ancestor of the chimpanzee, another (*P. major*) was identified as the ancestor of the gorilla (Pilbeam, 1969; Simons, 1967). But as further studies demonstrated the extremely primitive structure of these early

Miocene apes compared with that of living hominoids, it became certain that the radiation of the proconsulids antedates the radiation of the great apes (see, e.g., Ciochon and Corruccini, 1983). As discussed above, these putative chimpanzees and gorillas lacked almost all of the unique anatomical features that characterize living apes; they had been linked with the living species primarily on the basis of size and geography, rather than on the basis of unique, derived morphological similarities.

Both the primitive nature of the early Miocene apes and molecular predictions of the timing of hominoid evolution indicate that the evolutionary divergence of the lineages leading to the African great apes and to humans was probably some time in the later part of the Miocene, between 6 and 10 million years ago (Andrews, 1986). African fossil apes are extremely rare from this period. The best candidate is *Kenyapithecus*, which is known only from teeth, jaws, and a single maxillary fragment from Samburu Hills (Ishida *et al.*, 1984). None of these fossils can be clearly linked with chimpanzees, gorillas, or hominids, and the evolutionary history of chimpanzees and gorillas remains undocumented. In part, this reflects our lack of any substantial fossil record from western and central Africa, where these apes live today.

HOMINID ORIGINS   We can be almost certain that the earliest hominids evolved from some type of Miocene ape, but the identification of hominids among the various genera and species of fossil apes from that epoch has proved a fruitless exercise thus far. Widely cited as a Miocene hominid in earlier decades, *Ramapithecus* is now considered to be the same as *Sivapithecus* and more closely related to orangutans, as discussed above. All the same, the identification of orangutan-like features in *Sivapithecus* helped put the problem of hominid origins into a very different perspective. Many of the dental and gnathic features linking *Ramapithecus* or *Sivapithecus* with early hominids, such as thick molar enamel, robust jaws, and broad anterior lower premolars, are indeed shared similarities, but they are features that characterize many middle and late Miocene apes. As Martin (1986) has noted, the problem faced by paleoanthropologists is to identify the apes among the "dental hominids." As it turns out, the features that distinguished the earliest hominid from earlier apes are not the small teeth and large brain that are so distinctive of ourselves—these features came much later in human evolution—but rather the skeletal adaptations for bipedalism, particularly those of the pelvis. (We discuss hominid evolution in detail in Chapter 15.)

BIBLIOGRAPHY

MIOCENE EPOCH

Adams, C.G. (1981). An outline of Tertiary paleogeography. In *The Evolving Earth*, ed. L.R.M. Cocks, pp. 221–235. Cambridge: Cambridge University Press.

Bernor, R.L. (1983). Geochronology and zoogeographic relationships of Miocene Hominoidea. In *New Interpretations of Ape and Human Ancestry*, ed. R.L. Ciochon and R. Corruccini, pp. 21–66. New York: Plenum Press.

Bernor, R.L., Flynn, L.J., Harrison, T., Hussain, S.T., and Kelley, J. (1988). *Dionysopithecus* from southern Pakistan and the biochronology and biogeography of early Eurasian catarrhines. *J. Hum. Evol.* **17**:339–358.

Ginsburg, L. (1986). Chronology of the European pliopithecids. In *Primate Evolution*, ed. J.G. Else and P.C. Lee, pp. 47–58. Cambridge: Cambridge University Press.

Mein, P. (1986). Chronological succession of hominoids in the European Neogene. In *Primate Evolution*, ed. J.G. Else and P.C. Lee, pp. 59–70. Cambridge: Cambridge University Press.

Pickford, M. (1983). Sequence and environment of the Lower and Middle Miocene hominoids of western

Kenya. In *New Interpretations of Ape and Human Ancestry*, ed. R.L. Ciochon and R. Corruccini, pp. 421–440. New York: Plenum Press.

———. (1986). The geochronology of Miocene higher primate faunas of East Africa. In *Primate Evolution*, ed. J.G. Else and P.C. Lee, pp. 19–33. Cambridge: Cambridge University Press.

Whybrow, P.J. (1984). Geological and faunal evidence from Arabia for mammal "migrations" between Asia and Africa during the Miocene. In *The Early Evolution of Man*, ed. P. Andrews and J.L. Franzen, pp. 189–198. Senckenberg: Cour. Forsch. Inst.

AFRICAN EARLY AND MIDDLE MIOCENE APES

Andrews, P.J. (1978). A revision of the Miocene Hominoidea of East Africa. *Bull. Br. Mus. Nat. Hist. (Geol.)* **30**(2):85–224.

———. (1981). Species diversity and diet in monkeys and apes during the Miocene. In *Aspects of Human Evolution*, ed. C.B. Stringer, pp. 25–61. London: Taylor and Frances.

———. (1985). Family group systematics and evolution among catarrhine primates. In *Ancestors: The Hard Evidence*, ed. E. Delson, pp. 14–22. New York: Alan R. Liss.

Andrews, P.J., and Martin, L. (1987a). Cladistic relationships of extant and fossil hominids. *J. Hum. Evol.* **16**:101–118.

———. (1987b). The phyletic position of the Ad Dabtiyah hominoid. *Bull. Br. Mus. Nat. Hist. (Geol.)* **41**:383–393.

Andrews, P.J., and Simons, E.L. (1977). A new African Miocene gibbon-like genus *Dendropithecus* (Hominoidea, Primates) with distinctive postcranial adaptations: Its significance to origin of Hylobatidae. *Folia Primatol.* **28**:161–170.

Beard, K.C., Teaford, M.F., and Walker, A. (1986). New wrist bones of *Proconsul africanus* and *P. nyanzae* from Rusinga Island, Kenya. *Folia Primatol.* **47**:97–118.

Bernor, R.L., Flynn, L.J., Harrison, T., Hussain, S.T., and Kelley, J. (1988). *Dionysopithecus* from southern Pakistan and the biochronology and biogeography of early Eurasian catarrhines. *J. Hum. Evol.* **17**:339–358.

Bishop, W.W. (1967). The later Tertiary in East Africa—volcanics, sediments and faunal inventory. In *Background to Evolution in Africa*, ed. W.W. Bishop and J.D. Clark, pp. 31–56. Chicago: University of Chicago Press.

Davis, P.R., and Napier, J. (1963). A reconstruction of the skull of *Proconsul africanus* (R.S. 51). *Folia Primatol.* **1**:20–28.

Falk, D. (1983). A reconsideration of the endocast of *Proconsul africanus*: Implications for primate brain evolution. In *New Interpretations of Ape and Human Ancestry*, ed. R.L. Ciochon and R. Corruccini, pp. 239–248. New York: Plenum Press.

Feldesman, M.R. (1982). Morphometric analysis of the distal humerus of some Cenozoic catarrhines: The late divergence hypothesis revisited. *Am. J. Phys. Anthropol.* **59**:173–195.

Fleagle, J.G. (1975). A small gibbon-like hominid from the Miocene of Uganda. *Folia Primatol.* **24**:1–15.

———. (1983). Locomotor adaptations of Oligocene and Miocene hominoids and their phyletic implications. In *New Interpretations of Ape and Human Ancestry*, ed. R.L. Ciochon and R. Corruccini, pp. 301–324. New York: Plenum Press.

———. (1984). Are there any fossil gibbons? In *The Lesser Apes: Evolutionary and Behavioral Biology*, ed. D.J. Chivers, H. Preuschoft, N. Creel, and W. Brockelman, pp. 431–477. Edinburgh: Edinburgh University Press.

———. (1986). The fossil record of early catarrhine evolution. In *Major Topics in Primate and Human Evolution*, ed. B.A. Wood, L.B. Martin, and P. Andrews, pp. 130–139. Cambridge: Cambridge University Press.

Fleagle, J.G., and Kay, R.F. (1985). The paleobiology of catarrhines. In *Ancestors: The Hard Evidence*, ed. E. Delson, pp. 23–36. New York: Alan R. Liss.

Fleagle, J.G., and Simons, E.L. (1978). *Micropithecus clarki*, a small ape from the Miocene of Uganda. *Am. J. Phys. Anthropol.* **49**:427–440.

Gu, Y., and Lin, Y. (1983). First discovery of *Dryopithecus* in east China. *Acta Anthropol. Sinica* **2**(4):305–314.

Harrison, T. (1980). New finds of small fossil apes from the Miocene locality at Koru in Kenya. *J. Hum. Evol.* **10**:129–137.

———. (1986). New fossil anthropoids from the Middle Miocene of East Africa and their bearing on the origin of the Oreopithecidae. *Am. J. Phys. Anthropol.* **71**:265–284.

———. (1987). The phylogenetic relationships of the early catarrhine primates: A review of the current evidence. *J. Hum. Evol.* **16**:41–80.

Hill, A., and Odhiambo, I. (1987). New mandible of *Rangwapithecus* from Songhor, Kenya. *Am. J. Phys. Anthropol.* **72**:210.

Hopwood, A.T. (1933). Miocene primates from Kenya. *J. Linn. Soc. London, Zool.* **38**:437–464.

Ishida, H., Ishida, L., and Pickford, M. (1984). In *African Studies Monographs*, suppl. 3. Kyoto: Kyoto University Press.

Kay, R.F. (1977). Diets of early Miocene African

hominoids. *Nature (London)***268**:628–630.

Kelley, J. (1986). Species recognition and sexual dimorphism in *Proconsul* and *Rangwapithecus*. *J. Hum. Evol.* **15**:461–495.

Leakey, R.E., and Leakey, M.G. (1986a). A new Miocene hominoid from Kenya. *Nature* **324**:143–146.

———. (1986b). A second new Miocene hominoid from Kenya. *Nature (London)* **324**:146–148.

———. (1987). A new Miocene small-bodied ape from Kenya. *J. Hum. Evol.* **16**:369–387.

LeGros Clark, W.E., and Leakey, L.S.B. (1951). The Miocene Hominoidea of East Africa. In *Fossil Mammals of Africa. Br. Mus. Nat. Hist.* **1**:1–117.

LeGros Clark, W.E., and Thomas, D.P. (1951). Associated jaws and limb bones of *Limnopithecus macinnesi*. In *Fossil Mammals of Africa. Br. Mus. Nat. Hist.* **3**:1–27.

Li, C.-K. (1978). A Miocene gibbon-like primate from Shihhung, Kiangsu Province. *Vertebr. Palasiat.* **16**:187–192.

Napier, J.R., and Davis, P.R. (1959). The forelimb skeleton and associated remains of *Proconsul africanus*. In *Fossil Mammals of Africa. Br. Mus. Nat. Hist.* **16**:1–69.

Pickford, M. (1983). Sequence and environment of the Lower and Middle Miocene hominoids of western Kenya. In *New Interpretations of Ape and Human Ancestry*, ed. R.L. Ciochon and R. Corruccini, pp. 421–440. New York: Plenum Press.

———. (1986). Hominoids from the Miocene of East Africa and the phyletic position of *Kenyapithecus*. *Z. Morphol. Anthropol.* **76**:117–130.

Pilbeam, D.R. (1969). Tertiary Pongidae of East Africa: Evolutionary relationships and taxonomy. *Bull. Peabody Mus. Nat. Hist.* **31**:1–185.

———. (1972). Evolutionary changes in the hominoid dentition through geological time. In *Calibration of Hominoid Evolution: Recent Advances in Isotopic and Other Dating Methods Applicable to the Origin of Man*, ed. W.W. Bishop and J.A. Miller, pp. 369–380. Edinburgh: Scottish Academic Press.

———. (1984). The descent of hominoids and hominids. *Sci. Am.* **250**(3):84–96.

———. (1985). Patterns of hominoid evolution. In *Ancestors: The Hard Evidence*, ed. E. Delson, pp. 51–59. New York: Alan R. Liss.

Radinsky, L.B. (1975). The fossil evidence of anthropoid brain evolution. *Am. J. Phys. Anthropol.* **41**:15–28.

Rose, M.D. (1983). Miocene hominoid postcranial morphology: Monkey-like, ape-like, neither, or both? In *New Interpretations of Ape and Human Ancestry*, ed. R.L. Ciochon and R. Corruccini, pp. 405–420. New

York: Plenum Press.

———. (1988). Another look at the anthropoid elbow. *J. Hum. Evol.* **17**:193–224.

Simons, E.L. (1967). The earliest apes. *Sci. Am.* **217**:28–35.

———. (1987). New faces of *Aegyptopithecus* from the Oligocene of Egypt. *J. Hum. Evol.* **16**:273–289.

Simons, E.L., Andrews, P.J., and Pilbeam, D.R. (1978). Cenozoic apes. In *Evolution of African Mammals*, ed. V.J. Maglio and H.B.S. Cooke, pp. 120–146. Cambridge, Mass.: Harvard University Press.

Simons, E.L., and Pilbeam, D.R. (1965). Preliminary revision of the Dryopithecinae (Pongidae, Anthropoidea). *Folia Primatol.* **3**:81–152.

Walker, A.C., and Pickford, M. (1983). New postcranial fossils of *Proconsul africanus* and *Proconsul nyanzae*. In *New Interpretations of Ape and Human Ancestry*, ed. R.L. Ciochon and R. Corruccini, pp. 325–352. New York: Plenum Press.

Whybrow, P.J., and Andrews, P.J. (1978). Restoration of the holotype of *Proconsul nyanzae*. *Folia Primatol.* **30**:115–125.

EURASIAN FOSSIL APES

Andrews, P.J. (1983). The natural history of *Sivapithecus*. In *New Interpretations of Ape and Human Ancestry*, ed. R.L. Ciochon and R. Corruccini, pp. 441–464. New York: Plenum Press.

Andrews, P.J., and Cronin, J.E. (1982). The relationships of *Sivapithecus* and *Ramapithecus* and the evolution of the orangutan. *Nature (London)* **297**:541–546.

———. (1986). Fossil evidence on human origins and dispersal. *Cold Spring Harbor Symposia on Quantitative Biology* **51**:419–428.

Andrews, P., and Martin, L. (1987a). Cladistic relationships of extant and fossil hominoids. *J. Hum. Evol.* **16**:101–118.

Andrews, P.J., and Tekkaya, I. (1980). A revision of the Turkish Miocene hominoid *Sivapithecus meteai*. *Palaeontology* **23**:85.

Andrews, P.J., and Tobien, H. (1977). A new Miocene locality in Turkey with evidence on the origin of *Ramapithecus* and *Sivapithecus*. *Nature (London)* **268**:699–701.

Ankel, F. (1965). Der Canalis Sacralis als Indikator fur die Lange der Caudelregion der Primaten. *Folia Primatol.* **3**:263–276.

Barry, J.C., Jacobs, L.L., and Kelley, J. (1986). An early middle Miocene catarrhine from Pakistan with comments on the dispersal of catarrhines into Eurasia. *J. Hum. Evol.* **15**:501–508.

Bernor, R.L. (1983). Geochronology and zoogeographic relationships of Miocene Hominoidea. In *New Interpretations of Ape and Human Ancestry*, ed. R.L. Ciochon and R. Corruccini, pp. 21–66. New York: Plenum Press.

Bernor, R.L., Flynn, L.J., Harrison, T., Hussain, S.T., and Kelley, J. (1988). *Dionysopithecus* from southern Pakistan and the biochronology and biogeography of early Eurasian catarrhines. *J. Hum. Evol.* **17**:339–358.

Blainville, H.M.D. de (1839). Osteographie des Primates. In *Osteographie des Mammiferes*, I: *Primates et Secundates*. Paris: Bailliere.

Chopra, S.R.K. (1983). Significance of recent hominoid discoveries from the Siwalik Hills of India. In *New Interpretations of Ape and Human Ancestry*, ed. R.L. Ciochon and R. Corruccini, pp. 539–558. New York: Plenum Press.

Conroy, G.C. (1976). Hallucial tarsometatarsal joint in an Oligocene anthropoid, *Aegyptopithecus zeuxis*. *Nature (London)* **262**:684–686.

DeBonis, L., and Melentis, J. (1977). Les primates hominoides du Vellesian de Macedonia (Grece). Etude de la machoire infereure. *Geobios* **10**:849–885.

———. (1978). Les primates hominoides du Miocene superieur de Macedoine. *Ann. Paleontol. (Vert.)* **64**:185–202.

———. (1985). La place du genre *Ouranopithecus* dans l'evolution de Hominoides. *C. R. Acad. Sci.* **300**, *ser. II*, pp. 429–432.

———. (1987). Interet de l'anatomie naso-maxillaire pour la phylogenie des Hominidae. *C. R. Acad. Sci.* **304**, *ser. II*, pp. 767–769.

Fleagle, J.G. (1983). Locomotor adaptations of Oligocene and Miocene hominoids and their phyletic implications. In *New Interpretations of Ape and Human Ancestry*, ed. R.L. Ciochon and R. Corruccini, pp. 301–324. New York: Plenum Press.

———. (1984). Are there any fossil gibbons? In *The Lesser Apes: Evolutionary and Behavioral Biology*, ed. D.J. Chivers, H. Preuschoft, N. Creel, and W. Brockelman, pp. 431–477. Edinburgh: Edinburgh University Press.

Fleagle, J.G., and Kay, R.F. (1983). New interpretations of the phyletic position of Oligocene hominoids. In *New Interpretations of Ape and Human Ancestry*, ed. R.L. Ciochon and R. Corruccini, pp. 181–210. New York: Plenum Press.

Gervais, P. (1849). *Zoologie et Paleontologie francaises*. Paris: I. Bertrand.

Ginsburg, L. (1975). Le Pliopitheque des faluns Helvetiens de la Touraine et de l' Anjou. *Coll. Int. Cent. Nat. Rech. Sci.* **218**:877–885.

———. (1986). Chronology of the European pliopithecids. In *Primate Evolution*, ed. J.G. Else and P.C. Lee, pp. 47–58. Cambridge: Cambridge University Press.

Ginsburg, L., and Mein, P. (1980). *Crouzelia rhodanica*, nouvelle espece de primate Catarrhinien et essai sur la position systematique des Pliopithecidae. *Bull. Mus. Hist. Nat., Paris*, pp. 57–85.

Gu, Y., and Lin, Y. (1983). First discovery of *Dryopithecus* in east China. *Acta Anthropol. Sinica* **2**(4):305–314.

Harrison, T. (1986). A reassessment of the phylogenetic relationships of *Oreopithecus bambolii* Gervais. *J. Hum. Evol.* **15**:541–584.

Huxley, T.H. (1863). *Evidence as to Man's Place in Nature*. London: Williams and Norgate.

Jungers, W.L. (1987). Body size and morphometric affinities of the appendicular skeleton in *Oreopithecus bambolii* (IGF 11778). *J. Hum. Evol.* **16**:445–456.

Kay, R.F. (1981). The nut-crackers—a new theory of the adaptation of the Ramapithecinae. *Am. J. Phys. Anthropol.* **55**:141–151.

———. (1982a). Sexual dimorphism in Ramapithecinae. *Proc. Nat. Acad. Sci.* **79**:209–212.

———. (1982b). *Sivapithecus simonsi*, a new species of Miocene hominoid, with comments on the phylogenetic status of the Ramapithecinae. *Int. J. Primatol.* **3**(2):113–173.

Kay, R.F., and Simons, E.L. (1983). A reassessment of the relationships between later Miocene and subsequent Hominoidea. In *New Interpretations of Ape and Human Ancestry*, ed. R.L. Ciochon and R. Corruccini, pp. 577–624. New York: Plenum Press.

Kelley, J., and Pilbeam, D.R. (1986). The dryopithecines: Taxonomy, comparative anatomy, and phylogeny of Miocene large hominoids. In *Comparative Primate Biology*, vol. 1: *Systematics, Evolution, and Anatomy*, ed. D.R. Swindler and J. Erwin, pp. 361–411. New York: Alan R. Liss.

Kretzoi, M. (1975). New ramapithecines and *Pliopithecus* from the Lower Pliocene of Rudanbaya in north-eastern Hungary. *Nature (London)* **257**:578–581.

Lartet, E. (1856). Note sur un grand singe fossile qui se rattache au groupe des singes superieurs. *C. R. Acad. Sci. (Paris)* **43**:219–228.

LeGros Clark, W.E., and Leakey, L.S.B. (1951). The Miocene Hominoidea of East Africa. In *Fossil Mammals of Africa. Br. Mus. Nat. Hist.* **1**:1–117.

Lipson, S., and Pilbeam, D.R. (1982). *Ramapithecus* and hominoid evolution. *J. Hum. Evol.* **11**:545–548.

Martin, L. (1985). Significance of enamel thickness in hominid evolution. *Nature (London)* **314**:260–263.

———. (1986). Relationships among great apes and

humans. In *Major Topics in Primate and Human Evolution*, ed. B. Wood, L. Martin, and P. Andrews, pp. 161–187. Cambridge: Cambridge University Press.

Mein, P. (1986). Chronological succession of hominoids in the European Neogene. In *Primate Evolution*, ed. J.G. Else and P.C. Lee, pp. 59–70. Cambridge: Cambridge University Press.

Morbeck, M.E. (1983). Miocene hominoid discoveries from Rudabanya: Implications from the post cranial skeleton. In *New Interpretations of Ape and Human Ancestry*, ed. R.L. Ciochon and R. Corruccini, pp. 369–404. New York: Plenum Press.

Pilbeam, D.R. (1966). Notes on *Ramapithecus*, the earliest hominid, and *Dryopithecus*. *Am. J. Phys. Anthropol.* **25**:1–6.

———. (1968). The earliest hominids. *Nature* **219**:1335–1338.

———. (1970). *Gigantopithecus* and the origins of Hominidae. *Nature* **225**(5232):516–519.

———. (1972). *The Ascent of Man*. New York: Macmillan.

———. (1978). Rethinking human origins. *Discovery* **13**:2–9.

———. (1982). New hominoid skull material from the Miocene of Pakistan. *Nature (London)* **295**:232–234.

Pilbeam, D.R., Meyer, G.E., Badgley, C., and Lipschutz, B. (1980). Miocene hominoids from Pakistan. *Postilla* **181**:1–94.

Pilbeam, D.R., Meyer, G.E., Badgley, C., Rose, M.D., Pickford, M.H.L., Behrensmeyer, A.K., and Shah, S.M.I. (1977). New hominoid primates from the Siwaliks of Pakistan and their bearing on hominoid evolution. *Nature (London)* **270**:689–695.

Preuss, T.M. (1982). The face of *Sivapithecus indicus*: Description of a new, relatively complete specimen from the Siwaliks of Pakistan. *Folia Primatol.* **38**:141–157.

Sarmiento, E.E. (1987). The phyletic position of *Oreopithecus* and its significance in the origin of the Hominoidea. *Am. Mus. Nov.*, no. 2881, pp. 1–44.

Schwartz, J.H. (1986). *The Red Ape*. Boston: Houghton Mifflin.

Simons, E.L. (1961). The phyletic position of *Ramapithecus*. *Postilla* **54**:1–20.

———. (1977). *Ramapithecus*. *Sci. Am.* **236**:28–35.

Simons, E.L., and Chopra, S.R.K. (1969). *Gigantopithecus* (Pongidae, Hominoidea). A new species from northern India. *Postilla* **138**:1–18.

Simons, E.L., and Ettel, P.C. (1970). *Gigantopithecus*. *Sci. Am.* **222**(1):76–85.

Simons, E.L., and Fleagle, J.G. (1973). The history of extinct gibbon-like primates. *Gibbon and Siamang* **2**:121–148.

Simons, E.L., and Pilbeam, D.R. (1965). Preliminary revision of the Dryopithecinae (Pongidae, Anthropoidea). *Folia Primatol.* **3**:81–152.

Thenius, E. (1981). Bemerkungen zur taxonomischen und stammesgeschichtlichen Position der Gibbons (Hylobatidae, Primates). *Z. Saugetierkunde* **46**:232–241.

von Koenigswald, G.H.R. (1983). The significance of hitherto undescribed Miocene hominoids from the Siwaliks of Pakistan in the Senchenberg Museum, Frankfurt. In *New Interpretations of Ape and Human Ancestry*, ed. R.L. Ciochon and R.S. Corruccini, pp. 517–526. New York: Plenum Press.

Ward, S., and Brown, B. (1986). The facial skeleton of *Sivapithecus indicus*. In *Comparative Primate Biology*, vol. 1: *Systematics, Evolution, and Anatomy*, ed. D.R. Swindler and J. Erwin, pp. 413–452. New York: Alan R. Liss.

Ward, S.C., and Kimbel, W.H. (1983). Subnasal alveolar morphology and the systematic position of *Sivapithecus*. *Am. J. Phys. Anthropol.* **61**:157–171.

Ward, S.C., and Pilbeam, D.R. (1983). Maxillofacial morphology of Miocene hominoids from Africa and Indo-Pakistan. In *New Interpretations of Ape and Human Ancestry*, ed. R.L. Ciochon and R. Corruccini, pp. 211–238. New York: Plenum Press.

White, T.D. (1975). Geomorphology to paleoecology: *Gigantopithecus* reappraised. *J. Hum. Evol.* **4**:219–233.

Wolpoff, M.H. (1983). *Ramapithecus* and human origins: An anthropologist's perspective of changing interpretations. In *New Interpretations of Ape and Human Ancestry*, ed. R.L. Ciochon and R. Corruccini, pp. 651–676. New York: Plenum Press.

Woo, J.K. (1957). *Dryopithecus* teeth from Keiyun, Yunnan Province. *Vertebr. Palasiat.* **1**:25–32.

Wu, R. (1983). Hominid fossils from China and their bearing on human evolution. *Can. J. Anthropol.* **3**(2):207–214.

———. (1985). The cranium of *Ramapithecus* and *Sivapithecus* from Lufeng, China. In *The Early Evolution of Man*, ed. P. Andrews and J.L. Franzen, pp. 41–48. Senckenberg: Cour. Forsch. Inst.

———. (1987). A revision of the classification of the Lufeng great apes. *Acta Anthropol. Sinica* **6**:265–271.

Wu, R., and Pan, Y. (1984). A late Miocene gibbon-like primate from Lufeng, Yunnan Province. *Acta Anthropol. Sinica* **3**:193–200.

———. (1985). Preliminary observation on the cranium of *Laccopithecus robustus* from Lufeng, Yunnan, with reference to its phylogenetic relationship. *Acta Anthropol. Sinica* **4**(1):7–13.

Wu, R., Xu, Q., and Lu, Q. (1983). Morphological features of *Ramapithecus* and *Sivapithecus* and their phylogenetic relationships—morphology and comparisons of the cranium. *Acta Anthropol. Sinica* **2**(1):1–10.

Zapfe, H. (1958). The skeleton of *Pliopithecus (Epipliopithecus) vindobonensis* Zapfe and Hurzeler. *Am. J. Phys. Anthropol.* **16**:441–458.

———. (1960). Die Primatenfunde aus der Miozanen Spaltenfullung von Neudorf an der march (Devinzka nova ves), Tschechoslowakev. Mit. Anhang: Er Primatenfund aus dem Miozan von klein Hadersdorf in Niederoesterreich. *Schweiz. Pal. Abh.* **78**:4–293.

———. (1961). Ein primaten Fun aus der Miozanen Molasse von Oberosterreich. *Z. Morphol. Anthropol.* **51**(3):247–267.

EVOLUTION OF LIVING HOMINOIDS

Andrews, P. (1986). Fossil evidence on human origins and dispersal. *Cold Spring Harbor Symposia on Quantitative Biology* **51**:419–428.

Andrews, P., and Martin, L. (1987a). Cladistic relationships of extant and fossil hominoids. *J. Hum. Evol.* **16**:101–118.

Ciochon, R.L. (1983). Hominoid cladistics and the ancestry of modern apes and humans: A summary statement. In *New Interpretations of Ape and Human Ancestry*, ed. R.L. Ciochon and R. Corruccini, pp. 781–843. New York: Plenum Press.

Fleagle, J.G. (1976). Locomotion and posture of the Malayan siamang and implications for hominoid evolution. *Folia Primatol.* **26**:245–269.

Sarich, V. (1968). The origin of the hominids: An immunological approach. In *Perspectives on Human Evolution*, vol. 1, ed. S.L. Washburn and P.C. Jay, pp. 94–121. New York: Holt, Rinehart and Winston.

Simons, E.L., and Pilbeam, D.R. (1965). Preliminary revision of the Dryopithecinae (Pongidae, Anthropoidea). *Folia Primatol.* **3**:81–152.

Evolution of Gibbons

Fleagle, J.G. (1976). Locomotion and posture of the Malayan siamang and implications for hominoid evolution. *Folia Primatol.* **26**:245–269.

———. (1984). Are there any fossil gibbons? In *The Lesser Apes: Evolutionary and Behavioral Biology*, ed. D.J. Chivers, H. Preuschoft, N. Creel, and W. Brockelman, pp. 431–477. Edinburgh: Edinburgh University Press.

Evolution of Orangutans

Andrews, P.J., and Cronin, J.E. (1982). The relationships of *Sivapithecus* and *Ramapithecus* and the evolution of the orangutan. *Nature (London)* **297**:541–546.

Preuss, T.M. (1982). The face of *Sivapithecus indicus*: Description of a new, relatively complete specimen from the Siwaliks of Pakistan. *Folia Primatol.* **38**:141–157.

Evolution of African Apes

Andrews, P.J., and Cronin, J.E. (1982). The relationships of *Sivapithecus* and *Ramapithecus* and the evolution of the orangutan. *Nature* **297**:541–546.

Ciochon, R.L., and Corruccini, R., eds. (1983). *New Interpretations of Ape and Human Ancestry*. New York: Plenum Press.

Cronin, J.E. (1983). Apes, humans and molecular clocks: A reappraisal. In *New Interpretations of Ape and Human Ancestry*, ed. R.L. Ciochon and R. Corruccini, pp. 115–136. New York: Plenum Press.

Ishida, H., Ishida, L., and Pickford, M. (1984). In *African Studies Monographs*, suppl. 3. Kyoto: Kyoto University Press.

Simons, E.L. (1967). The earliest apes. *Sci. Am.* **217**:28–35.

Wolpoff, M.H. (1983). *Ramapithecus* and human origins: An anthropologist's perspective of changing interpretations. In *New Interpretations of Ape and Human Ancestry*, ed. R.L. Ciochon and R. Corruccini, pp. 651–676. New York: Plenum Press.

Hominid Origins

Kay, R.F., and Simons, E.L. (1983). A reassessment of the relationships between later Miocene and subsequent Hominoidea. In *New Interpretations of Ape and Human Ancestry*, ed. R.L. Ciochon and R. Corruccini, pp. 577–624. New York: Plenum Press.

Martin, L. (1986). Relationships among great apes and humans. In *Major Topics in Primate and Human Evolution*, ed. B. Wood, L. Martin, and P. Andrews, pp. 161–187. Cambridge: Cambridge University Press.

Wolpoff, M.H. (1983). *Ramapithecus* and human origins. An anthropologist's perspective of changing interpretations. In *New Interpretations of Ape and Human Ancestry*, ed. R.L. Ciochon and R. Corruccini, pp. 651–676. New York: Plenum Press.

Fossil Old World Monkeys

CERCOPITHECOID EVOLUTION

In the previous chapter we discussed the evolution of fossil apes, which are particularly well known from the earlier parts of the Miocene and become increasingly rare toward the end of that epoch. In this chapter we consider the fossil record of the modern success story of catarrhine evolution, the Old World monkeys. Although Old World monkeys first appear in the fossil record at approximately the same time as apes, the early Miocene, they are quite rare throughout that epoch and the major radiation of the group appears to have taken place much later. From the Pliocene to the present, Old World monkeys have an extensive fossil record from Africa, Europe, and Asia, including many parts of the world from which they are absent today (Fig. 14.1). The evolutionary history of this group is a much neglected aspect of primate evolution, largely because most workers have failed to realize the extraordinary diversity of this most successful radiation of modern catarrhines. However, with the recent discovery of new fossil monkeys from both the beginnings of the group and from their period of great diversity in the Plio-Pleistocene of Africa, Old World monkey evolution has become one of the most exciting areas in primate evolution.

Victoriapithecids: The Earliest Old World Monkeys

The first record of cercopithecoid monkeys comes from early Miocene deposits in northern and eastern Africa, but monkeys are relatively uncommon from this period. They are absent from many localities and only a few, very similar species are known. The lack of taxonomic diversity is likely to be a real phenomenon rather than the reflection of an incomplete fossil record, but the absence of cercopithecoids from many early Miocene localities is difficult to interpret. At some localities, early monkeys are as common as fossil apes; at other localities, they do not appear at all. This differential abundance of monkeys at early and middle Miocene sites seems to be related to environmental differences; the monkeys are more abundant in drier, more open habitats.

Like the early Miocene hominoids, early Old World monkeys are much more primitive than extant members of the same superfamily and cannot be placed conveniently in either of the modern subfamilies. Rather, they form a separate subfamily of more

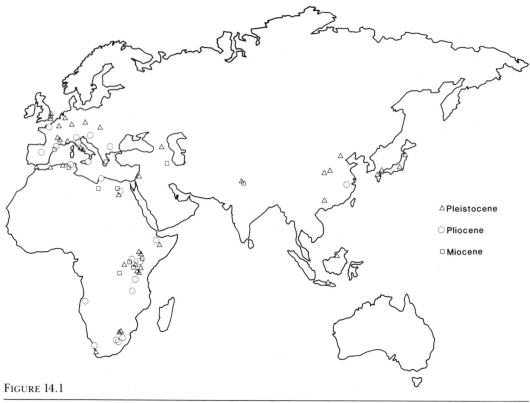

FIGURE 14.1

The modern Old World, showing fossil monkey localities from the Miocene, Pliocene, and Pleistocene.

primitive monkeys, the victoriapithecines, which preceded the divergence of colobines and cercopithecines and thus is also placed in a separate family. (Table 14.1).

Prohylobates and ***Victoriapithecus*** are very similar genera (Fig. 14.2). *Prohylobates* is known from the early Miocene of Egypt, Libya, and northern Kenya. *Victoriapithecus* is from the middle Miocene site on Maboko Island in Lake Victoria in Kenya. There is a single victoriapithecine tooth from the early Miocene site of Napak, in Uganda. Both are small- to medium-size monkeys (5–25 kg) and are known primarily from dental re-

mains. Like all later cercopithecoids, *Prohylobates* and *Victoriapithecus* have bilophodont lower molars, but their teeth are more primitive than those of later Old World monkeys and more like those of hominoids in that the upper molars frequently have a crista obliqua linking the metacone with the protocone and their lower molars often have a small hypoconulid (Fig. 14.3). Both of these dental features are present in primitive catarrhines and in apes but are absent in extant Old World monkeys. As in the propliopithecids, the last lower premolar has an expanded buccal face and the lower molars

TABLE 14.1
Infraorder Catarrhini
Family Victoriapithecidae
Subfamily VICTORIAPITHECINAE

Species	Body Weight (g)
Prohylobates (e. Miocene, N. and E. Africa)	
P. tandyi	7,000
P. simonsi	25,000
Victoriapithecus (?e.–m. Miocene, Kenya)	
V. macinnesi	7,000

have a very large base and a constricted occlusal surface. The trigonid is relatively short and the crown height is relatively low, as in colobines, but the molar cusps are relatively low, as in cercopithecines. Like all Old World monkeys, *Victoriapithecus* has sexually dimorphic canines. Overall, the dentition of these basal Old World monkeys is intermediate between that of the early catarrhines from the Oligocene and that of later Old World monkeys. The mandible is relatively deep, and the symphysis resembles that of later Old World monkeys in the position of the genioglossal pit but lacks the other characteristic features of either colobines or cercopithecines (Fig. 14.3; Benefit, 1985, 1987; Leakey, 1985).

The few skeletal elements of *Victoriapithecus* show the narrow articulation on the distal end of the humerus and the deep ulnar notch characteristic of living cercopithecoids. The limb bones are most similar to those of a small cercopithecine such as the vervet monkey, suggesting that *Victoriapithecus* was quadrupedal but not restricted to either arboreal or terrestrial substrates. The Maboko Island site, like most of the middle Miocene localities in Kenya, seems to have been an open woodland environment, where such locomotor abilities would be most appropriate.

Prohylobates and *Victoriapithecus* are distinctly more primitive than all later Old World monkeys. They are missing links between early catarrhines and modern cercopithecoids and provide clear evidence of

FIGURE 14.2

Lower jaws of *Prohylobates* and *Victoriapithecus*, from the early and middle Miocene of Kenya (courtesy of M. G. Leakey).

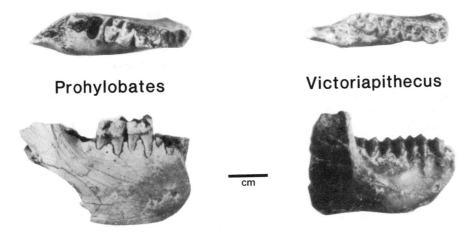

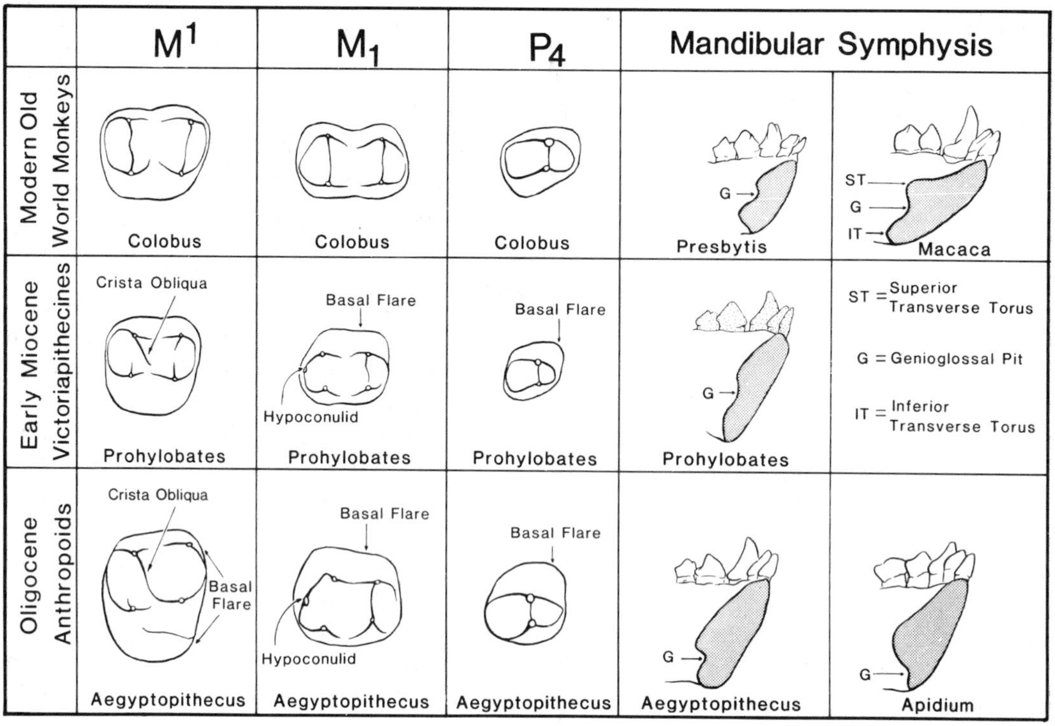

FIGURE 14.3

Dental and mandibular features of Oligocene anthropoids, early cercopithecoids, and modern cercopithecoids, showing the intermediate morphological features of *Victoriapithecus* and *Prohylobates*.

the sequence in which characteristic features of both subfamilies of extant cercopithecoids evolved. The retention of a trigon on the upper molars and a small hypoconulid on the lower molars in these genera confirms what dental anatomists have known for years—that the bilophodont teeth of Old World monkeys are derived from an ancestor with more apelike teeth (Butler, 1986). The question that remains unresolved is which group of primitive catarrhines is closest to the ancestry of cercopithecoids.

In the past, many authorities have argued that parapithecids are ancestral cercopithe-

coids on the basis of the bilophodont appearance of the lower molars in some species, especially *Parapithecus grangeri*. But it has recently become evident that parapithecids are much more primitive than any other Old World anthropoids. If parapithecids are uniquely ancestral to Old World monkeys, then many characteristic catarrhine (and anthropoid) features must have evolved independently in Old World monkeys and apes. In addition, the most cercopithecoid-like genus, *Parapithecus*, lacks permanent incisors, precluding it from ancestry of any later catarrhine. It therefore seems most unlikely that Old World monkeys evolved directly from parapithecids; rather, Old World monkeys and apes were derived from an early catarrhine that was more advanced than the parapithecids and similar to either the propliopithecids from Egypt or the proconsulids from East Africa.

Fossil Cercopithecids

After *Victoriapithecus* and *Prohylobates* there is a gap in the fossil record of nearly 10 million years with few fossil monkeys. However, in the latest Miocene and continuing through the Pliocene and Pleistocene, fossil monkeys are extremely abundant in fossil deposits throughout Africa and Eurasia. This radiation of monkeys was, for the most part, the same one that dominates living higher primate communities today, and all of the extinct forms can be readily grouped into the same subfamilies as living Old World monkeys.

Fossil Cercopithecines

Because many of the features that distinguish the living subfamilies of Old World monkeys are soft tissues, such as the sacculate stomachs of colobines or the cheek pouches of cercopithecines, there are potential hazards in assigning fossil monkeys to one family or another solely on the basis of dental and cranial remains. Nevertheless, extant cercopithecines can be distinguished from colobines by several dental and cranial features, including molars with long trigonids, higher crowns and relatively lower molar cusps, and skulls with longer snouts and narrower interorbital dimensions. Colobines have molars with shorter trigonids, lower crowns, and higher, more pronounced molar cusps, broader skulls with short snouts, narrow nasal openings, and a broad interorbital dimension (Fig. 6.3). These same features are used to identify fossil members of the two subfamilies, but the postcranial differences that characterize the living taxa do not so readily distinguish the fossils except for the tendency of cercopithecines to have longer thumbs and shorter digits than colobines. Fossil cercopithecines (Table 14.2) can be readily divided into four major groups: macaques, baboons and mangabeys, geladas, and guenons.

Macaques

The genus *Macaca* has the widest distribution of any nonhuman primate, extending from North Africa and Gibraltar in the west to Japan and the Philippines in Asia. Fossil macaques were even more widespread, especially in Europe and North Africa. Although they are quite abundant and widespread, most fossil macaques are strikingly similar to the extant genus, indicating that *Macaca* has retained a very conservative morphology over the last 5 million years or so (Delson and Rosenberger, 1984).

The earliest macaques are from latest Miocene or earliest Pliocene localities in Algeria, Libya, and Egypt and are known only from isolated teeth. ***Macaca prisca***, from the early Pliocene of southern France, is the earliest fossil cercopithecine in Europe. In the later Pliocene, macaques were widespread throughout much of North Africa and Europe (including Spain, France, Germany, Italy, the Netherlands, and Yugoslavia), and during the middle Pleistocene their range extended into Great Britain, southern Russia, and the Middle East. Most of these fossil populations cannot be distinguished in dental features from the living Barbary macaque, *M. sylvanus*, of Gibraltar

TABLE 14.2
Infraorder Catarrhini
Family Cercopithecidae
Subfamily CERCOPITHECINAE

Species	Body Weight (g)	Species	Body Weight (g)
Macaca (latest Miocene–Recent, N. Africa, Europe, Asia)		P. jonesi	19,000
M. sylvanus	—	P. whitei	30,000
M. prisca	—	P. antiquus	—
M. majori	—	P. ado	17,000
M. libyca	—	Dinopithecus (Pliocene, S. Africa)	
M. anderssoni	—	D. ingens	77,000
M. palaeindica	—	Gorgopithecus (Pleistocene, S. Africa)	
Procynocephalus (Pliocene, Asia)		G. major	41,000
P. wimani	—	Theropithecus (Plio-Pleistocene, Africa, ?Asia)	
P. subhimalayensis	—	(Simopithecus)	
Paradolichopithecus (Pliocene, Europe)		T. oswaldi	96,000
P. arvernensis	23,000	T. darti	—
Papio (Plio-Pleistocene–Recent, Africa)		(new subgenus)	
P. robinsoni	—	T. brumpti	50,000
P. izodi	—	T. baringensis	—
Cercocebus (Plio-Pleistocene, Africa)		T. quadrirostris	—
Parapapio (l. Miocene–e. Pleistocene, Africa)		Cercopithecus (Pliocene–Recent, Africa)	
P. broomi	23,000	unnamed species	—

and North Africa. The most distinctive fossil macaque, the Pliocene "dwarf macaque," **Macaca majori**, from the island of Sardinia, was about 5 to 10 percent smaller in dental dimensions than the living species.

In Asia, the earliest macaques were from the Pliocene of northern India and Pakistan. Macaques were also relatively common throughout most of the Pleistocene of China and Southeast Asia. The best known of the Asian species is **Macaca anderssoni** (= M. robusta), which was originally discovered among fossil teeth in Chinese drugstores and was also found in northern China at Zhoukoudian with fossil hominids. Most of

the fossil macaques from Asia are known only from isolated teeth and cannot be clearly distinguished from living species.

In addition to fossil representatives of the living *Macaca*, there are two genera of larger macaquelike monkeys from the late Pliocene and Pleistocene of Asia and Europe. **Procynocephalus** is a late Pliocene, Asian genus with one species from northern India and one from southern China. It has a macaque-like dentition and skull, and its baboonlike skeleton suggests locomotion resembling that of the more terrestrial macaques such as *M. nemestrina*. **Paradolichopithecus** is a similar baboonlike macaque from the Pliocene of

Europe. Like most cercopithecines, *Paradolichopithecus* seems to have sexually dimorphic canines, but it lacks any evidence of dimorphism in the cheek teeth or skull.

Baboons and Mangabeys

Macaques are the only cercopithecines to successfully colonize Europe and Asia. The other members of the subfamily are known almost totally from sub-Saharan Africa, where they remain abundant today. The most diverse group of fossil cercopithecines are the baboons and mangabeys, with over a dozen species since the late Miocene.

Fossil savannah baboons that are indistinguishable from the living genus *Papio* are known from the late Pliocene through much of the Pleistocene in eastern and southern Africa. Many of the fossil species and subspecies were much larger than the living species, but they are similar in dental, cranial, and skeletal morphology. Fossil mangabeys (*Cercocebus*) are also known from the late Pliocene and early Pleistocene of eastern and southern Africa, but most of the specimens are fragmentary and reveal little more than the presence of the genus.

Parapapio, from the late Miocene to early Pleistocene of eastern and southern Africa, is one of three genera of extinct baboons. In dental and cranial morphology it is the most primitive member of the baboon-mangabey group and is probably near the ancestry of both living genera. *Parapapio* is intermediate in size between mangabeys and savannah baboons, and it seems to have little sexual size dimorphism.

Dinopithecus is an extremely large (70–80 kg), sexually dimorphic baboon known mainly from the Swartkrans cave deposits (early Pleistocene) of South Africa. There are no skeletal remains assigned to the genus.

Gorgopithecus is a smaller (40 kg) baboon from South African Pleistocene deposits. Unlike *Dinopithecus*, *Gorgopithecus* seems to have little sexual dimorphism in the size of the cheek teeth, but otherwise it is probably very much like living savannah baboons.

Geladas

Theropithecus gelada, from the Ethiopian highlands, is the only living representative of a group of baboons that was much more successful and widespread during the Pliocene and Pleistocene. They were quite abundant in Africa, and apparently their range also extended as far as India. Like the living species, fossil *Theropithecus* has complex cheek teeth and a skull with a short deep face, presumably related to a dietary specialization on grass blades, seeds, and tubers. *Theropithecus* appears to have been the only predominantly folivorous cercopithecine. Unlike colobines, however, *Theropithecus* exploited this dietary niche on the ground by specializing on grass.

This group of baboons has long forelimbs and short phalanges, indicating terrestrial quadrupedalism. The extinct species seem to have the same digital proportions as extant geladas, with relatively long thumbs compared with the size of the index finger, and were probably manual foragers. The extinct species are generally much larger than the living gelada and show much more extreme dental, cranial, and skeletal specializations.

Theropithecus brumpti is an early species from the late Pliocene of East Africa that has a large anterior dentition (as in the living gelada) and extraordinary development of the zygomatic arches that must have given its face an extremely imposing appearance (Fig. 14.4). Its molars have the greatest development of shearing crests of any known cercopithecine, suggesting even more folivorous habits than the extant gelada.

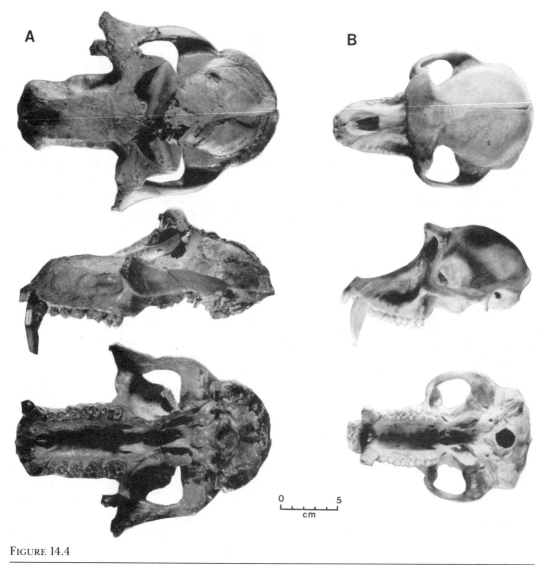

FIGURE 14.4

Skulls of (A) *Theropithecus brumpti* and (B) *Theropithecus gelada* (courtesy of Gerald Eck).

This species has been recovered from deposits indicating more forested environments. Its limbs also show greater similarities to the limbs of the forest-living mandrills than do the limbs of other *Theropithecus* species (Ciochon, 1986).

Theropithecus (Simopithecus) oswaldi was an enormous monkey that probably weighed as much as 100 kg and was extremely abundant in many East African Pliocene and Pleistocene sites (Fig. 14.5). Compared with *T. brumpti* it has greatly reduced, laterally compressed incisors and canines, large molar teeth, a short face, and very long limbs. The abundance of *T. oswaldi* at many fossil sites suggests that, like the living gelada,

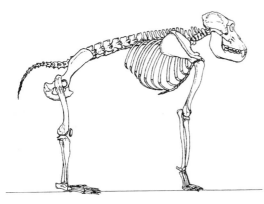

FIGURE 14.5

Theropithecus oswaldi skeleton.

TABLE 14.3
Infraorder Catarrhini
Family Cercopithecidae
Subfamily COLOBINAE

Species	Body Weight (g)
Mesopithecus (l. Miocene–Pliocene, Europe, W. Asia)	
M. pentelici	8,000
M. monspessulanus	5,000
Dolichopithecus (Pliocene, Europe)	
D. ruscinensis	18,000
Presbytis (l. Miocene–Recent, Asia)	
P. sivalensis	—
Rhinopithecus (e. Pleistocene–Recent, Asia)	
Colobus (l. Miocene–Recent, Africa)	
many undescribed species	—
C. flandrini	16,000
Libypithecus (l. Miocene–Pliocene, N. Africa)	
L. markgrafi	8,400
Microcolobus (l. Miocene, Africa)	
M. tugenensis	4,000
Cercopithecoides (Pliocene, Africa)	
C. williamsi	33,000
C. kimeui	—
Paracolobus (Plio-Pleistocene, Africa)	
P. chemeroni	35,000
P. mutiwa	—
Rhinocolobus (Plio-Pleistocene, Africa)	
R. turkanensis	21,000

these extinct baboons also lived in large herds. It has been suggested that they were preyed on by early hominids for food (Shipman *et al.*, 1981), and their extinction may well have been the result of human predation.

Theropithecus delsoni is a fossil gelada from the Pleistocene of northern India. This species is known only from a single tooth and is the only record of geladas outside Africa.

Guenons

Despite their abundance in sub-Saharan Africa today, guenons are very rare in the fossil record. There are *Cercopithecus* teeth from Pliocene and Pleistocene localities in Kenya and Ethiopia, but most of the material is fragmentary and has not been assigned to any particular species.

Fossil Colobines

In contrast with the cercopithecines, which are all relatively similar to extant genera, many fossil colobines from Miocene, Pliocene, and Pleistocene deposits are quite different from any living taxa and provide evidence of both a broader geographic

range and more diverse ecological adaptations in the extinct colobines (Table 14.3).

European Colobines

The oldest fossil colobine from Eurasia is ***Mesopithecus*** (Fig. 14.6). This langur-size monkey is known from many localities in the late Miocene through Pliocene of southern and central Europe. The genus ranged as far west as England and as far east as Iran. There are two species, *M. pentelici* (about 8 kg) and a younger, smaller species, *M.*

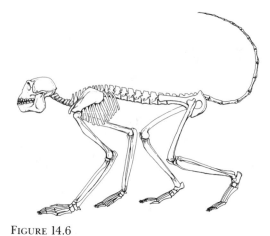

FIGURE 14.6

Mesopithecus skeleton.

monspessulanus (5 kg). *Mesopithecus* resembles living colobines in most dental and cranial features, including relatively small incisors, high-crowned cheek teeth, a deep mandible, a short face with large orbits, a narrow nasal opening, and a broad interorbital distance. It was probably a relatively folivorous monkey.

The limb skeleton of *Mesopithecus* resembles that of living colobines in having a relatively short thumb and a long tail. However, in the older species, *M. pentelici*, the limbs are more robust than in those of most living colobines and the digits relatively shorter, suggesting that it was partly terrestrial like the Hanuman langur of India. The localities that have yielded remains of this species seem to be characterized by woodland savannah environments (Delson, 1975). The later species, *M. monspessulanus*, is more like living colobines in its limb skeleton and also has been found in more wooded environments. Presumably it was more arboreal. Both species are sexually dimorphic and presumably lived in polygynous social groups.

Dolichopithecus is a European colobine that seems to be related to *Mesopithecus*; it was a Pliocene contemporary of the later species. Dentally it is similar to *Mesopithecus*, but it has a longer snout and a larger overall size (15–20 kg). It also is sexually dimorphic in tooth and skull size.

In its skeleton, *Dolichopithecus* has more extensive adaptations for terrestrial quadrupedalism than any other colobine. Its limb proportions and many of its joint articulations are baboonlike, and it has short, stout phalanges. The genus seems to have been associated with humid forests and probably foraged on the forest floor, a habitus that would have separated it ecologically from the sympatric, more arboreal *Mesopithecus* (Szalay and Delson, 1979).

It is not clear whether *Mesopithecus* and *Dolichopithecus* are more closely related to the living colobines of Africa or to those of Asia. There are few diagnostic features to link them unequivocally with either group, but their Eurasian distribution suggests closer affinities with the Asian langurs (Simons, 1970).

Asian Colobines

The fossil record of Asian colobines is extremely poor. A few late Miocene fossils have been assigned loosely to the genus *Presbytis*, but little is known of their anatomy or likely habits. Many living genera, including *Rhinopithecus* and several species of *Presbytis*, are known from Pleistocene deposits in China, India, and the islands of the Sunda Shelf. Unfortunately, these provide little information about the history of the group aside from documenting the presence of modern genera and species.

African Colobines

In contrast with Asia, Africa has an abundant record of fossil colobines, beginning in the late Miocene and extending into the Pliocene and Pleistocene. During this time there was an extensive radiation of African leaf-eating monkeys, many of which were unlike anything living today.

Microcolobus tugenensis was a small (about 4 kg) fossil colobine from the later Miocene of Kenya and one of the very few fossil monkeys from sub-Saharan Africa between 15 and 6 million years ago. It differs from later colobines and resembles *Mesopithecus* in having slightly lower molar cusps and more crushing surfaces on the lower premolars. It is also unusual among colobines in the shape of the mandibular symphysis. Both *Microcolobus* and *Mesopithecus* seem more primitive than all later colobines and probably preceded the modern radiations in Africa and Asia. In view of its small body size and less-developed shearing crests, it has been suggested that it was probably less folivorous than many later colobines.

Libypithecus markgrafi (Fig. 14.7), from Wadi Natrun in Egypt, was another small, late Miocene colobine. The species is known from a relatively complete skull and an isolated molar. The skull has a long snout compared with most extant colobines and well-developed sagittal and nuchal crests.

Some authors have suggested that it is closely allied with the European *Mesopithecus*; others have argued that it shows similarities to *Colobus* from sub-Saharan Africa. Because *Libypithecus* is known only from a skull, there is not suitable material for a direct comparison with *Microcolobus*.

There are latest Miocene or early Pliocene fossil colobines from Algeria, Libya, and Kenya which are often assigned to the genus ***Colobus***. Most of these monkeys are known only from isolated teeth or single jaws, and both their habits and their affinities with later forms are indeterminate at present. There are also many isolated teeth or jaws from the Pliocene and Pleistocene of East Africa which have been attributed to the living genus *Colobus* but have not been assigned to any particular species.

In addition to these *Colobus* fossils there is an impressive array of large extinct colobines from the Pliocene and earliest Pleistocene of southern and eastern Africa (Fig. 14.8). ***Cercopithecoides***, from the Pliocene and Pleistocene, has two species: *C. williamsi* (about 15 kg), from both southern and

FIGURE 14.7

Skull of *Libypithecus markgrafi*, a Pliocene colobine from Egypt.

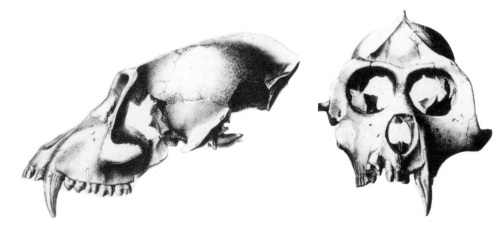

eastern Africa, and the larger *C. kimeui*, from eastern Africa. Both have relatively broad molars and a short-snouted skull associated with a relatively shallow, cercopithecine-like mandible (Fig. 14.8). Aside from canine differences, *Cercopithecoides* shows no evidence of sexual dimorphism in either the dentition or the skull. In the larger species, the broad molars have an inflated baboon-like appearance and are heavily worn on all of the individuals, suggesting a soft but perhaps gritty diet compared with that of most extant colobines.

The most striking adaptations of *Cercopithecoides* are in its limbs, which (if properly associated) resemble a terrestrial cercopithe-cine more than a typical colobine. *Cercopithecoides* was presumably a terrestrial forager and was particularly common in grassland environments.

Paracolobus is the largest colobine known and probably weighed over 30 kg. There are several species from the Pliocene of eastern Africa. *Paracolobus* has a longer face and deeper jaw than *Cercopithecoides* (Fig. 14.8). Dentally it is similar to living colobines, suggesting a largely folivorous diet. It has an intermembral index of 92, similar to that of the living proboscis monkey and red colobus. The skeleton indicates that *Paracolobus* was probably an arboreal quadruped.

Rhinocolobus turkanensis is another large

FIGURE 14.8

Skulls of various Plio-Pleistocene colobines and the extant *Colobus polykomos*. Note the greater size of the fossil monkeys.

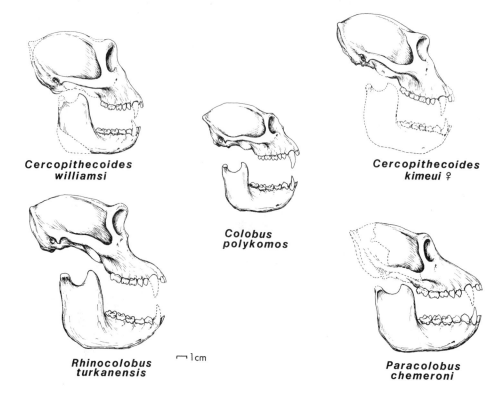

Cercopithecoides williamsi

Colobus polykomos

Cercopithecoides kimeui ♀

Rhinocolobus turkanensis

⌐ 1 cm

Paracolobus chemeroni

monkey from the later Pliocene and early Pleistocene of eastern Africa. It was slightly smaller than *Cercopithecoides* or *Paracolobus* and probably weighed about 20 kg. As the name indicates, *Rhinocolobus* has a pronounced snout on its relatively deep face (Fig. 14.8). Its dentition indicates a folivorous diet, and the few skeletal remains suggest that it was an arboreal monkey. It was common in woodland and gallery forest environments.

SUMMARY OF FOSSIL CERCOPITHECOIDS

The fossil record of Old World monkeys is quite different from that of the other major catarrhine group—the apes. For apes, we have abundant remains in the early Miocene and virtually nothing from the late Miocene to Recent. In contrast, Old World monkeys have a moderate fossil record of the early victoriapithecines from the early and middle Miocene and increasing numbers of fossil monkeys in the late Miocene through early Pleistocene. For apes, there are far more extinct genera and species than there are living taxa, and many of the extinct species are from extensive radiations that seem also to be largely extinct. In contrast, living monkeys far outnumber the extinct taxa, and many of the fossil monkeys seem to be part of the present-day radiation (Fig. 14.9).

Many authors have argued that the tem-

FIGURE 14.9

Cladogram of living (○) and fossil (●) Old World monkeys.

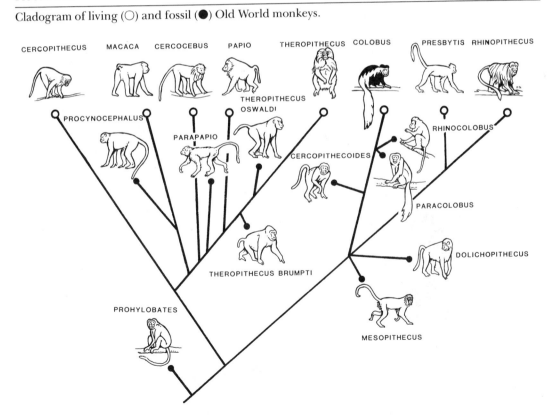

poral pattern of change in the relative abundance of monkeys and apes during the last 20 million years (Fig. 14.10) indicates an ecological replacement of early apes by Old World monkeys. It is equally likely, however, that this apparent change in the primate fauna reflects climatic changes during the Miocene of Africa and Europe rather than simply competition between monkeys and apes in a stable environment.

The earliest fossil monkeys, like the earliest fossil apes, provide evidence of intermediate stages in catarrhine evolution. The victoriapithecines and *Microcolobus* demonstrate that both colobines and cercopithecines preserve a mosaic of both primitive features from the earliest monkeys and also derived features unique to their respective subfamilies. The fossils expand our knowledge of Old World monkey evolution in several ways. They show that both colobines (*Mesopithecus* and *Dolichopithecus*) and cercopithecines (*Macaca* and *Paradolichopithecus*) ranged over much of Europe during the last five million years, and that *Theropithecus* was once found in Asia. The fossil record also suggests that the arboreal nature of most living colobines has not characterized all members of that subfamily. Both *Dolichopithecus* and *Cercopithecoides* were very terrestrial colobines.

A particularly striking feature of the cercopithecoid fossil record is the size difference between extinct and living monkeys. Many extinct colobines and cercopithecines were larger than related living genera. Like the extant Malagasy fauna, the living cercopithecoids are the smaller genera from the

FIGURE 14.10

Relative species diversity of hominoids and cercopithecoids during the past 20 million years in Africa. The diversity of monkeys has increased as the diversity of hominoids has decreased (after Andrews, 1986).

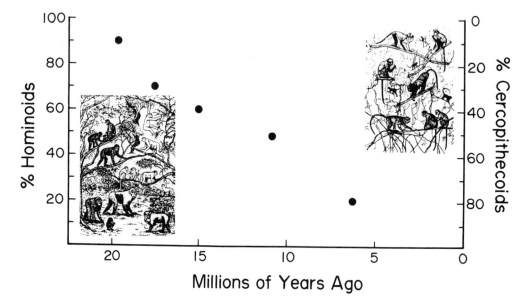

Pliocene and early Pleistocene. This Pleisto-cene extinction of relatively large species is a common phenomenon around the world that cannot clearly be attributed exclusively to either climatic changes or hominid hunting (Martin and Klein, 1984).

BIBLIOGRAPHY

EARLIEST OLD WORLD MONKEYS

Benefit, B. (1985). Dental remains of *Victoriapithecus* from the Maboko Formation. *S.V.P. News Bull.* **133**:21.

———. (1987). The molar morphology, natural history, and phylogenetic position of the middle Miocene monkey *Victoriapithecus*, and their implications for understanding the evolution of the Old World monkeys. Ph.D. Dissertation, New York University, New York.

Benefit, B.R., and Pickford, M. (1986). Miocene fossil cercopithecoids from Kenya. *Am. J. Phys. Anthropol.* **69**:441–464.

Butler, P.M. (1986). Problems of dental evolution in the higher primates. In *Major Topics in Primate and Human Evolution*, ed. B. Wood, L. Martin, and P. Andrews, pp. 89–106. Cambridge: Cambridge University Press.

Delson, E. (1975a). Evolutionary history of the Cercopithecidae. In *Approaches to Primate Paleobiology. Contributions to Primatology* vol. 5, ed. F.S. Szalay, pp. 167–217. Basel: Karger.

———. (1975b). Toward the origin of the Old World monkeys. *Actes CNRS Coll. Int.* **218**:839–850.

———. (1979). *Prohylobates* (Primates) from the early Miocene of Libya: A new species and its implication for cercopithecid origin. *Geobios* **12**:725–733.

Fourtau, R. (1918). *Contribution a l'etude des vertebres miocenes de l'Egypte*. Survey Dept., Ministry of Finance, Cairo.

Leakey, M.G. (1985). Early cercopithecids from Buluk, northern Kenya. *Folia Primatol.* **44**:1–14.

Pilbeam, D., and Walker, A. (1968). Fossil monkeys from the Miocene of Napak, northeastern Uganda. *Nature (London)* **220**:657–660.

Simons, E.L. (1969). Miocene monkey (*Prohylobates*) from north Egypt. *Nature (London)* **223**:687–689.

Simons, E.L., and Delson, E. (1978). Cercopithecidae and Parapithecidae. In *Evolution of African Mammals*, ed. V.J. Maglio and H.B.S. Cooke, pp. 100–119. Cambridge, Mass.: Harvard University Press.

von Koenigswald, G.H.R. (1969). Miocene Cercopithecoidea and Oreopithecoidea from the Miocene of East Africa. *Foss. Verts. Afr.* **1**:39–51.

FOSSIL CERCOPITHECINES

Fossil Macaques

Delson, E. (1980). Fossil macaques, phyletic relationships and a scenario of deployment. In *The Macaques: Studies in Ecology, Behavior and Evolution*, ed. D.G. Lindberg, pp. 10–30. New York: Van Nostrand.

Delson, E., and Rosenberger, A.L. (1984). Are there any anthropoid primate living fossils? In *Living Fossils*, ed. N. Eldridge and S.M. Stanley, pp. 50–61. New York: Springer Verlag.

Hooijer, D.A. (1962). Quaternary langurs and macaques from the Malay Archipelago. *Zool. Verhandl. Mus. Leiden* **55**:3–64.

———. (1963). Miocene mammalia of Congo. *Ann. Mus. Roy. Afr. Cent., ser. 8, Sci. Geol.* **46**:1–71.

Fossil Baboons and Mangabeys

Freedman, L. (1957). The fossil Cercopithecoidea of South Africa. *Ann. Transvaal Mus.* **23**:121–262.

———. (1965). Fossil and subfossil primates from the limestone deposits at Taung, Bolt's Farm and Witkrans, South Africa. *Paleontol. Afr.* **9**:19–48.

———. (1976). South African fossil Cercopithecoidea: A re-assessment including a description of new material from Makapandsgat, Sterkfontein and Taung. *J. Hum. Evol.* **5**:297–315.

Freedman, L., and Brain, C.K. (1972). Fossil cercopithecoid remains from the Kromdraai australopithecine site (Mammalia, Primates). *Ann. Transvaal Mus.* **28**(1):1–16.

Leakey, M.G., and Leakey, R.E.F. (1976). Further Cercopithecinae (Mammalia, Primates) from the Plio-Pleistocene of East Africa. *Foss. Verts. Afr.* **4**:121–146.

Maier, W. (1970a). Neue Ergebnisse der Systematik und der Stammesge schichte der Cercopithecoidea. *Z. Saugertierk.* **35**:193–214.

———. (1970b). New fossil Cercopithecoidea from the lower Pleistocene cave deposits of the Makapansgat limeworks, South Africa. *Paleontol. Afr.* **13**:69–108.

————. (1971). Two new skulls of *Parapapio antiquus* from Taung and a suggested phylogenetic arrangement of the genus *Parapapio*. *Ann. Sth. Afr. Mus.* **59**:1–16.

Simons, E.L., and Delson, E. (1978). Cercopithecidae and Parapithecidae. In *Evolution of African Mammals*, ed. V.J. Maglio and H.B.S. Cooke, pp. 100–119. Cambridge: Harvard University Press.

Fossil Geladas

Ciochon, R.L. (1986). The Cercopithecoid Forelimb: Anatomical Implications for the Evolution of African Plio-Pleistocene Species. Ph.D. dissertation, University of California, Berkeley.

Cronin, J.E., and Meikle, W.E. (1982). Hominid and gelada baboon evolution: Agreement between molecular and fossil time scales. *Int. J. Primatol.* **3**(4):469–482.

Eck, G. (1977). Diversity and frequency distribution of Omo group Cercopithecoidea. *J. Hum. Evol.* **6**:55–63.

Freedman, L. (1957). The fossil Cercopithecoidea of South Africa. *Ann. Transvaal Mus.* **23**:121–262.

Jolly, C.J. (1967). The evolution of the baboons. In *The Baboon in Medical Research*, vol. 2, ed. H. Vagtborg, pp. 427–457. Austin: University of Texas Press.

————. (1970). The large African monkeys as an adaptive array. In *Old World Monkeys*, ed. J.P. Napier and P.H. Napier, pp. 141–174. New York: Academic Press.

————. (1972). The classification and natural history of *Theropithecus* (*Simopithecus*) (Andrews, 1916), baboons of the African Plio-Pleistocene. *Bull. Brit. Mus. Nat. Hist., Geol.* **22**:1–122.

Leakey, M.G., and Leakey, R.E.F. (1973). Further evidence of *Simopithecus* (Mammalia, Primates) from Olduwai and Olorgesailie. *Foss. Verts. Afr.* **3**:101–120.

Maier, W. (1971). The first complete skull of *Simopithecus darti* from Makapansgat, South Africa, and its systematic position. *J. Hum. Evol.* **1**:395–405.

Shipman, P., Bosler, W., and Davis, K.L. (1981). Butchering of giant geladas at an Acheulian site. *Curr. Anthropol.* **22**(3):257–268.

Fossil Guenons

Eck, G., and Howell, F.C. (1972). New fossil *Cercopithecus* material from the lower Omo Basin, Ethiopia. *Folia Primatol.* **18**:325–355.

Leakey, M.G. (1976). Cercopithecoidea of the East Rudolf succession. In *Earliest Man and Environment in the Lake Rudolf Basin*, ed. Y. Coppens, F.C. Howell, G.Ll. Isaac, and R.E.F. Leakey, pp. 345–350. Chicago: University of Chicago Press.

FOSSIL COLOBINES

European Colobines

Aquirre, E., and Soto, E. (1978). *Paradolichopithecus* in La Puebla de Valverde, Spain: Cercopithecoidea in European Neogene stratigraphy. *J. Hum. Evol.* **7**:559–565.

Delson, E. (1975). Evolutionary history of the Cercopithecidae. In *Approaches to Primate Paleobiology. Contributions to Primatology* vol. 5, ed. F.S. Szalay, pp. 167–217. Basel: Karger.

Simons, E.L. (1970). The deployment and history of Old World monkeys (Cercopithecidae, Primates). In *Old World Monkeys*, ed. J.R. Napier and P.H. Napier, pp. 97–137. New York: Academic Press.

Szalay, F.S., and Delson, E. (1979). *Evolutionary History of the Primates*. New York: Academic Press.

Asian Colobines

Hooijer, D.A. (1962). Quaternary langurs and macaques from the Malay Archipelago. *Zool. Verhandl. Mus. Leiden* **55**:3–64.

Matthew, W.D., and Granger, W. (1923). New fossil mammals from the Pliocene of Szechuan, China. *Bull. Am. Mus. Nat. Hist.* **48**:563–598.

African Colobines

Birchette, M.G., Jr. (1982). The postcranial skeleton of *Paracolobus chemeroni*. Ph.D. Dissertation, Harvard University.

Freedman, L. (1957). The fossil Cercopithecoidea of South Africa. *Ann. Transvaal Mus.* **23**:121–262.

Leakey, M.G. (1976). Cercopithecoidea of the East Rudolf succession. In *Earliest Man and Environment in the Lake Rudolf Basin*, ed. Y. Coppens, F.C. Howell, G.Ll. Isaac, and R.E.F. Leakey, pp. 345–350. Chicago: University of Chicago Press.

————. (1982). Extinct large colobines from the Plio-Pleistocene of Africa. *Am. J. Phys. Anthropol.* **58**:153–172.

Leakey, M.G., and Leakey, R.E.F. (1973). New large Pleistocene Colobinae from East Africa. *Foss. Verts. Afr.* **3**:121–138.

Leakey, R.E.F. (1969). New Cercopithecidae from the Chemeron beds of Lake Baringo, Kenya. *Foss. Verts. Afr.* **1**:53–69.

Simons, E.L., and Delson, E. (1978). Cercopithecidae and Parapithecidae. In *Evolution of African Mammals*, ed. V.J. Maglio and H.B.S. Cooke, pp. 100–119. Cambridge, Mass.: Harvard University Press.

FOSSIL CERCOPITHECOIDS: GENERAL

Andrews, P. (1981). Species diversity and diet in monkeys and apes during the Miocene. In *Aspects of Human Evolution*, ed. C.B. Stringer, pp. 25–61. London: Taylor and Francis.

———. (1986). Fossil evidence on human origins and dispersal. *Cold Spring Harbor Symposia on Quantitative Biology* **51**:419–428.

Andrews, P., and Aiello, L. (1984). An evolutionary model for feeding and positional behavior. In *Food Acquisition and Processing in Primates*, ed. D.J. Chivers, B.A. Wood, and A. Bilsborough, pp. 422–460. New York: Plenum Press.

Delson, E., and Rosenberger, A.L. (1984). Are there any anthropoid primate living fossils? In *Living Fossils*, ed. N. Eldridge and S.M. Stanley, pp. 50–61. New York: Springer Verlag.

Fleagle, J.G., and Kay, R.F. (1985). The paleobiology of catarrhines. In *Ancestors: The Hard Evidence*, ed. E. Delson, pp. 23–36. New York: Alan R. Liss.

Martin, P.S., and Klein, R.G., eds. (1984). *Quaternary Extinctions: A Prehistoric Revolution*. Tucson: University of Arizona Press.

Hominids, the Bipedal Primates

PLIOCENE EPOCH

The short Pliocene epoch was a time of considerable faunal change in many parts of the world in association with several geographic rearrangements. The most prominent of these was the rise of the Panama land bridge between North and South America, which led to the exchange of faunas between those two previously separated continents. In the Old World, the Mediterranean Sea refilled at the beginning of the Pliocene after drying up in the latest Miocene. In general, sea levels were higher and temperatures were warmer in the early Pliocene than in the late Miocene. In primate evolution, the Pliocene is characterized by two major events: the spread of cercopithecoid monkeys throughout many parts of the Old World (see Chapter 14) and the first appearance of hominids (Fig. 15.1).

The separation of the lineages leading to

FIGURE 15.1

Map of fossil localities for *Australopithecus, Homo habilis,* and *Homo erectus.*

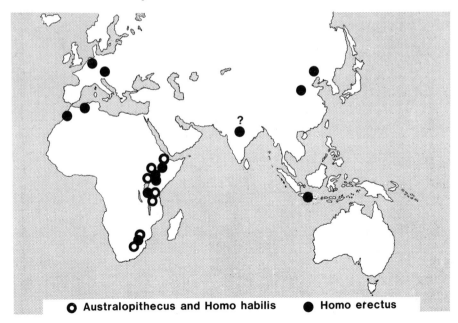

○ **Australopithecus and Homo habilis**   ● **Homo erectus**

the living African apes on the one hand and to humans on the other probably took place sometime in the late Miocene or earliest Pliocene (between about 10 and 4 million years ago). There are, however, only a few hominoid fossils from this period, and those that have been recovered are so fragmentary that it is difficult to determine if they are apes or hominids (Hill, 1985). At present, the earliest undoubted hominids come from the middle Pliocene.

Genus *Australopithecus*

Australopithecus ("southern ape"), from southern and eastern Africa, is the earliest and most primitive genus of hominid (Table 15.1). The earliest species, *A. afarensis*, is from sites approximately 4.5 to 3 million years old in Tanzania, Kenya, and Ethiopia, and the latest members of the genus extend well into the Pleistocene of eastern and southern Africa (Figs. 15.2, 15.3). *Australopithecus* species have big teeth and small brains compared with modern humans, but their size range is much like modern humans', with estimated body weights between 30 kg (the size of an Ituri pygmy) for the smallest individuals and 85 kg (the size of a small college football player) for the largest.

The species of *Australopithecus* differ considerably in cranial and dental anatomy. The earlier species are very similar to living apes and the latest species are quite specialized—

FIGURE 15.2

Geographic and temporal placement of early hominid sites in Africa (modified from White *et al.*, 1981).

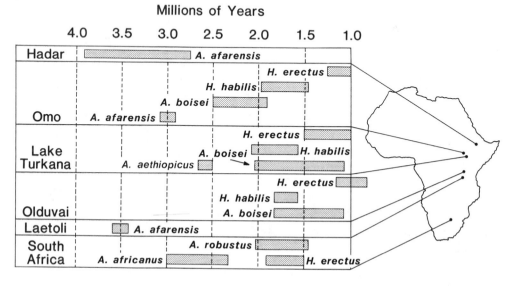

TABLE 15.1
Infraorder Catarrhini
Family HOMINIDAE

Species	Body Weight (g)
Australopithecus (Australopithecus) (Pliocene, ?Pleistocene, Africa)	
A. afarensis	50,000
A. africanus	46,000
Australopithecus (Paranthropus) (Plio–Pleistocene, Africa)	
A. robustus	50,000
A. boisei	50,000
A. aethiopicus	—
Homo (latest Pliocene to Recent, worldwide)	
H. habilis	40,000
H. erectus	50,000
H. sapiens	55,000 (40,000–70,000)

in some respects more specialized than early humans. Compared to living apes, all *Australopithecus* species have small incisors and canines relative to their body weight (Kay, 1985). The lower anterior premolar does not function as a sharpening blade for the upper canine. The molars of *Australopithecus* vary from large to extremely large and are characterized by thick to very thick enamel and bulbous cusps, features they share with the Miocene *Sivapithecus* (see Fig. 13.21). The mandible is thick and has a high ascending ramus.

Cranially, *Australopithecus* is more apelike than humanlike in proportions, with a large face and relatively small brain. On the other hand, the relatively short snout (associated with the reduced anterior dentition) and ventrally located foramen magnum of later species (associated with bipedalism) are very humanlike. Males frequently have large nu-

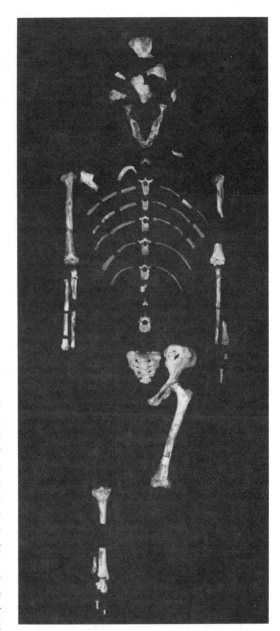

FIGURE 15.3

The skeleton of *Australopithecus afarensis*, "Lucy" (AL-288), from Hadar, Ethiopia. This is the most complete skeleton of an early hominid.

chal crests and sagittal crests that extend far forward. Details of brain morphology in *Australopithecus* have been debated since the first discovery of the genus. They are hindered by problems in estimating the body size of the different species and by the lack of clear impressions on the internal surface of the cranium. In general, it appears that their brains were relatively larger than those of other nonhuman primates but much smaller than those of later hominids or living humans. In external morphology their brains are generally apelike, with few human features (Falk, 1987).

There are isolated skeletal elements for several species of *Australopithecus*, and a relatively complete associated skeleton for the earliest species, *A. afarensis* (Fig. 15.3). Like all later hominids, *Australopithecus* was bipedal. This is evident from many aspects of its skeleton, including the relatively long legs, the short broad ilium, and the angulation of the knee joint. The reconstruction of bipedal habits for *Australopithecus* based on skeletal morphology was dramatically confirmed by a series of footprints preserved at Laetoli, Tanzania (Fig. 15.4). In spite of this, the early species of *Australopithecus* are more similar to living apes than to humans in many features of the skeleton, including the shoulder, the hand, the foot, and even details of the pelvis, femur, and tibia. The skeletal anatomy of *Australopithecus* is in many ways intermediate between those of living apes and humans, suggesting that these early hominids were both arboreal climbers and terrestrial bipeds (Fig. 15.5). The diversity in locomotor abilities among species of *Australopithecus* is difficult to determine because of a lack of skeletal material for many species and because of our weakness in creating models for interpreting a locomotor radiation with only one extant analogue, ourselves.

The systematics and biogeography of *Australopithecus* species are complicated by the same factors that cause confusion in the systematics of most other groups of fossil primates—inadequate dating of sites, fragmentary remains, sexual dimorphism, and differing taxonomic philosophies. Most current authorities recognize four or five species of *Australopithecus*. The two earlier, more primitive species are *A. afarensis* and *A. africanus*, from eastern and southern Africa, respectively. Two later, more specialized "robust" species are *A. robustus*, from southern Africa, and *A. boisei*, from eastern Africa. A hominid fossil ("The Black Skull," KNMWT 17,000) discovered by Richard Leakey and Alan Walker in West Turkana, Kenya, is probably a new species of "robust" *Australo-*

FIGURE 15.4

3.5 million-year-old footprints from Laetoli, Tanzania, presumably made by *Australopithecus afarensis* (photograph by P. Jones and T. White).

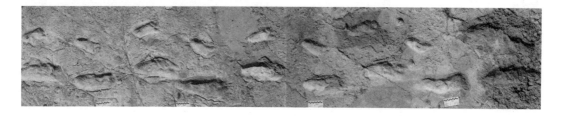

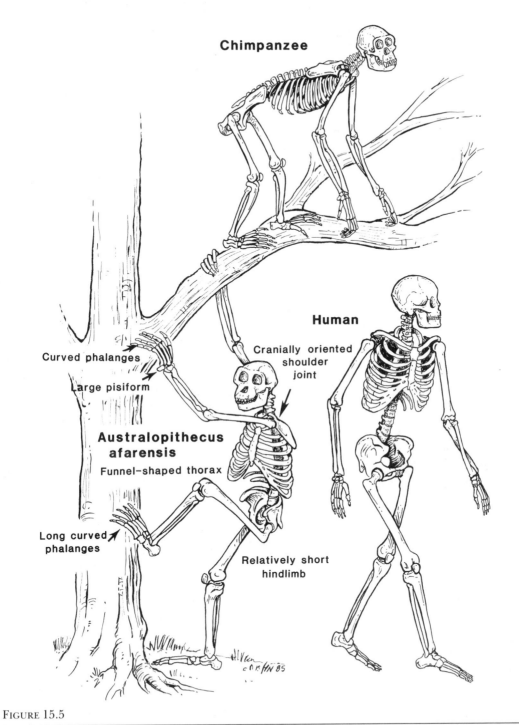

Chimpanzee

Human

Curved phalanges

Large pisiform

Cranially oriented
shoulder
joint

**Australopithecus
afarensis**

Funnel-shaped thorax

Long curved
phalanges

Relatively short
hindlimb

FIGURE 15.5

The skeletons of *Australopithecus afarensis*, *Pan troglodytes*, and *Homo sapiens*. Note the apelike features in *A. afarensis* that suggest climbing behavior.

pithecus, A. aethiopicus. These "robust" species (*robustus, boisei,* and *aethiopicus*) are often placed in a separate genus, *Paranthropus*, and there is increasing evidence that they represent a single lineage of hominids that went extinct about one million years ago (see Fig. 15.2 and 15.13). In this chapter, however, all of the early hominids are placed in a single genus, *Australopithecus*, with two separate subgenera: *Australopithecus* for *afarensis* and *africanus*, and *Paranthropus* for *robustus, boisei,* and *aethiopicus*. (In the Linnean system a subgenus is designated by placing the name in parentheses [*A. (Paranthropus)*]).

The phyletic relationships among these species and the issue of which species is closest to the origin of our own genus *Homo* will be discussed later in the chapter.

Australopithecus afarensis

Australopithecus afarensis (Fig. 15.6), from the middle and late Pliocene of Ethiopia, Tanzania, and Kenya, is the oldest and most primitive species of *Australopithecus*. It had a relatively long temporal span (4.5–2.8 million years). This early hominid was extremely sexually dimorphic in body size, with

FIGURE 15.6

A small *Australopithecus afarensis* group.

the smallest individuals weighing no more than 30 kg and the largest probably twice as much.

In dental proportions, *A. afarensis* is similar to a chimpanzee, with larger canines and incisors than modern hominids. The molars, however, are larger than those of living apes and have the low cusps and thick enamel also found in *Sivapithecus*. The relatively large anterior dentition suggests that this species was frugivorous, and the thick enamel indicates that nuts, grains, or hard fruit pits may have been part of its diet. Although one of the most sexually dimorphic in size of any primate species, *A. afarensis* has little canine dimorphism compared with living great apes—but more than modern humans.

The cranial anatomy of *A. afarensis* is known only from incomplete specimens, but it appears to be similar to that of living chimpanzees. This species has a longer snout and shallower face than later *Australopithecus* and an apelike nuchal region. The brain was small—the size of an orange. There is a sagittal crest both anteriorly, as in other australopithecines, and posteriorly, as in apes (Asfaw, 1987).

The skeletal anatomy of *A. afarensis* is better known than that of any other early hominid. One individual fossil from Hadar, "Lucy" (AL 288), is known from 40 percent of a skeleton, including large portions of almost all long bones (Fig. 15.3). In limb proportions, Lucy is intermediate between living chimpanzees and humans. Based on an estimated body weight of 30 kg, she has relatively short hindlimbs but forelimbs similar in length to those of a small human. Compared with a pygmy chimpanzee of the same size, she has relatively short arms but similar hindlimbs (Jungers, 1982).

The forelimb remains of Lucy, although humanlike in proportions, are more chimpanzee-like in other features (Fig. 15.5). The curved phalanges, large pisiform bone, and cranially oriented shoulder joint all suggest suspensory abilities for this early hominid, as do other chimplike features of the humerus and ulna (Susman *et al.*, 1984). The pelvis of *A. afarensis*, like that of all later hominids, has a short, broad ilium and a relatively short ischium, resembling that of bipedal humans more than that of any living ape in these features. Likewise, the distal part of the femur is strikingly humanlike in its valgus (knock-kneed) angulation. Many details of the ankle and foot, however, such as the relatively long, curved pedal phalanges, are more chimpanzee-like and suggest grasping behavior. Furthermore, even the more humanlike hindlimb elements are different in detail from those of all later hominids, suggesting that the bipedal locomotion of *A. afarensis* was different from that of extant humans (Fig. 15.6).

As with the dental and cranial remains of *A. afarensis*, there is considerable variability in both size and morphological detail among the skeletal remains attributed to this species. In general (but not always), the smaller bones tend to be more chimpanzee-like, while the larger ones tend to be more similar to those of living humans. The most widely held interpretation for this variation is that *A. afarensis*, like the larger living great apes, was characterized by sexual dimorphism in locomotor abilities. Perhaps the larger (male?) individuals were more terrestrial than the smaller (female?) ones.

Overall, *A. afarensis* is remarkably close to a missing link between later hominids and the living African apes in its dental, cranial, and skeletal morphology. Accordingly, this basal hominid is probably intermediate in many aspects of its behavior and also uniquely different from any living primate in many respects. In its diet, it was probably frugivorous and could also eat very hard objects such as seeds and nuts. It traveled bipedally on the ground but probably slept and perhaps foraged in the trees.

The social structure of *A. afarensis* is difficult to reconstruct with any confidence. The combination of little canine dimorphism with considerable body size dimorphism is unique among living primates but most like the pattern found among modern humans. Lovejoy (1981) and Kay (1981) have argued for monogamy from the canines, and Lovejoy has suggested that the size dimorphism reflects different foraging patterns and antipredator strategies for the two sexes. In Lovejoy's view, the more mobile (perhaps more terrestrial?) males relied on their larger size to deter predators, while the smaller females with offspring hid (or perhaps climbed trees). Similar sexual differences in antipredator behavior are characteristic of DeBrazza's monkey, a forest guenon with similar sexual dimorphism in body size that often lives in monogamous groups (see Chapter 6).

Others (Hrdy and Bennett, 1981) have suggested that the size dimorphism indicates a polygynous social structure for early hominids. In this case, the canine reduction might have nothing to do with social structure but would be related to dietary adaptations (Jolly, 1970). In view of the diversity in the social groups found among both extant apes and humans, it seems unlikely that we will ever be able to confidently reconstruct the social habits of early hominids by arguments from analogy.

Australopithecus africanus

Originally described by Raymond Dart (1925) from the limeworks at Taung in the Cape Province of South Africa, *Australopithecus africanus* is the original African apeman.

It is best known from the caves at Sterkfontein and Makapansgat and has long played an important role in our understanding of early human evolution (e.g., Tobias, 1983). Because the limestone caves are not amenable to radiometric dating, the absolute age of *A. africanus* is not precisely known, but faunal associations suggest that this species comes from deposits between 3.0 and 2.5 million years old (see Fig. 15.2).

Compared with *A. afarensis*, *A. africanus* has more similar-sized central and lateral upper incisors, and larger cheek teeth. The relatively smaller anterior dentition resulted in a shorter snout in *A. africanus*. The occipital region and the tympanic bones of the South African species are more like those of later hominids.

The skeleton of *A. africanus* is very similar to that of *A. afarensis*, with similar amounts of size dimorphism (McHenry, 1986).

Behaviorally, *A. africanus* was probably very similar to *A. afarensis*. Dental and cranial differences suggest that *A. africanus* was adapted for more powerful chewing than *A. afarensis* but had a softer, less gritty diet than the later *A. robustus* (see Figs. 15.8, 15.9).

Australopithecus robustus

The second South African apeman is *A. (Paranthropus) robustus* (Fig. 15.7), from the younger cave sites of Swartkrans and Kromdraii (estimated at 2–1 million years). There is evidence from the fauna that *A. robustus* lived (or at least died) in a drier, more open environment than did the earlier *A. africanus* (Brain, 1981a; Vrba, 1985). *Australopithecus robustus* was probably similar in size to *A. africanus* and was also sexually dimorphic in size.

Australopithecus robustus has smaller incisors and canines, larger cheek teeth with

FIGURE 15.7

Reconstruction of a group of *Australopithecus robustus* from Swartkrans in southern Africa. Faunal evidence suggests an open habitat for this species. Dental studies suggest an herbivorous, gritty diet, and anatomical studies of the hands and feet indicate bipedal locomotion and the possibility of tool use (courtesy of F. Grine).

Australopithecus africanus

Australopithecus robustus

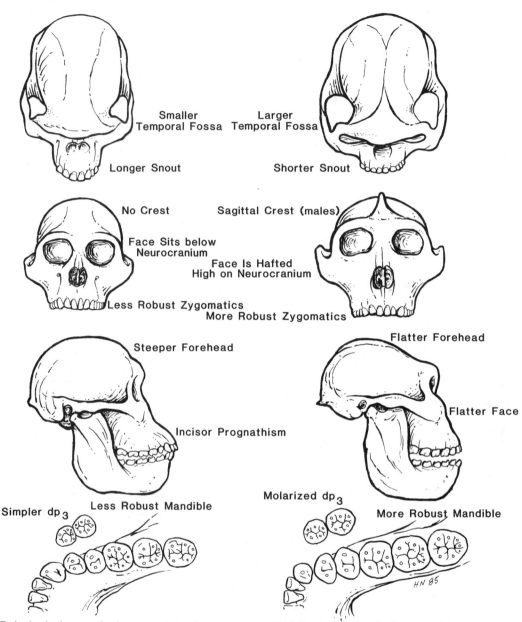

Smaller Temporal Fossa

Larger Temporal Fossa

Longer Snout

Shorter Snout

No Crest

Sagittal Crest (males)

Face Sits below Neurocranium

Face Is Hafted High on Neurocranium

Less Robust Zygomatics

More Robust Zygomatics

Steeper Forehead

Flatter Forehead

Incisor Prognathism

Flatter Face

Less Robust Mandible

Molarized dp_3

More Robust Mandible

Simpler dp_3

Relatively Larger Incisors and Canines

Relatively Smaller Premolars and Molars

Relatively Smaller Incisors and Canines

Relatively Larger Premolars and Molars

FIGURE 15.8

Cranial and dental features of *Australopithecus africanus* and *A. robustus*.

thicker enamel (see Fig. 13.21), and a thicker mandible than the earlier species (Fig. 15.8). These differences in dental morphology are associated with differences in both gross and microscopic tooth wear (Fig. 15.9). Individuals of *A. robustus* wore their teeth flatter and used more crushing than shearing. Their teeth are more heavily scratched and pitted than those of the earlier *A. africanus*, suggesting that their herbivorous diet contained harder, more resistant, and perhaps smaller food objects (Grine, 1981, 1986).

Australopithecus robustus has a shorter,

broader face with deeper zygomatic arches and a larger temporal fossa than seen in the skull of *A. africanus* (Fig. 15.8). The larger individuals (males?) have sagittal and nuchal crests. Like the molar differences, the cranial differences seem related to more powerful chewing in *A. robustus*.

The skeletal differences between *A. robustus* and *A. africanus* are more difficult to assess, because the skeleton of the large species is poorly known. A recent study suggests that *A. (Paranthropus) robustus* was more humanlike in both hands and feet than

FIGURE 15.9

Differences in dental wear between *Australopithecus africanus* and *A. robustus*. Wear facets on the molar teeth indicate that in *A. robustus* the chewing stroke was flatter than in *A. africanus*. The molar teeth of *A. robustus* also show more pits than do the molars of *A. africanus*.

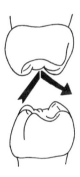

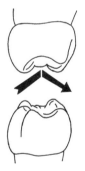

Australopithecus africanus   **Australopithecus robustus**

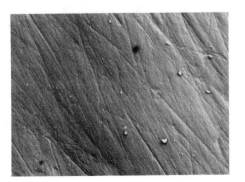

A. afarensis. The hand bones show evidence of manipulative abilities, suggesting that this species was capable of using and making tools (Susman, 1988a). The foot bones indicate that it was bipedal and less arboreal than *A. afarensis.*

The morphological differences between the earlier *A. africanus* and the temporally later *A. robustus* indicate different dietary adaptations rather than simply size differences in primates with similar habits. In dental and cranial features, the larger species was adapted for a tougher diet perhaps associated with a drier environment (Grine, 1981). This species apparently became extinct around 1 million years ago.

Australopithecus boisei

The dental and cranial features that characterize the South African *A. robustus* were developed even further by the "hyper-robust" *A. boisei,* from East Africa. This species, known from deposits between approximately 2.4 and 1 million years ago, was contemporaneous with members of our own genus, *Homo habilis* and *Homo erectus* (Walker and Leakey, 1978). *Australopithecus boisei* was similar in size to *A. robustus,* with an estimated body weight of about 50 kg. This species is sexually dimorphic in both size and cranial shape. Compared with *A. robustus, A. boisei* has smaller incisors and canines, absolutely larger cheek teeth, and a heavier mandible. The skull has an extremely broad, short face with a large temporal fossa between the flaring zygomatic arches and the relatively small brain. There are pronounced sagittal and nuchal crests in the large males.

Although there are few limb bones that can be definitely attributed to *A. boisei,* several very large forelimb bones from East African sites are often assigned to the species. These bones suggest suspensory abili-

ties (McHenry, 1973; Howell, 1978). Tools are often found in association with *A. boisei* in East Africa, but the existence of more advanced hominids (*Homo* sp.) from the same time span precludes any firm evidence that *A. boisei* made or used tools. Like *A. robustus, A. boisei* seems to have become extinct about 1 million years ago (Fig. 15.2).

Australopithecus aethiopicus

Walker and Leakey (Walker *et al.,* 1986) have recovered remarkable new *Australopithecus* material from West Turkana in Kenya dated at approximately 2.6 million years old. The best specimen (KNMWT 17,000, "The Black Skull") has the massive face of *A. boisei* combined with a relatively long snout, primitive cranial base like *A. afarensis,* and very large sagittal and nuchal crests. It appears to be another species of "robust" *Australopithecus* and is designated *A. aethiopicus.* Like many new fossil primates, it has caused considerable reevaluation of recent theories concerning the phyletic relationships among other early hominid species (Grine, 1988).

AUSTRALOPITHECINE ADAPTATIONS AND HOMINID ORIGINS

As the earliest hominid, *Australopithecus* can provide important clues to the adaptations associated with the origin of the human lineage and its divergence from that leading to living apes. Many people have attempted to reconstruct the habits of this basal hominid, and often these interpretations have been heavily colored by theoretical or personal views about human origins. On the one hand, many authors have probably been unduly influenced by the fact that there is only a single hominid living today and have reconstructed *Australopithecus* as a "Pliocene person" little different from modern humans. Others seem to have been overly influenced by their own views of the primitive aspects of human behavior and have seen this early hominid as a vicious killer apeman who bears little relationship to any living primate. With increased knowledge of nonhuman primate behavior, more complete and better-dated fossil remains, and a better appreciation of early hominid diversity, current reconstructions of the behavior of Pliocene hominids are more reliable than those put forth earlier in the century.

Australopithecines were medium-size hominoids with small incisors and canines, large cheek teeth, very robust jaws, and extremely large chewing muscles. In other mammals this combination of features is associated with an herbivorous diet of tough plant material. Later species of the genus evolved very small incisors and canines with even larger cheek teeth and thicker jaws, suggesting even more extreme adaptation for tough foods. All of the dental and cranial evidence indicates that these early hominids were herbivores. Contrary to earlier suggestions, recent estimates of australopithecine body size based on the limb skeleton suggest that, while all species show considerable sexual dimorphism, species means were all very similar, approximately 50 kg (see Table 15.1; Jungers, 1988; McHenry, 1988).

The teeth of *Australopithecus* also provide interesting information about growth and development in early hominids. One of the most characteristic features distinguishing humans from other primates is our slow development (see Figs. 2.26, 2.27) and long period of growth and maturation, usually associated with our reliance on learning. Studies of dental development indicate that in timing and sequence of dental development *A. africanus* was quite different from modern humans and more similar to that of African apes (Fig. 15.10; Bromage and Dean, 1985; Smith, 1986), while *Paranthropus* had a developmental pattern that resembled modern humans in some ways.

Relative to body size, the brains of *Australopithecus* species were no larger than the brains of living apes. There is also no evidence from either the external morphology of the brain or the shape of the skull base that *Australopithecus* had any greater capabilities for articulate speech than do living apes.

Australopithecus shares several distinctive features of its locomotor skeleton with *Homo sapiens*, indicating that it was a bipedal primate. Other similarities—to living pongids—indicate that the earliest hominids retained considerable abilities for arboreal locomotion such as vertical climbing, and that their bipedal gaits were probably noticeably different from humans'. It seems most likely that they climbed trees for foraging, sleeping, and escaping predators, but that they traveled bipedally on the ground. Arboreal foraging and terrestrial travel between food sources is a common behavioral pattern in a variety of living primates, including chimpanzees and pig-tailed macaques. A striking difference is that the bipedal hominids would have been able to use their hands for transporting food from place to place in a way that the quadrupedal apes and monkeys could not.

There is recent anatomical evidence that *A. (Paranthropus)* used stone tools. Bone and stone tools are often found in deposits with later (Pleistocene) species of *A. (Paranthropus)*, but the first appearance of stone tools coincides with the first appearance of more advanced hominids (*Homo habilis*), so there is always uncertainty as to whether *A. (Paranthropus)* or *Homo* is responsible (cf. Susman, 1988). As many authors have emphasized, it is certainly reasonable that they may have

Dental Development in *Australopithecus africanus*

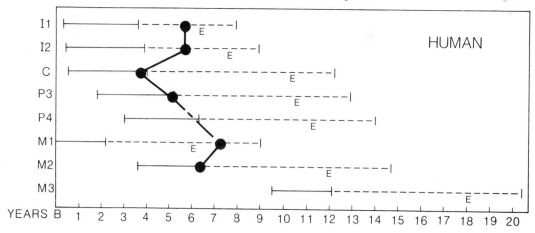

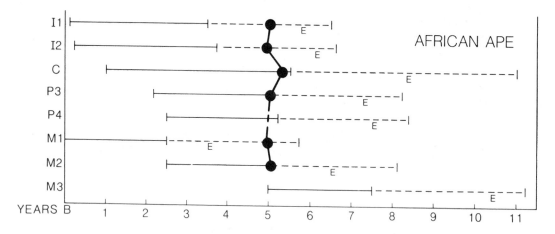

FIGURE 15.10

Comparison of the dental development in *Australopithecus africanus* (STW 151) and *Homo habilis* (KNMER 1590) with average patterns of dental development in humans and chimpanzees. For each tooth position, the solid lines represent the ages during which the tooth crown is formed and the dashed line indicates the time of root formation. Age of eruption is indicated by E. Thus a vertical line at any age gives the stage of development for each tooth. Compared with apes, humans are characterized by a slow development of all teeth, and the human canine develops relatively early (see Figs. 2.27, 2.28).

When the developmental stage of each tooth in the jaw of *A. africanus* is plotted on the human chart, the results are contradictory. The canine is most like that of a three-year-old child, but the first molar is more like that of a seven-year-old. All of the teeth are very similar in developmental stages to the conditions seen in a five-year-old chimpanzee, suggesting that *A. africanus* had an apelike rather than humanlike pattern of dental development. *Homo habilis* also shows an apelike rather than humanlike pattern of dental development.

428

Dental Development in *Homo habilis*

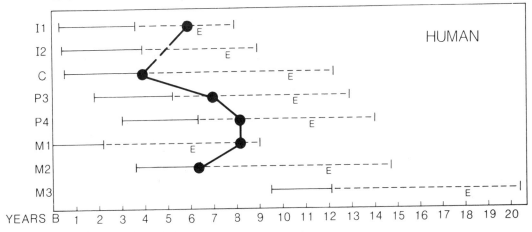

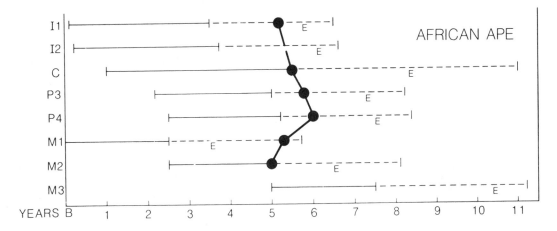

made and used some type of perishable tools, since these are used by both chimpanzees and later hominids; however, the current argument for tool use in *A. robustus* rests more explicitly on the presence of anatomical features related to manipulation in hand bones attributed to this species. Recovery of more complete remains of australopithecine skeletal material will certainly help clarify this issue.

Overall, the evidence suggests that australopithecines were rather nonhuman hominids. Many of their characteristic dental features, such as small canines and large flat molar teeth, were also present in earlier fossil apes such as *Sivapithecus*. Only the anatomical features indicative of bipedal walking and perhaps some manipulative abilities seem to separate these early hominids from other nonhuman primates. Most morphological features that characterize modern humans, such as an arched foot with an adducted hallux and a large brain, as well as our characteristically slow rates of growth and development, were not present in these primitive hominids. They were very much missing links between apes and people.

Humans differ from living apes in numerous morphological and behavioral features, and there has been a tendency in the study of human evolution to see all human features, including bipedalism, large brains, manipulative hands, tool use, and language, as integrally related into a single adaptive complex extending back to the origin of the hominid lineage (Darwin, 1871; Washburn, 1963). Such an approach was reasonable when there was only one known hominid, *Homo sapiens*, and no fossil record of more primitive ancestors lacking this complete

suite of features. But such completely integrated models provide no insight into the beginnings of the group—only a circular loop of explanations. The fossil record of *Australopithecus* provides direct evidence that the cluster of features characterizing living humans are not necessarily linked but rather evolved one by one (Fig. 15.11).

The evidence from *Australopithecus* indicates that hominids were bipedal well before they evolved brains appreciably larger than those of living apes and before they regularly made (and presumably used) stone tools. *Australopithecus* appears to have thrived for 3 or 4 million years as a biped without ever evolving, or for that matter needing, these novelties of the other hominid genus, *Homo*.

Like all other members of our order, *Australopithecus* was a primate that evolved a particular suite of adaptations that enabled

FIGURE 15.11

Appearance of major anatomical and behavioral features in the paleontological and archeological record of East Africa (modified from Harris, 1983).

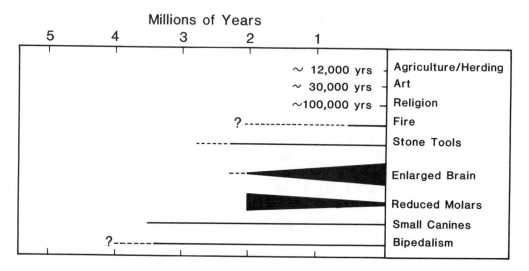

it to make a living in its particular time and place. The specific ecological factors that were critical for the evolution of bipedal locomotion in *Australopithecus afarensis* and thus the origin of the hominid lineage are the subject of considerable speculation (Fig. 15.12).

Some authors (Brace, 1979; Wolpoff, 1980) envision *Australopithecus* as a human-like creature. Like Darwin, they argue that bipedalism and canine reduction were linked with the regular use of tools. Because stone tools do not seem to be normally associated with *Australopithecus*, they posit that the earliest hominids must have used some type of wooden tools such as spears or clubs. Needless to say, such theories are virtually untestable.

FIGURE 15.12

Various theories of the origin of bipedal locomotion (courtesy of Jeanne Sept).

1 Carrying   1a Weapons and tools   1b Vegetable foods, water, and infants

2 Travelling between food trees   3 Feeding from bushes

4 Feeding on grass seeds   5 Provisioning family

In contrast, other theories attempt to explain the evolutionary divergence of hominids in a more nonhuman framework based on the nonhuman nature of *Australopithecus*. Zihlman and Tanner (1978) see considerable similarity between *Australopithecus* and living chimpanzees in many aspects of their behavior. They suggest that many of the traditionally accepted differences between hominids and apes are based on a misguided (and male-oriented) overemphasis on hunting as a characteristic hominid behavior. In modern hunter-gatherer communities, it is the gathering of plants by females that provides most of the food for subsistence, just as plants provide most of the food for living apes. In their view, we should pay more attention to the function of characteristic hominid features in a gathering rather than a hunting context. In such a context, bipedalism would enable females to more easily carry the extremely dependent hominid infants. Free hands would also enable gatherers to carry extra water, thus extending forage range into dry areas, and to carry surplus food in a way that living apes are generally unable. Zihlman and Tanner suggest that, if *Australopithecus* used tools, they were probably digging sticks, baskets, and bowls for water, not clubs and spears.

Another theory linking hominid origins with a unique foraging strategy is that of Rodman and McHenry (1980), who associate bipedalism with foraging in an open woodland habitat. In their view, the earliest hominids may well have fed in trees but adopted bipedal postures for traveling between trees. They suggest that bipedal walking rather than quadrupedal travel would be a more efficient way of moving. Other workers (Rose, 1976; Pilbeam, 1980; Wrangham, 1980) have suggested that the evolution of hominid bipedalism began as some sort of feeding posture enabling early hominids to feed on tall bushes or small trees, but like the open-habitat hypothesis this suggestion fails to explain why the other primates that regularly use bipedal postures have never adopted bipedal locomotion.

Several other recent theories directly address this problem. Jolly (1970) has argued that bipedalism in an herbivorous primate is only advantageous when the animal is feeding on small, evenly distributed objects such as nuts, grains, or small seeds. In this type of feeding situation, an individual's foraging efficiency is directly linked to the speed with which it can pluck and ingest food items. A squatting or partly bipedal animal with both hands free for foraging is best adapted to this type of diet. Furthermore, small hard objects such as seeds seem to be the type of foods that the *Australopithecus* dentition, with its broad, flat, thick-enameled molars, was designed for eating. Finally, Jolly argues that many foods of this nature are common in savannah environments, an open habitat in which bipedal locomotion seems most appropriate. Jolly's model of hominid origins, "the seed-eater" model, is based largely on the habits of the gelada baboon, an open-savannah, small-object-feeding baboon.

Lovejoy (1981) suggests that the major adaptive change distinguishing early hominids from their more apelike ancestors was in their reproductive capabilities. Living great apes, he argues, normally give birth at intervals of three to five years, largely because of the difficulties of caring for a large, slow-growing offspring. Humans, he suggests, generally have a much shorter spacing between successive offspring. This increased reproductive efficiency was made possible through a monogamous social system in which males provisioned their mates and offspring. In freeing the hands, bipedalism allowed the males to bring extra food back to the less widely foraging members of their family, a behavior unknown in other primates. Although Lovejoy notes that the teeth of *Australopithecus* clearly indicate a herbivorous diet, the advantage of bipedalism is not linked to any particular type of food, only to the ability to transport it. Lovejoy's suggestion that *Australopithecus* lived in a monogamous social system (in which male parental investment would be expected) is based on his view that there was little canine dimorphism. Other authors have suggested that the evidence of considerable sexual dimorphism in body size argues for a more polygynous social structure.

The major distinction between most of these theories of hominid origins and more traditional models (derived largely from Darwin) is that the former are designed to explain the evolution of a bipedal herbivorous primate with small canines based on the everyday parameters of a nonhuman primate way of life (feeding, travel, and reproduction) rather than on modern human behavior. Although they are somewhat speculative, they all represent a changing perspective on human origins. Today we see hominids as one of many peculiar radiations in primate history, the evolution of which should be explicable in terms of ecological adaptation. This is an important contrast to the more traditional views, in which nonhuman primates were, at best, stepping stones or, at worst, failed experiments on the road to humanity. This new view of hominid origins has come in part from an increased appreciation that *Australopithecus* was a more primitive, apelike hominid than earlier workers had suspected and one that was also uniquely specialized in ways that were adaptively distinct from both living apes and later hominids. The specialized nature of this early hominid has been particularly emphasized by the realization that for at least a million years *A. (Paranthropus)* was syn-

chronic and probably sympatric with a more human genus of hominid, *Homo*, which first appeared near the beginning of the Pleistocene epoch.

PHYLETIC RELATIONSHIPS OF EARLY HOMINIDS

There are several long-standing debates, and a few very new ones, concerning the phyletic relationships among the various australopithecine species and the origin of the genus *Homo* (Fig. 15.13). Most authorities agree that *A. afarensis* is the most primitive hominid species. *Australopithecus africanus* is very similar to *A. afarensis* but slightly more advanced in some cranial features (such as the structure of the petrosal part of the temporal bone) and is closer to the origin of the genus *Homo* (Skelton *et al.*, 1986).

The major problem concerning early hominid phylogeny revolves around the origin of the robust species relative to other species of *Australopithecus* and, later, humans. Various authorities have marshaled evidence to argue that the robust lineage is derived from *A. afarensis, A. africanus,* or an unknown ancestor intermediate between *A. africanus* and *Homo habilis*. The new skull from Kenya (WT 17,000) introduced yet another species into the sample, *A. aethiopicus,* and indicates that the robust australopithecines more likely evolved from a species like *A. afarensis* than from one like *A. africanus* (Grine, 1988). This new skull also demonstrated how little we currently know about the likely diversity of Pliocene hominids. *Australopithecus* was a very successful, diverse genus in the Pliocene and early Pleistocene, and further fossil finds will undoubtedly reveal even more surprises and lead to further questions concerning the phyletic relationships among the different species.

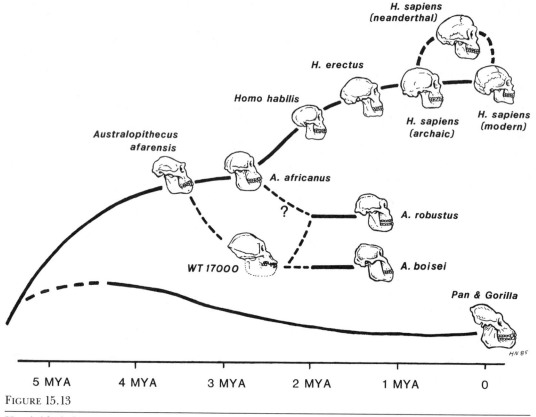

FIGURE 15.13

Hominid phylogeny, showing the temporal and likely phyletic positions of hominid species during the past 4 million years.

PLEISTOCENE EPOCH

The Pleistocene, from 2 million years ago until recent times, was characterized geologically by repeated glaciations of the Northern Hemisphere. The initial onset of dramatic cooling seems to have begun in the latest Pliocene (around 2.5 million years ago), and there is evidence of another extreme cooling after 1 million years ago. In Africa there is evidence of shifts in the flora and fauna during the Pleistocene which have been attributed to changes between relatively warm, wet climates and cooler, drier climates associated with these global events. In hominid evolution, this epoch saw the evolution of the genus *Homo*, which began in the latest Pliocene and continued to the present, and the extinction of the robust australopithecines approximately 1 million years ago. The correlation between these global climatic events and the major events in human evolution is a topic of considerable interest and controversy (e.g., Vrba, 1985).

Genus *Homo*

The fossil record of our own genus begins at the end of the Pliocene epoch, around 2 million years ago, and extends more or less continuously through the succeeding Pleistocene to the present. The three generally recognized species of *Homo* (*H. habilis, H. erectus,* and *H. sapiens*) seem to be drawn from a single, continuously evolving lineage that has been characterized by considerable geographic variation throughout its history (see Table 15.1). The timing and nature of the transitions between species of *Homo* are the subjects of long-standing debates.

Compared with *Australopithecus, Homo* is characterized by smaller molars and premolars and a more slender mandible. Throughout the evolution of the genus, there has been a trend toward reduction in the size of the cheek teeth. The anterior teeth, canines and incisors, are larger than those of the more specialized, robust species of *Australopithecus*. The cranium of *Homo* is characterized by a relatively larger brain size and smaller face than *Australopithecus*.

The skeleton of early species of *Homo* is not well known. Our genus seems to be characterized by a less beaked ilium and a larger femoral head than *Australopithecus*. The foot of *Homo* has shorter digits than those of more primitive hominids.

Homo habilis

The earliest and most primitive species, *Homo habilis* (Fig. 15.14), first appeared around 2 million years ago in the latest Pliocene and earliest Pleistocene of southern and eastern Africa, where it was contemporaneous with *A. (Paranthropus)*. As a species that is intermediate in many morphological features between earlier, more primitive *Australopithecus* and later, more advanced *H. erectus, H. habilis* has always been a difficult species for which to set morphological boundaries. Specimens regularly assigned to this species show considerable morphological variation and overlap with earlier and later species. Older fossils identified as *H. habilis* by one authority may be identified as *Australopithecus* by another. Younger fossils may be identified by one authority as *H. habilis* and by others as *H. erectus*. It is very much a transitional species, and one worker has questioned whether the transitional *H. habilis* fossils actually represent more than one species (Stringer, 1986).

Compared with *Australopithecus, H. habilis* has narrower premolars and first molars, a narrower mandible, a more coronal orientation of the petrous part of the temporal bone, and delayed eruption of the canines. The average cranial capacity is larger than that of the more primitive *Australopithecus* and smaller than that of most *H. erectus*, but individual specimens overlap with both taxa. The hand bones are more robust than those of later hominids and suggest that this species retained some suspensory abilities. The foot is more advanced than that of *Australopithecus* and resembles the foot of extant humans in most features, suggesting a similar bipedal gait (Susman, 1983). *Homo habilis* has relatively long forelimbs (Johanson *et al.,* 1987).

The first appearance of stone tools in Africa coincides roughly with the appearance of *H. habilis* in the fossil record, suggesting that this species used and made the artifacts (Fig. 15.10; Harris, 1983). The tools are crude choppers and scrapers, collectively called the Oldowan culture because of the original discovery at Olduvai Gorge, Tanzania (Fig. 15.15). Wear on the cutting edges indicates that these tools were used in a variety of activities, including butchering of small animals, trimming of leather, and

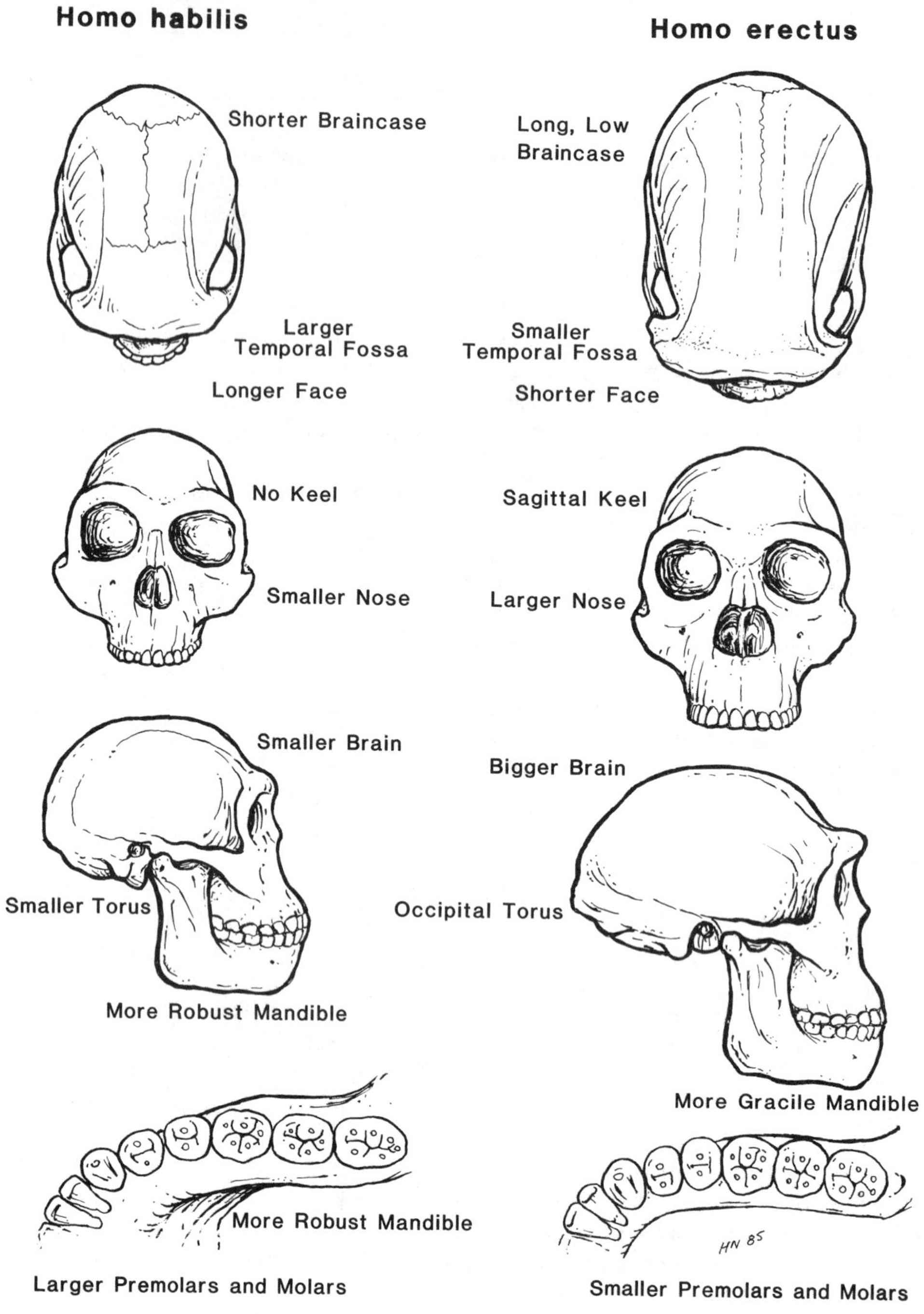

Homo habilis

Shorter Braincase

Larger Temporal Fossa

Longer Face

No Keel

Smaller Nose

Smaller Brain

Smaller Torus

More Robust Mandible

More Robust Mandible

Larger Premolars and Molars

Homo erectus

Long, Low Braincase

Smaller Temporal Fossa

Shorter Face

Sagittal Keel

Larger Nose

Bigger Brain

Occipital Torus

More Gracile Mandible

Smaller Premolars and Molars

HN 85

FIGURE 15.14

Cranial and dental characteristics of *Homo habilis* and *H. erectus*.

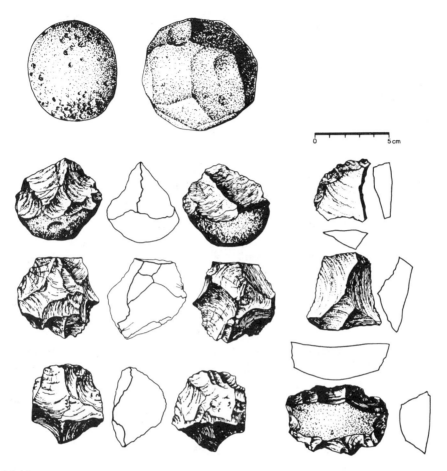

FIGURE 15.15

Primitive Oldowan tools (courtesy of K. Schick).

preparation of plant remains. However, recent work involving experimental manufacture and use of such primitive tools by present-day anthropologists indicates that the Oldowan "tools" may actually be the larger cores that were left in the process of producing smaller flakes (Toth, 1987). Although the core choppers were probably used for some activities, it is the smaller, razor-sharp flakes that are more effective in food processing.

Some of the broken animal bones found in association with Oldowan tools and *H. habilis* fossil remains show cut marks that appear to have been made by the stone tools. Furthermore, the concentrations of stone tools and broken bones suggest to some workers that animal parts were transported to the sites by hominids (Isaac, 1983), while others see both the cut marks and the accumulations as possible results of geological processes (Binford, 1981). Isaac (1983) has argued that these concentrations indicate the emergence of some type of home base or "central-place foraging" and possibly food sharing among early humans. Whether the animal parts are the result of hunting or scavenging activities cannot be determined. It seems likely that the role of meat eating and hunting in early hominids has been overemphasized in past archaeology. Unfortunately, plant foods leave few fossilized remains, so it is not possible to reconstruct the relative proportions of meat and plant material in the diet of *H. habilis*.

At several sites in East Africa, *H. habilis* appears in the latest Pliocene in association with *A. robustus* or *A. boisei*, and it seems likely that our genus evolved from an earlier, more primitive species of *Australopithecus*. There is debate over which species of *Australopithecus* is closer to the origin of the human lineage, but it seems most likely that the divergence of the lineage leading to *Homo* took place in Africa during the middle of the Pliocene epoch, from something more advanced than *A. africanus*.

Homo erectus

Homo erectus first appeared in the early Pleistocene of Africa about 1.6 million years ago. It was a relatively long-lived species. Fossils assigned to this species are known from sites in Africa and Asia until the middle Pleistocene and possibly also from Europe, although hominid remains are less common than tools. The geographic distribution of *H. erectus* exceeds that of any other primate prior to its time (Fig. 15.1).

Compared with *H. habilis* and *Australopithecus*, *H. erectus* has still smaller cheek teeth and a more slender mandible, in keeping with the general trend of tooth reduction within the genus (Fig. 15.14). Brain size is significantly larger than in earlier hominids,

with an average of about 900 cc for the species. The cranium of *H. erectus* is characterized by very thick bones, a long, low vault with sagittal keeling, projecting brow ridges, and a prominent occipital torus. The face of this species was relatively broad and had a large nasal opening. *Homo erectus* seems to have had a mean body size similar to or slightly larger than that of australopithecines, but with less sexual dimorphism (Rightmire, 1986).

As with *H. habilis*, there is debate over the morphological and temporal boundaries of this evolving, geographically variable taxon. Some authors argue that *H. erectus* shows almost no morphological change from its first appearance in the early Pleistocene until its replacement by *H. sapiens* in the middle Pleistocene (e.g., Rightmire, 1981, 1985), but more thorough analyses of the fossils with larger samples clearly indicate morphological change in *H. erectus* with time. The earlier fossils resemble *Australopithecus* and *Homo habilis*, and the later ones approximate *H. sapiens* (Wolpoff, 1984).

Until quite recently there were very few nonpathological remains of the skeleton of *H. erectus*. Brown *et al.* (1985) have reported a nearly complete skeleton of a young *H. erectus* boy that lived approximately 1.6 million years ago on the west side of Lake Turkana in Kenya (Fig. 15.16). *Homo erectus* was much larger in stature than earlier hominids, with limb bones as large as or larger than those of many living people. The young (12-year-old) male from the early Pleistocene had an estimated adult height of nearly six feet. The limb proportions are similar to those of *H. sapiens*, but most of the limb bones are more robust. The chest is more conical, as in apes, and the femoral neck is long, as in *Australopithecus*, but the femoral head is large, as in modern humans.

In Africa, *H. erectus* remains are associated

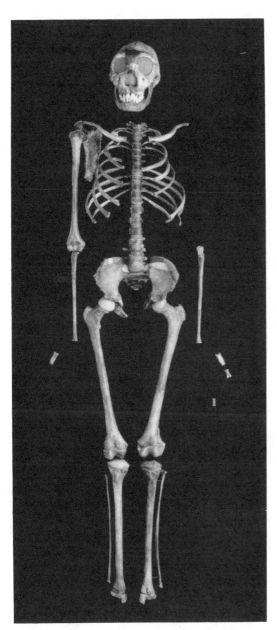

with Acheulian hand axes; in Asia the species is found with more primitive chopping tools, similar to earlier Oldowan artifacts. *Homo erectus* is the first fossil primate with a substantial archeological record. The species developed a wide range of stone implements for different purposes, many of which are still manufactured and used today. Archeological sites attributed to *H. erectus* are widespread and diverse. Some seem to have been camps, others were sites of animal kills, and others were butchering sites. The Zhoukoudian site in northern China (from 460,000 to 230,000 years ago) indicates that *H. erectus* used fire, and some of the later sites show evidence of simple structures. The variation in size of the camps suggests a social organization of individual families that sometimes camped and presumably foraged alone and at other times joined with other families—a social structure similar to that of living hunter-gatherers.

Homo erectus were hunters that successfully preyed on a variety of medium and large mammals including elephants, ungulates, and deer. At later sites, the archeological evidence indicates that they exploited virtually all available animals in the area. Like both their primate forebears and living hunter-gatherers, *H. erectus* probably relied on plant parts of some sort for most of their diet. There are remains of berries at Zhoukoudian and other sites. As with other hominids, this part of the diet of *H. erectus* is very difficult to reconstruct, and our view of their subsistence behavior is certainly distorted by an overemphasis on hunting because of abundant animal bones.

FIGURE 15.16

The 1.6 million-year-old skeleton of a *Homo erectus* boy from West Turkana, Kenya (from Leakey and Walker, 1985) (photograph: David L. Brill © National Geographic Society).

Homo sapiens

The first fossils attributed to our own species, *Homo sapiens*, come from the middle Pleistocene of Africa and Europe. From this

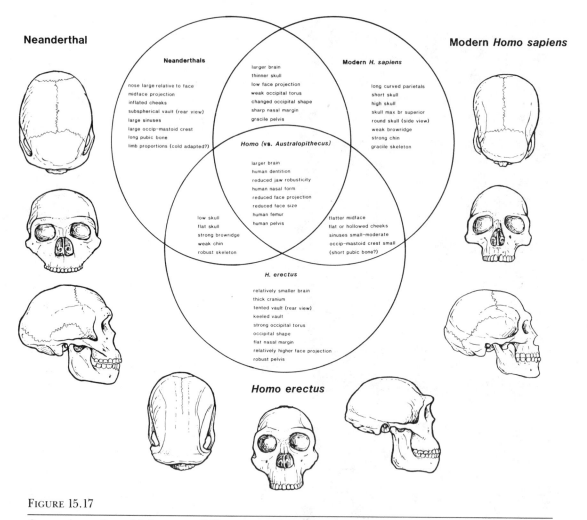

FIGURE 15.17

Comparison of cranial features of *Homo erectus*, Neandertals, and modern *H. sapiens* (from Stringer, 1985).

period there are many fossils that also seem to be intermediate between *H. erectus* and *H. sapiens*, suggesting that the transition was a gradual one. *Homo sapiens*, like *H. erectus*, has always been a widespread, geographically variable species, but it differs from *H. erectus* in many features (Fig. 15.17).

Compared with *H. erectus*, our species has smaller cheek teeth, a more slender mandi-

ble, and less sexual dimorphism. We have thinner cranial bones and a larger brain, averaging about 1,300 cc, which is housed in a large globular cranium without pronounced marking on the nuchal region. Our facial skeleton is less protruding and more gracile.

A major feature of our skeleton is the decreased robustness of the limb bones compared with those of *H. erectus*. A further gracility of the limbs also distinguishes more recent *H. sapiens* from earlier members of the species. The markedly decreased robustness of the skeleton in modern *H. sapiens* in the late Pleistocene seems to be related to the continued replacement of physical exertion with technological skill.

This technological ability, which involves not only the modification and design of inanimate objects such as stone, wood, and metals but also domestication of many plant and animal species, is an outstanding characteristic of *H. sapiens*. Thus the ecological adaptations of our species, unlike those of *H. erectus*, cannot be characterized as uniformly "hunter-gatherer." Although the earliest *H. sapiens* were probably similar to *H. erectus* in their habits, later populations specialized on fish, shellfish, small mammals, and undoubtedly many other foods before developing the sophisticated habits of agriculture and animal husbandry in the very recent past. Likewise, the development of food preparation by cooking, a striking departure from our primate heritage, has been a major contributor to the continued dental reduction that characterizes our species.

The technological skill of our species has enabled us to exploit virtually all available habitats on earth, from the more traditional woodlands and savannahs to tropical forests, oceanic islands, and the arctic. *Homo sapiens* considerably extended the geographic range of *H. erectus*, successfully colonizing areas that had previously been beyond the range of catarrhines, such as North America and South America, as well as islands such as Madagascar that had never been colonized by higher primates, and Australia, which had never seen a primate. We are still expanding our range.

Our own species has been studied more thoroughly than any other primate species,

and many temporal and geographic differences in populations of *H. sapiens* during the past 300,000 years are well documented— even though the relationships among them are poorly understood. The earliest members of *H. sapiens*, from the middle Pleistocene of Europe and Africa (and perhaps also some younger fossils from Asia), share many primitive features with *H. erectus*. In the early part of the late Pleistocene, Europe was populated by a distinctive group of humans, the Neandertals, which were characterized by very large brain size, inflated faces, and limb proportions with short distal elements compared with living humans (Fig. 15.17). Anatomically modern *H. sapiens*, similar to living populations, are known from Africa approximately 100,000 years ago and the Middle East and only about 40,000 years ago in Europe (Fig. 15.18). It is debated whether the current distribution of modern humans throughout the world has come about through replacement of more archaic populations or by gradual evolution of modern types from more archaic types on different continents. Most recent analyses of both fossil and molecular data support a replacement model (e.g., Bräuer, 1984; Cann *et al.*, 1987; Stringer and Andrews, 1988).

HUMAN PHYLOGENY

The major issues in human phylogeny concern whether the various populations and species of fossil humans are part of a single lineage or whether there have been multiple lineages at any one time—some of which became extinct while others gave rise to later species. This issue has been raised for virtually every transition between species in hominid evolution (Fig. 15.13).

When our knowledge of early hominid evolution was much less complete than it is

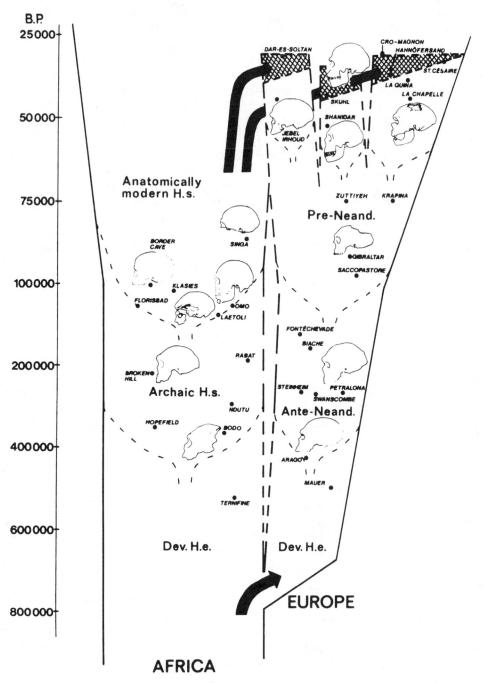

FIGURE 15.18

Diagram of temporal placement of human cranial remains from Africa and Europe (from Bräuer, 1984b). Bräuer supports a replacement model, with modern *Homo sapiens* evolving in Africa and migrating to Europe approximately 40,000 years ago.

today, several authors (notably Wolpoff, 1971) argued that the hominid adaptive niche was so broad that there could never be more than one hominid alive at any one time and that all hominid evolution was unilineal. The basic premise of this **single-species hypothesis** was that culture defined the adaptive niche of hominids, and that competitive exclusion precluded two culture-bearing primates from existing at the same place at the same time. However, new fossils have shown that for approximately 1 million years in the early Pleistocene of both East Africa and South Africa there were at least two coexisting hominids—one species of *A. (Paranthropus)* (either *A. robustus* or *A. boisei*) and one species of *Homo* (either *H. habilis* or *H. erectus*). Why *A. (Paranthropus)* became extinct approximately one million years ago is unclear.

The issue of unilineal versus multilineal evolution continues to be a consideration with respect to *H. erectus*. Many authorities (Eldredge and Tattersall, 1975; Tattersall and Eldredge, 1977; Wood, 1985) suggest that *H. sapiens* evolved directly from an early Pleistocene hominid such as *H. habilis*, which has a thin, domed braincase, rather than from *H. erectus*, with its low flat cranial vault. This view is countered by the many authors who emphasize that the middle Pleistocene hominids grouped as *H. erectus* have much greater morphological variation than is normally appreciated and seem to bridge the gaps in both time and morphology between *H. habilis* and *H. sapiens*. Thorne and Wolpoff (1981) have revived and reanalyzed earlier observations by Weidenreich (1946) and suggested that many regional characteristics seen in *H. erectus* fossils from Africa and Asia persist through later *H. sapiens* to modern populations. In their view, "*Homo erectus*" fossils should be considered part of *H. sapiens*, and the evolution of *H. sapiens*

from "*H. erectus*" a more cosmopolitan event, with *sapiens* genes spreading through the "*erectus*" gene pool in such a way that local features were preserved. Most other authorities argue that the transition from *H. erectus* to *H. sapiens* occurred only once, probably in Africa, and that then *H. sapiens* replaced *H. erectus* (e.g., Stringer and Andrews, 1988).

The most controversial aspect of the evolution of *H. sapiens* is whether the Neandertals from the upper Pleistocene of Europe contributed to the evolution of anatomically modern *H. sapiens* on that continent or were replaced by the earliest anatomically modern *H. sapiens*—the Cro-Magnon people (Fig. 15.18; see Smith and Spencer, 1985; Trinkaus, 1986). Many studies indicate that anatomically modern *H. sapiens* was present in Africa nearly 100,000 years ago when there were only Neandertals in Europe. Moreover, new dates indicate that modern *H. sapiens* may have been in the Middle East before Neandertals (Valladas *et al.*, 1988). All of these chronological data support morphological studies which argue that Neandertals were a uniquely derived group of fossil humans, unlikely to be ancestral to modern *H. sapiens* (Stringer and Andrews, 1988).

HUMANS AS AN ADAPTIVE RADIATION

In some respects it seems inappropriate to discuss what is probably a single genus (*Homo*) of time-successive species as an adaptive radiation. Yet, from a primate perspective, the ecological diversity and geographic spread of fossil and living humans are their most striking feature. In nonhuman primates and almost all other organisms, adaptive diversity associated with the exploitation of diverse environments and geographic dispersion are usually accompanied by morphological diversity and ultimately speciation. Living, and presumably fossil, humans have evolved minor differences such as skin color or various blood polymorphisms in association with different environments, and perhaps other persistent differences due to random genetic drift and geographic isolation. Since at least the early Pleistocene, however, humans have ranged throughout Africa, Asia, and Europe with relatively little morphological differentiation in different environments (cf. Tattersall, 1986). Like macaques, humans are an example of what one scientist has called "specialized generalists" (Rose, 1983).

The evolutionary specializations that have permitted humans to exploit such a wide range of environments in many different ways are those that characterize the genus—the large brain that facilitates learning abilities and memory, our uniquely proportioned hand with its very mobile thumb, and our uniquely shaped vocal tract, which in conjunction with our brain permits a wide range of linguistic communication. Together, these features facilitated the technological abilities that more directly enabled humans to exploit such diverse habitats with a uniform body.

BIBLIOGRAPHY

GENERAL

Coppens, Y., Howell, F.C., Isaac, G.Ll., and Leakey, R.E.F. (1976). *Earliest Man and Environments in the Lake Rudolph Basin: Stratigraphy, Paleoecology and Evolution*. Chicago: University of Chicago Press.

Day, M.H. (1986). *Guide to Fossil Man*. Chicago: University of Chicago Press.

Delson, E. (1985). *Ancestors: The Hard Evidence*. New York: Alan R. Liss.

Grine, F.E. (1988). *Evolutionary History of the Robust Australopithecines*. Hawthorne, N.Y.: Aldine.

Isaac, G.Ll., and McCown, E.R. (1976). *Human Origins: Louis Leakey and the East African Evidence*. Menlo Park, Ca.: Benjamin-Cummings.

Jolly, C.J., (1978). *Early Hominids of Africa*. New York: St. Martin's.

Sigmon, B.A., and Cybulski, J.S. (1981). *Homo erectus: Papers in Honor of Davidson Black*. Toronto: University of Toronto Press.

Smith, F.H., and Spencer, F. (1984). *The Origins of Modern Humans: A World Survey of the Fossil Evidence*. New York: Alan R. Liss.

Tobias, P.V. (1985). *Hominid Evolution: Past, Present and Future*. New York: Alan R. Liss.

Toth, N., and Schick, K.D. (1986). The first million years: The archeology of protohuman culture. In *Advances in Archaeological Method and Theory*, vol. 9, ed. M.B. Schiffer, pp. 1–96. Orlando, Fla.: Academic Press.

GENUS *AUSTRALOPITHECUS*

Asfaw, B. (1987). The Belohdelie frontal: New evidence of early hominid cranial morphology from the Afar of Ethiopia. *J. Hum. Evol.* **16** (6/7).

Brace, C.L. (1979). Biological parameters and Pleistocene hominid lifeways. In *Primate Ecology and Human Origins: Ecological Influences on Social Organization*, ed. I.S. Bernstein and E.O. Smith, pp. 263–289. New York: Garland STPM Press.

Brain, C.K. (1981a). *The Hunters or the Hunted? An Introduction to African Cave Taphonomy*. Chicago: University of Chicago Press.

———. (1981b). Hominid evolution and climatic changes. *S. Afr. J. Sci.* **77**:104–105.

Bromage, T.G., and Dean, M.C. (1985). Re-evaluation of the age at death of immature fossil hominids. *Nature (London)* **317**:525–527.

Clark, J.D., Asfaw, B., Assefa, G., Harris, J.W.K., Kurashina, H., Walter, R.G., White, T.D., and Williams, M.A.J. (1984). Paleoanthropological discoveries in the Middle Awash Valley, Ethiopia. *Nature (London)* **307**:423–428.

Conroy, G.C., and Vannier, M.W. (1987). Dental development of the Taung skull from computerized tomography. *Nature (London)* **329**:625–627.

Dart, R.A. (1925). *Australopithecus africanus:* The manape of South Africa. *Nature (London)* **115**:195–199.

Darwin, C. (1871). *The Descent of Man and Selection in Relation to Sex*. London: Murray.

Day, M.H. (1986). Bipedalism: Pressures, origins and modes. In *Major Topics in Primate and Human Evolution*, ed. B.A. Wood, L. Martin, and P. Andrews, pp. 188–202. Cambridge: Cambridge University Press.

Falk, D. (1987). Hominid paleoneurology. *Annu. Rev. Anthropol.* **16**:13–30.

Grine, F.E. (1981). Trophic differences between 'gracile' and 'robust' Australopithecines: A scanning electron microscope analysis of occlusal events. *S. Afr. J. Sci.* **77**:203–230.

———. (1986). Dental evidence for dietary differences in *Australopithecus* and *Paranthropus:* A quantitative analysis of permanent molar microwear. *J. Hum. Evol.* **15**:783–822.

———. (1988). *Evolutionary History of the Robust Australopithecines*. Hawthorne, N.Y.: Aldine.

Harris, J.W.K. (1983). Cultural beginnings: Plio-Pleistocene archaeological occurrences from the Afar, Ethiopia. *Afr. Archaeol. Rev.* **1**:3–31.

Hill, A. (1985). Early hominid from Baringo District, Kenya. *Nature (London)* **315**:222–224.

Holloway, H.L. (1983). Human brain evolution: A search for units, models and synthesis. *Can. J. Anthropol.* **3**:215.

Howell, F.C. (1978). Hominidae. In *Evolution of African Mammals*, ed. V.J. Maglio and H.B.S. Cooke, pp. 154–248. Cambridge, Mass.: Harvard University Press.

Hrdy, S.B., and Bennett, W. (1981). Lucy's husband: What did he stand for? *Harvard Magazine*, (July–August) p. 7.

Isaac, G. (1978a). Food sharing and human evolution. *J. Anthropol. Res.* **34**:311–325.

———. (1978b). The food sharing behavior of protohuman hominids. *Sci. Am.* **238**:90–108.

———. (1983). Aspects of human evolution. In *Evolution from Molecules to Men*, ed. D.S. Bendall, pp. 509–543. Cambridge: Cambridge University Press.

———. (1984). The archaeology of human origins: Studies of the Lower Pleistocene in East Africa 1971–1981. In *Advances in World Archaeology*, vol. 3, ed. F. Wendorf and A. Close, pp.1–87. Orlando, Fla.: Academic Press.

Jolly, C.F. (1970). The seed-eaters: A new model of hominid differentiation based on a baboon analogy. *Man* **5**:5–28.

Johanson, D.C., and Edey, M. (1981). *Lucy: The Beginnings of Humankind*. New York: Simon and Schuster.

Johanson, D.C., and White, T.D. (1979). A systematic assessment of early African hominids. *Science* **203**:321–330.

Johanson, D.C., White, T.D., and Coppens, Y. (1978). A new species of the genus *Australopithecus* (Primates: Hominidae) from the Pliocene of eastern Africa. *Kirtlandia*, no. 28, pp. 1–14.

Jungers, W.L. (1982). Lucy's limbs: Skeletal allometry and locomotion in *Australopithecus afarensis*. *Nature (London)* **297**:676–678.

———. (1988). New estimates of body size in australopithecines. In *Evolutionary History of the Robust Australopithecines*, ed. F.E. Grine. Hawthorne, N.Y.: Aldine.

Kay, R.F. (1981). The nut-crackers: A new theory of the adaptations of the Ramapithecinae. *Am. J. Phys. Anthropol.* **55**:141–151.

———. (1985). Dental evidence for the diet of *Australopithecus*. *Ann. Rev. Anthropol.* **14**:315–342.

Kimbel, W.H., White, T.D., and Johanson, D.C. (1984). Cranial morphology of *Australopithecus afarensis:* A comparative study based on a composite reconstruction of the adult skull. *Am. J. Phys. Anthropol.* **64**:337–388.

Leakey, M.D., and Hay, R.L. (1979). Pliocene footprints in the Laetoli beds at Laetoli, northern Tanzania. *Nature (London)* **278**:317.

Lovejoy, C.O. (1981). The origin of man. *Science* **211**:341–350.

Lovejoy, C.O., Johanson, D.C., and Coppens, Y. (1982). Hominid lower limb bones recovered from the Hadar Formation: 1974–1977 collections. *Am. J. Phys. Anthropol.* **57**:679–700.

Mann, A. (1975). *Paleodemographic Aspects of the South African Australopithecines*. Philadelphia: University of Pennsylvania Press.

McHenry, H.M. (1973). Humerus of robust *Australopithecus*. *Science* **182**:396.

———. (1975). Fossils and the mosaic nature of human evolution. *Science* **190**:425–431.

———. (1986). The first bipeds: A comparison of the *A. afarensis* and *A. africanus* postcranium and implications for the evolution of bipedalism. *J. Hum. Evol.* **15**:177–191.

———. (1988). New estimates of body weight in early hominids and their significance to encephalization and megadentia in robust australopithecines. In *Evolutionary History of the Robust Australopithecines*, ed. F.E. Grine. Hawthorne, N.Y.: Aldine.

Olson, T. (1981). Basicranial morphology of the extant hominoids and Pliocene hominids: The new material from the Hadar Formation and its significance in early human evolution and taxonomy. In *Aspects of Human Evolution*, ed. C.B. Stringer, pp. 99–128. London: Taylor and Francis.

———. (1985). Cranial morphology and systematics of the Hadar Formation hominid and *"Australopithecus" africanus*. In *Ancestors: The Hard Evidence*, ed. E. Delson, pp. 102–119. New York: Alan R. Liss.

Oxnard, C.E. (1975). *Uniqueness and Diversity in Human Evolution*. Chicago: University of Chicago Press.

Pilbeam, D.R. (1980). Major trends in human evolution. In *Current Arguments on Early Man*, ed. L.K. Konigsson, pp. 261–285. Oxford: Pergamon.

Rodman, P.S., and McHenry, H.M. (1980). Bioenergetics and the origin of hominid bipedalism. *Am. J. Phys. Anthropol.* **52**: 103–106.

Rose, M.D. (1976). Bipedal behavior of olive baboons (*Papio anubis*) and its relevance to an understanding of the evolution of human bipedalism. *Am. J. Phys. Anthropol.* **44**:247–261.

Senut, B. (1981). *L'Humerus et ses articulations chez les Hominides Plio-Pleistocene*. Paris: CNRS.

Skelton, R.R., McHenry, H.M., and Drawhorn, G.M. (1986). Phylogenetic analysis of early hominids. *Curr. Anthropol.* **27**:21–43.

Smith, B.H. (1986). Dental development in *Australopithecus* and early *Homo*. *Nature (London)* **323**:327–330.

Stern, J.T., Jr., and Susman, R.L. (1983). The locomotor anatomy of *Australopithecus afarensis*. *Am. J. Phys. Anthropol.* **60**:279–317.

Susman, R.L. (1983). Evolution of the human foot: Evidence from Plio-Pleistocene hominids. *Foot Ankle* **3**:365–376.

———. (1988a). Evidence for tool behavior in *Paranthropus robustus* from Member I, Swartkrans. *Science* **240**:781–784.

———. (1988b). New postcranial remains from Swartkrans and their bearing on the functional morphology and behavior of *Paranthropus robustus*. In *Evolutionary History of the Robust Australopithecines*, ed. F.E. Grine. Hawthorne, N.Y.: Aldine.

Susman, R.L., Stern, J.T., Jr., and Jungers, W.L. (1984). Arboreality and bipedality in the Hadar hominids. *Folia Primatol.* **43**:113–156.

Tobias, P.V. (1967). *Olduvai Gorge*, vol. 2: *The Cranium and Maxillary Dentition of Australopithecus (Zinjanthropus) boisei*. Cambridge: Cambridge University Press.

———. (1980). A survey and synthesis of the African hominids of the late Tertiary and early Quaternary periods. In *Current Arguments on Early Man*, ed. L.K. Konigsson, pp. 86–113. Oxford: Pergamon.

———. (1983). Hominid evolution in Africa. *Can. J. Anthropol.* **3**:163–190.

———. (1985). *Hominid Evolution*. New York: Alan R. Liss.

Vrba, E.S. (1975). Some evidence of chronology and paleoecology of Sterkfontein, Swartkraans and Kromdraai from the fossil Bovidae. *Nature (London)* **254**:301–304.

———. (1985). Ecological and adaptive changes associated with early hominid evolution. In *Ancestors: The Hard Evidence*, ed. E. Delson, pp. 63–71. New York: Alan R. Liss.

Walker, A., and Leakey, R.E.F. (1978). The hominids of East Turkana. *Sci. Am.* **239**(2):54–66.

Walker, A., Leakey, R.E., Harris, J.M., and Brown, F.H. (1986). 2.5 Myr *Australopithecus boisei* from west of Lake Turkana, Kenya. *Nature* **322**:517–522.

Washburn, S.L. (1963). Behavior and human evolution. In *Classification and Human Evolution*, ed. S.L. Washburn. Chicago: Aldine.

White, T.D., Johanson, D.C., and Kimbel, W.H. (1981). *Australopithecus africanus:* Its phyletic position reconsidered. *S. Afr. J. Sci.* **77**:445–470.

Wood, B.A., and Chamberlain, A.T. (1986). *Australopithecus:* Grade or clade? In *Major Topics in Primate and Human Evolution*, ed. B.A. Wood, L. Martin, and P. Andrews, pp. 220–248. Cambridge: Cambridge University Press.

Wrangham, R.W. (1980). Bipedal locomotion as a feeding adaptation in gelada baboons and its implications for hominid evolution. *J. Hum. Evol.* **9**:329–332.

Wolpoff, M. (1980). *Paleoanthropology*. New York: Knopf.

Zihlman, A., and Tanner, N.M. (1978). Gathering and the hominid adaptation. In *Female Hierarchies*, ed. L. Tiger and H. Fowler, pp. 163–194. Chicago: Beresford Book Service.

GENUS *HOMO*

Andrews, P., and Franzen, J.L., eds. (1984). *The Early Evolution of Man with Special Emphasis on Southeast Asia and Africa. Cour. Forschr. Inst. Senckberg* **69**.

Binford, L. (1981). *Bones: Ancient Men and Modern Myths*. Orlando, Fla.: Academic Press.

Bräuer, G. (1984a). A craniological approach to the origin of anatomically modern *Homo sapiens* in Africa and implications for the appearance of modern Europeans. In *The Origins of Modern Humans: A World Survey of the Fossil Evidence*, ed. F.H. Smith and

F. Spencer, pp. 327–410. New York: Alan R. Liss.

———. (1984b). The "Afro-European sapiens hypothesis," and hominid evolution in East Asia during the late Middle and Upper Pleistocene. In *The Early Evolution of Man with Special Emphasis on Southeast Asia and Africa*, ed. P. Andrews and J.L. Franzen, *Cour. Forschr. Inst. Senckberg* **69**:145–165.

Brown, F., Harris, J.R., Leakey, R.E.F., and Walker, A. (1985). Early *Homo erectus* skeleton from west Lake Turkana, Kenya. *Nature* **316**:788–792.

Cann, R.L., Stoneking, M., and Wilson, A.C. (1987). Mitochondrial DNA and human evolution. *Nature (London)* **325**:31–36.

Delson, E. (1985). Late Pleistocene human fossils and evolutionary relationships. In *Ancestors: The Hard Evidence*, ed. E. Delson, pp. 296–300. New York: Alan R. Liss.

Eldredge, N., and Tattersall, I. (1975). Evolutionary models, phylogenetic reconstruction and another look at hominid phylogeny. In *Approaches to Primate Paleobiology*, ed. F.S. Szalay, pp. 218–242. Basel: Karger.

Harris, J.W.K. (1983). Cultural beginnings: Plio-Pleistocene archaeological occurrences from the Afar, Ethiopia. *Afr. Archaeol. Rev.* **1**:3–31.

Howells, W.W. (1976). Explaining modern man: Evolutionists versus migrationists. *J. Hum. Evol.* **5**:577–596.

———. (1980). *Homo erectus*—who, when and where: A survey. *Yrbk. Phys. Anthropol.* **23**:1–23.

Isaac, G. (1983). Aspects of human evolution. In *Evolution from Molecules to Men*, ed. D.S. Bendall, pp. 509–543. Cambridge: Cambridge University Press.

Johanson, D.C., Masao, T.T., Eck, G.G., White, T.D., Walter, R.C., Kimbel, W.H., Asfaw, B., Manega, P., Ndessokia, P., and Suwa, G. (1987). New partial skeleton of *Homo habilis* from Olduvai Gorge, Tanzania. *Nature* **327**:205–209.

Leakey, R.E.F., and Walker, A. (1976). *Australopithecus, Homo erectus*, and the single species hypothesis. *Nature* **261**:572–574.

———. (1985). *Homo erectus* unearthed. *Nat. Geographic* **168**(5):624–629.

Potts, R., and Shipman, P. (1981). Cutmarks made by stone tools on bones from Olduvai Gorge, Tanzania. *Nature* **291**:577–580.

Rak, Y. (1986). The Neanderthal: A new look at an old face. *J. Hum. Evol.* **15**:151–164.

Rightmire, G.P. (1981). Patterns in the evolution of *Homo erectus*. *Paleobiology* **7**:241–246.

———. (1985). The tempo of change in the evolution of mid-Pleistocene *Homo*. In *Ancestors: The Hard Evidence*, ed. E. Delson, pp. 255–264. New York: Alan R. Liss.

———. (1986). Body size and encephalization in *Homo erectus*. *Anthropos* **23**:139–150.

Rose, M.D. (1983). Miocene hominoid postcranial morphology: Monkey-like, ape-like, neither, or both? In *New Interpretations of Ape and Human Ancestry*, ed. R.L. Ciochon and R. Corrucini, pp. 405–420. New York: Plenum Press.

Shipman, P. (1986). Scavenging or hunting in early hominids: Theoretical framework and tests. *Am. Anthropol.* **88**:27–43.

Sigmon, B.A., and Cybulski, J.S. (1981). *Homo erectus: Papers in Honor of Davidson Black*. Toronto: University of Toronto Press.

Smith, F.H., and Spencer, F.I., eds. (1985). *The Origins of Modern Humans*. New York: Alan R. Liss.

Stringer, C.B. (1985). Middle Pleistocene hominid variability and the origin of late Pleistocene humans. In *Ancestors: The Hard Evidence*, ed. E. Delson, pp. 289–295. New York: Alan R. Liss.

———. (1986). The credibility of *Homo habilis*. In *Major Topics in Primate and Human Evolution*, ed. B. Wood, L. Martin, and P. Andrews, pp. 266–294. Cambridge: Cambridge University Press.

Stringer, C.B., and Andrews, P. (1988). Genetic and fossil evidence for the origin of modern humans. *Science* **239**:1263–1268.

Susman, R.L. (1983). Evolution of the human foot: Evidence from Plio-Pleistocene hominids. *Foot and Ankle* **3**(6):365–376.

Susman, R.L., and Creel, N. (1979). Functional and morphological affinities of the subadult hand (O.H. 7) from Olduvai Gorge. *Am. J. Phys. Anthropol.* **51**:311–332.

Susman, R.L., and Stern, J.T. (1982). Functional morphology of *Homo habilis*. *Science* **217**:931–934.

Tattersall, I. (1986). Species recognition in human paleontology. *J. Hum. Evol.* **69**:131–143.

Tattersall, I., and Eldredge, N. (1977). Fact, theory and fantasy in human paleontology. *Am. Sci.* **65**:204–211.

Thorne, A.G., and Wolpoff, M.H. (1981). Regional continuity in Australasian Pleistocene hominid evolution. *Am. J. Phys. Anthropol.* **55**:337–349.

Toth, N. (1987). The first technology. *Sci. Am.* **256**:112–121.

Toth, N., and Schick, K.D. (1986). The first million years: The archeology of protohuman culture. In *Advances in Archaeological Method and Theory*, vol. 9, ed. M.B. Schiffer, pp. 1–96. Orlando, Fla.: Academic Press.

Trinkaus, E. (1986). The Neandertals and modern human origins. *Annu. Rev. Anthropol.* **15**:193–218.

Trinkaus, E., and Howells, W.W. (1979). The Neanderthals. *Sci. Am.* **241**:118–133.

Valladas, H., Reyss, J.L., Joron, J.L., Valladas, G., Bar-Yosef, D., and Vandermeersch, B. (1988). Thermoluminescence dating of Mousterian "Proto-Cro-Magnon" remains from Israel and the origin of modern man. *Nature (London)* **331**:614–616.

Weidenreich, F. (1946). *Apes, Giants and Man.* Chicago: University of Chicago Press.

Wolpoff, M.H. (1971). Competitive exclusion among Lower Pleistocene hominids: The single species hypothesis. *Man* **6**:601–614.

———. (1980). *Paleoanthropology.* New York: Knopf.

———. (1984). Evolution in *Homo erectus:* The question of status. *Paleobiology* **10**(4):389–406.

Wood, B. (1985). Early *Homo* in Kenya, and its systematic relationships. In *Ancestors: The Hard Evidence*, ed. E. Delson, pp. 206–214. New York: Alan R. Liss.

Wu, R., and Olsen, J.W., eds. (1985). *Palaeoanthropology and Palaeolithic Archaeology in the People's Republic of China.* New York: Academic Press.

Patterns in Primate Evolution

PRIMATES AND EVOLUTIONARY THEORY

Throughout the preceding chapters we have examined primates—their phyletic relationships and ecological and behavioral adaptations—in more or less chronological order, family by family. Now that we have outlined and described the primate radiations, we are in a position to look for general trends. How can we characterize primate evolution as a whole? Are there repetitive patterns in the evolution of this order? With a good account of primate history and phylogeny at hand,

we can also begin to examine theoretical questions about evolutionary processes. How do the various theories of evolutionary mechanisms, the theories of speciation and of species extinction, fit the primate evidence? The fossil record of primates is probably as complete as that of any group of mammals, and it has certainly been more thoroughly studied. It is, then, particularly appropriate for such investigations.

Primate Adaptive Radiations

One of the most striking features of the primate fossil record is the extraordinary diversity of extinct forms, not just isolated species and genera but major radiations of families. There are approximately fifty genera and two hundred species of living primates. Nearly twice as many fossil species have been discovered and described, and many more remain to be uncovered. The vast majority of primate taxa that have ever lived are now extinct.

Unfortunately, comparing the numbers of living species with the total number of extinct species cannot give us a good indication about how the present diversity of primates compares with that in the past. Living primates are from a single slice in

time over a broad geographic area, while our knowledge of the fossil record is derived from samples of very restricted geographic areas and relatively long periods of time. More significant, there is virtually no paleontological record from the areas in which primate diversity is greatest today—the Amazon Basin, the Zaire Basin, and Southeast Asia.

As an alternative to comparing the taxonomic diversity of primates in the past with that of today, we can compare the morphological and reconstructed ecological diversity of selected primate faunas from the past with what we find among primate faunas today. In this way we can at least speculate about how the types of adaptations exploited by the primate faunas of the past were similar to or different from those characterizing the living radiations.

Body Size Changes

As we discussed in Chapter 8, body size is closely correlated with many aspects of a species' ecology, including diet and locomotion, and is also an easy parameter by which to compare species. Comparing the size distributions of living and fossil primate faunas gives us some indication of the adaptive diversity of the groups and the extent to which they seem to occupy similar ecological niches in a very broad sense. Figure 16.1 compares the range of body sizes occupied by various extinct primate groups with that found among living primates.

Several patterns are evident in the comparisons. First, there has been considerable diversity in size among primates throughout

FIGURE 16.1

Body size distribution of prosimians and Old World anthropoids through time (redrawn and modified from Covert, 1986; Fleagle and Kay, 1985).

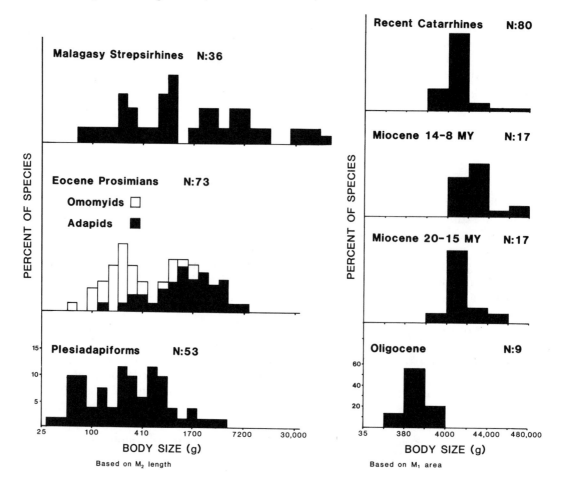

their evolution. This size diversity is certainly associated with a considerable ecological diversity, as we discussed in earlier chapters.

Like many other groups of mammals, primates seem to have increased in size during the past 65 million years. This overall size increase for primates during their evolution is reflected in two different aspects of the distributions. First, very tiny primates were relatively common in the Paleocene and Eocene, but they are missing from most Oligocene to Recent faunas (Fleagle, 1978; Covert, 1986). Second, Miocene through Recent primates include very large species that are unknown from earlier periods (Fleagle and Kay, 1985). Both of these size changes suggest that the adaptive space occupied by primates has shifted through time and that primates of the past showed different ecological adaptations from those found among living primate species. In this regard, the size changes corroborate the indications from specific morphological features, such as teeth and limbs (discussed in the following section).

Dietary Diversity

Although there is evidence of ecological diversity (especially in diet) throughout primate evolution, the expression of adaptations among different groups has probably varied considerably. A frugivorous plesiadapid was probably not much like a frugivorous adapid in many of its nondental adaptations, just as a frugivorous gibbon is very different in many aspects of its biology and foraging strategy from a frugivorous leaf monkey (e.g., Chivers, 1980). The detailed differences in foraging strategy that have been documented for extant primates in earlier chapters of this book are, of course, far beyond the realistic scope of paleontological studies, but we can see some general trends in the adaptive diversity of primate radiations through time. On the basis of dental morphology, it seems most likely that the plesiadapiforms were predominantly insectivorous and frugivorous, probably with some gum specialists. There are few species that show indications of extensive adaptation for folivory. In contrast, the fossil prosimians from the Eocene of North America and Europe include many folivorous species (especially among the adapids), as well as others adapted for insectivorous (especially the omomyids) and frugivorous diets (see Covert, 1986). Of the dietary diversity of prosimians between the Oligocene and the present we can conclude little—the fossil record is too poor to permit reliable assessments for the past 40 million years.

We do have evidence of broad dietary changes among Old World anthropoids during the past 30 million years (Fig. 16.2). Oligocene higher primates, known only from the rich Fayum deposits of Egypt, show a primate fauna of predominantly frugivorous anthropoids. There are no species that show dental adaptations to folivory comparable to those of many modern leaf eaters. In the early part of the Miocene epoch there are a few species with dentitions suggesting folivory, but it seems that most of the species were frugivorous. (The contemporaneous early Miocene lorises and galagos were presumably frugivores, insectivores, and gum eaters like their modern relatives.) By the later part of the Miocene there were more higher primates with teeth suggesting folivorous habits (e.g., *Mesopithecus*, *Microcolobus*, and *Oreopithecus*), and the proportions of frugivores to folivores was comparable to that among living Old World anthropoids.

Locomotor Diversity

Temporal changes in primate locomotor habits are very difficult to document in the paleontological record because of the rarity

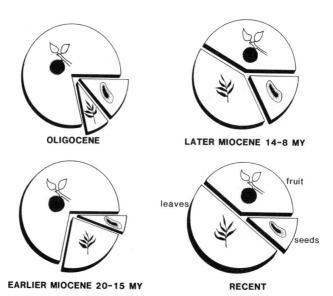

FIGURE 16.2

Changes in dietary diversity of Old World higher primates over the past 30 million years. Oligocene anthropoids were predominantly frugivorous, but in subsequent epochs folivorous and seed-eating species have become more common (modified from Fleagle and Kay, 1985).

FIGURE 16.3

Changes in substrate use of Old World higher primates during the past 30 million years. Oligocene anthropoids were arboreal, but in later epochs terrestrial species have become common (modified from Fleagle and Kay, 1985).

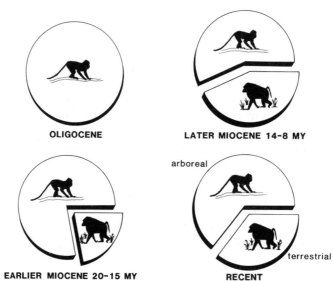

of fossil skeletons. Still, there seem to be some general patterns (Figs. 16.3, 16.4). The few skeletal remains of plesiadapiforms, particularly the ankle and the claws, suggest arboreal habits, but the lack of an opposable hallux and the presence of long curved claws indicate that their arboreal behavior was qualitatively different from that practiced by Eocene to Recent primates. They were probably not leapers. In contrast, the Eocene fossil prosimians are similar to extant prosimians in general skeletal anatomy. Most seem to have been arboreal quadrupeds and quadrupedal leapers (see Covert, 1986). Many of the diagnostic features of modern primates that first appear in the early Eocene seem to reflect an adaptation to arboreal leaping (Dagosto, 1988). On the other hand, there is no indication among the Eocene prosimians of the specialized vertical clinging behaviors that characterize living indriids or tarsiers. There is also no evidence of either terrestrial species or large species with extreme suspensory abilities, such as those found in *Palaeopropithecus*.

Old World higher primates show patterns of locomotor change over the past 30 million years. Oligocene higher primates were arboreal, platyrrhine-like species, and all of the skeletal remains indicate leaping and quadrupedal habits. There is no evidence of terrestrial species or of either suspensory species or specialized clingers. However, the first appearance of hominoids and cercopithecoids in the early Miocene is associated with evidence of more terrestrial and more suspensory species. The few remains from the last half of the Miocene indicate the presence of essentially modern locomotor adaptations in the fossil monkeys and apes. The most extreme locomotor specializations found among living higher primates seem

FIGURE 16.4

Changes in arboreal locomotor habits of Old World higher primates during the past 30 million years. Oligocene anthropoids were all arboreal quadrupeds and leapers. Since the early Miocene, there have been a variety of suspensory species (modified from Fleagle and Kay, 1985).

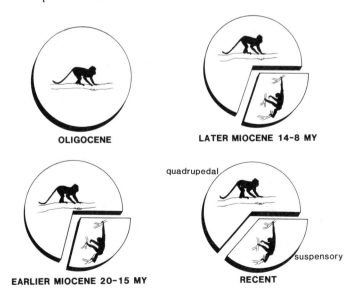

to have evolved even more recently. The skeletal specializations of gibbons associated with brachiation are unknown prior to the Pleistocene, and human bipedal locomotion seems to have evolved during the past 4 million years.

The patterns in body size, diet, and locomotion described in the previous paragraphs are interrelated. It is not surprising that the first appearance of relatively large higher primates in the early Miocene is associated with evidence of more folivorous and terrestrial habits, since folivory and terrestriality are functionally linked with relatively large size. In addition, many of the changes in size and adaptation that we see in the fossil record are clearly associated with the appearance or disappearance of particular taxonomic groups (Fig. 16.5). The major taxonomic groups of living and fossil primates often have characteristic adaptive features that permit them to exploit a unique array of resources. Thus, changes in the "primate" adaptive zone through time may be linked with the appearance or disappearance of particular groups of primates. For example, the disappearance of tiny primates in the fossil record reflects the extinction of the plesiadapiforms and the omomyids, and

FIGURE 16.5

Changes in the taxonomic abundance of different groups of Old World higher primates during the past 30 million years. In the Oligocene, parapithecids were the most common anthropoids and propliopithecids were slightly less common. The earlier Miocene was characterized by an abundance of fossil apes; the proconsulids were the most diverse group. In the later Miocene, cercopithecoids and pongids became more common. Recent Old World higher primate communities are dominated by cercopithecoid monkeys (modified from Fleagle and Kay, 1985).

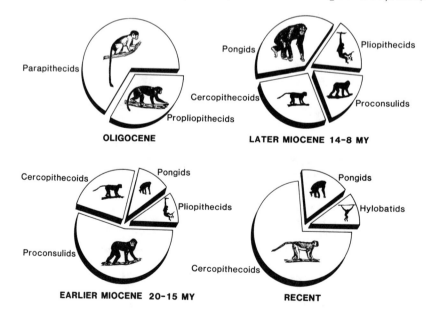

the increase in folivory among higher primates since the earlier Miocene is associated with the radiation of Old World monkeys.

Patterns in Primate Phylogeny

In the previous chapters, we considered the evolution of the major groups of both living and fossil primates one at a time, with particular consideration of their adaptive diversity. It is also interesting to compare the evolutionary history of these different primate groups during the past 65 million years. Theoretically, there are many different evolutionary patterns we might expect to find in primate evolution. One pattern

would be a whole series of distinct, long-lived lineages. Alternatively, we might find a series of evolutionary radiations succeeding one another in time, or maybe one slowly replacing another.

As should be no surprise, the record of primate evolution shows evidence of all of these possible evolutionary patterns in various groups at various times. At a gross level, the major pattern seems to be one of succeeding radiations—an initial radiation of plesiadapiforms in the Paleocene, followed by a radiation of prosimians at the beginning of the Eocene, and finally the radiation of anthropoids beginning in the Oligocene (Fig. 16.6). It is important to remember, however, that this is a summary of our knowledge of all primates rather than an

FIGURE 16.6

The major primate radiations of the Cenozoic era, showing the successive appearance of different radiations in the Paleocene, Eocene, and

Oligocene through Recent (modified from Gingerich, 1986).

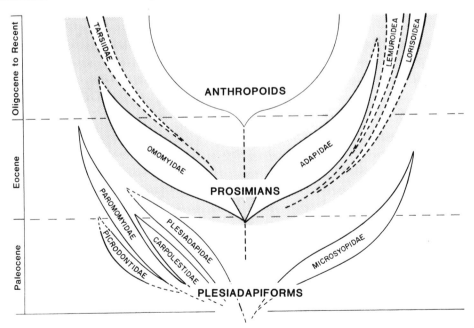

account of the events that took place in many different places simultaneously. Only from western Europe do we actually have a fossil record in which all three radiations succeed one another, and the anthropoids never seem to have been very diverse there. In other continental areas, one or more of these major radiations is absent from the fossil record, either because the animals were never there or because we have not uncovered the fossils of some particular period. Higher primates other than humans have never successfully invaded North America as far as we know; Asia has no clear record of plesiadapiforms; only higher primates are known from South America; and on Madagascar, prosimians were the only primates until the recent arrival of humans. Many of the global patterns that we see in the primate fossil record, then, reflect our available sample of fossil primates more than the actual timing or biogeography of the evolution of particular taxa.

If we examine evolutionary radiations of subfamilies and families of primates within restricted geographic areas, we can probably obtain a less distorted, but more parochial view of evolutionary changes in primate history. Again, we find a diversity of evolutionary patterns. Among the platyrrhine monkeys of South America we see evidence of many distinctive, relatively old lineages—none of which has ever been very diverse (see, e.g., Rosenberger, 1984). The history of higher primate evolution in the Old World seems to have been very different, with a succession of very different anthropoids in the Oligocene, parapithecids and propliopithecids; early Miocene, proconsulids; and late Miocene to Recent, cercopithecoid monkeys (Fig. 16.5).

An analysis at the generic level shows similar heterogeneity. There are a few primate genera that seem to have persisted for

tens of millions of years with very little change: *Tarsius* (Simons and Bown, 1985; Ginsburg and Mein, 1987), *Aotus* (Setoguchi and Rosenberger, 1987), and *Macaca* (Delson and Rosenberger, 1984). Other genera have undergone dramatic morphological changes in a relatively short time, the most notable being *Homo*.

Primate Evolution at the Species Level

One of the most hotly debated issues in evolutionary biology today is the same one that preoccupied Darwin—the origin of species. In Darwin's day the major issue was over the mechanism leading to evolutionary change and the appearance of new species. Darwin (1859) resolved this issue with his "discovery" and description of natural selection. Current debate is over the tempo of evolutionary change—whether new species appear by gradual modification of earlier types or through rapid changes in form (Eldredge and Gould, 1972; Smith, 1983).

There are several periods in primate evolution for which the fossil record is sufficiently well sampled over a long period of time that questions of this nature can be fruitfully examined. Two of the best examples come from western North America, where P.D. Gingerich, T.M. Bown, and K.D. Rose have carefully documented the evolutionary history of fossil mammals through a long, continuous series of late Paleocene and early Eocene sediments in northern Wyoming. Gingerich (e.g., Gingerich, 1976, 1979, 1985) has studied evolutionary change in two lineages of early primates—the late Paleocene plesiadapids and the early Eocene adapids (see Fig. 11.6). He found extensive evidence of gradual morphological change in both lineages through time, and most of the "new" species are clearly the result of the gradual modification of earlier forms. For the few instances in which a new species appears abruptly, it is impossible to determine whether this abrupt appearance is the result of rapid, discontinuous change from another local form, immigration from another area, or an absence of linking forms because of missing fossils. This is one of the major difficulties in testing theories of evolutionary change with fossil evidence. A rec-

ord of gradual change is positive evidence for gradual evolution, but a record of discontinuous change can be interpreted as the result of several very different phenomena.

More recently, Bown and Rose (1987) have produced extraordinarily detailed documentation of evolutionary change within lineages of early Eocene omomyid prosimians in northern Wyoming (Figs. 16.7, 16.8). Charting gradual change in many aspects of

FIGURE 16.7

Changes through time in the lower dentition of a lineage of fossil prosimians from the early Eocene of Wyoming. Note the loss of P_2 and the gradual

change in the size and shape of P_3 and P_4 (modified from Bown and Rose, 1987).

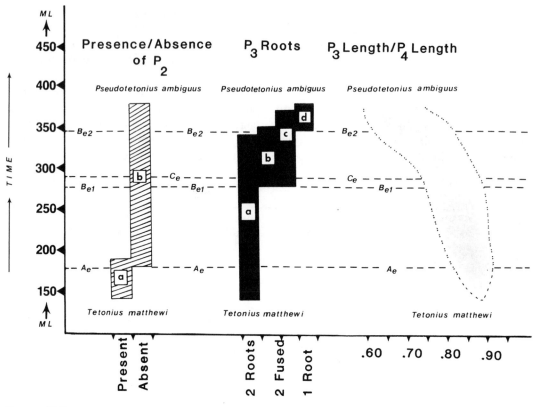

FIGURE 16.8

Changes in the size of P_3/P_4, the number of roots on P_3, and the presence or absence of P_2 in an evolving lineage of early Eocene omomyids. The morphological changes take place at different rates and at different times in the stratigraphic sequence. Because of the mosaic nature of mor-phological change in this lineage, identification of the boundary between the older species, *Tetonius matthewi*, and the younger species, *Pseudotetonius ambiguus*, must be based on an arbitrary choice of criteria (modified from Bown and Rose, 1987).

dental morphology, including reduction in size and loss of teeth, changes in the size and shape of cusps, and changes in the number and size of tooth roots, they show that the transition from one paleospecies to another within a lineage is characterized by changing frequencies of the diagnostic features within intermediate populations, but also that the morphological differences that characterize the end products rarely change at the same rate. Thus, they argue, identification of any species-specific morphology is most often only possible when one lacks the intermediate forms. For a very good fossil record, the identification of species boundaries becomes arbitrary, depending on which morphological feature is used. *Tetonius matthewi*, for example, gradually changes into *Pseudoteto-nius ambiguus* in the early Eocene deposits of northern Wyoming. However, the diagnostic features of the latter appear at different levels in the stratigraphic section and at different rates. Loss of P_2 occurs relatively low in the geological section, reduction of the roots of P_3 takes place in a series of steps, and changes in the size of P_3/P_4 take place gradually. Depending on which criterion or combination of criteria is used to define *Pseudotetonius ambiguus*, one can place the taxonomic transition at almost any time within the evolutionary lineage. There is no abrupt appearance of a new species.

Another period of primate evolution for which there has been extensive debate over the tempo of evolutionary change is the past 2 million years of hominid evolution. Rightmire (1981) argues that for over a million years during the first half of the Pleistocene human evolution was static. During this period, he claims, *Homo erectus* showed no significant morphological change—a conclusion that has been challenged by several workers (Cronin *et al.*, 1981; Levinton, 1982; Wolpoff, 1984). Cronin and his colleagues (1981) argue that the past 3 million years of hominid evolution show no evidence of morphological leaps and question the evidence of long periods of stasis. The distinct hominid species, they suggest, are all linked by intermediate specimens that bridge most of the morphological gaps between taxa. In their view, the morphological gaps in hominid phylogeny are not evidence for evolutionary jumps but rather reflect temporal and geographic gaps in the fossil record. Like the studies of early primates, the studies of hominid evolution indicate that there is more convincing evidence for gradual evolution and migrations than for a model of repeated abrupt evolutionary change.

Primate Extinctions

Over the past 65 million years many new primate species have appeared and slightly fewer have become extinct. In many cases, a species disappeared by evolving into an animal with a different morphology which we recognize as a new species; this phenomenon is called **pseudoextinction**. In other cases, species and lineages disappeared leaving no descendants. Is it possible to identify any common factors that have caused whole groups of primates to become extinct? Three primary reasons are commonly given to account for the major extinctions in the primate fossil record: climatic changes, competition from other primates or other mammals, and predation.

Climatic Changes

Living primates are for the most part tropical animals, with only a few hardy genera and species found in temperate climates. Most primates are arboreal animals; only baboons and humans have successfully abandoned forested areas for a life in open savannahs. Thus it is not surprising that climatic change has frequently been put forth to explain the disappearance of a primate group (Fig. 16.9). Gingerich (1986) in particular argues that climate has played a major role in the extinction of many primate groups from northern continents, including the adapids and omomyids of Europe at the end of the Eocene and the European hominoids at the end of the Miocene. Increased aridity has been suggested as a contributor to the disappearance of the proconsulids in the middle Miocene of Africa (Pickford, 1983). A more unusual case is the extinction of most archaic primates in the early Eocene of North America and Europe. Since the plesiadapiforms thrived during the temperate climates of the Paleocene, it has been suggested that the more tropical Eocene climates were too warm for them and led to their extinction.

Competition

Since climatic changes in the fossil record are invariably associated with the appear-

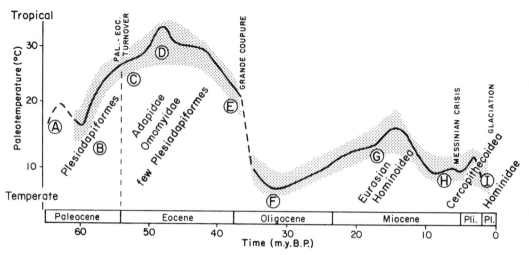

FIGURE 16.9

Temperatures during the Cenozoic, with major events in the primate fossil record of the Northern Hemisphere marked: A, appearance of *Purgatorius*; B, radiation of plesiadapiforms; C, appearance of adapids and omomyids; D, height of primate diversity in Europe and North America; E, decline and extinction of primates in Europe and North America; F, nadir of primate diversity in northern continents; G, appearance of hominoids in Europe and Asia; H, disappearance of northern hominoids, extensive ecological radiation of cercopithecoids, and emergence of hominids; I, evolution and dispersal of humans (from Gingerich, 1986).

ance of new groups of primates and other mammals, we must attempt to distinguish the effects of climate per se from the effects of ecological competition with other species. The extinction of the plesiadapiforms, for example, coincides roughly with the radiation of both rodents and early prosimians (Fig. 16.10), and the decline of proconsulids in the middle and late Miocene of Africa is associated with an increased abundance of cercopithecoid monkeys (see Fig. 14.10). It is always extremely hard to determine whether a new group contributes to the extinction of earlier groups through direct competition for resources or merely fills the gap left by their extinction, and careful analysis is required to sort out the alternatives (see Maas *et al.*, 1988). In the case of the extinction of

plesiadapiforms, it appears that competition with rodents may well have contributed to the demise of the group, whereas the radiation of the prosimians occurred after the decline or disappearance of most plesiadapiforms and so seems a less likely cause of extinction.

Predation

A cause of primate extinctions that seems well documented, in at least one case, is human predation. The disappearance of the large diurnal lemurs from the fauna of Madagascar clearly postdates the appearance of humans on the island and is almost certainly the result of human hunting and habitat destruction (Dewar, 1984). The

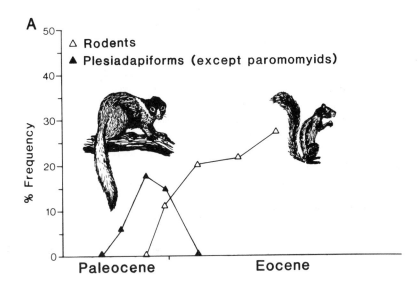

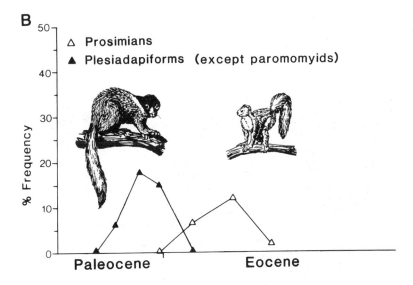

FIGURE 16.10

Abundance of plesiadapiforms in the fossil record from the Paleocene and Eocene of North America compared with the abundance of early rodents (A) and fossil prosimians (B). Note that the decline of plesiadapiform abundance corresponds to the initial appearance and dramatic explosion of fossil rodents in this period, suggesting that the appearance of rodents was primarily responsible for the extinction of plesiadapiforms. In contrast, the radiation of fossil prosimians occurred primarily after the major decline in plesiadapiforms, suggesting that the prosimians may have exploited some aspect of the ecological space occupied by plesiadapiforms without being primarily responsible for the extinction of the plesiadapiforms. Percent frequency based on minimum number of individuals (MNI) (modified from Maas *et al.*, 1988).

lemur extinction on Madagascar was a relatively recent event, dating to the last 5,000 years, but it is not unlikely that human predation was also responsible for the extinction of most of the large monkeys from East Africa during the early and middle Pleistocene and for the disappearance of the orangutan from mainland Asia (Ciochon, 1988).

Limiting Primate Extinctions

We often have difficulty reconstructing the events that led to the extinction of particular species, genera, and lineages of primates from earlier periods. In contrast, the factors that will probably lead to the extinction of many primate species alive today are relatively easy to identify: habitat destruction, hunting, and live capture for the pet or research markets (Mittermeier and Cheney, 1986). Of these, habitat destruction due to our expanding human population is the greatest threat. More than 90 percent of all primate species live in the tropical forests of Asia, Africa, and South and Central America, and their fate is intimately linked with

FIGURE 16.11

Eight of the primates closest to extinction today. Left to right: top row, the indris from Madagascar, the mountain gorilla from central Africa, and the aye-aye from Madagascar; middle row, the muriqui from Brazil, and the lion-tail macaque from India; bottom row, the golden monkey from China, the golden lion tamarin from Brazil, and the pygmy chimpanzee from Zaire (courtesy of Stephen Nash, World Wildlife Fund-U.S.).

the future of these forests. There are endangered primates in virtually every part of the tropical world, but the greatest threats are in Madagascar and eastern Brazil, where faunas of unique primates with limited geographic ranges are under extreme pressure from rapidly growing populations that need the same resources of forests and land (Fig. 16.11).

With an expanding global human population and the developmental demands of many parts of the tropical world, it is inevitable that many primate populations and species will disappear in the coming decades. The goal of conservationists is to limit the losses as much as possible. This involves "1) protecting areas for particularly endangered and vulnerable species; 2) creating large national parks and reserves in areas of high primate diversity or abundance; 3) maintaining parks and reserves that already exist and enforcing protective legislature in them; 4) creating public awareness of the need for primate conservation and the importance of primates as both a national heritage and a resource; ... 5) determining ways in which people and other primates can coexist in multiple-use areas" (Mittermeier and Cheney, 1986, p. 488). Only such a broad approach will ensure that future generations are able to appreciate the living remnants of 65 million years of primate adaptation and evolution.

BIBLIOGRAPHY

PRIMATE ADAPTIVE RADIATIONS

Chivers, D.J., ed. (1980). *Malayan Forest Primates: Ten Years' Study in the Tropical Forest*. New York: Plenum Press.

Covert, H.H. (1986). Biology of early Cenozoic primates. In *Comparative Primate Biology*, Vol. 1: *Systematics, Evolution, and Anatomy*, ed. D.W. Swindler and J. Erwin, pp. 335–359. New York: Alan R. Liss.

Dagosto, M. (1988). Implications of postcranial evidence for the origin of euprimates. *J. Hum. Evol.* **17**(1/2).

Fleagle, J.G. (1978). Size distributions of living and fossil primate faunas. *Paleobiology* **4**:67–76.

Fleagle, J.G., and Kay, R.F. (1985). The paleobiology of catarrhines. In *Ancestors: The Hard Evidence*, ed. E. Delson, pp. 23–36. New York: Alan R. Liss.

PATTERNS IN PRIMATE PHYLOGENY

Delson, E., and Rosenberger, A.L. (1984). Are there any anthropoid primate living fossils? In *Living Fossils*, ed. N. Eldredge and S.M. Stanley, pp. 50–61. New York: Springer-Verlag.

Gingerich, P.D. (1984). Primate evolution. In *Mammals: Notes for a Short Course*, ed. T.W. Broadhead, pp. 167–181. Knoxville: University of Tennessee, Dept. of Geological Sciences.

Ginsburg, L., and Mein, P. (1987). *Tarsius thailandica nov. sp.*, premier Tarsiidae (Primates, Mammalia) fossile d'Asie. *C. R. Acad. Sc. (Paris), t. 304, Serie II*, no. 19, pp. 1213–1215.

Rosenberger, A.L. (1984). Fossil New World monkeys dispute the molecular clock. *J. Hum. Evol.* **13**:737–742.

Setoguchi, T., and Rosenberger, A.L. (1987). A fossil owl monkey from La Venta, Colombia. *Nature (London)* **326**:692–694.

Simons, E.L., and Bown, T.M. (1985). *Afrotarsius chatrathi*, first tarsiiform primate (?Tarsiidae) from Africa. *Nature (London)* **313**:475–477.

EVOLUTION OF PRIMATE SPECIES

Bown, T.M., and Rose, K.D. (1987). Patterns of dental evolution in early Eocene anaptomorphine primates (Omomyidae) from the Bighorn Basin, Wyoming. Paleontological Society Memoir no. 23 *J. Paleontol.* **61**:1–62.

Cronin, J.E., Boaz, N.T., Stringer, C.B., and Rak, Y. (1981). Tempo and mode in hominid evolution. *Nature (London)* **292**:113–122.

Darwin, C. (1859). *On the Origin of Species by Means of Natural Selection, or the Preservation of Favoured Races in the Struggle for Life*. London: John Murray.

Eldredge, N., and Gould, S.J. (1972). Punctuated equilibria: An alternative to phyletic gradualism. In *Models in Paleobiology*, ed. T.J.M. Schopf, pp. 82–115. San Francisco: Freeman, Cooper.

Gingerich, P.D. (1976). Cranial anatomy and evolution of early Tertiary Plesiadapidae (Mammalia, Primates). *Contr. Mus. Paleontol., Univ. Michigan* **15**.

———. (1979). Paleontology, phylogeny, and classification: An example from the mammalian fossil record. *Systematic Zool.* **28**:451–464.

———. (1985). Species in the fossil record: Concepts, trends, and transitions. *Paleobiology* **11**:27–41.

Levinton, J.S. (1982). Estimating stasis: Can a null hypothesis be too null? *Paleobiology* **8**:307.

Rightmire, G.P. (1981). Patterns in the evolution of *Homo erectus. Paleobiology* **7**:241–246.

Rose, K.D., and Bown, T.M. (1986). Gradual evolution and species discrimination in the fossil record. In *Vertebrates, Phylogeny, and Philosophy*, ed. K.M. Flanagan and J.A. Lillegraven, pp. 119–130. *Contr. Geol., Univ. Wyoming, spec. paper no.* 3.

Smith, J. Maynard. (1983). Current controversies in evolutionary biology. In *Dimensions of Darwinism*, ed. M.J. Grene, pp. 273–286. Cambridge: Cambridge University Press.

Wolpoff, M.H. (1984). Evolution in *Homo erectus*: The question of stasis. *Paleobiology* **10**:389–406.

PRIMATE EXTINCTIONS

Ciochon, R.L. (1988). *Gigantopithecus*: The king of all apes. *Animal Kingdom* **91**:32–39.

Dewar, R.E. (1984). Extinctions in Madagascar: The loss of the subfossil fauna. In *Quaternary Extinctions: A Prehistoric Revolution*, ed. P.S. Martin and R.G. Klein, pp. 574–593. Tucson: University of Arizona Press.

Gingerich, P.D. (1986). *Plesiadapis* and the delineation of the order Primates. In *Major Topics in Primate and Human Evolution*, ed. B. Wood, L. Martin, and P. Andrews, pp. 32–46. Cambridge: Cambridge University Press.

Maas, M.C., Krause, D.W., and Strait, S.G. (1988). Decline and extinction of Plesiadapiformes (?Primates: Mammalia) in North America: Displacement or replacement? *Paleobiology* **14**(4).

Maas, M.C., Strait, S.G., and Krause, D.W. (1987). The decline and extinction of plesiadapiform primates in North America. *Am. J. Phys. Anthropol.* **72**:228.

Marsh, C.W., and Mittermeier, R.A., eds. (1987). *Primate Conservation in the Tropical Rain Forest. Monographs in Primatology*, vol. 9. New York: Alan R. Liss.

Mittermeier, R.A., and Cheney, D.L. (1986). Conservation of primates and their habitats. In *Primate Societies*, ed. B.B. Smuts, D.L. Cheney, R.M. Seyfarth, R.W. Wrangham, and T.T. Struhsaker, pp. 477–490. Chicago: University of Chicago Press.

Pickford, M. (1983). Sequence and environments of the lower and middle Miocene hominoids of western Kenya. In *New Interpretations of Ape and Human Ancestry*, ed. R.L. Ciochon and R.S. Corruccini, pp. 421–439. New York: Plenum Press.

Walker, A. (1967). Patterns of extinction among the subfossil Madagascan lemuroids. In *Pleistocene Extinctions: The Search for a Cause*, ed. P.S. Martin and H.E. Wright, Jr., pp. 425–432. New Haven and London: Yale University Press.

abduction movement of a limb or part of a limb away from the midline of the body.

absolute dating determination of the age, in years, of a fossil or fossil site, usually on the basis of the amount of change in radioactive elements in rocks.

adaptation process whereby an organism changes in order to survive in its given environment; or, a specific new characteristic that enables survival.

adaptive radiation a group of closely related organisms that have evolved morphological and behavioral features enabling them to exploit different ecological niches.

adduction movement of a limb or part of a limb toward the midline of the body.

age-graded group a social group with several adult females and several adult males who differ in social and reproductive status according to age. Thus, with time, an age-graded group can change from a one-male reproductive system to a multi-male system.

allometry the relationship between the size and shape of an organism, or, more broadly, the relationship between an organism's size and various aspects of its biology, such as morphology, ecology, or behavior; also, the study of such relationships.

alloparenting (aunting) assistance in care of infants and juveniles by individuals other than parents.

allopatry the absence of overlap in the geographical range of two species or populations.

arboreal living in trees.

arboreal quadrupedalism mode of locomotion in which the animal moves along horizontal branches with a regular gait pattern involving all four limbs.

articulation a joint between two or more bones.

basal metabolism the energy requirements of an animal at rest.

bilateral symmetry a type of developmental shape in which right and left sides of the organism are mirror images of one another.

bilophodonty a condition of the molar teeth in which the mesial and distal pairs of cusps form ridges or lophs.

biomass the sum of the weights of the organisms in a particular area.

bipedalism mode of locomotion using only the hindlimbs, usually alternately rather than together.

brachiation arboreal locomotion in which the animal progresses below branches by using only the forelimbs.

buccal the cheek side of a tooth.

bunodont (teeth) having low, rounded cusps.

canopy a layer of forest foliage that is laterally continuous and usually distinct vertically from other layers. A tropical forest often has one or more distinct canopies.

carnivore an animal that eats primarily the flesh of other animals; also often used to refer to the mammalian order Carnivora (which includes cats, dogs, skunks, raccoons, and bears).

cathemeral active intermittently throughout the twenty-four-hour day rather than active only during the day (diurnal) or only during the night (nocturnal).

clade a group composed of all the species descended from a single common ancestor; a monophyletic group.

cladogram branching tree diagram used to represent phyletic relationships.

conspecific belonging to the same species.

convergent evolution the independent evolution of similar morphological features from different ancestral conditions. The wings of bats and birds are an example of convergent evolution.

core area the part of a group's home range that is used most intensively.

cranial capacity the volume of the brain, usually determined by measuring the volume of the inside of the neurocranium.

crepuscular active primarily during the hours around dawn and dusk.

cryptic hidden; not normally visible.

day range the distance a group of animals travels during a single day.

deciduous dentition the milk teeth or first set of teeth in the mammalian jaw. The deciduous dentition is replaced by the permanent dentition.

dental eruption sequence the order in which the different teeth erupt or come into use.

dental formula a notation of the number of incisors, canines, premolars, and molars in the upper and lower dentition of a species. In humans, the adult dental formula is $\frac{2.1.2.3}{2.1.2.3}$.

derived feature a specialized morphological (or behavioral) characteristic that departs from the condition found in the ancestors of a species or group of species.

diagnostic distinguishing or characteristic, as the diagnostic features of a group of organisms.

diurnal active primarily during daylight hours.

dorsal toward the back side of the body; the opposite of ventral.

dizygotic twins (fraternal twins) twins that develop from two fertilized eggs or zygotes. This contrasts with monozygotic (or identical) twins, which develop from a single fertilized egg or zygote.

eclectic coming from many sources, as an eclectic diet.

ecological niche the complex of features (such as diet, forest type preference, canopy preference, activity pattern) that characterize the position a species occupies in the ecosystem.

ecology the study of the relationship between an organism and all aspects of its environment; or, all aspects of the environment of an organism which affect its way of life.

emergent trees the trees in a tropical forest which extend above the relatively continuous canopy.

endocast an impression of the inside of the cranium, often preserving features of the surface of the brain.

extant living, as opposed to extinct.

extension a movement in which the angle of a limb joint increases.

exudate a substance, such as gum, sap, or resin, which flows from the vascular system of a plant.

faunal correlation determination of the relative ages of different geological strata by comparing the fossils within the strata and assigning similar ages to strata with similar fossils; a method of relative dating.

faunivore an animal that eats primarily other animals (includes insectivores and carnivores).

fission-fusion society a type of social organization in which individuals regularly form small subgroups for foraging but from time to time also join together in larger groups; the variation in grouping usually depends on the type of food.

flexion a movement in which the angle of a limb joint decreases; the opposite of extension.

folivore an animal that feeds primarily on leaves.

foraging strategy the behavioral adaptations of a species related to its acquisition of food items.

frugivore an animal that feeds primarily on fruit.

gallery forest a forest along a river or stream.

genotype the genetic make-up of an organism.

gracile relatively slender or delicately built.

grade a level or stage of organization, or a group of organisms sharing a suite of features (either primitive or derived) that distinguishes them from more advanced or more primitive animals but does not necessarily define a clade.

gradistic classification a classification in which organisms are grouped according to grade or level of organization rather than according to ancestry or phylogeny.

graminivore an animal that eats primarily grains; often also used to describe an animal that eats seeds.

gregarious living in regular social groups; contrasted to solitary living.

grooming the cleaning of the body surface by licking, biting, picking with fingers or claws, or other kinds of manipulation.

growth allometry the relationship between size and shape during the growth (or ontogeny) of an organism.

holophyletic group a taxonomic group of organisms which has a single common ancestor and which includes all descendants of that ancestor.

home range the area of land that is regularly used by a group of animals for a year or longer.

homologous having the same developmental and evolutionary origin. The bones in the hands of primates and the wings of bats are homologous.

infanticide the killing of infants.

insectivore an animal that eats primarily insects (and other invertebrates); also used to refer to the mammalian order Insectivora (which includes shrews, moles, and hedgehogs).

insertion the attachment of a muscle or ligament farthest from the trunk or center of the body.

intermembral index a measure of the relative length of the forelimbs and hindlimbs of an animal: humerus plus radius length $\times$ 100 divided by femur plus tibia length.

interspecific allometry the relationship between size and shape among a range of different species; for example, a comparison between mouse and elephant.

ischial callosity a fatty sitting pad on the ischium of all Old World monkeys and gibbons.

Kay's threshold the body weight (approximately 500 grams) that is roughly the upper size limit of predominantly insectivorous primates and the lower size limit of predominantly folivorous primates.

knuckle-walking a type of quadrupedal walking, used by chimpanzees and gorillas, in which the upper body is supported by the dorsal surface of the middle phalanges of the hands.

kyphosis dorsally convex curvature of the back.

life history parameter a characteristic of the growth and development of an organism such as the length of gestation, timing of sexual maturity, length of reproductive period, or lifespan.

locomotion movement from one place to another.

lordosis ventrally convex curvature of the back.

mandible jawbone.

mandibular symphysis the joint between the

right and left halves of the mandible. In humans and other higher primates, this joint is fused.

monogamy   a social system based on mated pairs and their offspring.

monophyletic group   a taxonomic group of organisms which has a single common ancestor.

morphology   the shape of anatomical structures.

multi-male group   a group of animals in which several adult males and several adult females are reproductively active.

natural selection   a nonrandom differential preservation of genotypes from one generation to the next which leads to changes in the genetic structure of a population.

neoteny   the retention of the features of a juvenile animal of one species in the adult form of a different species.

neotropics   the tropical regions of North America, Central America, and South America.

nocturnal   active primarily during the night.

noyau   a type of social organization in which adult individuals have separate home ranges; ranges of individuals do not overlap with those of other individuals of the same sex, but they do overlap with ranges of individuals of the opposite sex.

olfaction   the sense of smell.

one-male group   a social group containing several reproductively active females but only one reproductively active male.

ontogeny   the development of an organism from conception to adulthood.

organ of Jacobson   an organ for chemical reception found in the anterior part of the roof of the mouth of many vertebrates.

paleomagnetism   study of the magnetism of rocks that were formed in earlier time periods.

More broadly, the study of changes in the earth's magnetic fields during geological time.

palmar   pertaining to the palm side of the hand.

parallel evolution   independent evolution of similar (and homologous) morphological features in separate lineages.

paraphyletic classification   a classification in which a taxonomic group contains some, but not all, of the members of a clade.

phyletic classification   a classification in which taxonomic groups correspond to monophyletic groups.

phyletic gradualism   a model of evolution in which change takes place slowly in small steps, in contrast to the punctuated equilibrium model.

phylogeny   the evolutionary or genealogical relationships among a group of organisms.

plantar   pertaining to the sole of the foot.

polyandry   a type of social organization in which there are two or more reproductively active males and a single reproductively active female.

polygyny   any type of social organization in which one male mates with more than one female.

prehensile   capable of grasping; for example, the prehensile tail of some platyrrhine monkeys.

primary rain forest   rain forest characterized by the later stages of the vegetational succession cycle.

primitive feature   a behavioral or morphological feature that is characteristic of a species and its ancestors.

procumbent   inclined forward, protruding, as in the procumbent incisors of some primates.

prognathism   prominence of the snout.

pronation   rotation of the forearm so that the palm faces dorsally or downward; the reverse movement from supination.

punctuated equilibrium a model of evolution in which change takes place primarily by abrupt genetic shifts, in contrast to phyletic gradualism.

quadrumanous four-handed; as in quadrumanous climbing, in which many suspensory primates use their feet in the same manner that they use their hands.

ramus the vertical part of the mandible, often called the ascending ramus.

relative dating a determination of whether a fossil or fossil site is younger or older than other fossils or sites, usually through study of the stratigraphic position or evolutionary relationships of the fauna; contrasts with absolute dating.

reproductive strategy an organism's complex of behavioral and physiological features concerned with reproduction. The reproductive strategy of oysters, for example, is characterized by the production of large numbers of offspring and no parental care, whereas the reproductive strategy of humans is often characterized by production of relatively few offspring and investment of large amounts of parental care by both parents.

reproductive success the contribution of an individual to the gene pool of the next generation.

sagittal crest a bony ridge on the top of the neurocranium formed by the attachment of the temporalis muscles.

saltation leaping; either a type of locomotion, or a description of rapid evolutionary change characterized by a lack of intermediate forms, i.e., "leaping" from one distinct species to another.

savannah a type of vegetation zone characterized by grasslands with scattered trees.

schizodactyly grasping between the second and third digits of the hand rather than between the pollex (thumb) and second digit.

secondary compounds poisons produced by plants which exist in leaves, flowers, etc., and deter animals from eating them.

secondary rain forest rain forest characterized by immature stages of the succession cycle, commonly found on the edges of forests, along rivers, and around tree falls.

sexual dichromatism the condition in which males and females of a species differ in color.

sexual dimorphism any condition in which males and females of a species differ in some aspect of their nonreproductive anatomy such as body size, canine tooth size, or snout length.

single-species hypothesis the theory that there has never been more than one hominid lineage at any time because all hominids are characterized by culture and thus all occupy the same ecological niche.

speciation appearance of new species.

subfossil recently extinct, often from historical time periods. Some prosimians from Madagascar, for example, have become extinct in the past thousand years.

supination rotation of the forearm such that the palmar surface faces anteriorly or upward; the reverse movement from pronation.

suspensory behavior locomotor and postural habits characterized by hanging or suspension of the body below or among branches rather than walking, running, or sitting on top of branches.

suture a joint between two bones in which the bones interdigitate and are separated by fibrous tissue. The joints between most of the bones of the skull are sutures.

sympatry overlap in the geographical range of two species or populations.

systematics the science of classifying organisms and the study of their genealogical relationships.

taphonomy study of the processes that affect the remains of organisms from the death of the organism through its fossilization.

taxonomy   the science of describing, naming, and classifying organisms.

terrestrial   on the ground.

terrestrial quadrupedalism   four-limbed locomotion on the ground.

territory   part of a home range that is exclusive to a group of animals and is actively defended from other groups of the same species.

tooth comb   a formation of the lower incisors into a comblike structure for grooming.

tympanic bone   the bone that forms the bony ring for the eardrum.

type specimen   a single designated individual of an organism which serves as the basis for the original name and description of the species.

understory   the part of a forest that lies below the canopy layers.

valgus   an angulation of the femur such that the knees are closer together than the hip joints; "knock-kneed."

ventral   toward the belly side of an animal; the opposite of dorsal.

vertical clinging and leaping   a type of locomotion and posture in which animals cling to vertical supports and move by leaping between these vertical supports.

woodland   a vegetation type characterized by discontinuous stands of relatively short trees separated by grassland.

Genera in boldface contain extant members.

SUBORDER PLESIADAPIFORMES

Family Microsyopidae
Palaechthon
Plesiolestes
Talpohenach
Torrejonia
Palenochtha
Berruvius
Navajovius
Micromomys
Tinimomys
Niptomomys
Uintasorex
Microsyops
Arctodontomys
Craseops
Alveojunctus

Family Plesiadapidae
Pronothodectes
Nannodectes
Plesiadapis
Chiromyoides
Platychoerops

Family Carpolestidae
Elphidotarsius
Carpodaptes
Carpolestes

Family Saxonellidae
Saxonella

Family Paromomyidae
Paromomys
Ignacius
Phenacolemur
Elwynella
Arcius

Family Picrodontidae
Picrodus
Zanycteris
Draconodus

Family *incertae sedis*
Purgatorius

SUBORDER PROSIMII

INFRAORDER LEMURIFORMES

Superfamily Lemuroidea
Family Lemuridae
Lemur
Varecia
Pachylemur
Hapalemur

Family Lepilemuridae
Subfamily Lepilemurinae
Lepilemur
Subfamily Megaladapinae
Megaladapis

Family Indriidae
Subfamily Indriinae
Avahi
Propithecus
Indri
Mesopropithecus
Subfamily Archaeolemurinae
Archaeolemur
Hadropithecus
Subfamily Palaeopropithecinae
Palaeopropithecus
Archaeoindris

Family Daubentoniidae
Daubentonia

471

Superfamily Lorisoidea
  Family Cheirogaleidae
    Microcebus
    Mirza
    Cheirogaleus
    Phaner
    Allocebus
  Family Galagidae
    Otolemur
    Galago
    Galagoides
    Euoticus
    Progalago
    Komba
  Family Lorisidae
    Perodicticus
    Arctocebus
    Nycticebus
    Loris
    Mioeuoticus
    Nycticeboides

INFRAORDER ADAPIFORMES

  Family Adapidae
    Subfamily Notharctinae
      Cantius
      Copelemur
      Notharctus
      Smilodectes
      Pelycodus
    Subfamily Adapinae
      Donrussellia
      Protoadapis
      Europolemur
      Periconodon
      Agerinia
      Caenopithecus
      Pronycticebus
      Cercamonius
      Cryptadapis
      Microadapis
      Anchomomys
      Adapis
      Leptadapis
      Mahgarita

    Subfamily Sivaladapinae
      Indraloris
      Sivaladapis
      Sinoadapis
    Subfamily *incertae sedis*
      Azibius
      Panobius
      Hoanghonius
      Lushius

INFRAORDER TARSIIFORMES

  Family Tarsiidae
    Tarsius
    Afrotarsius

INFRAORDER OMOMYIFORMES

  Family Omomyidae
    Subfamily Anaptomorphinae
      Teilhardina
      Anemorhysis
      Chlororhysis
      Pseudotetonius
      Absarokius
      Anaptomorphus
      Tetonius
      Trogolemur
      Aycrossia
      Strigorhysis
      Gazinius
      Steinius
      Loveina
    Subfamily Omomyinae
      Arapahovius
      Omomys
      Chumashius
      Ourayia
      Shoshonius
      Washakius
      Utahia
      Hemiacodon
      Dyseolemur
      Stockia
      Macrotarsius
      Uintanius
      Jemezius
      Rooneyia
      Ekgmowechashala

Subfamily Microchoerinae
  Nannopithex
  Pseudoloris
  Necrolemur
  Microchoerus
Subfamily *incertae sedis*
  Altanius
  Kohatius

SUBORDER ANTHROPOIDEA

INFRAORDER *INCERTAE SEDIS*

Amphipithecus
Pondaungia
Oligopithecus

INFRAORDER PARAPITHECOIDEA

Family Parapithecidae
  Qatrania
  Apidium
  Parapithecus

INFRAORDER PLATYRRHINI

Superfamily Ceboidea
  Family Cebidae
    Subfamily Cebinae
      Cebus
      Saimiri
      Neosaimiri
    Subfamily Aotinae
      Aotus
      Tremacebus
      Callicebus
      Homunculus
  Family Atelidae
    Subfamily Pitheciinae
      Pithecia
      Chiropotes
      Cacajao
      Mohanamico
      Cebupithecia
    Subfamily Atelinae
      Alouatta
      Stirtonia
      Lagothrix
      Brachyteles
      Ateles

Family Callitrichidae
  Subfamily Callitrichinae
    Callimico
    Saguinus
    Leontopithecus
    Callithrix
    Cebuella
    Micodon

Family *incertae sedis*
  Subfamily *incertae sedis*
    Branisella
    Dolichocebus
    Soriacebus
    Xenothrix

INFRAORDER CATARRHINI

Superfamily Hominoidea
  Family Propliopithecidae
    Propliopithecus
    Aegyptopithecus
  Family Pliopithecidae
    Pliopithecus
    Crouzelia
    Laccopithecus
  Family Proconsulidae
    Proconsul
    Limnopithecus
    Dendropithecus
    Simiolus
    Rangwapithecus
    Micropithecus
    Dionysopithecus
    Platydontopithecus
  Family *incertae sedis*
    Turkanapithecus
    Afropithecus
    Kenyapithecus
  Family Oreopithecidae
    Nyanzapithecus
    Oreopithecus
  Family Hylobatidae
    Hylobates

Family Pongidae
  Subfamily Dryopithecinae
    Dryopithecus
    Lufengpithecus
  Subfamily Ponginae
    Pongo
    Sivapithecus
    Gigantopithecus
  Subfamily Gorillinae
    Graecopithecus
    Gorilla
    Pan

Family Hominidae
  Australopithecus
  Homo

Superfamily Cercopithecoidea
  Family Victoriapithecidae
    Subfamily Victoriapithecinae
      Victoriapithecus
      Prohylobates

  Family Cercopithecidae
    Subfamily Cercopithecinae
      Macaca
      Procynocephalus
      Paradolichopithecus
      Cercocebus
      Parapapio
      Papio
      Mandrillus
      Dinopithecus
      Gorgopithecus
      Theropithecus
      Cercopithecus
      Allenopithecus
      Miopithecus
      Erythrocebus

Subfamily Colobinae
  Mesopithecus
  Dolichopithecus
  Microcolobus
  Libypithecus
  Cercopithecoides
  Paracolobus
  Rhinocolobus
  Colobus
  Piliocolobus
  Procolobus
  Presbytis
  Simias
  Nasalis
  Pygathrix
  Rhinopithecus

Page numbers of illustrations are indexed in italic.

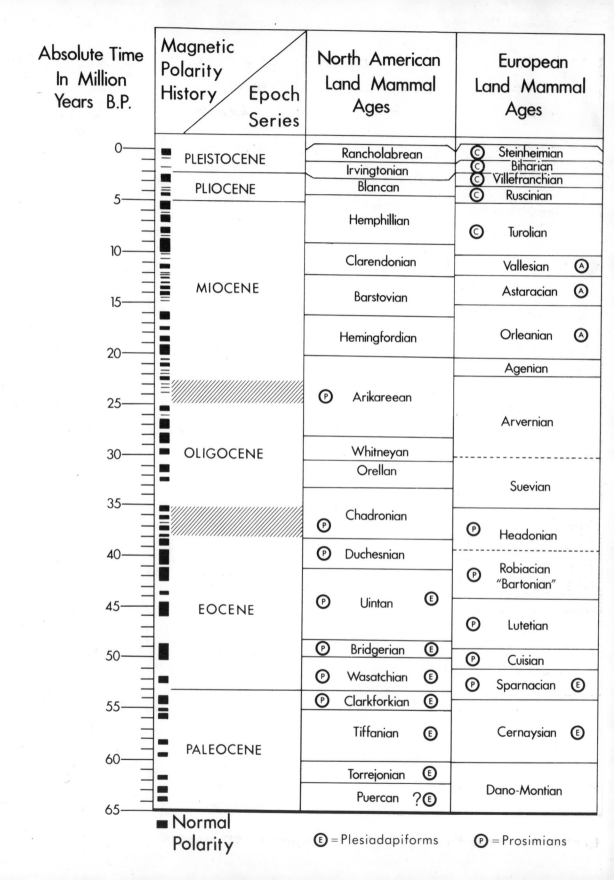